NUTRITION AND ENVIRONMENTAL HEALTH—Volume I: The Vitamins
Edward J. Calabrese

NUTRITION AND ENVIRONMENTAL HEALTH—Volume II: Minerals and Macronutrients
Edward J. Calabrese

SULFUR IN THE ENVIRONMENT, Parts I and II
Jerome O. Nriagu, Editor

COPPER IN THE ENVIRONMENT, Parts I and II
Jerome O. Nriagu, Editor

ZINC IN THE ENVIRONMENT, Parts I and II
Jerome O. Nriagu, Editor

CADMIUM IN THE ENVIRONMENT, Parts I and II
Jerome O. Nriagu, Editor

NICKEL IN THE ENVIRONMENT
Jerome O. Nriagu, Editor

ENERGY UTILIZATION AND ENVIRONMENTAL HEALTH
Richard A. Wadden, Editor

FOOD, CLIMATE AND MAN
Margaret R. Biswas and Asit K. Biswas, Editors

CHEMICAL CONCEPTS IN POLLUTANT BEHAVIOR
Ian J. Tinsley

RESOURCE RECOVERY AND RECYCLING
A. F. M. Barton

QUANTITATIVE TOXICOLOGY
V. A. Filov, A. A. Golubev, E. I. Liublina, and N.A. Tolokontsev

ATMOSPHERIC MOTION AND AIR POLLUTION
Richard A. Dobbins

INDUSTRIAL POLLUTION CONTROL—Volume I: Agro-Industries
E. Joe Middlebrooks

BREEDING PLANTS RESISTANT TO INSECTS
Fowden G. Maxwell and Peter Jennings, Editors

NEW TECHNOLOGY OF PEST CONTROL
Carl B. Huffaker, Editor

THE SCIENCE OF 2,4,5-T AND ASSOCIATED PHENOXY HERBICIDES
Rodney W. Bovey and Alvin L. Young

PALEOCLIMATE ANALYSIS AND MODELING

PALEOCLIMATE ANALYSIS AND MODELING

Edited by

ALAN D. HECHT

Director, National Climate Program Office
National Oceanic and Atmospheric Administration

A Wiley-Interscience Publication

JOHN WILEY & SONS

NEW YORK CHICHESTER BRISBANE TORONTO SINGAPORE

Library of Congress Cataloging in Publication Data

Paleoclimate analysis and modeling.

(Environmental science and technology, ISSN 0194-0287)
"A Wiley-Interscience publication."
Includes index.
1. Paleoclimatology. 2. Climatic changes. I. Hecht,

Alan D. 1944- . II. Series.
QC884.P35 1985 551.6 84-22175
ISBN 0-471-86527-3

Printed in the United States of America

10 9 8 7 6 5 4 3 2 1

CONTRIBUTORS

Eric J. Barron, National Center for Atmospheric Research, Boulder, Colorado

R. G. Barry, World Data Center-A for Glaciology (Snow and Ice), Cooperative Institute for Research in Environmental Sciences and Department of Geography, University of Colorado, Boulder, Colorado

William R. Boggess, Laboratory for Tree-Ring Research, University of Arizona, Tucson, Arizona

R. L. Burk, Golder Associates, Seattle, Washington

S. P. Harrison, School of Geography, University of Oxford, Oxford, England

John Imbrie, Department of Geological Sciences, Brown University, Providence, Rhode Island

Helmut E. Landsberg, University of Maryland, College Park, Maryland

David M. Meko, Laboratory for Tree-Ring Research, University of Arizona, Tucson, Arizona

William F. Ruddiman, Lamont-Doherty Geological Observatory of Columbia University, Palisades, New York

Barry Saltzman, Department of Geology and Geophysics, Yale University, New Haven, Connecticut

Charles W. Stockton, Laboratory for Tree-Ring Research, University of Arizona, Tucson, Arizona

F. A. Street-Perrott, School of Geography, University of Oxford, Oxford, England

Minze Stuiver, Quaternary Research Center, University of Washington, Seattle, Washington

Thompson Webb III, Department of Geological Sciences, Brown University, Providence, Rhode Island

To Dunya, Jennifer, and Gregory,
who have weathered all the storms

FOREWORD

The more we have learned about the climate of the earth in bygone eras, the more we have come to appreciate the wondrous capacity of climate to vary and change. Perhaps just as wondrous, in its own way, is the story of how we have managed to establish the climates of the past, and the ingenious techniques that have been developed to reconstruct the chronology of past changes of climate.

Paleoclimatology is the study of past climates. Since its inception well over a century ago when Louis Agassiz, in 1837, convinced the contemporary scientific world that scratches found on boulders in the Swiss Alps had been caused by ancient glaciations, paleoclimatology has blossomed into a major frontier of knowledge about our planet and its history. Today, scientists can probe the past in sometimes remarkable detail by extracting cores from trees, lake beds, ice sheets, and ocean bottom sediments and by applying to these a wide range of geophysical, geochemical, or other experimental techniques to derive a diverse range of quantitative and reliably dated indices of past climatic changes. From these indices they can develop reconstructions of the global-scale patterns of climate as they existed in past epochs, for example, at the time the continental ice sheets were near their greatest size and extent about 18,000 years ago. Or they can map the variations of climate over time to clarify the behavior of the atmosphere, the oceans, and the ice sheets in the course of a climatic change, and the relationships between these various domains of the "climate system" to help elucidate the physical mechanisms—even the primal causes— of climatic change.

The fruits of paleoclimatic studies are many and by no means lacking in relevance to practical, modern day concerns about climate. They help us to understand the capacity of climate to vary, the rates of change, and the degree of consistency and regularity. They also illuminate the causes of climatic changes. Such information is essential not only to comprehend past climatic changes but to frame intelligent assessments of the possibilities for future climatic change. Because the modern meteorological record is so short, it can adequately reflect neither the full range of climatic variability that is possible under present day environmental conditions nor the "slow physics" involved in longer-term climatic changes. Paleoclimatic studies provide the longer historical perspective needed to assess both. Much of what we can now deduce about the capacity of climate to vary in the future is based on unverified theory as articulated through experiments with computer models of the global climate system. Paleoclimatic studies provide a framework for testing the validity of such models under a wider range of environmental conditions than otherwise possible, to elucidate the mechanisms of climatic change and the feasibility of predicting future climate.

At the present time, the science of paleoclimatology is advancing rapidly on many interdisciplinary fronts. Documentation of new research in the field, already crowding traditional disciplinary journals, is being reported more frequently in newly established journals, as well as in conference proceedings volumes, of an appropriately cross-disciplinary scope.

There are widening perceptions among paleoclimatologists that now is a good time to take stock of our current knowledge in the field. Most particularly, an assessment is needed of the databases upon which our information about past climates has evolved, and of the techniques by which that information can be refined, from each of several principal categories of climate-recording media. This book is a response to that need. It is without precedent as a scholarly and richly documented appraisal of the nature, quality, and information content of the basic data from which most inferences about past climates and their changes have been drawn. The book is written by a stellar team of scientists, each an eminent authority in his or her field and each a seasoned working-level practitioner in the topic of his or her chapter. Its publication represents a milestone in paleoclimatology. It is destined to serve as a valuable sourcebook for generations of working scientists and students to come.

J. Murray Mitchell
Senior Research Climatologist
National Oceanic and
Atmospheric Administration

SERIES PREFACE

Environmental Science and Technology

The Environmental Science and Technology Series of Monographs, Textbooks, and Advances is devoted to the study of the quality of the environment and to the technology of its conservation. Environmental science therefore relates to the chemical, physical, and biological changes in the environment through contamination or modification, to the physical nature and biological behavior of air, water, soil, food, and waste as they are affected by man's agricultural, industrial, and social activities, and to the application of science and technology to the control and improvement of environmental quality.

The deterioration of environmental quality, which began when man first collected into villages and utilized fire, has existed as a serious problem under the ever-increasing impacts of exponentially increasing population and of industrializing society. Environmental contamination of air, water, soil, and food has become a threat to the continued existence of many plant and animal communities of the ecosystem and may ultimately threaten the very survival of the human race.

It seems clear that if we are to preserve for future generations some semblance of the biological order of the world of the past and hope to improve on the deteriorating standards of urban public health, environmental science and technology must quickly come to play a dominant role in designing our social and industrial structure for tomorrow. Scientifically rigorous criteria of environmental quality must be developed. Based in part on these criteria, realistic standards must be established and our technological progress must be tailored to meet them. It is obvious that civilization will continue to require increasing amounts of fuel, transportation, industrial chemicals, fertilizers, pesticides, and countless other products; and that it will continue to produce waste products of all descriptions. What is urgently needed is a total systems approach to modern civilization through which the pooled talents of scientists and engineers, in cooperation with social scientists and the medical profession, can be focused on the development of order and equilibrium in the presently disparate segments of the human environment. Most of the skills and tools that are needed are already in existence. We surely have a right to hope a technology that has created such manifold environmental problems is also capable of solving them. It is our hope that this Series in Environmental Sciences and Technology will not only serve to make this challenge more explicit to the established professionals, but that it also will help to stimulate the student toward the career opportunities in this vital area.

Robert L. Metcalf
Werner Stumm

CONTENTS

PALEOCLIMATE ANALYSIS AND MODELING

1

PALEOCLIMATOLOGY: A RETROSPECTIVE OF THE PAST 20 YEARS

Alan D. Hecht

Director, National Climate Program Office, NOAA
Washington, D.C.

The nineteenth and twentieth centuries have seen the development of unifying theories explaining the evolution of the biosphere and lithosphere. Together these theories of evolution and plate tectonics provide dynamic models for interpreting the history of the earth. In a stimulating editorial in *Quaternary Research* in 1974, the late Richard Foster Flint[1] suggested that the next great synthesis will be to elucidate the dynamics of the atmosphere, explaining how the complex interactions of oceanic and atmospheric circulation control global climate. To deduce such a synthesis requires globally distributed quantitative data which reflect, at least, the temperature and precipitation history of the earth on a wide range of time scales.

Today such data are obtained by satellite, meteorological stations, radiosondes, and other means. The global meteorological network recording modern atmospheric data has grown steadily over the past 20 years and while gaps exist, particularly in the Southern Hemisphere, sufficient data are now available for use in numerical models that simulate the main characteristics of the global circulation (figure 1a–f). Modern data for the ocean, mainly derived from merchant ships, are concentrated in the Northern Hemisphere. Figure 2, for example, shows the distribution of bathythermographic data collected between October and December 1980. A global ocean data set of all oceanographic data from the files of the National Oceanographic Data Center are shown in figures 3a and b.

But climate is a time-transgressive phenomenon being the average state of the atmosphere over periods of 25 to 30 years or more. To reconstruct the history of climate requires data older than the existing meteorological and oceanographic files. For historic times, Lamb[2] and others have summarized from numerous historical records the variability and trends of climate. The French historian LeRoy Ladurie[3]

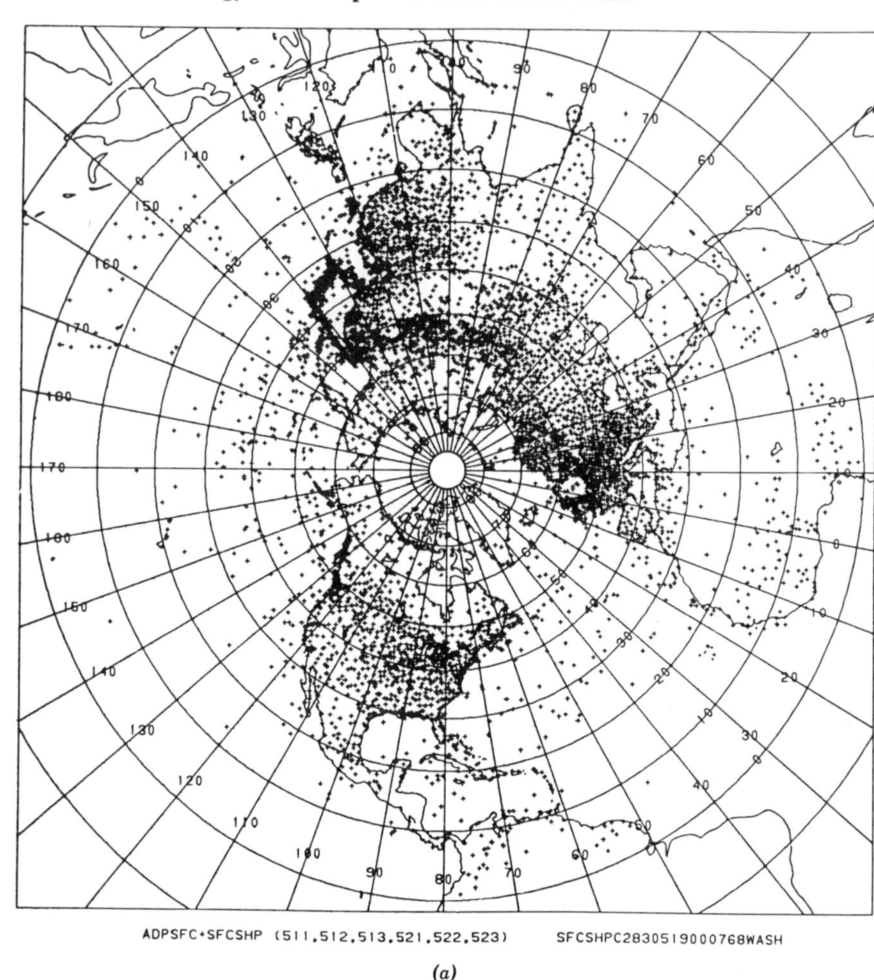

(a)

Figure 1. (*a*) Distribution of surface land and ocean temperatures for Northern Hemisphere reported daily to National Meteorological Center (NOAA) Washington, D.C. (Date of observation, May 19, 1983, O GMT.) Courtesy of National Meteorological Center. (*b*) Same as *a*, but for Southern Hemisphere. (*c*) Distribution of air temperature and wind data collected by radiosonde and reported daily to National Meteorological Center Washington, D.C. (May 19, 1983). Courtesy of National Meteorological Center. (*d*) Same as *c*, but for Southern Hemisphere. (*e*) Distribution of surface winds determined by satellite for Northern Hemisphere, May 19, 1983. Courtesy of National Meteorological Center. (*f*) Same as *e*, but for Southern Hemisphere.

from his analysis of diverse historic records such as those of wine growers in France and of famines in Europe, has reconstructed the pattern of climatic change in Europe over the past millennium. Helmut Landsberg, who has for many years studied weather diaries and historic records, describes in the next chapter how climate information is deduced from these sources.

Over longer time scales, the history of climate must be deduced from biologic and geologic records such as the variations of the widths of rings in trees, the record of pollen changes preserved in lakes and bogs, and the changes in marine flora and

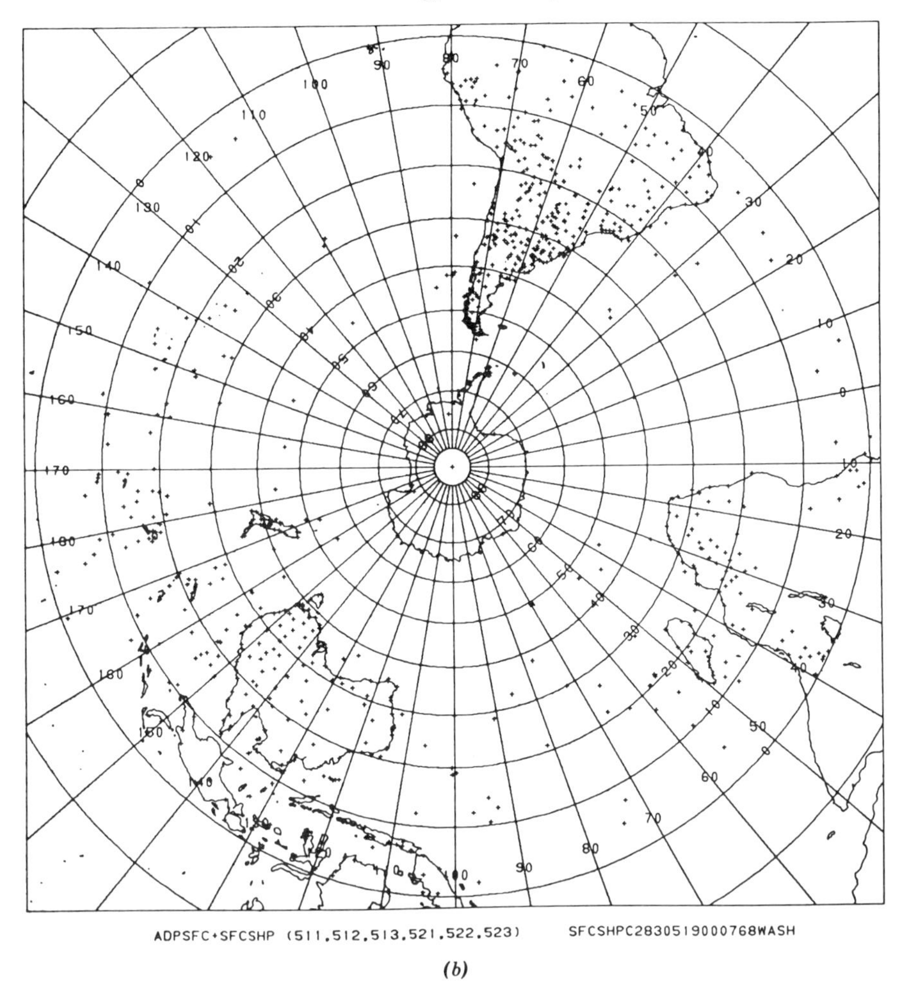

(b)

Figure 1. *(Continued)*

fauna in the deep sea. These data can be interpreted in quantitative ways and provide detailed climate information for at least the last million years of geologic time.

It is the record of quantitative biological and geologic data covering the last million years that is the main subject of this book. Earth scientists see climate change on many different time scales and much has happened in the past 20 years in earth and related sciences to bring us closer to the synthesis imagined by Flint in 1974. Though hardly a complete list, the following achievements have served to open broader opportunities in paleoclimate research and represent the major advances of the past 20 years.

Major Advances In Paleoclimatology, 1960–1982

1. Development of isotopic geochemical techniques, and establishment of an accurately dated stratigraphy for the last 780,000 years.

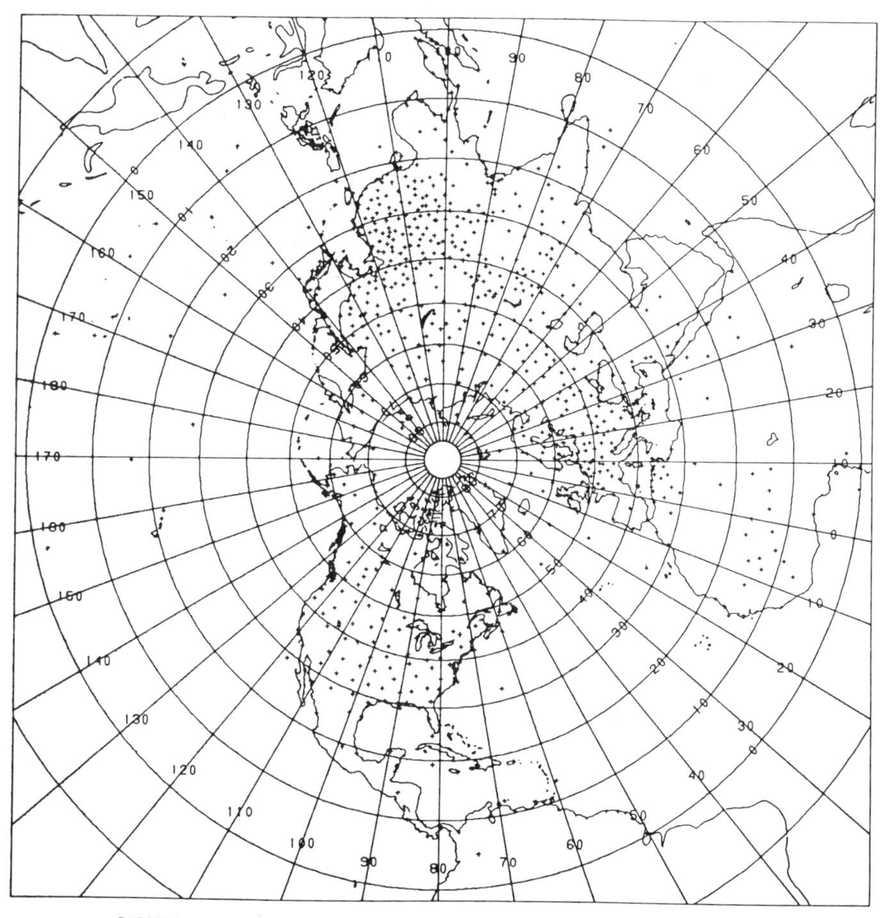

(c)

Figure 1. (*Continued*)

2. Development of methodologies for quantitative reconstruction of past ocean and continental temperatures.
3. Reconstruction of land-sea-ice geography during time of the last maximum glaciation, about 18,000 years ago.
4. Development of atmospheric and ocean general circulation models and simulation of ice age climate.
5. Development of theory of plate tectonics and reconstruction of past continental and oceanic positions.
6. Development of ice core drilling and analysis techniques and evidence of past fluctuations of atmospheric CO_2.
7. Documentation of orbital effects on climate; development of ice sheet climate models.

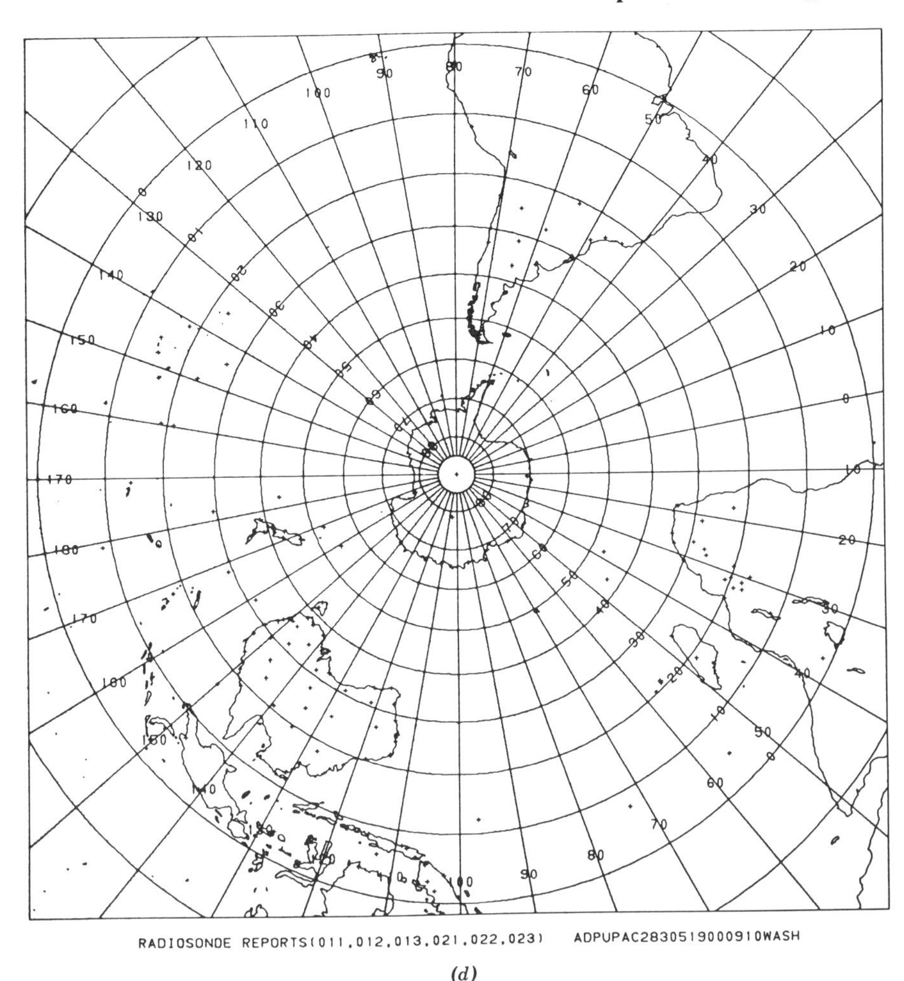

(d)

Figure 1. (*Continued*)

ISOTOPIC PALEOCLIMATOLOGY

While paleoclimatology was rapidly developing during the 1960s it was given major impetus from the application of a newly developed geochemical technique. Harold Urey hypothesized in 1947[4] that changes in the oxygen isotopic composition of carbonate-secreting organisms could be used as a paleotemperature indicator. He suggested that the fractionation of stable isotopes was dependent on temperature and that if such isotopes could be measured in fossil shells, they would yield an estimate of the temperature of the water in which the organism grew. "The temperature coefficient for the abundance of the oxygen isotope in calcium carbonate," wrote Urey,[5] "makes possible a new thermometer of great durability which may have been buried in the rocks for hundreds of millions of years after recording the temperature of some past geological epoch and then having remained unchanged

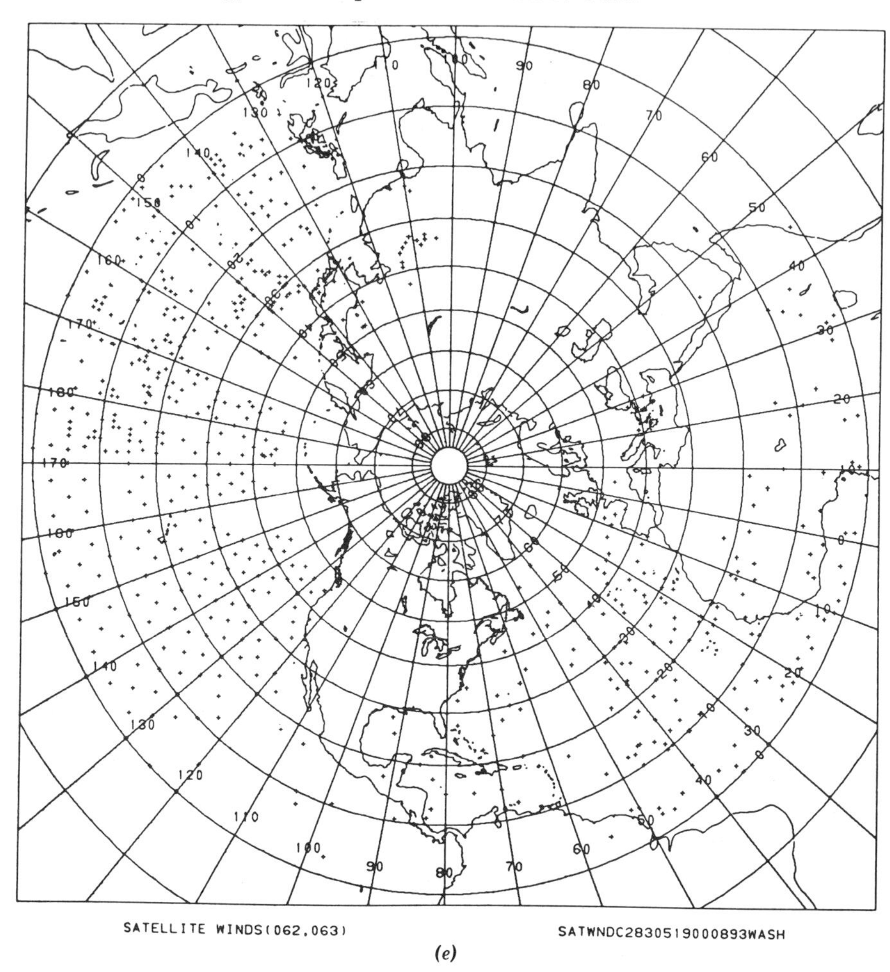

(e)

Figure 1. (*Continued*)

to the present." However, based on the capability of mass spectrometers available at the time, the most accurate paleotemperature that could be obtained was about ±6°C. The problems of reducing this temperature estimate seemed so large that few people volunteered to help Urey develop the idea. The first people who worked with him were Charles McKinney, an engineer, John McCrea, a graduate student, and Samuel Epstein, a young post-doctoral geochemist from McGill University. Urey and this group worked fulltime and improved the precision of the mass spectrometer by a factor of 10 in less than one year. This was the necessary first step before the technique could be developed. Later McKinney and McCrea were replaced by Ralph Bachsbaum, a biologist, and Heinz Lowenstam, a geologist, who turned their attention to determining the oxygen isotopic fractionation in shells of mollusks grown in tanks. Theory predicts that the isotopic fractionation measured

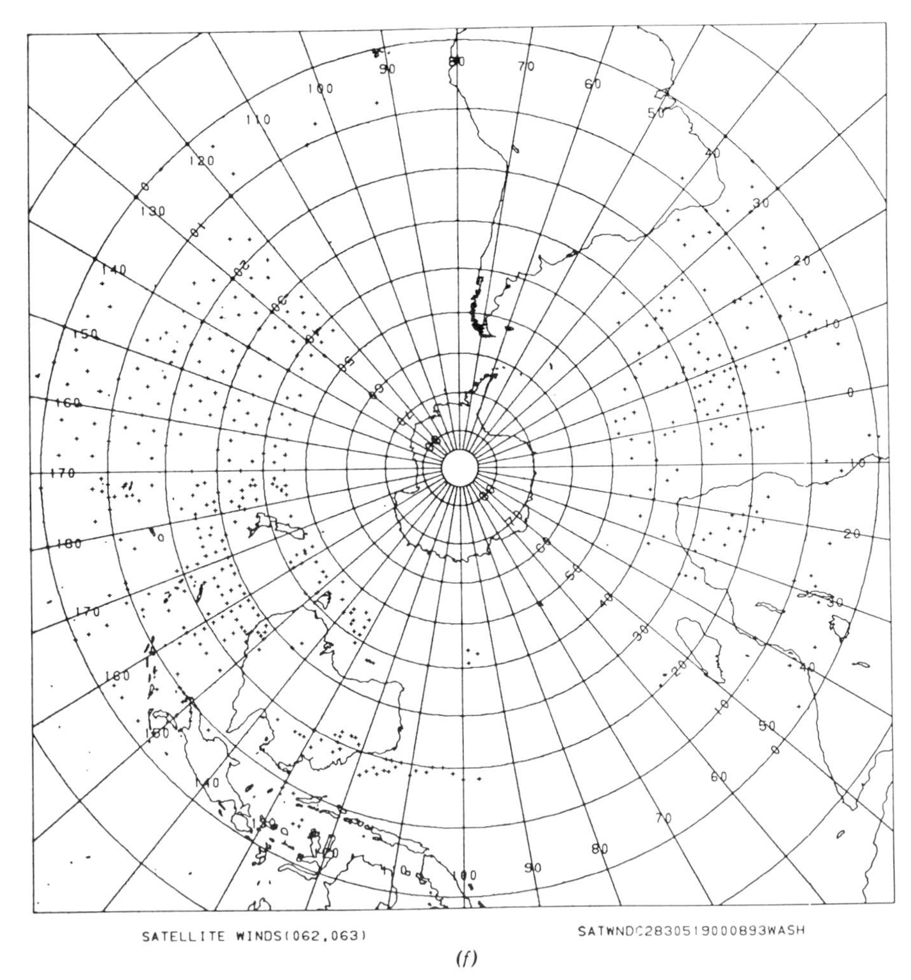

(*f*)

Figure 1. (*Continued*)

in carbonate shells is largely a function of the isotopic composition of the water in which growth occurs and the temperature of the water. However, the enormous chemical problems involved in developing the technique of extracting oxygen isotopic gas from the carbonate shells took about two years to solve. After this time a series of papers was published which documented the analytical techniques and which showed that reliable estimates of the temperatures of shell growth could be obtained.[6,7] The techniques described in these early papers have not changed much since.

In 1950 Cesare Emiliani joined Urey at the University of Chicago and began to examine the remains of foraminifera preserved in ocean cores being collected for the first time by a newly developed piston corer. He postulated that these microorganisms with carbonate shells floating both at the surface and living on the ocean

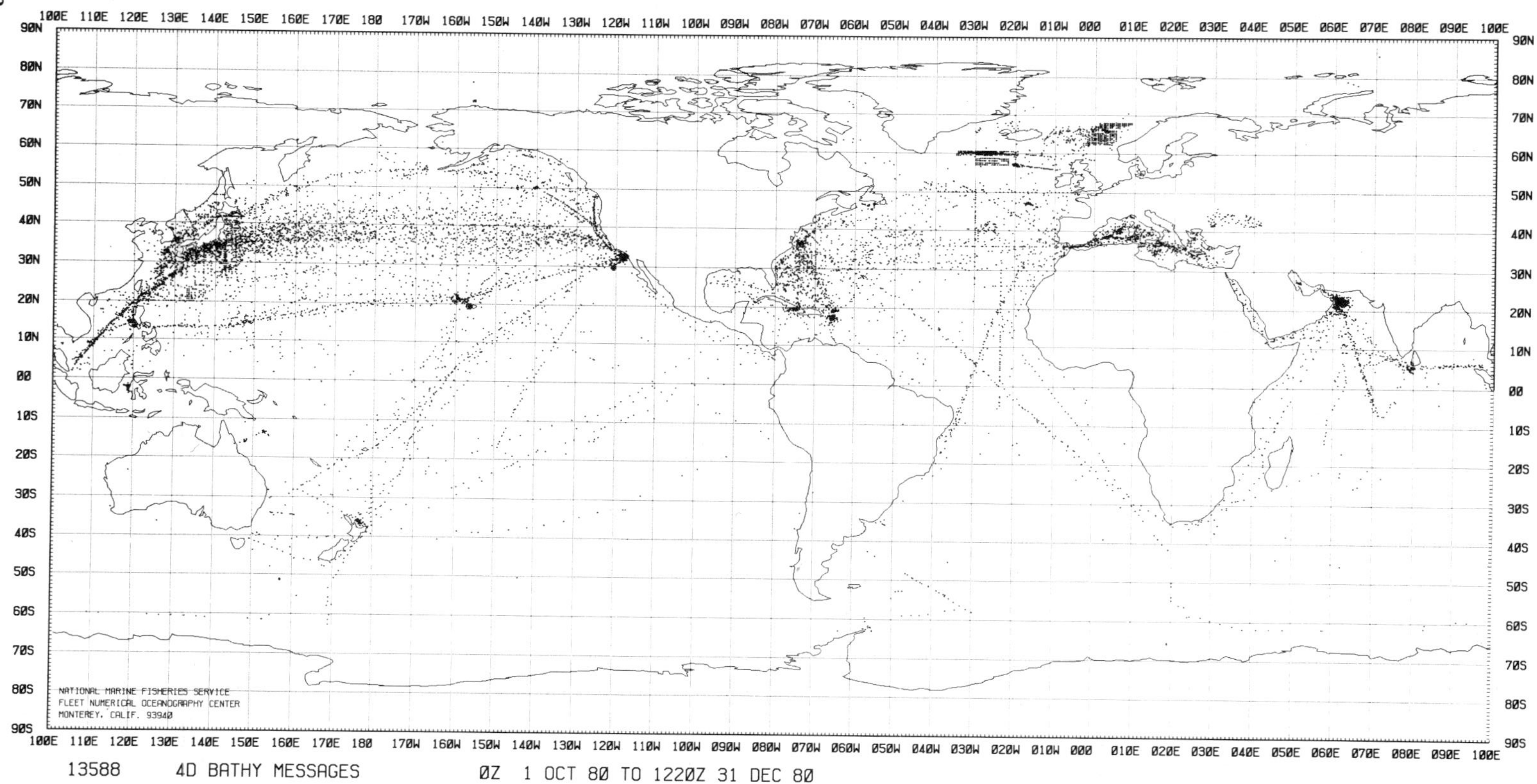

Figure 2. Distribution of radio Bathy messages October to December 1980.

bottom could provide major evidence of past ocean temperatures. Emiliani, who was later to make significant contributions to paleoclimatology, began his new research by first documenting that living foraminifera did in fact preserve an accurate record of ocean temperatures. In 1955 he published[8] his first evidence which showed a close relationship between observed ocean temperatures and calculated temperatures using the paleotemperature equation.

The oxygen isotopic composition in the shells of foraminifera preserved in deep-sea cores is a function of both changes in ocean temperature and oxygen isotopic compositions of the water in which the animals lived. Emiliani[9] calculated, from many indirect sources, that nearly 80 percent of the isotopic change observed between glacial and inter glacial planktonic foraminifera was due to ocean temperature changes and that the remaining amount reflected changes in the isotopic composition of the water. The changes in water composition were due in large part to the waxing and waning of the continental ice sheets. Shackleton[10] alternately proposed that the ice volume changes were more controlling than the temperature ones and that the isotopic record was, in fact, a record of changes in continental volumes of ice. Critical evidence in this interpretation of the isotopic record was given by Shackleton and Opdyke[11] who showed similar isotopic fluctuations for both planktonic and benthonic foraminifera from the same ocean core. If temperature were the dominant factor in controlling isotopic values, then the isotopic record observed in benthonic foraminifera would show low variability due to the general stability of bottom water temperatures. The fact that the isotopic record of benthonic foraminifera revealed large variations implied a large contribution from changes in water chemistry.

Oxygen isotopic analyses have come to be a major tool of paleoclimatic research but not exactly in the sense first suggested by Urey. For the period of time when the earth was glaciated, oxygen isotopic ratios have come to be used as a measure of the changing chemistry of the oceans and atmosphere reflecting mainly the waxing and waning of the polar and subpolar ice sheets. Because ocean waters mix on a time scale of 300 to 700 years, the isotopic records from different ocean basins can be closely correlated. This record provides the standard stratigraphic framework for interpreting Pleistocene climate history. Absolute stratigraphy for the past 800,000 years has recently been developed based on radiometric dating and a simple model in which the isotopic record is considered as the response of a single-exponential system being forced by variations in obliquity and precession of the earth's orbit.[12] For other, nonglaciated times of the earth, the oxygen isotopic record can usually be interpreted as a paleotemperature indicator and can be used to deduce quantitatively the temperature history of the earth over the past 100 million years. As shown in chapter 10, the oceanic isotope record can be related to the time scale of orbital changes, a fact which allows for the first time a consistent and accurate chronology of first order climate variations over the last million years.

Variations in the isotope record also provide fundamental clues to the dynamics of the climate system. The isotopic record is predominately saw-tooth shaped[13,14] and dominated by periods associated with changes in the earth's orbit.[15] Figure 4 shows that isotope maxima and minima are similar in magnitude and duration,

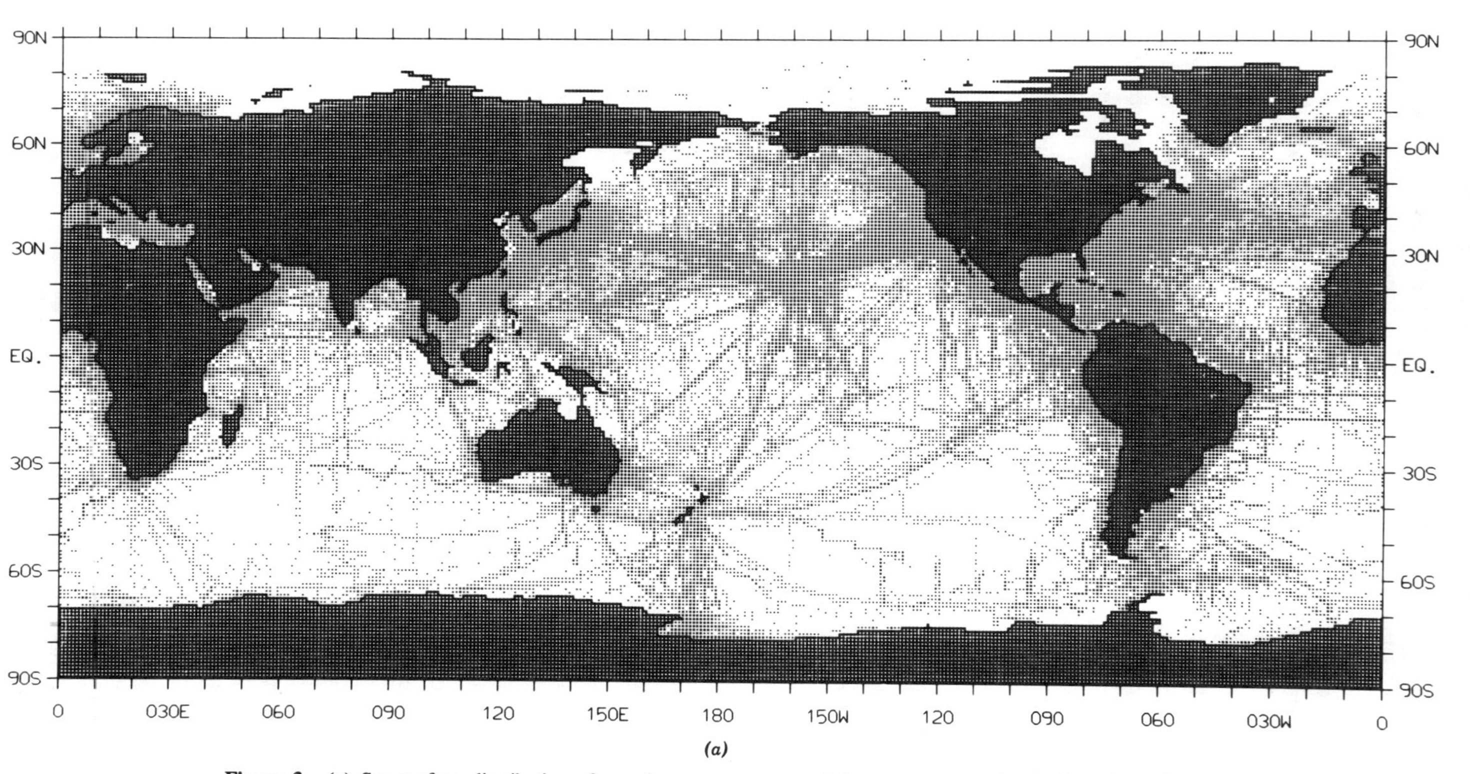

Figure 3. (*a*) Sea surface distribution of one-degree squares containing temperature observations from file of NODC for Northern Hemisphere winter (February, March, April). A small dot indicates a one-degree square containing 1–4 observations; a large dot indicates 5 or more observations. Data compiled by Sydney Levitus and published as NOAA Professional Paper 13, Climatological Atlas of the World Ocean, 1982, reprinted with permission of author. (*b*) Same as *a*, but for Northern Hemisphere summer (August–October).

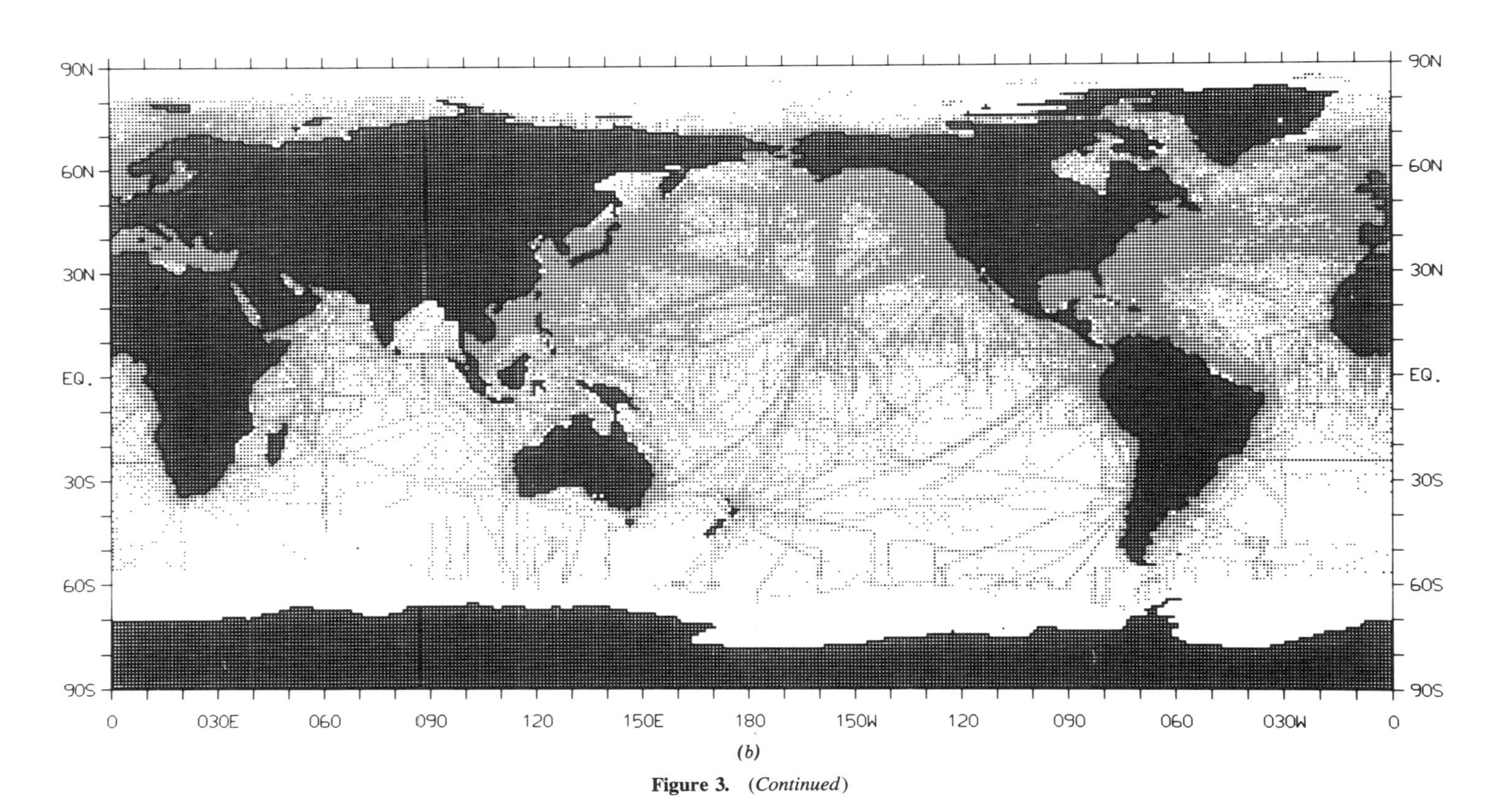

(*b*)

Figure 3. (*Continued*)

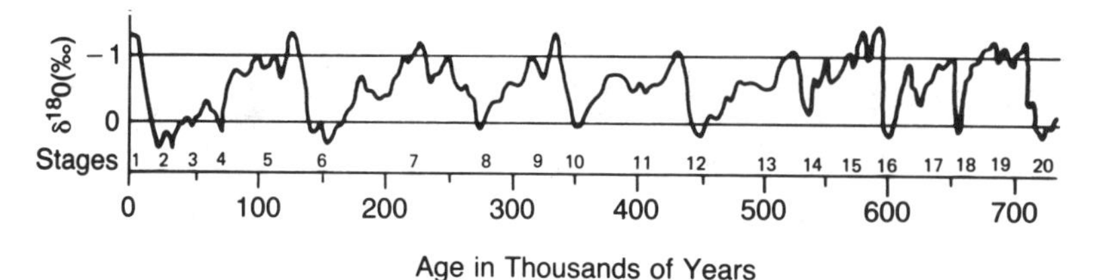

Figure 4. Composite δO^{18} curve showing sawtooth pattern and well defined limits of maximum and minimum values. Numbers refer to intervals of glacial and non-glacial periods. Reproduced from "Climate in Earth History," 1982, with permission from National Academy Press, Washington, D.C. 20418.

although ice ages are characterized by long periods of glacial build-up and rapid terminations. The isotope record reveals many periods of rapid climate change which are corroborated by other geologic data[16] and which can often be modeled as step-function inputs.[14,17] The isotope record is, of course, not a direct response to any single forcing. The record reflects feedback responses (in part distorted due to sediment mixing by biologic organisms) involving the atmosphere, ocean, cryosphere, and land. Yet this record remains a fundamental data set for testing models and theories of climate change.

Application of stable isotope studies to noncarbonate marine fossils has been hindered by complicated biological factors and no reliable paleotemperature relationship has yet been established using such organisms.

Applications of stable isotopes to other biological data such as plants and tree rings have been advanced by many studies. Use of tree ring isotope chemistry to obtain paleoclimate information is a new area of research with potentially wide application. Initial studies (as reviewed in chapter 3) are promising but have not yet provided unequivocal paleoclimatic data. Correlations of variations in the stable isotopes of carbon, oxygen, and hydrogen have been made with climate data derived from tree ring variations.[18-20] Isotope variations appear to be related to several environmental factors. For example, C^{13} variations appear related to changes in both precipitation and temperature.[21] These factors cannot yet be separated in interpreting the carbon variations without help from other methods. Temperature, humidity, and soil water isotope chemistry are factors that effect variations in oxygen isotopic composition of trees, and no unambiguous paleoclimate information has as yet been interpreted from these data. However, it appears that variations in hydrogen isotopes are unaffected by humidity and it may thus be possible to isolate temperature effects using both hydrogen and oxygen isotopes.[22] Variations in hydrogen isotopes appears to offer the most promising lead in paleoclimate studies derived from tree rings.[23]

In the 1960's the application of isotopic analyses to paleoclimatic problems began a quantification process which in the early 1970s was further advanced by the application of multivariate statistical techniques to geological and biological data. By such statistical procedures, terrestrial and marine geologic data were transformed to quantitative climatic data.

QUANTITATIVE RECONSTRUCTIONS OF PAST CLIMATES

In 1970 John Imbrie, a marine geologist at Brown University; Reid Bryson, a meteorologist at the University of Wisconsin; and Harold Fritts, a biologist and expert in dendroclimatology at the University of Arizona, shared common research objectives. Although they would not all meet together until 1974, each knew the other's work. Bryson, for example, took Imbrie's classic study of factor analysis[24] and with the help of Tom Webb and John Kutzbach applied it to their studies of pollen and vegetation changes in the midwest. This group finally met together with the others at the first conference on "transfer functions" at the University of Wisconsin in 1974. Transfer functions became a unifying theme in quantitative paleoclimatic studies—namely to treat biological and geologic data in such a way as to yield quantitative estimates of past climate. The methodologies are based on the assumption that biological changes in species distribution, diversity, and growth are related to climatic factors. Empirical equations can be written which express these relationships. In the case of tree ring variations, annual growth is related to temperature and precipitation for both current growth years and previous ones. It is also assumed that these climatic-biologic relationships are independent of evolutionary changes on the time scale of the life of a species. The statistical techniques used are now well developed[25–27] and collectively they provide a quantitative insight into climate history not previously possible with geologic data.

Imbrie and Kipp[25] pioneered the development of transfer-function methodology as applied to microorganisms preserved in deep-sea cores (foraminifera, radiolaria, and coccolithaphorid). This methodology, as described in the next section and in chapter 5 provides the basis for reconstructing sea surface temperatures during the Quaternary.

Transfer-function methods have been applied to a variety of paleoclimate data. Past patterns of seasonal temperature and precipitation for the United States for the past 300 to 500 years have been reconstructed from tree ring data (chapter 3). These data supplement the existing network of instrumental data and extend it back in time by 300 to 500 years. From such data, which can be resolved to annual events, Fritts[28] was able to reconstruct the general patterns of temperature and precipitation over the United States for each season back to nearly A.D.1600. Validation of these reconstructions can be obtained from historical data. For example, the network of trees in the western United States showed a weather of unusual severity during the summer and winter of 1849 when pioneers traveled west to exploit the newly discovered gold in California. The tribulations of the pioneers are clearly documented in the historical accounts of their expedition.[29]

The value of tree-ring data as an extension of the instrumental record is illustrated by the reconstruction of annual precipitation for a 300-year-old tree in Iowa (figure 5). The reconstructed record shows that over the last 300 years there were five decades comparable in dryness to the dust bowl decade of the 1930s. Four of these occurred prior to the instrumental record, and evidence of them is found only in tree ring data.[30] Other examples of the use of tree ring data for climate reconstruction are given in chapter 3. Similar statistical techniques can be applied to

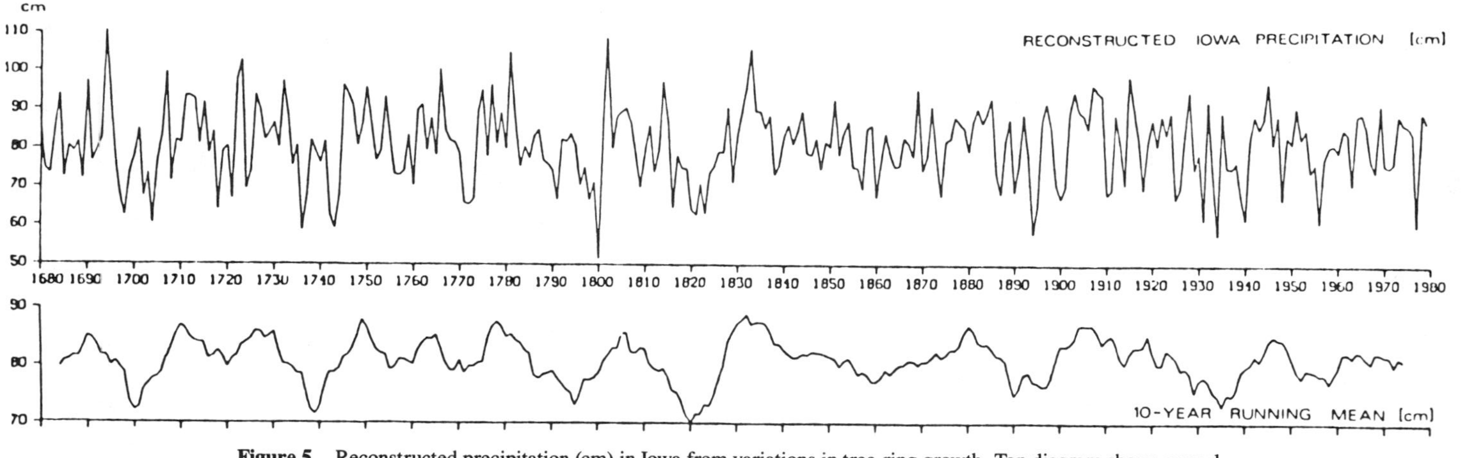

Figure 5. Reconstructed precipitation (cm) in Iowa from variations in tree-ring growth. Top diagram shows annual values; bottom diagram shows 10-year running mean. From ref. 30; reprinted with permission of author and American Meteorology Society, Boston, Mass.

pollen data thereby permitting detailed climate reconstructions for many areas of the Northern Hemisphere over the past 10,000 years (chapter 4).

RECONSTRUCTION OF ICE AGE GEOGRAPHY

Reconstruction of conditions prevailing over the earth at the culmination of the last ice age (18,000 years B.P.) is a major scientific accomplishment. Synoptic climate reconstructions can be developed for time intervals that have an adequate regional or global database and which can be well dated. A reconstruction of the ice age earth has only been possible because of the global stratigraphy provided by oxygen iosotopic records together with the development of quantitative methods of interprcting fossil biota preserved in deep-sea cores. In the program known as CLIMAP (Climate Long range Investigation and Mapping)[31,32] the oxygen isotope stratigraphy was determined for hundreds of ocean cores. In each core the datum for the inferred maximum ice volume was determined and correlated from one ocean core to another. The faunal composition of these horizons was determined and from these data an estimate was given for surface temperatures. Land vegetation and geography were estimated from biological-geological features.

The reconstruction of ice sheet extent during the last glacial maximum was compiled from nearly a century of field work. However, geologic evidence does not permit an unambiguous reconstruction and two models of ice sheet extent could be developed (see chapter 6). The maximum and minimum models of ice sheet extent reflect alternate interpretations of available geologic data. Uncertainties in the geologic data include how extensive were marine-based ice sheets and how widespread were ice sheets in large areas of Northern Canada and Asia.[33]

The above biological and geological data were compiled to reconstruct an ice age earth. Such an analysis provides sufficient quantitative data for atmospheric scientists to model the general circulation of the atmosphere at a time when the climate was vastly different from today.

Analysis of the ice age reconstruction (for both an average August and an average February) and comparison of the ice age seasons with those of today provide important views of the climate system during two extreme conditions (glacial and interglacial maximum) and insight into possible mechanisms that drive the climate system.

The major conclusions of the CLIMAP Program suggest the following:

1. During the Last Glacial Maximum several huge land-based ice sheets that were as much as 3 km thick existed in the Northern Hemisphere. In the Southern Hemisphere the most striking contrast with today was the greater extent of sea ice. Global sea level was lowered by at least 100 m.

2. Global average sea surface temperatures (SSTs) associated with the ice age maximum were about 1.7°C cooler in August and 1.4°C cooler in February.

3. The average surface albedo of the last glacial maximum was higher than today mainly due to expanded glaciers and sea-ice fields. The albedo of nonglaciated

areas was also higher than today due to lower density of vegetative cover and to contrasting characteristics of glacial soils.

4. The ice age ocean was characterized by steepening of thermal gradients along frontal systems, an equatorial displacement of subtropical convergences in the Southern Hemisphere, a large shift in path of the Gulf Stream, which flowed due east at 40°N during glacial maximum; nearly stable positions and temperatures of the central gyres; and increased upwelling along the zones of equatorial divergence.

The ice age climate reconstruction provides a new perspective on the ice age earth. Most previous studies largely based on terrestrial data from mid-latitudes emphasized the growth of the ice sheets and compression of climatic zones toward the equator. The oceanic data show a compression of climatic zones toward the centers of the subtropical oceanic gyres which form the thermally and geographically stable part of the climate system. Surrounding them were areas of modest cooling in the equatorial and boundary currents with areas of greater cooling occurring in higher latitudes.

GENERAL CIRCULATION MODELING—ICE AGE SIMULATION

Models of the general circulation of the atmosphere attempt to simulate the complex interactions of the atmosphere, ocean, cryosphere, and land, all of which must be included in the climate system. Models can be ranked in a hierarchy with respect to both type and degree of resolution.[34] Simple models are used to compute the spatial distribution at equilibrium of a single key climatic parameter—usually temperature—as a function of latitude, elevation in the atmosphere, or both. Such models are concerned primarily with changes in gross climatic conditions that occur when the near-equilibrium between energy gain and loss of the climate system is changed to some other equilibrium state. These models are simple in design but give useful information about the sensitivity of the climate system to changes in variables such as solar radiation.

Simple models are also useful in interpreting the results of more complex models. The most elaborate models depict several climatic quantities for all three spatial dimensions and indicate their variations in time. These models are commonly referred to as "general circulation models" (GCMs). They specify an average future or past climate from given initial and boundary conditions.

A major advance of the past 20 years has been the development of GCMs, and their application to a number of problems such as simulation of the ice age earth and prediction of future climate with increased levels of CO_2 in the atmosphere (chapter 8). The value of paleoclimate simulation experiments is that they are a test of the validity of the model when called to simulate a climate vastly different from today. The GCM is also an important tool for diagnostic studies and is useful in identifying important physical factors underlying climate change in general.

Using the boundary conditions of sea surface temperature, ice sheet topography sea level lowering, and surface albedo assembled by CLIMAP for the ice age about

18,000 years B.P., February and August paleoclimates were simulated with GCMs developed at three major laboratories: The National Center of Atmospheric Research (NCAR), Oregon State University (OSU), and the Geophysical Fluid Dynamics Laboratory (GFDL). Results of the latter two models are more directly comparable because the NCAR model used boundary conditions different from that of OSU and GFDL.[35-37] A comparison of the output of these models is given by Heath.[38] A more recent simulation has been done by Gates (unpublished) using the revised and final CLIMAP data set.[32]

Aside from the boundary condition mentioned above, the other conditions used in the Gates' model experiment which were different during the ice age (relative to today) were a sea level lowering of 150 m and a lowering of CO_2 level in the atmosphere from 320 ppm (as used in control experiment) to 220 ppm. The change of CO_2 level in the atmosphere during the ice age is based on recent observations of CO_2 levels determined from ice cores.[39,40]

The ice age simulation shows in both August and February a zonally averaged cooling of the surface air temperature at all latitudes with a maximum cooling of about 10°C at northern latitudes (figure 6). The Globally averaged cooling is about 3.4°C which is less than previously estimated.[37] In tropical regions, average temperature change is only about 2°C.

Ice age precipitation rates show a decrease at most latitudes from 3 percent in February to 9 percent in August. There is a narrowing of the subtropical dry zone near 14° North in the ice age and a southward displacement of the zonally averaged mid-latitude precipitation maximum from 34°N to 26°N. The ice age is thus cooler and drier over most of the globe. Predicted land temperatures agree in a general way with estimated temperatures derived from geologic data. These atmospheric modeling experiments provide a first order picture of global climate given changes in boundary conditions. A coupled ocean-atmosphere model would provide a more realistic picture of climate change, but such modeling experiments remain a problem for the future.

There exists a comparable spectrum of ocean models suitable for study of ocean processes and climate.[41] Ocean GCMs are able to simulate the general surface temperature distribution, although there are regional errors of several degrees Celsius.[42] Knowledge of the vertical circulation of the ocean remains limited and is difficult to model. Realistic models of seasonal or longer climate changes will require an interactive atmosphere and ocean at least to include the upper 70 m of water. Several varieties of coupled atmosphere and ocean models are being developed.[41] These models are at present complex and require excessive computer time, consequently paleoclimatic simulation with coupled ocean-atmosphere GCM, is not likely to occur in the immediate future.

PLATE TECTONICS AND CLIMATE MODELING

The development of the theory of plate tectonics provides a unifying model for interpreting the evolution of the earth over the past hundreds of millions of years. Changes in the amount and location of land and sea have likely affected global

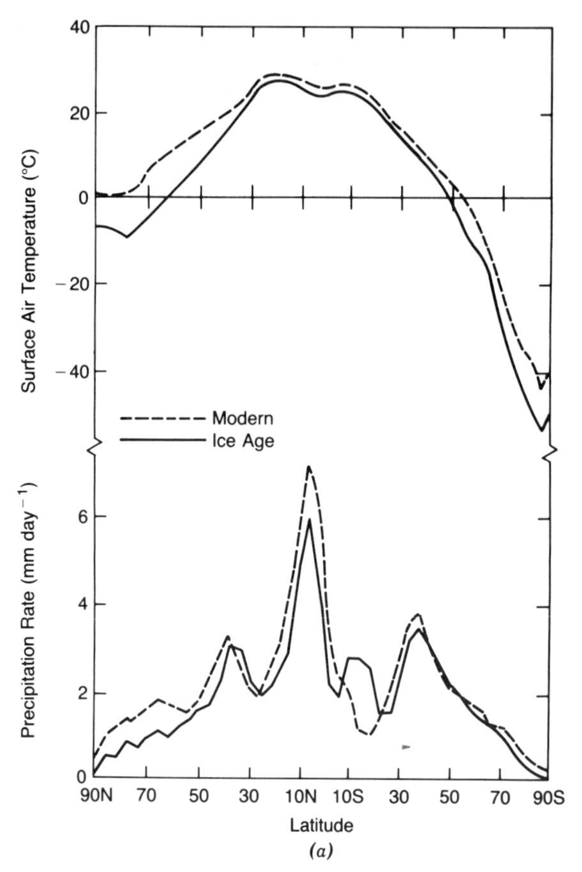

Figure 6. (*a*) August zonal average surface air temperature (above) and precipitation rate (below) simulated by Oregon State University General Circulation Model (GCM) with both modern and ice age boundary conditions. After W. L. Gates unpublished. (*b*) Same as *a* except for February.

climate patterns. The question remains as to whether or not such geographic changes in themselves account for the observed over all global cooling during the Tertiary Period. Barron et al.[43] used a zonally averaged energy-balanced model developed by Thompson and Schneider[44] to investigate the mechanisms that could explain the warm and equatable climate that appears to have characterized the mid-Cretaceous Period (about 100 million years ago). The model results suggest that changes in geography were important in bringing about climate change. However, the meridional distribution of Cretaceous temperature could not be successfully simulated with this model until additional feedback processes involving cloud and meridional heat transport were included. Barron et al.[45] have also produced a series of paleogeographic reconstructions at 20 million year intervals from 180 million years ago to the present. Continents with ancient shorelines and oceans are reconstructed from paleomagnetic and geologic data. These reconstructions provide a framework for modeling climatic responses and for determining the extent to which these

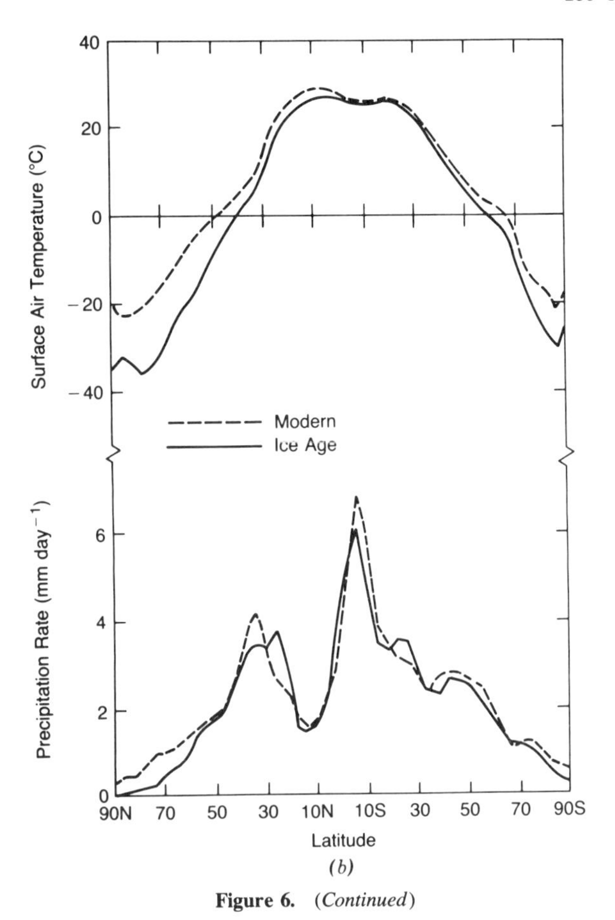

Figure 6. (*Continued*)

geographic changes account for observed climatic changes. Recent modeling of such climate reconstructions is discussed in chapter 9. Modeling of pre-Pleistocene climate change with large scale models is in its infancy. Despite limitations in deriving quantitative paleoclimatic data for pre-Pleistocene times, there are numerous opportunities for modeling which may provide new insight into climate change on longer time scales.

ICE CORES

In 1966 the U.S. Army Cold Regions Research and Engineering Laboratory (CRREL) succeeded in drilling a core through the Greenland Ice Sheet at Camp Century. The isotopic and geochemical data from this core provided for the first time a direct window on the history of climate of the polar regions.[46] Moreover, the success of the drilling program suggested the possibility that very high resolution records extending back many thousands of years could be obtained in other polar regions.

The papers published by Dansgaard and his colleagues[46,47] between 1969 and 1971 (1) established the general pattern of the response of the (northern) polar regions over the last interglacial-glacial cycle; (2) showed that the oxygen-isotopic ratio in ice sheets was in fact far lower during the last glacial interval; (3) provided a detailed record of such Late Quaternary events as the "little ice age" and the pre-Holocene; and (4) seemed to provide a continuous, dated chronology of climatic events for the last glaciation. The famous Camp Century profile was published in Flint's textbook in 1971[48] and became a standard climate series against which other climatic records could be compared.[49]

An overview of the potential and problems of ice core research since 1966 is found in chapter 6. Since 1971 much of the effort in studying these cores has gone into developing a reliable chronology. As a result the chronology of the upper part of several ice cores, representing approximately the past 20,000 to 30,000 years, has been substantially improved. Some progress has also been made in interpreting the climatic significance of proxy records recorded in ice cores. Within the last several years parameters other than temperature have been reconstructed from ice core data. Electrical resistivity measurements apparently reflect the influx of volcanic aerosols; microparticle concentrations reflect atmospheric dust loading; and Na measurements may reflect the influx of sea salts. Recent work by Delmas et al.[39] and Neftel et al.[40] has also raised the possibility that estimates of past atmospheric CO_2 concentration can be obtained from air bubbles trapped in ice cores. This work has documented for the first time the fluctuations of CO_2 levels in the atmosphere over the past 20,000 years.

ORBITAL EFFECTS

It is now apparent that changes in the geometry of the earth's orbit have an effect on the earth's climate. Croll in 1864[50] and Milankovitch in 1930[51] each proposed a theory of climate dependent on orbital changes as a prime factor in affecting the distribution of solar radiation incident on the earth's surface. Since first proposed, there had been conflicting geologic evidence to support these theories. In 1961, A. Nairn edited the first modern paleoclimate monograph[52] and in the introduction wrote "the effect of the varying distance between the earth and the sun from perihelion to aphelion, the basis of Croll's theory of the origin of ice ages is not now thought to be a significant factor in climate." Compelling evidence to the contrary has since become available from better dating of ocean sediments, which provide a continuous and reliable time series for analysis and application of spectral techniques to identify the crucial frequencies in deep sea sediments.

Hays et al.[15] showed by means of spectral analysis that three climatically sensitive parameters in deep sea sediment cores have the same frequencies as the major orbital ones, namely: about 100,000 years (eccentricity), about 41,000 years (obliquity) and about 21,000 years (precession). Subsequent studies of ocean cores from the North Atlantic (chapter 5) show that the strength of the orbital influence varies as a function of latitude, as predicted by the Milankovitch theory. For example, the 21,000 year precession rhythm is very strong and easily identified in ocean cores near 40°N latitude. The intensity of this spectral signature in ocean cores decreases significantly northward and is negligible at latitudes higher than 50–55°N. However, at higher latitudes the obliquity rhythm (41,000 years) is strongest in ocean cores and decreases in intensity toward lower latitudes.[53,54] It is thus apparent that different parts of the spectral frequency will affect different parts of the climate system. Because of recent advances in collecting and dating ocean cores[11,12] and in climate theory, it is now possible to focus future research on the mechanism by which different parts of the climate system respond to changes in radiation boundary conditions.[17] Imbrie and Imbrie summarize recent work in modeling climatic response to orbital variations. They point out that except for studies of the annual cycle, there is no other problem in climate dynamics where the primary external forcing functions are known so precisely. Since both the temporal and spatial structure of orbital changes can be specified exactly, a means is now available to compare geological observations with predictions derived from physically based models of climate.

It is also now apparent that orbital changes may have a pronounced effect on low latitude climate as well. About 9000 years ago the amount of solar radiation reaching the earth's surface was approximately 7 percent greater in summer and 7 percent less in winter across low and middle latitudes of both hemispheres (figure 7). While the annual average solar radiation at 9000 year B.P. is close to the present value, the seasonal radiation differences were significantly increased. Kutzbach[55] has taken this change in the radiation boundary condition at the top of the earth's atmosphere as input in a low-resolution GCM to investigate the effect of orbital parameter changes on climate. Kutzbach's analysis shows that the amplified seasonal cycle of solar radiation, coupled with sea-surface temperatures close to the modern ocean, create an intensified monsoon circulation over Africa and India. The model predicts increased rainfall over Africa which is compatible with geologic data compiled by Street Perrot (chapter 7). This model experiment provides further documentation of the important role orbital changes play in affecting the earth's climate.

Ice sheet variations are the principal feature of the Ice Age. It is evident from the time series and spectral analyses of the oxygen isotopic record (figure 4) that ice sheets decay more rapidly than they grow and that growth and decay of ice sheets is caused by changes in solar radiation due to changes in the earth's orbit. While the response of the ice sheets to radiation changes has been explored in several studies and is discussed in chapters 6 and 8, the processes in the earth's climate system (ocean, atmosphere, and cryosphere) which translate the insolation anomalies into climate change are largely unknown.

It is also apparent from spectral analysis of oxygen isotopic record[11,14] that a frequency of 100,000 years (eccentricity) is dominant. However, this frequency in the isotopic record cannot be simulated as a linear response to insolation forcing.[17,56–58] Recent ice modeling studies suggest that ice volume changes with pe-

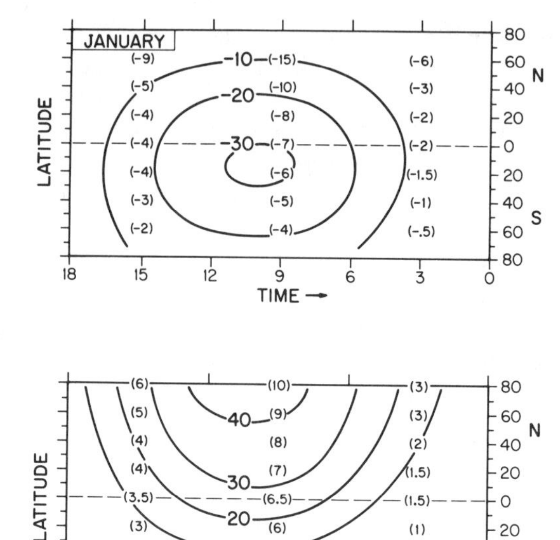

Figure 7. Incoming solar radiation differences (in W/m^2) at the top of the earth's atmosphere, expressed as past values minus present value, for the past 18,000 years; time scale in thousands of years before present. Top, January; bottom, July. Numbers in parentheses at 3-, 6-, and 15,000 years before present are the differences in percent, 100 × (past-present)/present. Monthly values refer to the average radiation over a 30° arc of celestial longitude.[55] Calculations by Peter Guetter of the University of Wisconsin-Madison; unpublished.

riods near 100,000 years result from nonlinear response of the ice sheet itself and the lithosphere-asthenosphere response to the ice sheet loading.[58,59]

SUMMARY

The past 20 years have been characterized by development of new techniques for collecting and analyzing biological and geologic data. There has been a rapid expansion in data collecting and in establishing reliable time scales. Global data sets for certain time periods are within reach. New opportunities are now open for documenting climate history on several time scales and in testing models of climate change.[60] While the first goal of paleoclimatic research is to describe variations in climate that span intervals beyond the range of instrumental data, the ultimate goal is to identify the physical causes of climate change. Speculative ideas about the nature and causes of climate variations have always been plentiful. Recent ad vances in this science suggest that the next decade will be a period when observations and modeling experiments will make it possible for the first time to test some of the ideas of climate change.

ACKNOWLEDGMENTS

J. M. Mitchell (NOAA), Stephen C. Porter (University of Washington), John Kutzbach (University of Wisconsin), and Richard Poore (USGS) reviewed this article and greatly improved its quality and clarity. Illustrations have been graciously provided by William Bonner and Sydney Levitus (NOAA), John Kutzbach (U. Wisconsin), and W. L. Gates (Oregon State University). The manuscript was typed by Wendy McGhee Graham.

REFERENCES

1. Richard F. Flint, "Three theories in time," *Quat. Res.*, **4,** 1-9 (1974).
2. H. H. Lamb, *Climate Present, Past and Future,* Vol. 2, Methuen, London, 1977.
3. E. LeRoy Ladurie, *Times of Feast, Times of Famine: A History of Climate Since the Year 1000,* Doubleday, Garden City, N.Y., 1977.
4. H. C. Urey, "The thermodynamic properties of isotopic substances," *J. Chem. Soc.*, 562–581 (1947).
5. H. C. Urey, "Oxygen isotope in nature and in the laboratory," *Science,* **108,** 489–496 (1948).
6. S. Epstein, R. Buchsbaum, H. A. Lowenstam, and H. C. Urey, "Revised carbonate—water isotopic temperature scale," *Bull. Geol. Soc. Am.*, **64,** 1315–1326 (1953).
7. H. A. Lowenstam and S. Epstein, "Paleotemperatures of the Post-Aptian Cretaceous as determined by the oxygen isotope method, *J. Geol.*, **62,** 207–248 (1954).
8. C. Emiliani, "Depth habitats of some species of pelagic forminifera as indicated by oxygen isotope ratios," *Am. J. Sci.*, **52,** 149–158 (1954).
9. C. Emiliani, "Pleistocene temperatures," *J. Geol.*, **63,** 538–578 (1955).
10. N. Shackleton, "Oxygen isotope analyses and pleistocene temperatures re-assessed," *Nature,* **215,** 15–17 (1967).
11. N. Shackleton and N. D. Opdyke, "Oxygen isotope and paleomagnetic stratigraphy of Equatorial Pacific Core V28-238: Oxygen isotope temperatures and ice volumes on a 10^5 year and 10^6 year scale," *Quat. Res.*, **3,** 39–55 (1973).
12. J. Imbrie, J. D. Hays, D. G. Martinson, Andrew McIntyre, A. C. Mix, J. J. Morley, N. G. Pisias, W. L. Prell, and N. J. Shackleton, "The Orbital Theory of Pleistocene Climate: Support from a Revised Chronology of the Marine O^{18} Record." In A. Berger, ed., *Milankovitch and Climate,* D. Reidel, Dordrecht, 1984, 269–305.
13. W. S. Broecker and J. van Donk, "Insolation changes, ice volume and the O^{18} record in deep sea cores," *Rev. Geophys. Space Phys.*, **8,** 169–198 (1970).
14. W. S. Berger, "Climatic Steps in Ocean History." In *Climate in Earth History,* National Academy of Sciences, Washington, D.C., 1982, pp. 43–54.
15. J. D. Hays, J. Imbrie, and N. Shackleton, "Variations in the earth's orbit: Pacemaker of the ice age," *Science,* **194,** 1121–1132 (1976).
16. G. Woillard and W. G. Mook, "Carbon 14 dates at Grande Pile: Correlation of land and sea chronologies," *Science,* **215,** 159–161 (1982).
17. J. Imbrie and J. Z. Imbrie, "Modeling the climate response to orbital variations," *Science,* **207,** 943–953 (1980).
18. S. Epstein, C. J. Yapp, and J. H. Hall, "The determination of the D/H ratio of non-exchangeable hydrogen in cellulose extracted from aquatic and land plants," *Earth Plan. Sci. Lett.*, **30,** 241–251 (1976).

19. S. Epstein and C. Yapp, "Isotope tree thermometers," *Science,* **266,** 477–478 (1977).

20. R. L. Burk and M. Stuiver, "Oxygen isotope ratios in trees reflect mean annual temperature and humidity," *Science,* **211,** 1417–1419 (1981).

21. P. O. Tans, "Past atmospheric CO_2 levels and the C^{13}/C^{12} ratios in tree rings," *Tellus,* **32,** 268–283 (1980).

22. J. W. C. White, "The relationship between non-exchangeable hydrogen of tree ring cellulose and the source waters for tree sap," *Proc. Int. Meet. Stable Isot. Tree Ring Res.*, U.S. Department of Energy, Washington, D.C., 58–65, (1980).

23. C. J. Yapp and S. Epstein, "Climatic implication of D/H ratios of meteoric water over North America (9500–22,000 yr BP) as inferred from ancient wood cellulose C-H hydrogen," *Earth Plan. Sci. Lett.*, **34,** 333–350 (1977).

24. J. Imbrie, "Factor and vector analysis programs for analyzing geologic data," Tech. Rpt. 6, Office of Naval Research Task No. 389-135 (1963).

25. J. Imbrie and N. Kipp, "A New Micropaleontological Method for Quantitative Paleoclimatology: Application to a Late Pleistocene Caribbean Core." In K. K. Turekian, ed., *Late Cenozoic Glacial Ages,* Yale University Press, New Haven, 1971, pp. 71–181.

26. Harvey Sachs, T. Webb III, and D. R. Clark, "Paleoecological transfer functions," *Ann. Rev. Earth Plan. Sci.*, **5,** 159–178 (1977).

27. T. Webb III and D. R. Clark, "Calibrating micropaleontological data in climatic terms: A critical review," *Ann. N.Y. Acad. Sci.*, **228,** 93–118 (1977).

28. H. C. Fritts, *Tree Rings and Climate;* Academic, New York, 1976.

29. M. P. Lawson, "The climate of the great american desert, reconstruction of the climate of western interior U.S., 1800–1850," University of Nebraska Study N.S. 46, Lincoln (1947).

30. D. N. Duvick and T. J. Blasing, "A dendroclimatic reconstruction of annual precipitation in Iowa since 1680," *Water Resour. Res.*, **17,** 1183–1189 (1981).

31. CLIMAP, "The surface of the Ice-Age," *Science,* **191,** 1131–1137 (1976).

32. CLIMAP, "Seasonal reconstruction of earth's surface at the last glacial maximum," *Geol. Soc. Am.*, Map and Chart Series MC-36 (1981).

33. G. H. Denton and T. J. Hughes, eds., *The Last Great Ice Sheets,* John Wiley and Sons, New York, 1981.

34. Kenneth H. Bergman, A. D. Hecht, and Stephen H. Schneider, "Climate models," *Phys. Today,* October, 45–51 (1981).

35. J. Williams, R. G. Berry, and W. M. Washington, "Simulation of the atmospheric circulation using the NCAR global circulation model with Ice-Age boundary conditions," *J. Appl. Meteorol.*, **13,** 305–317 (1974).

36. S. Manabe and D. C. Hahn, "Simulation of tropical climate of an Ice Age," *J. Geophys. Res.*, **82,** 3889–3912 (1977).

37. W. L. Gates, "The numerical simulation of Ice Age climate with general circulation model," *J. Atmos. Sci.*, **33,** 1844–1873.

38. G. Ross Heath, "Simulations of a glacial paleoclimate by three different atmospheric general circulation models," *Palaeogeogr. Palaeoclim. Palaeocol.*, **26,** 291–303 (1979).

39. R. J. Delmas, J. -M. Ascencio, and M. Legrand, "Polar ice evidence that atmospheric CO_2 20,000 yr BP was 50% of present," *Nature,* **284,** 155–157 (1980).

40. A. Neftel, H. Oeschger, J. Schwander, B. Stauffer, and R. Zumbrunn, "Ice core sample measurements give atmospheric CO_2 content during the past 40,000 years," *Nature,* **295,** 220–222 (1982).

41. U.S. Committee for the Global Atmospheric Research Program, *Ocean Models for Climate Research: A Workshop,* National Academy of Sciences, Washington, D.C., 1980.

42. W. L. Gates, "Paleoclimate Modeling—A Review with Reference to Problems and Prospects for

the Pre-Pleistocene." In *Climate in Earth History,* National Academy of Sciences, Washington, D.C., 1982, pp. 26–42.

43. Eric J. Barron, Starley L. Thompson, and Stephen H. Schneider, "An ice-free creteceous results from climate model simulators," *Science,* **212,** 501–508 (1981).
44. Starley L. Thompson and Stephen H. Schneider, "A seasonal zonal energy balance climate model with an interactive lower layer," *J. Geophys. Res.*, **84,** 2401–2414 (1979).
45. Eric Barron, C. G. A. Harrison, James L. Sloan II, and William W. Hay, "Paleogeography, 180 million years ago to the present," *Eclogae. Geol. Helv.*, **74** (2), 443–470 (1981).
46. W. S. Dansgaard, S. J. Johnsen, J. Miller, and C. C. Langway Jr., "One thousand centuries of climate record from Camp Century on the Greenland Ice Sheet," *Science,* **166,** 377–381 (1969).
47. W. S. Dansgaard, S. J. Johnsen, H. B. Clausen, and C. C. Langway Jr., "Climate Record Revealed by the Camp Century Ice Core," In K. K. Turekian, ed., *The Late Cenozoic Glacial Ages,* Yale University Press, New Haven, 1971, pp. 43–56.
48. Richard F. Flint, *Glacial and Quaternary Geology,* Wiley, New York, 1971.
49. W. Dansgaard, S. J. Johnson, N. Reeh, N. Gundestrup, H. B. Clausen, and C. U. Hammer, "Climatic changes, Norsemen and modern man," *Nature,* **255,** 24–28 (1975).
50. J. Croll, "On the eccentricity of the earth's orbit, and its physical relations to the glacial epoch," *Philosophical Mag.*, **33,** 119–131 (1867).
51. M. Milankovitch, "Kanon der Erdbestrahlung und seine Anwendung auf das Eiszeitenproblem," Royal Serb. Acad. Spec Pub 133, Belgrade. [English translation published in 1969 by Israel Program for Scientific Translations.]
52. A. Nairn, *Descriptive Paleoclimatology,* Interscience, New York, 1961.
53. W. F. Ruddiman and A. McIntyre, "Oceanic mechanisms for amplification of the 23,000 year ice volume cycle," *Science,* **212,** 617–627 (1981).
54. W. F. Ruddiman and A. McIntyre, "The mode and mechanism of the last deglaciation: Oceanic evidence," *Quat. Res.*, **16,** 125–134 (1981).
55. J. Kutzbach and B. L. Otto-Bliesner, "The sensitivity of the African-Asian monsoonal climate to orbital parameter changes for 9000 years B.P. in a low-resolution general circulation model," *J. Atmos. Sci.*, **39,** 1177–1188 (1982).
56. D. Pollard, "An investigation of the astronomical theory of the Ice Age using a simple climate ice sheet model," *Nature,* **272,** 233–235 (1978).
57. M. Ghil and H. LeTreut, "A climate model with cryodynamics and geodynamics," *J. Geophys. Res.*, **86,** 5262–5270 (1981).
58. G. E. Birchfield, J. Weertman, and A. T. Lunde, "A paleoclimate model of northern hemisphere ice sheets," *Quat. Res.*, **15,** 126–142 (1981).
59. J. Oerlemans, "Glacial cycles and ice sheet modeling," *Clim. Change,* **4,** 353–374 (1982).
60. A. D. Hecht, ed., "Paleoclimatic research: Status and opportunities," *Quat. Res.*, **12,** 6–17 (1979).

2

HISTORIC WEATHER DATA AND EARLY METEOROLOGICAL OBSERVATIONS

Helmut E. Landsberg

University of Maryland,
College Park, Maryland

It has been common practice in climatology to use only the so-called *official* observations. These have been made by calibrated instruments and trained observers at the weather stations of state weather services or by volunteer cooperative observers at substations. These observations are made under uniform rules which have become more and more standardized on a worldwide basis since establishment of the International Meteorological Organization in 1873 (called the World Meteorological Organization since 1951). Only shortly before that time did the various national weather services originate. (In the United States, the Weather Bureau was created by an act of the Forty-First Congress in 1870). The observations of these services are usually subjected to quality control and deposited in central archives. In recent years much of this information has been placed in formats that can be processed by computers. This makes it easy for modelers and others to study climatic fluctuations, but it essentially restricts them to the last 100 years at most. In a post-Pleistocene history of about 10,000 years, that represents only about 1 percent of recent climatic history. It neglects much available material that can throw light on climatic developments and, occasionally, leads to premature judgments on climatic history.

It will be the purpose of this chapter to explore other sources (both instrumental and noninstrumental) of meteorological information, inspect their validity, and assess their use for the study of climate and its variations. It must be emphatically stated at the outset that only a very limited amount of such information has as yet been discovered and even less has been analyzed.

CHRONOLOGIES

There are many interesting sources in which weather is casually or systematically mentioned. Many of these concern weather events that have been thought to have been influential in a historical event. There are literally dozens of these. There are cases from classical periods well reported by Neuman.[1,2] There are the essays of Ludlum on weather and the battles of the American Revolution.[3] The Russian winters which defeated Napoleon and Hitler are too well known to need further elaboration. There was the escape of the German battle cruisers Scharnhorst and Gneisenau (Stöbe, 1978)[4] during World War II. There was that critical forecast for the invasion of France (Stagg, 1971).[5] And what schoolchild does not know about the defeat of the great Spanish Armada in 1588 by the wind? Much could be added to this list; unfortunately these are individual events which give us little insight into climatic history which depends on reasonably gap-less information (see also, Dettwiller, 1981).[6]

Somewhat better in this respect are weather events which have impact on human life and activities. Major catastrophic weather systems always enter the chronicles. Thus we have records of intense, damaging hurricanes (or typhoons) in many parts of the world. In the Caribbean, most large storms have been noted since the early eighteenth century. Certainly when thousands of people or even hundreds are killed, or when vessels founder or sink, the facts are noted. Yet there are problems. Such information does not include off-shore tropical storms and hence early statistics are not comparable to present ones. Thus we cannot draw any conclusions about whether or not storms have increased or decreased in number. Nor is the information adequate for us to study the annual frequency. In the North Atlantic and Caribbean this information may now span a 100 years but only since the advent of satellites, just a little over two decades ago, has the global tropical storm frequency been reasonably well established.

Almost any city, especially in Europe, has a chronicle. Many of these faithfully record events—often for centuries. Lightning strikes that ignited fires are often mentioned. Church steeples in the early days were frequent targets until lightning rods became commonplace. Little or no climatic information can be gleaned from such records. Much ado has been made about the climatological significance of such agricultural information as grape harvesting dates, quantity and quality of wine (LeRoy Ladurie and Boulant, 1981).[7] Similarly, wheat prices have been cited as indices of climate. These concepts are mostly unacceptable as evidence for year to year variations of climate. A single late frost can wipe out a grape crop. Harvest dates while correlated with summer temperatures also have high correlations with other weather events. Wine quality is a function of late summer and early fall sunshine—a piece of climatic information of little significance in isolation. Wheat or other commodity prices are too much influenced by carry-overs of supplies, hoarding, speculation, or trade systems to be reliable climatic indices.

Somewhat better are records of floods on rivers and their dates. Such information has been kept in many settled regions, sometimes for centuries. In Europe the information on high water levels has been marked on riparian structures (figure 1).

Figure 1. High water marks on Main River at the Toll Tower in Frankfurt, Germany. Such flood inscriptions are common in Europe.

Although the flood height may not be comparable to contemporaneous floods because of river regulation, the interval between major floods on a drainage system may well convey some information. The time of year when major floods occur also has relevance. It indicates if melting ice and snow or excessive rainfall was a cause of a flood.

Records of drought are usually even more informative. Droughts are generally a slowly building-up phenomenon. They are usually fairly widespread geographically in contrast to a flood which can be caused by a severe local storm. Although old records generally indicate the existence of drought conditions by crop failure, there is almost always other corroborating evidence such as notice of low river stand; impeding navigation; and drying out of streams, wells, and springs. In hydraulic civilizations, as principally represented by China, centuries of careful stream observations have been kept. They reflect floods, droughts, and conditions in between (Wittfogel and Teng, 1949;[8] Central Meteorological Research Bureau of China, 1977[9]). Drought frequencies and durations in earlier centuries can be compared with current conditions.

Other catastrophic events also enter the chronicles. Thus major inundations in coastal areas as a result of gale-driven waves figure in the European lowlands. In mountainous regions, avalanches take their toll and are sorrowfully recorded. None of these cases that have extensive human impact, however, mean much for climatic history.

PHENOLOGICAL AND OTHER DATABLE EVENTS

In many areas records have been kept of certain events in the local plant growth. Some of these have cultural or religious significance. The best known are the Japanese cherry blossoms. The date of full bloom is determined by the antecedent temperatures of late winter or early spring (Arakawa, 1956;[10] Sekiguti, 1970[11]). The usual measure for the condition prompting progress of plant growth are the accumulated degree days (or degree hours). These are the daily (or hourly) sums of degrees by which the temperature exceeds a specific threshold for a given species of tree or shrub. Schnelle (1981)[12] has recently published all long phenological observations for Europe. An example are the dates of leafing of the horse chestnut in Geneva, Switzerland, for which a long series of observations exists, beginning in 1808 (Lauscher and Lauscher, 1981).[13] Comparisons with the local temperatures, maintained for over 200 years for early spring temperatures, indicate that linear regressions were able to explain about 50 percent of the variance in the leafing dates. This is about what has been noted for the fruit trees and Japanese cherries. From current simultaneous phenological and meteorological statistical relations, for example, a regression between heat sums in spring and blossom time is established. This relation can then be used if a series of phenological observations exists prior to the meteorological observations to estimate the pertinent temperature for the period before meteorological observations started. This will, of course, yield only approximations for a limited period of the year.

A similar procedure can be used for the dates of freezing of rivers, bays, or lakes. Another classical study of Arakawa (1954)[14] pointed the way when he compiled freezing dates for Lake Suva in Japan back to the winter of 1474–75. Gray (1974)[15] compared the recent dates with the simultaneous winter temperatures at Tokyo, and derived a relationship which was then used to estimate winter temperatures of earlier years. The calculated linear regression between freezing date and the Tokyo winter temperature after a few statistical corrections (Landsberg and Kaylor, 1977),[16] explains about 26 percent of the variance. An identical technique was used to reconstruct early Leningrad (St. Petersburg) winter temperatures from available Baltic sea ice cover, to be discussed below. In these cases, again, the temperature of only a single season can be reconstructed.

TRAVEL LOGS

Often considerable climatic insight can be gained from travel logs of early explorers. Created in the days before instruments were available, these are very qualitative—but nonetheless informative. Often the emphasis is on storms or other untoward happenings but some very astute comments survive. It may be instructive in this context to quote from a 1635 account of the Rev. Andrew White, S. J.,[17] of conditions in Maryland, chartered in 1632 to Cecil Calvert. In modern English we read:

> The temper of the air is very good and agrees well with the English, as appeared at their first coming thither, when they had no houses to shelter them and their people were forced, not only to labor during the day but took turns to watch by night. Yet their health was exceedingly well: In summer it is hot as in Spain, and in winter there is frost and snow, but it seldom lasts long; this last winter [presumably 1633–34] was the coldest that had been known in many years; but the year before those was scarcely any sign of winter. . . . The winds there are variable; from the south comes heat, gusts and thunder; from the north, or northwest, cold weather, and in winter, frost and snow; from the east and southeast, rain.

One can only describe this as a very astute summary of the seasonal conditions in Maryland's lowlands. The comparison with Spain in summers is indeed very apt. The Madrid summer temperature in recent years has been about 23°C, its highest maximum 39°C. Compare this with Annapolis with a mean summer temperature of 24°C and a maximum of 39°C. The description of the weather accompanying various winds also closely corresponds to current conditions. The account, though very brief, does not indicate that there have been any radical changes in the area in the last 350 years.

There are a great many other travel reports about other parts of the world. Climatologists have generally ignored them. In the United States much interesting information is contained in the diaries of Mason and Dixon during their surveying activities in Pennsylvania and Maryland, 1763–1768 (American Philosophical Society, 1969).[18] Aside from frequent references to periods of cloudy weather which interfered with the star observations essential for their surveying activities, there are some consistent temperature observations for the period of June 21 to about October 2 of 1767. These were taken at 2 P.M. and obviously approximate daily maxima. The location is not absoutely certain from the text but seems to refer to New Town on the Chester River in Maryland. A maximum of 102°F (38.9°C) is specially noted at 4 P.M. on June 30, 1767, with the comment: "The height of the Fahrenheit thermometer hung in the shade on the north side of a house standing on a hill about three miles eastward of Mr. Harland's." The listed values for early afternoon temperatures look entirely reasonable and fit easily within current observational data in that area during those months.

More elaborate observations, with daily comments on the weather, are contained in the Journals of the Lewis and Clark Expedition, 1804–1806 (Coues, 1893).[19]

The explicit instructions by President Thomas Jefferson, himself a veteran weather observer, given in 1803 directed the following observations:

> Climate, as characterized by the thermometer, by the proportion of rainy, cloudy, and clear days; by lightning, hail, snow; by the access and recess of frost; by the winds prevailing at different seasons; the dates at which particular plants put forth or lose their flower or leaf; times of particular birds, reptiles, or insects.

At the end of their report Lewis and Clark appended a "Meteorological Register" with twice daily notes on wind and weather. Thermometer readings were given until September 1805. The location of the readings shifted with the progress of the expedition and hence the information remains a grab sample. Although such data for times and regions (where climatic material is scarce) will give some insight, the most one can get out of these is a feel for their conformity to or departure from observations collected in a later period.

GENERAL DIARIES

Far more useful for climatological purposes are diaries containing weather information. Some of these personal accounts may only contain casual mention of notable weather events; others make a point of frequent or even daily comment. A second category separately kept for entering weather information is especially valuable. But the occasional notations are not worthless. They often can serve to corroborate other information about a season. Thus, references to late and early frosts or severe hail storms may confirm crop losses. References to ice fairs on rivers bear on winter intensities in the same way that notations that a river was fordable on foot serve as corollary information on droughts.

There are similarly valuable hints in material kept as records of commercial companies on shipping conditions or weather happenings affecting navigation. A number of such records have been usefully exploited for climatological purposes. A notable example are the reconstructions of winter temperatures prior to regular meteorological observations at De Bilt in the Netherlands. Among the basic data were the number of days the canal between Haarlem and Leiden was frozen, in part based on records of barge trips since 1634. This is a century before meteorological observations became available for De Bilt in 1735. Using later observations, a regression between the number of days the canal was frozen and the mean winter temperatures was established. This relation is shown in figure 2, according to van den Dool et al. (1977).[20] The root mean square error for the period of reconstruction is estimated at 1.1 °C. Considering the fact the winter temperatures in De Bilt have varied in the past between −3.0 and 5.5 °C, this is a very satisfactory result and certainly indicates that no particularly mild or severe winters of the reconstruction period have been missed.

A somewhat different analysis has been performed on freezing dates for ports on Hudson Bay, 1714–1871 (Moodie and Catchpole, 1975;[21] Catchpole and Moodie,

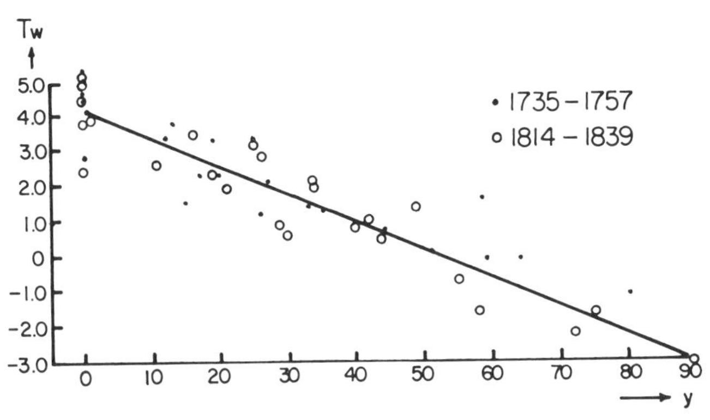

Figure 2. DeBilt winter temperatures derived from freezing observations on the Dutch canals (after van den Dool et al.[20]) (*y* number of days frozen). Reproduced with permission of author and D. Reidel Publishing Co.

1978[22]). This analysis has been based on the observations given in the Hudson Bay Company's post journals. The authors of this work point to the problem of interpretation of the wording found in the journals. These were kept by different employees and occasionally contain ambiguous wording. They refer to the appearance of ice and water. This requires inquiry into what has been referred to as "content analysis" of such documents. Although there is little ambiguity about dates when a bay or harbor is completely frozen, the freezing and break-up stages are expressed in varying terms and comparisons of dates at various posts become more shaky. Still this kind of information for a period when climatic data in northern North America are essentially nonexistent is quite valuable.

The important question raised by this work permeates all evaluations of diaries. Were the concepts in earlier years the same as nowadays? Did the words mean the same thing? Many of the earlier diaries, especially the European ones, were written in Latin. Do the translations of terms convey the correct interpretations? Is there any useful information in expressions such as "severe cold," "cold," "mild," "pleasant," "warm," "hot," "sultry," and others often found in diaries? It is much easier to deal with notations such as "ice or frozen water," "fog," "snow," "rain," "thunder," "lightning." Most of these terms do not suffer degradation when translated from Latin or other foreign language. There is little doubt about the event the diarist describes.

An example of a general diary with frequent weather notes will illustrate the value to climatology of information gathered from a long-kept diary. This was written by Joshua Hempstead from 1712 to 1758 in New London, Connecticut.[23] He was a carpenter who noted the weather character almost daily. This is not unusual among inhabitants of New England where the weather is often inclement and frequently changing. The notes make it easy to establish the frequency of various weather types. This was done for the frequency distribution of the number of annual days with snow in the 45 winters of Hempstead's observations. A comparison of that frequency distribution with recent observations taken at nearby

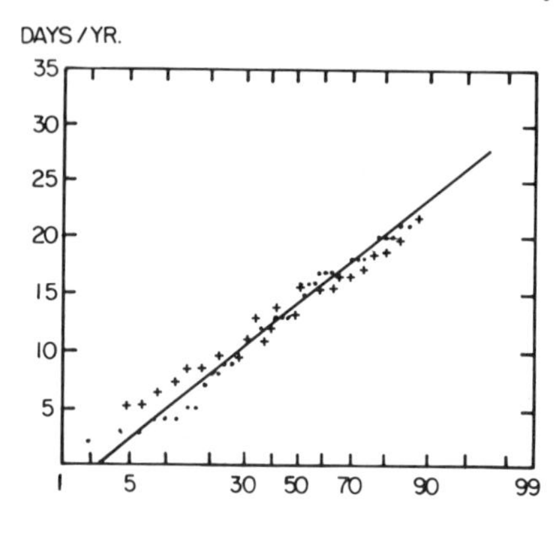

Figure 3. Probability of annual days with snow at New London, Connecticut (dots), as compiled from Joshua Hempstead's diary[23] (1712–1758), compared with present nearby climatic station at Groton, 1948–1977 (crosses).

Groton between 1948 and 1977 show almost identical conditions, as shown in figure 3, where the old and the recent data were plotted on probability paper. In contrast, rain frequency, as gathered from the diary, was noted only on 80 days per year; current data indicate about 102 days of rain per year at that place. The deficit in earlier years does not point to a drier climate in the eighteenth century but rather indicates that small nocturnal rainfalls probably escaped the attention of the diarist. For snowfalls the evidence is usually still there in the morning after a snow falls at night.

There are many more such diaries in the possession of private persons, historical societies, and archives. Their potential for climatological studies remains to be exploited.

WEATHER DIARIES

From at least the fourteenth century onward there are diaries specifically devoted to weather observations—at least in the western world. What may be available in the eastern countries is at present unknown. The earliest such medieval document is a manuscript in the Bodleian Library. The author was William Merle who kept a weather journal from January 1337 to January 1344.[24] These observations were made primarily at Driby in Lincolnshire (where Merle was rector of the church) and also at Oxford. The material reads much like our familiar almanacs. Thus November 1342 is described as follows (in translation from the Latin script, according to Lawrence, 1972): "2nd, heavy fall of snow. 3rd and 4th slight frost. 5th thaw. 6th wind, with rain. 7th stronger W wind than on the 6th. 8th and 9th, slight frost, with hoarfrost. 10th cold with frost, but after mid-day there was no cold until the 15th when there was a slight frost with hoarfrost. 20th, 22nd, 23rd light rain. (21st) strong SW wind. (22nd) wind but not so strong, and before the 21st was W wind. 25th and 26th, light rain. On the last day there was slight frost, with ice and hoarfrost."[25]

There have been a number of attempts to compare the seven years of Merle's observations with current data. Mortimer (1981)[26] also tried to verify some of the observations by records in contemporary chronicles, which seem, in general, to support Merle's observations. Also the material again indicates that these are not radically different from what modern observations reveal. There are also indications in Merle's work that an interpretation of weather events in astrological terms was sought for. This tendency persisted for centuries.

The next weather record, also from Europe, was found in Basel. It contains a daily record from 1399 to 1400. The observer and place of observations is unknown but they originated most likely in northern Switzerland or nearby areas of France (Thorndike, 1966;[27] Frederick et al., 1966;[28] Klemm, 1969[29]). Even with the uncertainty of locality the observations of winds and weather are quite compatible with current observations of the weather in that region.

From the sixteenth century onward, weather observations from Europe multiply rapidly. This was facilitated by the availability of printed calendars which provided writing space. These also contained for each date lunar phases and planetary constellations. The astronomers of the sixteenth and seventeenth centuries, including such luminaries as Tycho Brahe and Johannes Kepler (Frisch, 1868)[30] kept weather diaries in the hope that they might relate to the position of heavenly bodies and thus permit long-range forecasting. Vestiges of this belief persist today.

Flohn (1979)[31] has given a list of Central European weather records of this type for a year or longer (Zürich, 1545–46, 1550–76; Hven, 1582–97; Fürstenfeldbruck, 1587–93; East Friesland, 1588–1613; Linz, 1617–26; Kassel-Rotenburg, 1623–50). These have all been transcribed but little critical analytical work has been done. Flohn has investigated the manuscripts from Ingolstadt (December 1508 to September 1518) and Eichstädt (April 1513 to December 1531), both in Bavaria. He concludes that summer precipitation frequency might have been slightly higher than in modern times. For winter conditions he gives us the following table:

	Ingolstadt (1508–1518)	Eichstädt (1513–1531)	Weissenburg (1901–1930)
Last snowfall	April 6	April 7	April 21
First snowfall	October 27	October 14	November 11
Snow-free interval	203 days	190 days	200 days

Weissenburg is a nearby representative modern station. One must marvel at the very comparable values over a 400-year time span.

Flohn also uses a regression between the snowfall share in winter with the mean winter temperature in Central Europe, developed from modern data

$$T_w = 4.97 - 9.0\ S_f$$

where T_w = mean winter temperature in °C
S_f = ratio of solid to liquid preciptation cases

This simple relation explains about 80 percent of the variance. If this is applied to the data of 1508 to 1513 at the two Bavarian stations, the derived winter temper-

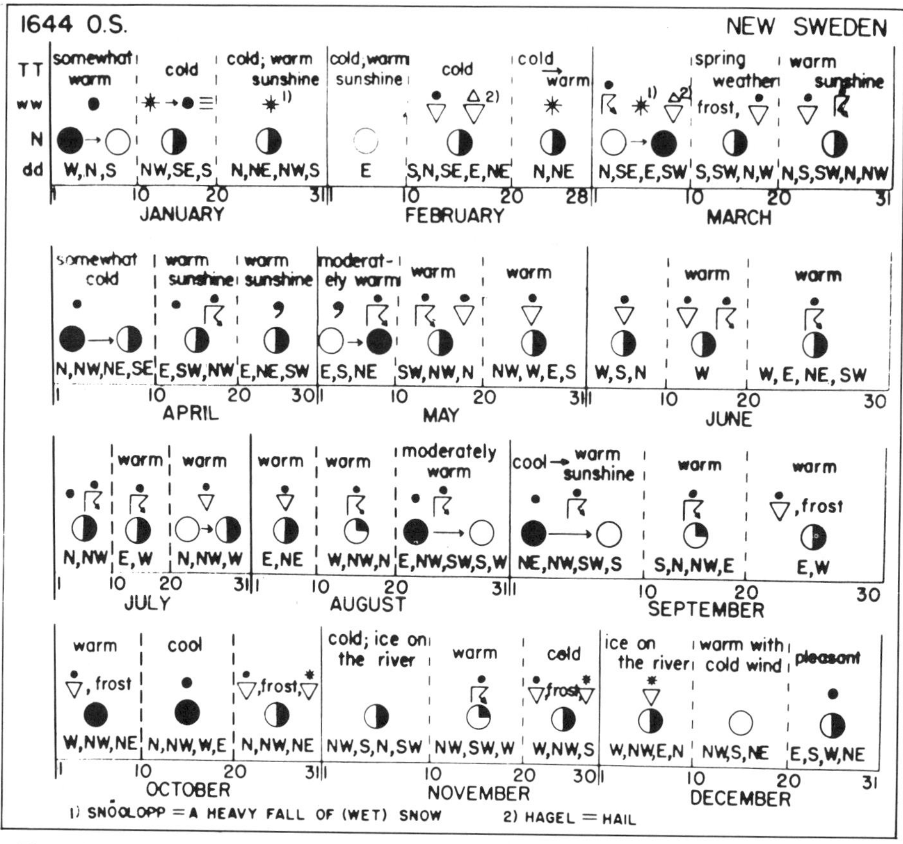

Figure 4. Summary of non-instrumental weather observations by John Companius Holm at Old Swedes Fort (now Wilmington, Delaware) in 1644, as compiled by J. M. Havens,[32] using modern weather symbols designating Temperature (TT), weather (ww), cloud-cover (N), wind direction (dd).

atures lie only 0.2°C below current values, a statistically insignificant value. In the area of the United States we have one early systematic weather record for 1644–45 made by John Campanius Holm (1702), the Chaplain of the Swedish occupation force at old Swedes Fort (now Wilmington, Delaware). Havens (1955)[32] evaluated these observations and represented them in the common current weather symbols (figure 4). In England a search for diaries has yielded some results already (Kington, 1980).[33]

There is still much to be done on the known calendar weather notations of the pre-instrumental period. Hellmann (1926)[34] lists no less than 42 pre-instrumental weather diaries between 1481 and 1700. All of these are from Europe, mostly the central parts. Klemm (1973, 1976, 1979)[35–37] made further searches for systematic weather observations prior to 1700 in the German area. The prejudice of climatologists against the usefulness of this type of information is gradually subsiding and one may hope that their contents will soon be exploited. Similarly an imperative task exists to search for such material in the archives of other nations. Particular targets should be Spanish and Portuguese records for the great period of exploration and the reports from Latin America to their erstwhile colonial masters. Another basic source that requires systematic exploration is the archive of the Vatican. There must be casual or systematic reports from missionaries and bishops from many parts of the world. This must be expected because missionaries of the Catholic Church were often admitted to otherwise xenophobic societies as astronomers, seismologists, or meteorologists. In fact, in later years, meteorological services were started by Jesuits, for example, in parts of China, Madagascar and the Phillipines. In China, where many meticulous records were kept, undoubtedly much relevant material still exists, as indicated by Wang (1980).[38] There is no reason why a reasonable reconstruction of climatic fluctuations from about 1500 onward for much of the northern hemisphere, and perhaps the western area of the southern hemisphere cannot be made from a variety of historical documents.

SHIP LOGS

A notable source of information, of an oceanwide nature, are ships' logs. Since the time of Columbus, these logs have given us information on severe storms encountered at sea. They yielded the first information on the trade winds and the doldrums. They led to the first description and interpretation of the effect of the deflecting force of the earth's rotation (Halley, 1686;[39] Hadley, 1735[40]). The dependence of sailing ships on wind and currents was recognized early. The value for navigational planning of earlier observations from past trips of clipper ships was substantial. Thus merchant and naval fleets kept very meticulous records. In areas of high traffic density, such as the North Atlantic, comparison of earlier and present data has been of great value for analysis of climatic fluctuations. Systematic collections and evaluations of such data by M. F. Maury of the U.S. Navy begun in the 1840s led to the publication of a work on the climatic conditions of the explored oceans.[41] First published in 1855, this book went through many editions and brought world fame to its author. Maury in 1853, with the help of the U. S. State Department, arranged for a conference of seafaring nations to agree on internationally uniform standards of weather reporting by ships.

Such reports were systematically collected. They had reliable wind directions and estimates of wind force according to a scheme devised by Admiral Sir Francis Beaufort (Kinsman, 1969;[42] Crutcher, 1975[43]). They also gave estimates of wave height and of barometric pressure. Lamb (1977)[44] pioneered the use of such data to estimate circulation changes in the North Atlantic. Other ship reports on ice conditions in the North Atlantic, near Iceland, Greenland, and the northern shores of the USSR, have been frequently cited in support of various hypotheses of climatic change. But the vagaries of ice drift and the lack of systematic data on sea-ice thickness makes many interpretations doubtful.

A major contribution of early marine observations has been the compilation of storm tracks at sea, especially of tropical cyclones. The ship observations were a decisive factor in affirming the so-called law of storms which explained the rotary motion of winds around storm centers. A study of the ship reports enabled Reid (1849)[45] not only to contribute decisively to this dynamic problem but also to trace the first track chart of major Caribbean and North Atlantic storms from 1780 to 1847, with extensive quotations from ships logs, including storms from the Indian, Chinese, and Southern Hemisphere areas. References to climatic conditions in many parts of the world and their relations to storms made Reid's book a very important contribution to marine climatology.

There is little doubt that the classical ship observations can readily serve to reconstruct storm tracks. This has not been done for many storms, except those which had some other historical significance. One of these was the tropical storm which hit the Crimea in the middle of November 1854. It wreaked havoc with the French and British fleets during the Crimean War with Russia. It essentially stopped the war. The ships logs, permitted retrospective analysis of the storm using wind and barometer observations. The latter are used more readily when observed on ships at sea level than from early land stations, which are often difficult to reduce to sea level (Landsberg, 1954).[46] Although many early marine observations from ships logs exist, there have been only limited efforts to reconstruct oceanic storm tracks, principally because of the prohibitive cost to collect the material from such a large number of sources.

EARLY INSTRUMENTAL RECORDS (GENERAL)

Instrumental records of meteorological variables began principally in the middle of the seventeenth century—rain gauges were known and used since antiquity. However, no quantitative records seem to have survived. Even though it is known that almost modern-appearing rain gauges were in existence in the fifteenth century in Korea (figure 5), it took three more centuries before interpretable data became available.

Middleton has traced the invention of the meteorological instruments in general (1969)[47] and the developments of barometers (1964)[48] and thermometers (1966)[49] in particular. The early history of the thermometer is a bit clouded, but the group of Florentine scientists sponsored by Ferdinand II of Tuscany was certainly the first to use this instrument routinely. They also sponsored the manufacture of the instrument, encouraged its distribution to other learned groups, and tried to obtain information of readings in various places. Unfortunately their attempts to organize a network of such observations failed for a number of reasons, not the least of which was the lack of uniformity. These will be discussed in more detail below.

The barometer, invented by Torricelli in 1644, was more successful from the beginning, in spite of its greater bulk. The main reason was probably the fact that its fluctuations were at an early date recognized as reflecting weather changes. This was probably the cause of its enormous popularity in the British Isles where weather changes rapidly. Even though its rise or fall was not an infallible prognosticator, every manor soon had one of these instruments. They became elaborate works of

Figure 5. Ancient Korean rain gauge introduced by King Sejong of the Yi dynasty in 1441. The receiver was reconstructed in bronze 1837. The inscription dates from 1770.

art (Goodison, 1968).[50] There was no ambiguity of calibration; the height of the mercury level above the reservoir represented the weight of the atmosphere above. But for absolute values, a correction was needed: a reduction to a standard temperature to counteract the mercury's thermal expansion and the effect of a station's elevation above sea level. The gravity correction caused by the deviation of the earth from perfect sphericity was a minor wrinkle. It took well into the nineteenth century before complete standardization of barometric readings was established. Thus, even where observations of the seventeenth and eighteenth centuries exist, they are difficult to use in a climatic context even though the day-to-day changes reflect the passages of the various weather systems. The popularity of the barometer on shipboard has already been mentioned. No elevation correction was needed and information was principally used to discover rapid falls of pressure, indicating that the ship was approaching a storm and thus warning the crew to take precautions.

The quest among scientists of the seventeenth century to learn more about the weather was lively and it was clear to them that little knowledge could be acquired by single station observations. The latter still served those who searched for astrological connections, which soon centered on lunar phases. But for progress there was a need to establish procedures which would assure comparability. There was also agreement that the instruments needed scales. Only for wind direction observations was there a universal convention, namely the compass rose, and since antiquity it was practice to label a wind by the direction it came from (figure 6). This makes it possible to use notations of wind directions in weather diaries without much hesitation even if no equipment was used. But the wind vanes, often exposed

Figure 6. Renaissance wind vane, reproduced from old woodcut; dial at left represents wind rose on base plate with the ancient Greek wind names.

on church steeples or watchtowers, were so universal that even such observations going back to the Middle Ages are of use if they were systematically kept.

EARLY WESTERN INSTRUMENTAL RECORDS (THERMOMETERS)

The early development of the thermometer in the seventeenth century is somewhat clouded. Its invention has been attributed to a number of persons. Various liquid-in-glass versions evolved, although air thermometers with an index liquid were also used (D'Alancé, 1688)[51] (figure 7). The expansion coefficients of the different liquids used was uncertain and the art of glass blowing was not sufficiently advanced to assure uniform bore for the tubes holding the liquid. But probably the worst deficiency was the lack of uniform calibration. In the beginning the thermometers were related to human sensation scales. Thus some of them (in various languages) are simply labeled: "hot," "warm," "moderate," "cool," "cold," "very cold." We now know that such sensations are not solely governed by temperature but at the lower end also by wind and at the upper end also by humidity. It was not until the beginning of the eighteenth century that uniformly sized capillaries were introduced and that either mercury or colored ethyl alcohol became the thermometer fluids.

Even though some instrumental series of thermometer notations from the seventeenth century have survived, only a few can be interpreted in modern terms. Only two of these early series can be tied in with modern observations, as indicated

Figure 7. Seventeenth century thermometer, exposed outdoors, representing the large variety of early liquid-in-glass thermometers with arbitrary calibrations (from D'Alancé[51]).

below. Those taken in Pisa (1657) and Florence (1658), Italy, remain curiosities and so do some early Stockholm notes. A set of data taken by Rudolph Jakob Camerarius (1665–1721) in Tübingen (West Germany) has been re-evaluated by Lenke (1961).[52] Although Camerarius did make observations until 1717 only the interval from July 1, 1691, to June 30, 1694, was completely published (Camerarius, 1696).[53] Camerarius used a Florentine thermometer, 15 cm long and a bulb of 3 cm. The scale had 40 degrees but they were not evenly spaced. It was exposed near a shaded window and thus would only react very slowly to outdoor changes. Yet by an ingenious procedure Lenke was able to obtain an approximate calibration. He assumed that the shape of the frequency distribution of the observations is not altered in various observational series at the same locality. This procedure has to be applied separately for various times and seasons of observations. With the help of current data he inferred not only the values of the thermometer scale used but also quite reasonable values for monthly values during the three years. Such a record is only of limited worth for climatological purposes.

This leaves us up to now only two long climatological series reaching back into the seventeenth century. One is that for central England, which Manley (1974)[54] painstakingly reconstructed back to 1659. It is based on a variety of sources, some of them on data kept at Oxford by such notables as Robert Boyle and John Locke. The latter not only commented on the small effort required to jot down a daily line about the weather and the instrumental readings, but urged prompt publication of the data. The learned philosopher was two centuries ahead of the times. The central England series has been much discussed and variously analyzed (Mason, 1976,[55] Schönwiese, 1978[56]). Aside from the quasi-biennial oscillation and a long period of 90 to 100 years (attributed by some to the solar Gleissberg rhythm), there is little evidence of any notable climatic change in central England over the past three centuries.

The second long European series covers the interval from 1680 to 1980 by Dettwiller (1981).[57] He comments that although observations in Paris were made since 1664 by de LaHire, the uncertainty of calibration made it impossible to use the earliest data. But actual usable observational data did not become available for Paris until 1757. However, using the central England data and observations at de Bilt (Netherlands), where meteorological observations started in 1706 (Labrijn, 1944)[58] enabled Dettwiller to use regression methods to obtain reasonable monthly and yearly temperature values for the location of Paris-Montsouris back to 1680. It is interesting to assess the magnitude of errors in such reconstructions. Dettwiller evaluated them as indicated in table 1.

For the eighteenth century there is a gradual increase in usable temperature series. These begin in Berlin in 1700 (Brumme, 1978, 1980)[59,60] and, with use of other locations, in de Bilt (Netherlands) since 1706. Interrupted records are available from Ulm since 1710 and from Copenhagen, Denmark, since 1756 (Horrebow, 1780).[61] From Stockholm, Sweden, we have continuous temperature records since 1756 (Hamberg, 1906).[62]

In North America there is a barometric record from Boston for 1725–1726 but the earliest thermometer readings are from 1731 in Germantown near Philadelphia, Pennsylvania, for 1731–32 (Kirch, 1737;[63] Havens, 1953[64]). They were soon fol-

Table 1. Estimated Error of Annual Temperature Values for Paris-Montsouris in Various Intervals (Dettwiller, 1981)

Interval	Error Estimate °C
1680–1705	0.4
1706–1756	0.35
1757–1776	0.2 (maximally)
1777–1967	0.1 (maximally)
1948–1980	0

lowed in 1738 by the observations of John Lining (1743,[65] 1748[66]) in Charleston, South Carolina, and by data from Cambridge, Massachusetts, and Nottingham, Maryland, 1753 (Brooke, 1760a and b).[67,68]

By statistical methods the early American temperature observations from New England, New York, Pennsylvania, Maryland, Virginia, and South Carolina were reduced to a long temperature series, centered on Philadelphia, starting in 1731 (Landsberg et al., 1968).[69] There is still a chance to discover new data that may improve this series and the known fact that thermometers had been brought to America by 1715 lets this work stand as "preliminary." A very complete account of all early European data is found in v. Rudloff (1967).[70]

EARLY EASTERN EUROPEAN RECORDS (TEMPERATURE)

As far as is now known, there are no usable weather records for the vast realms of Russia until Czar Peter I opened a window to the West. With the establishment of the Academy of Sciences came an influx of scholars from Western Europe. Actually they began meteorological work in the capital on the Neva in 1725. An earlier record made by what appears to be two English observers (attributed to the Rev. Tho. Consett, Derham, 1733,[71]) starting in November 1724, is essentially uninterpretable. But much of the data observed, presumably by Academicians F. Chr. Meier, I. G. Leitman, and G. W. Kraft, since December, 1725 seem to be usable for comparison with present data. These data end in 1732 but then a new seies starts in 1751 for the old Russian capital of St. Petersburg (now Leningrad). It is not impossible that as yet unevaluated material exists for the 1732–1750 gap as it is known that Academician Kraft (to 1744) and his successor Academician J. A. Braun continued the observations.

The reconstruction of the early St. Petersburg temperature records is an instructive case history of a salvage operation of old data. The basic material bears no identifying marks as regards the observers, place of observation, or instruments used. There are two manuscripts, judging by the different handwriting, not all readily legible. The first is in German script, the second mostly in Latin with some notations in German. This probably explains why this material lay dormant for 250 years in the archives of the Russian Academy and later in the Main Geophysical Observatory in Leningrad.

Although there were also observations of barometric pressure (in unintelligible units), weather, and wind, emphasis here will be on the temperature observations. The first set of observations covered the period from December 1, 1725 (Julian calendar) to July 1728. The calibration of the thermometer was in terms of "cold" degrees and "warm" degrees. A common practice to obtain a fixed point on these unknown thermometers is to obtain temperature readings when the simultaneous weather observations indicate a transition from rain to snow. This usually indicates a temperature near the freezing point. From a number of cases, this placed 0°C near 21° of the unknown scale. This value pointed to a Florentine thermometer of second construction. Much effort was exerted in later years of the eighteenth century to compare various thermometer scales (Martine, 1740;[72] Cotte, 1774;[73] van Swinden, 1778;[74] Luz, 1781[75]). They all agree closely on this value but a later calibration of a stock of a second variety of (Florentine) thermometers in 1829 by G. Libri (Khrgian, 1959),[76] indicated the freezing point at 13.5°. Brumme (1978[59], 1980[60]) in her reconstruction of the early Berlin temperature records by the Kirch family, arrived at a regression between the readings on the Berlin (T_B) Florentine thermometer and the Celsius scale

$$T_C = 0.67T_B - 13.33. \tag{1}$$

This approximated very closely our derivation for the first St. Petersburg thermometer readings ($T_{P(1)}$)

$$T_C = 0.7T_{P(1)} - 14. \tag{2}$$

This indicates that 1°C is equivalent to 0.67° on the Florentine scale.

The thermometer used in the second interval of early St. Petersburg observations was not as readily identified. The freezing point was located at about 46° on the unknown scale, again identified by the weather transitions from rain to snow. By use of the technique developed by Lenke on the persistence of the shape of the frequency distribution of individual observations, it could be determined that 1°C corresponded to about $3\frac{1}{3}$° on the unknown scale, now designated as $T_{P(2)}$. This is the interval on an old Fahrenheit thermometer whicch preceded his later, much better known scale. The intriguing item about the second St. Petersburg scale was its upside down character, that is, the lower the temperature, the higher the reading. A number of the old thermometer makers adopted a very high temperature value as zero and increased their reading with decreasing temperatures, the reason being that in this way negative numbers were avoided. The first version of the Celsius scale (1742)[77] had the boiling point at zero and the freezing point at 100. A scale developed evidently with Russian conditions in mind by the Frenchman DeLisle (1738)[78] also was "upside down." Its boiling point was at 0° and the freezing point at 150°.

The second scale encountered in the St. Petersburg manuscripts yielded a conversion to Celsius degrees

$$T_C \approx 15 - 0.33T_{P(2)}. \qquad (3)$$

If the available observations are converted to °C for the 63 months that had enough data to permit an approximation of monthly means, the data fall quite comfortably within or near the range of contemporary observations, as taken from the World Weather Records for the last 100 years (figures 8 and 9). This is more evidence for the theory that European temperatures have not changed radically in recent centuries (Landsberg, 1981).[79]

An independent check on the winter temperatures (December–February) was attempted by use of proxy data. For this purpose the Baltic sea ice observations, which date back to 1720, were used (Betin and Preobrajenski, 1962).[80] This series has also been reproduced by Lamb (1977).[81] From the observed winter temperatures at St. Petersburg-Leningrad in the 87 year interval 1871/72 to 1956/57 and the

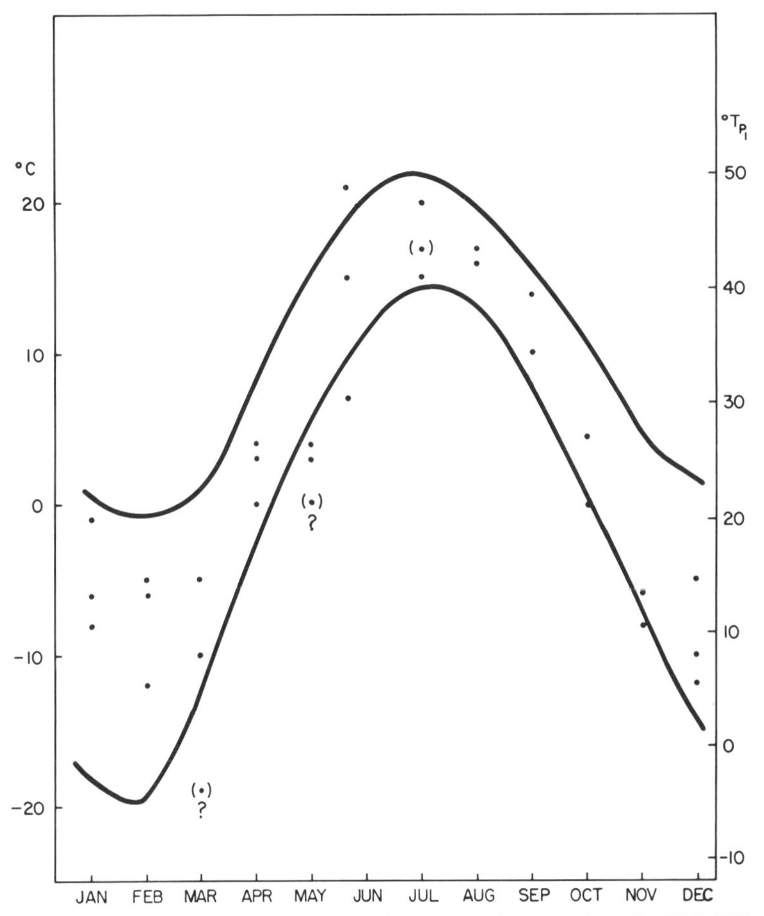

Figure 8. Reconstructed monthly temperatures for St. Petersburg (Leningrad), 1725–1728, in °C. (Right scale original calibration of Florentine thermometer in use at the Imperial Academy of Science.) Solid lines represent envelope of highest and lowest monthly values observed 1871–1970.

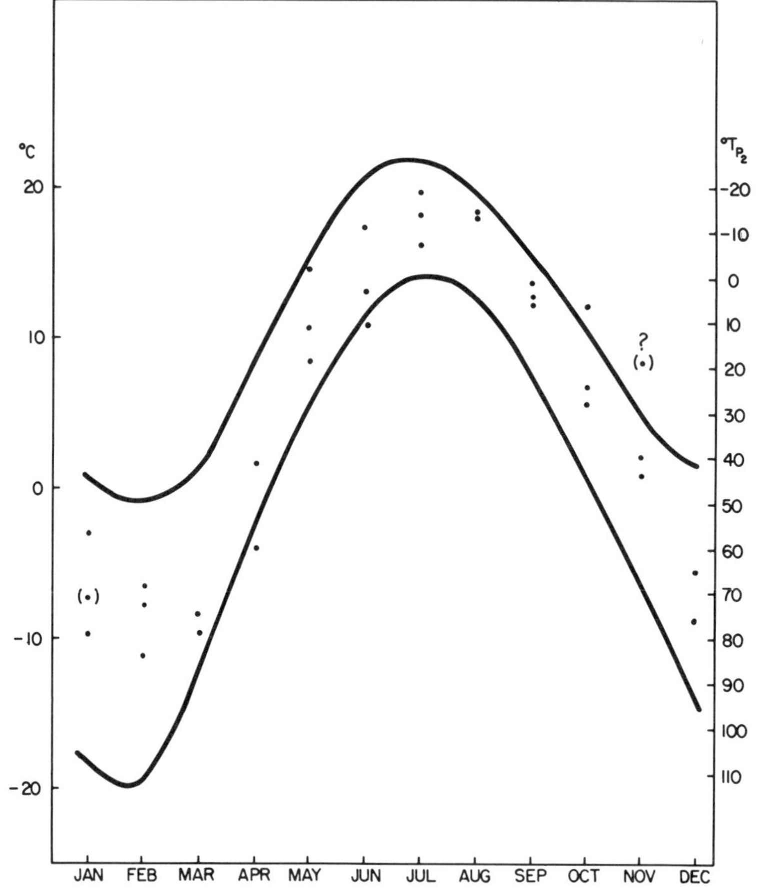

Figure 9. Same as in Figure 8.8, except for 1728–1731 and a thermometer of unknown make with an inverted scale.

Baltic sea ice cover, a regression was established, as follows:

$$-T_{LW} = (B_{sic} + 57)/36 \tag{4}$$

where T_{LW} = Leningrad winter temperature, °C
B_{sic} = Baltic sea ice cover, km^2 × 10^3

The correlation coefficient is +0.83, explaining about 69 percent of the variance at a better than 0.01 level of significance. The data pairs are shown in figure 10, which shows a clear breakdown of the relation when the Baltic is completely frozen. Still 76 percent of the estimates in the comparison interval were within ±2°C, which is about the error one has to expect for such reconstructions from proxy data.

When the whole sea ice data set is used for the interval from 1720 to 1871, a total of 152 years, the estimated mean winter temperature for the city is −8.5°C with a standard deviation of 2.7°C. For the later years, 1872 to 1957, the mean is

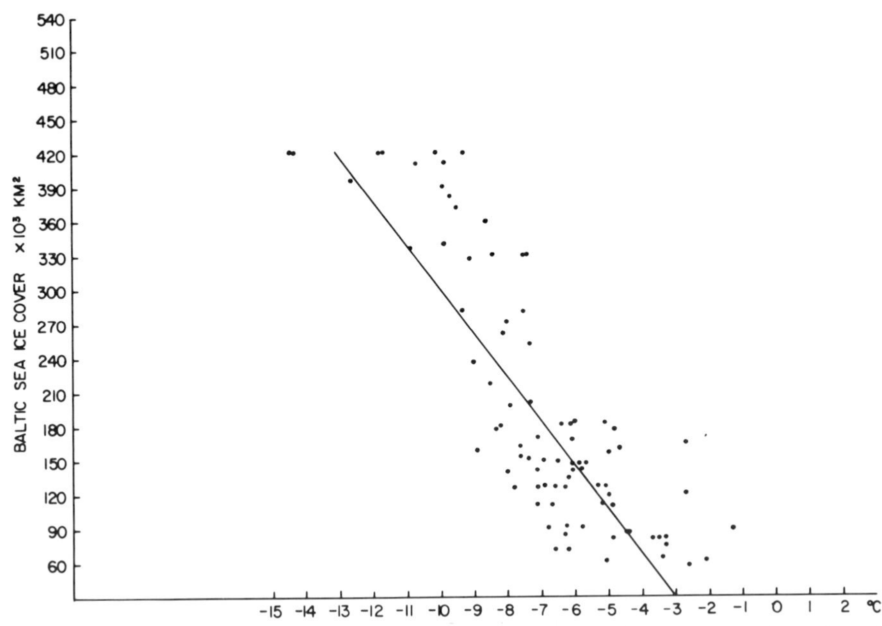

Figure 10. Relation between the extent of Baltic sea ice cover and the winter temperature in Leningrad for the period of simultaneous observations, 1872–1957.

−7.0°C with a standard deviation of 2.6°C. The current (1931 to 1960) mean is −6.6°C. These values may reflect the warming in the early part of the twentieth century and the current value may contain an unknown rise caused by urban influences.

For the reconstructed winter, data from the 1725–1731 interval and a summary table given for 1743–1745 (obtained from the Main Geophysical Observatory, without identification of the observer) are compared with the calculated values. The results are shown in table 2.

Table 2. Comparison of Reconstructed Winter Temperatures for St. Petersburg with Simulated Values Obtained from Baltic Sea Ice Cover

Winter	Observed T°C	Calculated T°C	Difference °C Observed-Calculated
1725/26	−8.3	−11.2	+2.9
1726/27	−5.3	− 6.3	+1.0
1727/28	−8.0	− 8.7	+0.7
1729/30	−9.0	− 3.9	−5.1
1730/31	(−8.0)	−10.0	+2.0
1743/44	−7.8	− 5.4	−2.4
1744/45	−8.1	−10.4	+2.3

Only one value, the winter 1729/30, appears to be quite out of line. The others fall within the same limits established by the test series. But it must be emphasized again that both the reconstructed observations and the simulations are undoubtedly afflicted with large inaccuracies.

THE EARLY THERMOMETER TANGLE

The thermometer, much simpler and cheaper than the more elaborate barometer, was a popular article beyond the circle of natural philosophers. In the eighteenth century a large number of models emerged. Each maker made up his own scale, although there is good evidence that since 1710 (Melin, 1710)[81] there were advocates of centesimal calibrations. Actually only three scales survived the eighteenth century. They were those of Fahrenheit (second version), who divided the range from freezing to boiling point into 180 degrees. (His zero point was the lowest then artifically attainable temperature), the Réaumur scale (1730)[82] with 80° between freezing and boiling point (based on the coefficient of expansion of mercury), and the inverted Celsius temperature scale (see, for example, v. Oettingen).[83]

But the various thermometer makers, some of them scientists, others instrument artisans, were not necessarily swayed by the plain logic of using the simple calibration marks of freezing point and boiling point (at sea level pressure) but proceeded to produce a welter of scales. Some comparisons were provided by Martine, 1744,[72] Cotte, 1744,[73] and van Swinden, 1778.[74] Evidently not all scales have survived. Khrgian (1959)[76] states that there were 60 of them. Middleton (1966)[49] cogently remarked that different batches of thermometers, even from one instrument maker, showed discrepancies. The interest in air temperatures early in the eighteenth century was greatly stimulated by an intense winter in 1708/09. It was extremely cold in eastern, central, and western Europe. The Baltic was frozen and the Danish astronomer Rømer (Meyer, 1913)[84] reported from Copenhagen the lowest temperatures ever recorded. The young instrument maker Fahrenheit, hailing from the Baltic city of Danzig (now Gdansk), is alleged to have fixed the lowest point on his thermometer according to the lowest temperature there. Much speculation by the contemporary natural philosophers centered around the question of whether or not there could be any lower natural temperature.

Thus many thermometers, different in shape, rationale, and scales, were manufactured. Fahrenheit's were the first to achieve uniformity and to accept reasonably well reproducible calibration points. They were the sal ammoniac-ice mixture at zero and the boiling point of water at sea level at 210 degrees (with 100° erroneously interpreted as normal human body temperature). Soon thereafter the French Marquis de Réaumur (1713)[82] designed his thermometer, using alcohol as fluid. He adopted the freezing and boiling points of water as the fixed points. He determined the coefficient of expansion of alcohol, setting the column length at 1000 at the freezing point and at 1080 at the boiling point of water. This set the rationale of the 80 degree Réaumur scale from freezing to boiling of water.

After another dismal European winter (1739/40, Celsius (1742)[77] adopted the centesimal scale for the freezing to boiling point of water thermometer expansion.

There is some question if he deserves priority because Christin (1745)[85] had already suggested the same division and about the same time in an obscure place. After that notorious winter, when the Baltic was frozen over again and Academician Krafft (1741)[86] reported on the curious ice house on the frozen Neva, standardization of thermometers advanced rapidly.* Comparative calibrations were made (van Swinden, 1778;[74] Luz, 1781[75]). In the nineteenth century, only the Fahrenheit, Réaumur, and Celsius scales survived. In the twentieth century, for meteorological and other scientific work, only the Celsius scale is in use.

For work in climatic history it remains essential to be aware of the calibrations of all the thermometers for which such information could be obtained. There is hope that more interpretable temperature observations may yet be discovered. Every addition will give us more insight into climatic fluctuations in historical time. Further, every measured value is another, if a small, step away from speculation. Up to now no less than 36 scales (not including the Celsius scale itself) have been noted and there is some evidence that others may have existed (Khrgian, 1959.[76] After an earlier tentative publication of 35 calibrations (Landsberg, 1981,[79] J. Dettwiller called attention to a 36th which was contained in a paper by Renou (1876).[87] These comparisons are shown in table 3, from a variety of sources, cited earlier. They show our estimates of the freezing points and the conversion to °C. Others have reported different freezing points, due to the uncertainties noted above. A quote from Luz (1781),[75] in translation, is interesting here: "The calibration of Florentine alcohol thermometers was so arbitrary that a comparison of, say, ten of them would have shown not one to agree with any of the others." It must be emphasized that there remain the uncertainties of deviations of individual thermometers because of errors in the manufacturers' calibrations. One of the earliest thermometers was that of de LaHire, used in 1664. From the work by Cotte (1774)[73] our table gives the freezing point as 30°LH but Dettwiller called attention to the fact that Renou placed it at 31.25°LH. Such discrepancies occur throughout the comparisons of early thermometer scales. Thus one can at best hope for a reliability of about 1°C for the monthly means obtained from these observations.

EARLY PRECIPITATION OBSERVATIONS

With regard to precipitation records, there are many qualitative references throughout recorded history. They are couched in terms of floods and droughts. We have already referred to river or stream flow data (see section on Chronologies). But quantitative information has not become available until the end of the seventeenth century. This refers to precipitation at Kew near London, England. There are two notable contributions. One refers to Kew since 1697 (Wales-Smith, 1971),[88] the other to 11 regions, covering nearly all of England since 1725 (Craddock, 1976).[89]

*It may be of some interest here to quote Academician Krafft on the lowest minima for Europe collected for him for that winter in °C. 1739 Dec. 30: Danzig −23°, Frankfurt −43°(?), Hamburg −22°, Hague −23°; Dec. 31: Haarlem −18°, Wittenberg −23°. 1740 Jan. 25: St. Petersburg −34°; Jan. 27: Berlin −23°; Feb. 5: Uppsala −28°; Feb. 14: Basel −20°, Leipzig −29°, Weimar −24°. Without date London −22°. It might also be noted here that Dettwiller (1981)[57] judges the year 1740 as the coldest observed in the last 300 years in Paris.

Table 3. Some Thermometer Scales Used in the Eighteenth Century Compared to °C[a]

Thermometer	Freezing Point (0°C)	1°C Difference	Approximate Conversion ($Tx \rightarrow Tc$)
Allgöwer	−30°Al	3.75°Al	0.26°Al + 8
Amontons	50.5°Am	5°Am	0.2°Am − 10.3
Barnsdorf	7°B	0.625°B	1.6°B − 11.2
Brisson_I	$0°\text{Br}_I$	$0.87°\text{Br}_I$	$1.15°\text{Br}_I$
Brisson_{II}	$0°\text{Br}_{II}$	$0.83°\text{Br}_{II}$	$1.21°\text{Br}_{II}$
Camararius	13°Ca	0.6°Ca	1.7°Ca − 22.1
Christin	0°Chr	1°Chr	°Chr = °C
De Luc_I	$0°\text{DL}_I$	$0.8°\text{DL}_I$	$1.25°\text{DL}_I$
De Luc_{II}	$0°\text{DL}_{II}$	$0.65°\text{DL}_{II}$	$1.53°\text{DL}_{II}$
du Crest	~10°dC	1.2°dC	0.83°dC − 8.3
Crucquius	1070°Cr	6.75°Cr	0.15°Cr − 160.5
Edinburgh_I	$8.2°\text{E}_I$	$5.47°\text{E}_I$	$0.18°\text{E}_I - 1.5$
Edinburgh_{II}	$15°\text{E}_{II}$	$0.82°\text{E}_{II}$	$1.22°\text{E}_{II} - 18.3$
Fahrenheit (old)	30°σF	3.33°σF	0.3°σF − 9
Fahrenheit (new)	32°F	1.8°F	0.56°F − 18
"Florentine"$_I$	$13.5°\text{Fl}_I$	$2.22°\text{Fl}_I$	$0.45°\text{Fl}_I - 6.1$
"Florentine"$_{II}$	$20°\text{Fl}_{II}$	$0.67°\text{Fl}_{II}$	$0.67°\text{Fl}_{II} - 13.33$
Fowler	34°Fo	1.33°Fo	0.75°Fo − 25.5
Fricke	33°Fr	2.81°Fr	0.36°Fr − 11.7
Hales	0°H	2.0°H	0.5°H
Hanow (old)	$30.0°\text{Hn}_I$	$1.65°\text{Hn}_I$	$0.61°\text{Hn}_I - 18.3$
Hanow (new)	$40.0°\text{Hn}_{II}$	$1.65°\text{Hn}_{II}$	$0.61°\text{Hn}_{II} - 24.4$
Kirch	7.0°K	0.45°K	2.22°K − 15.5
La Hire	30°LH	1.44°LH	0.44°LH − 13.2
de L'Isle	150°dL	1.5°dL	100 − 0.67°dL
Messier	0°M	0.853°M	1.172°M
Newton	0°N	3.125°N	0.32°N
"Paris"	25°Pa	0.69°Pa	1.45°Pa − 36.3
Patrice	32°Pt	0.875°Pt	1.14°Pt − 36.5
Poleni	47.5°Po	0.22°Po	4.5°Po − 214
Réaumur	0°R	0.8°R	1.25°R
Richter	18°Ri	0.82°Ri	1.22°Ri − 22
"Roy. Soc." (Hauksbee)	77°RS	3.125°RS	30.8 − 0.4°RS
"Royale" (Soumille)	0°Ro	0.6°Ro	1.67°Ro
Schalch	24°Sch	1.2°Sch	0.56°Sch − 13.4
Suchodolez	24°Su	0.9°Su	0.42°Su − 10.1

[a]The Celsius scale is used here in its present form, not in the inverted way first given by Celsius (1742).

From Zürich, Switzerland, there are interpretable rainfall records since 1708 (Pfister, 1978).[90] From that part of the world there are also some rather interesting data on the duration of snow cover for the outgoing seventeenth and early eighteenth centuries (Pfister, 1977).[91] Since 1715 there is a series in the Netherlands for Hoofddorp-Zwanenburg published by Labrijn (1945).[56] An interesting footnote to history of rainfall observations is the fact that Sir Christopher Wren, principally noted as an architect, constructed the first recording rain gauge in 1663. In Italy a long series of precipitation without gap is now available for Padua (Polli, 1943).[92] In North America the oldest precipitation records date back to 1738 when John Lining started observations at Charleston, South Carolina.[65,66]

At a very early stage rain gauges were developed in the Far East. There the significance of the monsoonal precipitation for crops, especially rice, can not escape attention. Thus we find records of raingauges in Korea as early as the fifteenth century (Wada, 1911).[93] No records from these early times have survived but an uninterrupted record for Seoul, Korea, has been reconstructed by Arakawa, (1944).[94] By the end of the eighteenth century, a substantial number of rainfall records were available in Europe. A few are available from North America. In India the first series did not begin until 1813 (Madras).[95] In South America and Australia, records back to the middle of the nineteenth century are available; for the Near East a record back to 1846 (Rosenan, 1955)[96] exists for Jerusalem. Continuous records for Africa are scarce although early data in St. Louis in Senegal exist since 1738 (Nicolas, 1959).[97] It is no exaggeration to state that it was not until the twentieth century before a reasonably complete climatological picture of precipitation for the world's land areas could be established. It is now (1980) estimated that around 170,000 rainfall stations are in operation. The dearth of information in the earlier centuries necessitates reliance on proxy data, and great caution in the interpretation of the existing instrumental observations must be advised.

OTHER TYPES OF WEATHER OBSERVATIONS

Although information on climatic conditions obtained for temperature and precipitation, either by instrumental observations or inferred from proxy data is a mainstay of paleoclimatic analysis, other elements should not be entirely neglected. We have already seen the usefulness of frequencies of events such as snowfall, but many other similar eye observations contained in sundry old sources can be utilized. Aside from precipitation, clouds are most frequently recorded, in part because they hinder observations of the sun during the day and the stars at night. Astronomical observations, of course, date back to antiquity; certainly since the sixteenth century, the heyday of astrology, the skies were eagerly watched. Thus observatory records often indicate the sky conditions. Attempts have also been made to infer them from reports of meteorite sightings which require a clear sky. Yet this is somewhat shaky evidence which has to be accumulated for decades.

Astronomical records often contain sightings of sunspots and of auroras. Climatologically these are of importance because of the claims that solar activity has some influence on climatic fluctuations (Eddy, 1977).[98] (This topic will be touched upon again later. Here only the significance of such observations for prevalent weather conditions will be discussed.) Quite revealing in this connection are sunspot sightings with the naked eye. These can only be made when solar brightness is sufficiently reduced by obscuring material in the air, usually dust. Most of these observations come from the Far East, China and Korea (Willis et al., 1980).[99] The combination of good record keeping and vast dust clouds blowing off the enormous Loess deposits and the interior Chinese deserts have given us these observations. They show a distinct seasonal variation for 139 sightings from the first to the seventeenth century of sunspots (figure 11). This reflects the annual variation of wind speed, with an early spring maximum and a midsummar minimum. This is still the annual wind pattern of recent periods. It is questionable whether any other inferences of a climatic nature can be drawn from these sightings.

Wind observations have also been made since antiquity. They were related by the Greek physicians to the outbreak of various diseases and the sailors of the various nations surrounding the Mediterranean were keen observers of the air currents. However, little but generalities have survived. For interpretable observations one has to wait for the outgoing Middle Ages. Much useful material has been preserved in England and H. H. Lamb (1972) has diligently compiled and analyzed such data for London and eastern England since the fourteenth century. From the frequency of various wind components he has drawn inferences about the circulation patterns in the eastern North Atlantic for various decades from about 1350 to 1960. Some of the information had to be interpolated. But the share of Southwesterly components varied in that interval from about 19 to 35 percent of all wind observations (figure 12). In an area so close to one of the major centers of action in the general circulation—the Icelandic Low—this must represent notable interdecadal fluctuations. It must, however, be stated that sensual observations of wind direction are often biased. The direction gustiness of wind, even if a wind vane is available, makes it difficult to gauge the average compass point the wind comes from. This

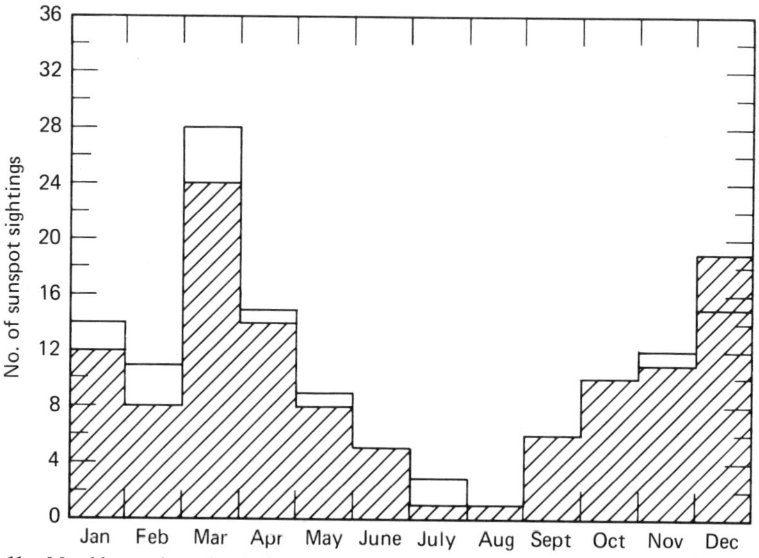

Figure 11. Monthly number of naked-eye sightings of sunspots in China from 28 B.C. to A.D. 160. Cross-hatched part refers to those with reliable dating, open part is uncertain in dating. Such sightings are only possible with large amounts of wind-blown dust suspended in the air. (From Willis et al., 1980.)[99]

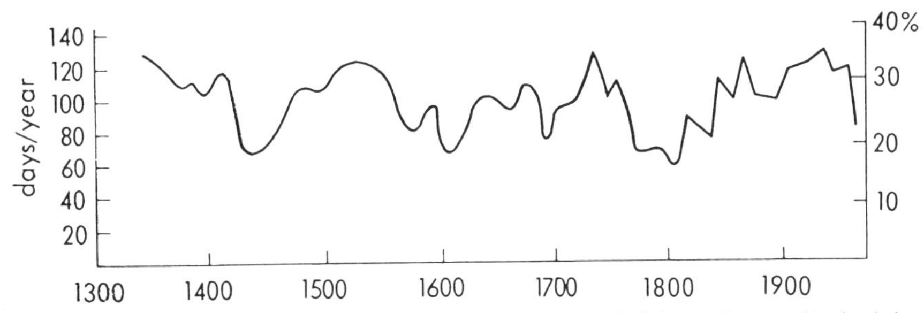

Figure 12. Frequency, in days per year, of southwesterly surface winds in southeastern England since 1340, by decades. Reproduced with permission of author.[124]

applies not only the old observations but is still applicable at present. Such observations show a distinct bias toward major directions.

Evaluation of the St. Petersburg (Leningrad) weather diaries for 1725–1732, show the predominance of westerly directions. The three components SW, W, NW share in winter 53 percent, in spring 40 percent, in summer 46 percent, and in autumn 54 percent of all observations. In this series it is evident that frequency of calms is usually underestimated. The interval of the St. Petersburg observations shows also a fairly high frequency of westerly components as in Lamb's analysis for England. They still predominate in Leningrad during the most recent decades.

Most weather diaries contain wind direction observations and, although frequency analyses are useful, there are other particulars which can be gleaned from these, especially if more than one observation per day is available. From windshifts it is possible to reconstruct the frequency of frontal passages. These become, of course, more certain if other corollary weather information is available, such as barometer and thermometer readings, onset and cessation of precipitation, and changes in cloud cover.

Frequency counts of various weather types may permit comparisons with current conditions but often are disappointing. Two examples will illustrate the potential complications. The first is again drawn from the St. Petersburg (Leningrad) diaries. These are shown in table 4.

The agreement of the number of cloudy days over two centuries apart must be

Table 4. St. Petersburg Weather Types, 1725–1732 and Current Data (days/year)

Type	Diaries	Current Data
Cloudy	174	172
Precipitation	96	194
Thunderstorms	3	15
Fog	4	57
Gale	3	2.4

viewed with caution. It may well be fortuitous. The difference in number of days with precipitation is probably ascribable to lack of equipment in the early series, where probably many nocturnal rainfalls escaped attention. Although it is uncertain how "gale" was defined in the eighteenth century, one might accept the frequency as real. In case of thunderstorms the discrepancy is startling. It seems fairly hard to miss thunder and lightning. A discrepancy in definition in this case is rather unlikely. This is more likely in the case of fog. The old observers probably only noted down cases of dense fog with sharply reduced visibility, whereas modern practice might include events with visibilities up to 1 km. It is also quite possible that increased urbanization and industrialization has raised the number of fog days.

A second example can be drawn from the records of the Rev. Anton Pilgram (1788)[100] in Vienna, Austria. He summarized observations from 1765 to 1784, a total of twenty years. The annual variation of various elements is shown in table 5. Pilgram interestingly enough normalized all his months to 30 days but the table has rounded all frequencies, where appropriate, to full days. The annual variation of the various elements looks entirely reasonable and is consonant with current conditions. But in individual elements there are discrepancies in the frequencies. Thus the old work reports only 8 thunderstorms per year, compared with about 26 per year at present. Is it a matter of definition again? Some of the old chronicles report only thunderstorms when cloud to ground lightning caused some damage, whereas today audible perception of even a distant thunder is recorded. In case of fog frequency, Pilgram notes 74 per year but in modern times only 52 per year are observed. This may have a number of causes. One is again the definition. Another could be the location of the place of observation. In Pilgram's day this was undoubtedly the densely settled inner city with many domestic fires, while the modern observatory is on an elevation in the far more open suburban areas. For precipitation the old records show 84 days per year, the modern series gives for days with ≥ 1 mm a total of 96 days per year. Here again the omission of nocturnal rain may have reduced the earlier information.

It remains to add a brief comment on other phenomena shown in the diaries. There is frequent mention of hail because of its damaging effect on crops. Here again any frequencies are generally biased toward the cases that had an impact

Table 5. Frequency of Weather Character in Vienna, Austria, 1765–1784, after Pilgram (1788).[100]

Weather	Days per Month and Year												
Type	Jan.	Feb.	Mar.	Apr.	May	June	July	Aug.	Sep.	Oct.	Nov.	Dec.	Year
Clear	4	5	8	9	11	9	9	12	11	8	3	4	93
Changeable	8	11	12	12	13	14	15	12	10	10	9	4	130
Cloudy	18	14	10	9	6	7	6	6	9	12	18	22	137
Rain	2	8	5	9	8	9	9	8	7	7	8	4	84
Snow	6	6	3	1	0.1	—	—	—	—	—	3	5	24
Thunder	0.4	—	0.2	0.6	1	2	2	2	0.4	0.1	0.1	0.1	8
Fog	11	12	9	4	1	0.5	0.5	2	2	7	12	13	74

while others are not recorded. Similar problems exist with observations of frost. From the context it is often quite clear that not a slight, white ice crystal deposit is referred to but an event that caused damage. In contrast to frost, references to ice formation are far more reliable and useful.

EARLY METEOROLOGICAL NETWORKS

It was recognized by the early meteorologists of the seventeenth century that better insight into weather and climate would be gained by observations from many locations. The effort by the Florentine Academy to start such a network was abortive because of the nonuniform equipment, lack of procedures and instructions, and difficulties of communications.

In 1723 the Secretary of the Royal Society of London, Jacob (James) Jurin, M.D., tried to remedy the situation by issuing a call for meteorological observations to be made under a uniform scheme.[101] He gave very explicit instructions as to how a meteorological journal should be kept. Barometric pressure, temperature, weather, and wind direction should be listed. He also drew up a scheme of estimating wind speeds, an early forerunner of the Beaufort scale. Force 1 was to indicate the lightest winds barely moving leaves and 4 the most violent winds, with 2 and 3 somewhere between these extremes, and 0 reserved for complete calm. Rain amounts were also included in the scheme, as was a sample table (figure 13). The proposal, although extremely meritorious at the time, had two major flaws. One was the recommendation to use instruments by a London instrument maker, Francis Hauksbee, Jr. Unfortunately this artisan's thermometer calibrations seemed to have been very haphazard. The freezing point variously described at somewhere between 65 and 77. Worse yet was Jurin's sugggestion to hang the thermometer in an unheated north room instead of an outdoor location. The lag of such an indoor location, even though radiation protected, distorted temperature variations to such an extent as to vitiate most observations made according to this scheme. Jurin requested also that the journals be sent to the Royal Society for publication but very few actually appeared in the Philosophical Transactions, often apparently in condensed form. Jurin's exposure scheme was promptly criticized. Johann Weidler is credited by Middleton (1966)[49] to have indicated in 1729 the need for outdoor exposure in the shade.

There was some dormancy in the efforts to organize meteorological observations. Toward the end of the eighteenth century, a better constellation arose. Much improved instruments had become available and a dedicated scientist with a powerful sponsor started a new network. It all started at the capital of Karl Theodor, Prince Elector of the Palatinate, who was an ardent supporter of the arts and sciences. In Mannheim was a major cultural center of the period and its science academy was founded in 1763. One of its most active members was the court chaplain Johann Jacob Hemmer (1733–1790). Hemmer, who was also a notable physicist, persuaded the Prince Elector in 1779 to create a meteorological division of the Mannheim Academy. This was established in 1780 as Societas Meteorologica Palatina in 1780. The Elector's decree charged it with the establishment of a worldwide network of

Diarii Forma.

Dies & Hora 1723. Nov. St. V.	Barom. alt. dig.dec	Therm alt. gr. dec.	Vent.	Tempeſtas.	Pluvia. dig.dec
1. 8 *a. m.*	29.75	49 . 6	S. W. 1	Cœlum nubibus obduct.	0.035
4 *p. m.*	29.56	47 . 3	S. W. 2	Imbres interrupti. Sol pervices intercurrens	0.043
2. 7 ½ *a. m.*	29.24	48 . 5	S. 1	Pluvia fere perpetua	0.725
3. 9 *a. m.*	29.95	49 . 7	N. 1	Cœlum ſudum	0.032
5 *p. m.*	30. 4	49 . 2	N. 1	Cœlum ſudum	0.000
4. 7 *a. m.*	29. 9	47 . 0	S. W. 1	Nubes ſparſæ	0.000
10	29. 7	46 . 2	S. W. 2	Imbres intercurrentes	0.103
12	29. 4	45 . 0	S. 3	Cœlum nubibus undique fere tectum	0.050
3 *p. m.*	28. 8	46 . 0	S. 4	Nubes ſparſæ	0.000
5	28. 6	47 . 2	S. W. 4	Eadem Cœli facies	0.000
7	28. 9	48 . 0	S. W. 2	Pluit	0.000
9	28. 9	48 . 2	0	Pluvia fere perpetua	0.305
5. 7 *a. m.*	29. 7	53 . 4	N. E. 1	Sudum. Gelu.	0.250

Figure 13. Scheme of weather observations proposed by the Secretary of the Royal Society, J. Jurin, in 1723.[101]

meteorological observations, directed his diplomatic representatives to deliver instruments to foreign correspondents, and the return of data through diplomatic channels. It appointed Hemmer as secretary and provided funds for the procurement of equipment. Hemmer calibrated all meteorological and magnetic instruments (figure 14*a,b*); issued instructions which provided for three observations per day; and sent invitations to national academies, learned societies, and individual scientists to participate in the observations. He got more responses than he could furnish with equipment (with some notable exceptions, including the Royal Society, which remained silent).

The scheme might have been unsuccessful had it not been for the collection of the returns. Many of these came from Hemmer's learned ecclesiastical brethren and were carefully filled out according to the basic instructions. All the returns were meticulously examined and edited by Hemmer. Most important, they were *published* (Societas Meteorologica Palatina, 1783–1795).[102] Eventually 39 localities sent in their data to Mannheim. Most of the European countries were represented (except England). There were three Russian stations, one in Greenland, and two in Massachusetts. There were 12 volumes of *Ephemerides* (figure 15) published between 1783 and 1795, with all daily observations. Some of them were edited by Hemmer's successor, Medical Counselor Güthe. They cover observations from 1781

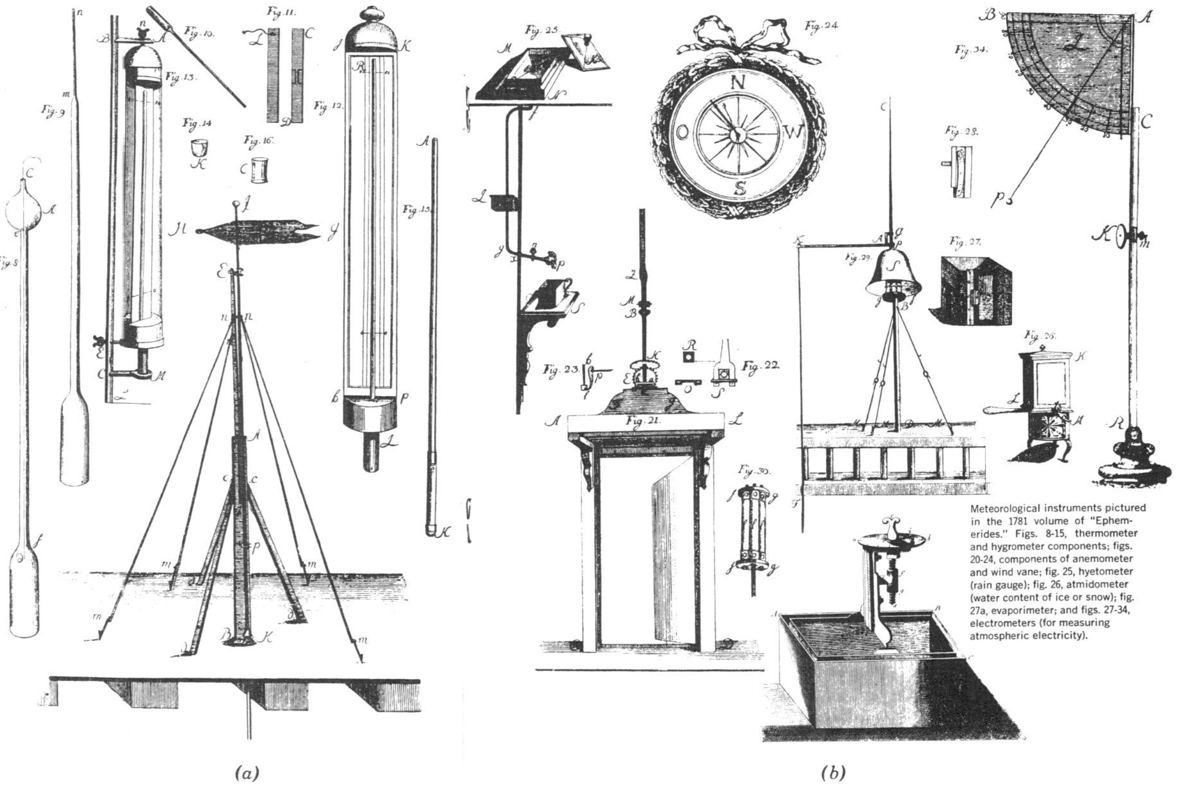

(a) (b)

Figure 14. Illustrations of instruments used by the Palatine Meteorological Society in its network (From Societas Meteorologica Palatina, Ephemerides 1781).[102] (*a*) Figures 8–15 show parts and assemblies of thermometers and hygrometers. Figure 16 a wind vane. (*b*) Figures 25, 26, and 27 show rain gauges and evaporimeter, the other figures are for atmospheric electric and magnetic observations.

OBSERVATIONES ROMANAE

Horae obfervationis ordinariae 7 mat. 2 pom. 9 vefp.

Tyberis obfervatur circa meridiem. In columna Meteor. Signum + indicat *infenfibilis.*

J a n u a r i u s.

Dies	Barom.	Th. juxta barom. fufpenf.	Th. libero aëri ex pof.	Hygr.	Declin.	Ventus.	Pluvia.	Evap.	Tyberis.	Luna.	Coeli fac.	Meteora.
	dig. lin. dec.	gr. dec.	gr. dec.	gr dec.	gr. min.	direct. vires.	part. lin. [illegible]	lin. integrae cum [illegible]	ped. dig			
1	27, 8, 2 8, 1 9, 9	9, 6 8, 9 9, 0	7, 1 10, 9 7, 9	23, 8 31, 5 30, 0	16, 37 40 39	N N O 2 N 2 N 2			-0, 1	♌ 5 h. 25 m. mane.	= = a. cin. ∞- a. t. ☉	
2	28, 0, 3 0, 9 1, 8	8, 8 8, 6 8, 7	4, 6 7, 8 4, 6	39, 3 50, 4 44, 5	16, 40 41 40	N N W 1 N 2 N N O 2		2, 15	-0, 2 2/3	♌	☉ ☉ ☉	
3	28, 2, 5 2, 1 1, 5	8, 2 8, 4 8, 5	1, 0 6, 1 4, 7	36, 0 34, 8 29, 9	16, 38 41 38	N N O 1 N N O O S O		1, 35	-0, 2	♍ 2h. 6m. vefp.	☉ = a. l. ☉	
4	28, 0, 7 1, 1 2, 0	8, 4 8, 3 8, 5	4, 8 9, 5 5, 9	25, 4 36, 3 37, 6	16, 38 42 40	N N O 1 N 1 N N O	437	1, 5	-0, 3 3/4	♍	☉ ☉ ☉	⁞ Hora 1/2 mat.
5	28, 1, 6 1, 1 0, 9	9, 1 8, 2 8, 3	3, 3 6, 9 4, 8	33, 7 36, 4 33, 4	16, 39 42 39	N N N O N 1		0, 46	-0, 6	♎ 6 h. 25 m. vefp.	⋲ a. fafc. ∞- a. t. fafc. ⋲ cin. obf.	⁙ hora 7 mat.
6	27, 11, 7 10, 8 10, 9	8, 1 8, 4 8, 6	5, 9 10, 3 7, 0	30, 6 30, 2 25, 6	16, 40 41 38	N S W 1 N O 1		0, 43	-0, 6 1/2	♎ ☾ 11h. 28m vefp.	∞- r. cin. ∞- a. cin. ☉	⁙ hor. 7 mat. rara.
7	28, 0, 4 0, 8 1, 1	8, 2 8, 2 8, 3	4, 5 8, 3 6, 6	24, 1 28, 4 27, 1	16, 40 43 39	N N W		0, 39	-0, 8	♏ 9 h. 51 m. vefp.	⊕ cin. l. = a. ⊕ obf.	⁙ hor. 7 mat.
8	27, 11, 3 9, 9 10, 8	8, 2 8, 2 8, 3	6, 8 8, 7 7, 0	25, 1 25, 3 24, 4	16, 40 43 38	S 1 S S O N 1	252	0, 38	-0, 8 2/3	♏	= = cin. r. ∞- a. l. ☉	⁞ h. 4 mat. ⁞ h. 7 mat. ⁞ h. 1. 1/2 pom.

Figure 15. Sample table of observations from its network published annually by the Societas Meteorologica Palatina in its Ephemerides.[102]

to 1792. The war disturbances of the French revolution terminated the activities of the Mannheim Academy. But many of the stations then active continued to the present day. Two of them are in exactly the same locality as they were at that time. One is Prague (Praha) and the other is in Hohenpeissenberg in Bavaria (Belohlavek, 1977;[103] Attmannspacher, 1981[104]). Observations at many of the other localities of the network have also continued since but with station shifts; other had later interruptions in the records.

The history of the Societas Meteorologica Palatina has been extensively sketched (Kington, 1974;[105] Rigby, 1973;[106] Cappel, 1980a,b;[107,108] Landsberg, 1980;[109] Lingelbach, 1980[110]). The data have also been widely used. They were decisive for the history of climatology. Humboldt (1817)[111] was the first to use these data for an initial attempt to construct an isothermal chart of the northern hemisphere. Brandes (1820)[112] used the data to come up with the first synoptic analysis of moving weather systems. Kämtz (1832)[113] prepared mean values of the data and, in connection with other sources, used it for the first comprehensive description of hemispheric climate in his textbook on meteorology.

The idea of network observations was actually debated before the enterprise of the Societas Meteorologica Palatina. A local network for Badenia had been advocated by Boeckmann (1778)[114] and instituted, but only one year of data was published (Boeckmann, 1780).[115] In France, upon the initiative of Cotte (1774),[73] the

Socieété Royale de Médecine started a network of 31 stations in 1776. The data were published by Cotte in the annual volumes of the *Histoire de la Société de Médicine*. This effort came to an abrupt halt in 1786 as a result of the French Revolution. The wars following this event also ended the existence of the Mannheim Society in 1795. In Europe several decades elapsed before organized meteorological networks, with regular data publications, returned to the scene.

The impression, prevalent since classical times, that weather influences health, gave the impetus to the next network. This came about by an order of the Surgeon General of the U.S. Army, James Tilton, in 1814. This made it a duty of the post surgeons to keep a diary of the weather. This resulted in centralized procurement of instruments and issuance of uniform instructions (Landsberg, 1964).[116] It might be noted that this network was the first to be established under governmental authority. There was great sympathy for weather observation in the United States. Both George Washington and Thomas Jefferson were ardent observers of the weather (McAdie, 1894;[117] Hodge, 1980[118]).

The first observations of the military network came from Benjamin Waterhouse, M.D., Hospital Surgeon at Cambridge, Massachusetts, covering the months from March to June 1816 (figure 16). Starting with 1818, under the vigorous aegis of Surgeon General Joseph Lovell, returns flowed regularly to the Army Medical Department in Washington. The data were published at regular five year intervals (see, e.g., U.S. Surgeon General's Office, 1826;[119] Lawson, 1840[120]). The fortunate circumstance was that Army forts were widely scattered through the country: from Maine to Florida to the unsettled areas of the Midwest and even to the Pacific Northwest. An example was Fort Snelling (now St. Paul), Minnesota, where records began in 1819 (Baker, 1960).[124] Many of the data were summarized by Forry (1842)[122] and again by Blodget (1857).[123] The latter also used data from the New York and Pennsylvania networks established at secondary schools. The Army records of the early nineteenth century fortunately have survived in the National Archives, while some of the others have been destroyed in fires or lost. The quality of most of these early records in the United States is high and quite equivalent to the later Weather Bureau records.

INTERPRETATION AND USE OF EARLY WEATHER OBSERVATIONS

With a few exceptions, such as the work of Lamb (1963,[124] 1966,[125] 1969[126]), interest in and use of early documentation of weather has been restricted to the last few years. The lure of readily obtainable and generally reliable data collected by the various national weather services has been too strong. Their main advantage, especially since the end of World War II, has been the fairly broad geographical coverage and the three-dimensional surveillance of the atmosphere. Starting with the International Geophysical Year (1957/58) and the advent of meteorological satellites (1960), the coverage has included all polar areas and become truly global.

The problem with these new materials is that two or three decades are simply inadequate to assess climatic fluctuations or establish trends. This applies particularly to extreme values. Older items of information can add valuable information

Explanation of the characters
● Fair
÷ Mostly fair
–⊏ Few clouds
I Mostly cloudy

Meteorological Journal.
Cambridge, March, 1816.

Explanation of the characters
== Cloudy
ŦŦ Rain
‡ Snow

Day	Barometer 7 a.m.	Barometer 2 p.m.	Barometer 9 p.m.	Thermometer 7 a.m.	Thermometer 2 p.m.	Thermometer 9 p.m.	Face of Sky 7 a.m.	Face of Sky 2 p.m.	Face of Sky 9 p.m.	Winds 7 a.m.	Winds 2 p.m.	Winds 9 p.m.	
1	30 27	30 26	30 41	33	47	31	●	==	==	N	E 1	E	
2	30 28	30 27	30 19	33	38	33	==	==	==	NE	NE	NE 2	Ŧ Ŧ Ŧ
3	30 14	30 14	30 23	35	40	37	==	==	==	N	E	E	Ŧ
4	30 27	30 10	29 99	32	35	33	==	==	==	NE	NE	E	‡ Ŧ
5	30 24	30 35	30 37	30	38	27	÷	÷	÷	NW 2	NW 2	W	
6	30 38	31 33	30 29	30	35	33	==	I–	÷	E	SW	W	
7	30 27	29 97	29 74	23	43	32	÷	I–	==	W	S	SE	
8	29 79	29 95	30 08	31	25	8	I–	–⊏	●	W	NW	NW 3	
9	30 13	30 12	30 25	0	27	17	●	●	●	NW 2	NW 2	NW 2	
10	30 30	30 39	30 38	15	39	25	●	●	●	NW 2	NW 1	NW 2	
11	30 50	30 54	30 41	19	36	25	●	●	–=	W	E	SW	
12	30 27	30 16	37 17	30	42	23	==	==	÷	E	E	N	
13	30 27	30 24	30 10	13	47	30	●	=	–=	SW	E	E	
14	29 76	29 75	30 03	38	48	32	==	I–	–=	S	SW	W	
15	30 34	30 35	30 37	14	37	18	÷	÷	●	NW 2	NW	NW	
16	30 15	30 00	30 00	15	38	34	●	I–	=	SW 2	SW 3	SW	
17	30 05	30 12	30 61	33	48	10	I–	–=	●	SW	W 1	NW 3	
18	30 95	30 95	30 92	51	20	15	●	●	=	NW 2	NW 0	NW	
19	30 75	30 52	30 38	20	32	20	==	==	==	E	NE	NE	‡ ‡ ‡
20	30 17	30 95	29 92	21	35	32	==	==	●	NW	NW	W 2	
21	30 08	30 08	30 18	25	40	27	÷	=	–=	W	SW 0	W	
22	30 36	30 39	30 44	16	34	15	●	–=	●	NW	NW	W	
23	30 44	30 26	30 12	20	48	28	I–	–⊏	==	N	N	SW	
24	30 13	30 11	30 11	32	43	28	I–	–⊏	●	W	NW	W	
25	29 97	30 14	30 30	34	40	28	÷	●	●	W 2	NW	W	
26	30 06	29 85	29 87	35	56	34	I–	–=	÷	W	W 0	W	
27	29 90	29 85	29 72	32	48	38	÷	–=	–=	W	E	E	
28	29 45	29 50	29 78	43	52	38	I–	=	÷	S	W	SW 2	
29	30 00	30 22	30 44	38	34	20	I–	÷	●	N 2	W	NW	
30	30 64	30 55	30 53	20	30	32	–=	I–	I–	N	E	E	
31	30 45	30 37	30 27	31	40	35	==	==	==	NE	NE	E	

Figure 16. Table from first set of weather observations made by post surgeons under orders from the Surgeon General of the Army. These are for March 1816 made at Cambridge, Massachusetts, under the direction of Dr. Benjamin Waterhouse. From Landsberg, 1964.[116]

for diagnostic studies of climate. The major problem is to judge the reliability of the information. It has already been outlined for some elements what the difficulties are. Early instrumental records are afflicted by uncertainties of calibration and exposure of instruments, as well as times of observation. For these reasons the interpretation of such data must assume considerable errors, especially for material collected prior to the inauguration of the Mannheim network. Eye and event observations are usually more reliable even into the period of the weather diaries of the fifteenth and sixteenth centuries. Their main drawback is the fact that they usually do not constitute a 24-hour weather watch and nocturnal events are often missed. The exploitation and interpretation of logs of astronomical observatories remains a task for the future.

In the preceding paragraph it was assumed that instrumental and other weather observations were made fairly regularly and cover a reasonable interval of time. In contrast, extreme, damaging, or strange events are rare and require a different scrutiny. The earlier the date of such reports the more difficult the assessment of reality. Content analysis becomes very essential then. This is often complicated by the language problems. Not only did the languages change but often manuscript material is not always easy to decipher. Another problem is vagueness which applies both to dates and to expression. This is a characteristic of most early chronicles. An example are the quotations from the ancient Chinese chronicles (Wang, 1979)[127] such as: "Severe drought," "flood," "thunder struck temple." Unless one has a fairly complete chronological listing of such events for a specific locality, one must treat this as anecdotal. Where such good records exist, analysis of frequency or frequency of intervals is possible. There have been some attempts at such chronicles (e.g., Weikinn, 1958).[128] Chronicles are generally not flawless, yet once they are available it is easy to amend them when new information is found.

If one deals with historical documents, interpretation and corroboration becomes very important (Herlihy, 1981).[129] For some weather events corroborative evidence can be readily found. This is, for example, usually possible for drought, which is never a local phenomenon but generally widespread. Spring floods resulting from snow melt also usually affect many localities along the course of major rivers. Ice on large rivers, in bays, or on lakes can be verified from a number of sources. However, flash floods on small streams, hail, thunderstorms, and tornados are, with few exceptions, isolated phenomena and in earlier times there are rarely multiple sources for corroboration.

Greatest caution has to be exercised when dealing with information that is only indirectly related to weather. There is a vast literature dealing with quantity and quality of grain, fruit, and other harvests. But weather is only one of many factors affecting these and there is a variety of weather factors that can play a role. Although precipitation is usually the most important weather element for crop yield, temperature, sunshine, freeze dates, and hail can be quite influential. Sorting out what caused a good or poor crop is not easy. A particular pitfall is to apply correlations of yield with weather factors derived from recent data to events centuries ago when farm practices were radically different. In this connection it must be remembered that often the weather influence is not on the corp plant itself but on a pest affecting the plant, an uncontrollable circumstance in earlier times.

Perhaps the most widely committed error in interpretation of climatic information in a data-sparse era is extrapolation from one place to another; or, more commonly, the use of locally acquired information to large areas. Teleconnections can, of course, be useful adjuncts to interpretations but they cannot be stretched too far. Especially when one is confronted with some exceptional conditions, it is well known that broad-scale circulation anomalies exist which bear little relation to connections prevalent in more common patterns. This is not meant to preclude the use of data from nearby stations for estimates of missing data at another station because correlations are usually high, especially for temperature data.

Other potential misinterpretations concern the advances and retreats of mountain glaciers. These may indeed be related to changes in circulation patterns but advances are not necessarily an indication of cooling and retreats may not mean warming. The height of the glacier plays a role and it is possible for lower-level glaciers to advance and higher-level glaciers to retreat at the same time. A multiplicity of factors can be involved. A common scenario for advancing glaciers is warmer winters with more snowfall, followed by cooler summers with less melting, with essentially constant annual temperature.

There have also been climatological interpretations of entirely nonphysical recorded facts of history. Among these are famines. The long chain of reasoning assumes that these are due to poor crops which, in turn, are caused by bad weather. Similarly removed are such elements as commodity prices. Here again the facile explanation is: the grain price is high, the crop was poor, hence the weather must have been the culprit. There is nothing said about possible pest and rodent damage, nothing about carryover, and nothing about speculation. The same criticism applies to the interpretation of widespread epidemics. While it is known that weather will affect the human body, the carriers of disease, and the etiological agents, it is only one of a large number of other factors, principally of immunological and nutritional variety. Migrations also must be viewed with considerable circumspection as indicators of climate. Economical, political, and religious factors claim equal importance for moves of people. For all these proxy elements, independent evidence must be adduced to implicate weather or climate.

One serious handicap for a broader reconstruction of climate in the period prior to about 1800 has been the preponderance of information from Europe. For some periods it is so overwhelming in volume and detail that it has at times dominated interpretations of hemispheric or even global climatic fluctuations. It has certainly dominated the discussions of the so-called Little Ice Age. This term was introduced in 1939 by François E. Matthes, a geologist.[130] He meant it to designate the whole of the Holocene following the warm period of the hypsithermal (Climatic Optimum) about 5000 to 6000 years B.P. Unfortunately others have applied this to the interval of about A.D., 1550 to 1800, where it is a misnomer because *continental* glaciation did not increase nor was there any sustained low global temperature. That interval was undoubtedly cooler than the preceding Middle Ages and, particularly, the warm interval in the early decades of the twentieth century. Any reasonable resolution of available information shows fluctuations of a similar nature as indicated by current instrumental observations. It would be far better to expurgate the scientifically misleading (even if journalistically appealing) term "Little Ice Age," because of the fact that the interval* so designated was not uniformly cold in space or time. It is much preferable to designate cool decades as *katathermal* periods and warm decades, such as the 1900 to 1940 interval, as *anathermal* periods.

*The literature is not even consistent about what interval is designated as "Little Ice Age": for example, 1550–1700, eighteenth and nineteenth centuries, 1550–1850, fifteenth through nineteenth centuries. (Lamb, 1977;[44] Pittock et al., 1978;[134] Schönwiese, 1979;[135] Berger, 1981[136]).

In a reconstruction of northern hemisphere temperatures for the interval between 1579 and 1880, covering a major portion of the cool interval, this can be clearly seen (Groveman and Landsberg, 1979).[131] This reconstruction was based on the recognition that a selection of thermal records in higher latitudes reflected fairly faithfully the temperature fluctuations of the northern hemisphere (Landsberg et al., 1978).[132] Appropriate regression between observed temperatures for the whole hemisphere and selected northern stations indicated a good agreement by such an abbreviated scheme. Using a number of long temperature series for the nineteenth and eighteenth centuries, and Manley's Central England series (back to 1658 and before that tree ring and freeze proxy data), a first approximation for annual hemisphere temperature departures from average was obtained. The uncertainties increase, of course, toward the early years. A summary of this reconstruction by decades is shown in figure 17. It is clearly seen that there were alternations of warmer and cooler periods of varying length and amplitudes. Particularly notable was the sharp drop of temperatures from a warm interval around 1750 to a very cold era from about 1800 to 1820. From the 1881–1975 northern hemisphere temperature calculation of Borzenkova et al. (1976),[133] also shown in the figure, the warm period of 1920–1940 stands out.

The spatial differences are also readily demonstrable. The Alps offer some examples. In the Austrian Alps the glaciers advanced rapidly in the sixteenth century; in the western Alps the glaciers had their greatest extent in 1850 (LeRoy Ladurie, 1971);[7] in Iceland, Dragnajökull glacier in the North reached its maximum extent about 1760 but Vatnajökull in the south-central area culminated in 1850 (Björnson, 1979).[137] In the northern foreland of the Alps, freezing of upper Lake Constance shows the lowest winter temperatures in the fifteenth and sixteenth centuries; see figure 18 (Steinjans, 1976).[138] Japanese studies place principal cold periods during the first half of the fifteenth century and they designate 1750–1850 as their "Little Ice Age" (Yamamoto 1967,[139] 1971[140]).

There is a concept in some recent writings that the coldest excursion of the "Little Ice Age" coincided with the so-called Maunder Minimum of sunspots, 1645–1715 (Eddy, 1976,[141] 1977,[98] 1981[142]). Although this seems to have been an interval of low solar activity, there is ample evidence that the usual sunspot rhythm

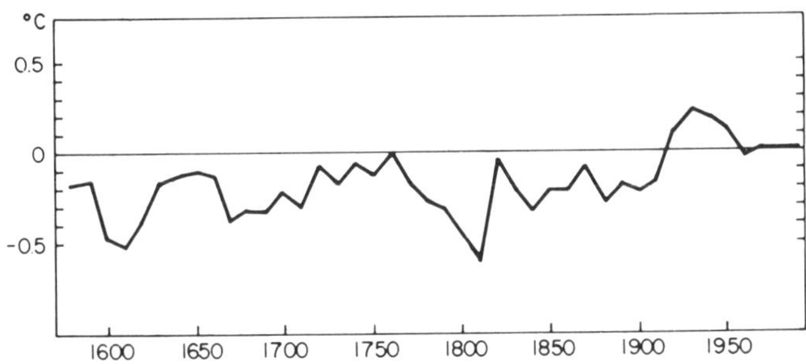

Figure 17. Decadal values of Northern Hemisphere temperatures, reconstructed by Groveman and Landsberg.[131]

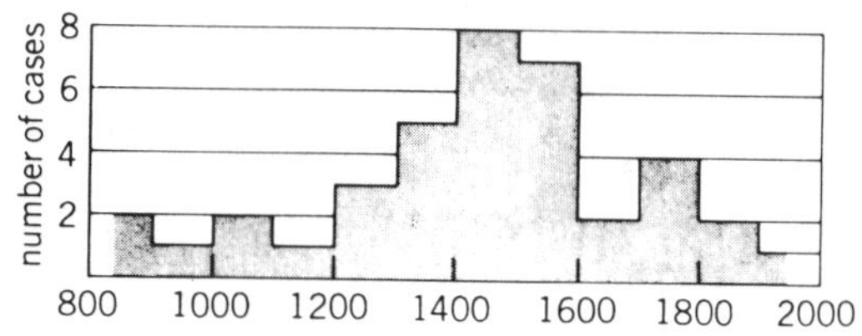

Figure 18. Number of cases per century when Upper Lake Constance was frozen, from data by Steinjans.[138]

was maintained. Lack of auroral sightings has been cited as proof but a principal source of auroral observations, compiled by Fritz (1873)[144] is quite deficient (Schröder, 1979;[145] Landsberg, 1980[143]). There were notable sunspot observations in 1672 at Paris (see, for example, Bion 1751).[146] Chinese eye observations also occurred in that interval (Xu, 1979;[147] Cullen, 1980[148]). The solar rhythm, albeit weakened, seemed to have shown the usual maxima and minima (Gleissberg, 1977;[149] Gleissberg and Darnboldt, 1979;[150] Botley, 1979[151]). The main contention that the Maunder interval was especially cold is not borne out by climatic analyses. Although 1670–1680 was hemispherically a cold decade and some years between 1690 and 1700 were very cold in Europe (as was the winter 1708/09), there were colder decades such as 1605–1615 before the "Maunder Minimum" and 1805–1815 afterwards. There is also new evidence that the sun emanates more energy at times of low activity than when there is turmoil visible on its disk (Agee, 1982),[152] an interpretation that seems to be borne out by recent satellite observations of solar flux.

One can also be reasonably certain that continued search in diaries and chronicles will turn up further observations not in the printed record yet. There are many sources waiting to be tapped.

ACKNOWLEDGMENTS

The author is indebted for helpful comments and corrections on the thermometer scale conversions to Dr. Karl Keil (Deutscher Wetterdienst, ret.) and Mr. Jacques Dettwiller (Météorologie Nationale). He is also grateful to Ms. K. Mesztenyi for typing the manuscripts, to Ms. Clare Villanti for graphics, and to the Climate Dynamics Program of the National Science Foundation for supporting part of this work through Grant ATM 80-18510.

REFERENCES

1. J. Neumann, "Great historical events that were significantly affected by the weather; 3, The cold winter of 1657–58, the Swedish army crosses Denmark's frozen sea areas, *Bull. Am. Meteorol. Soc.*, **59,** 1432–1437 (1975).
2. J. Neumann, Great historical events that were significantly affected by the weather; 2, The year leading to the revolution of 1789 in France, *Bull. Am. Meteorol. Soc.*, **58,** 163–168 (1977).

3. D. M. Ludlum, *Early American Winters*, Vol. 1, 1604-1820, and Vol. 2, 1821-1870, American Meteorological Society, Boston, 1966, 1967.
4. W. Stöbe, "Forecasting for the escape of Scharnhorst und Gneisenau," *Meteorol. Mag.*, **107,** 321-338 (1978).
5. J. M. Stagg, *Forecast for Overlord*, Norton, New York, 1971.
6. J. Dettwiller, "Pages d'historie emportées par le vent," *Cols Bleus, Hebdomadaire de la Marine et des Arsenaux*, **1680,** 6-12 (17 Octobre 1981).
7. E. LeRoy Ladurie, *Times of Feast, Times of Famine, A History of Climate Since the Year 1000*, Doubleday, Garden City, N.Y., 1971.
8. K. Wittfogel and T. Chiao-Sheng, "History of Chinese society: Liao," *Am. Philos. Soc.*, **752**, 907-1145 (1949).
9. Central Meteorological Research Bureau of China. "A study of Occurrences of Flood and Drought in the Last 500 Years in Northern and Northeastern China (translated title)." In *Collected Papers on Climatic Change and Long-Range Weather Forecasts*, Peking, 1977, pp. 164-170.
10. H. Arakawa, "Climatic change as revealed by the blossoming dates of the cherry blossoms in Kyoto," *J. Meteorol.*, **13**, 599-600 (1956).
11. T. Sekiguti, "Historical dates of Japanese cherry festivals since the 8th century and her climatic changes," *Jpn. Prog. in Climatol.*, 38-45 (1970).
12. F. Schnelle, "Beiträge zur Phänologie Europas IV - lange phänologische Beobachtungsreihen in West - Mittel - und Osteuropa," *Ber. d. Dt. Wetterdienstes*, **158**, (1981).
13. A. und F. Lauscher, "Vom Einfluss der Temperatur auf die Belaubung der Rosskastanie nach den Beobachtungen in Genf. seit 1808," *Wetter und Leben,* **33**, 103-112 (1981).
14. H. Arakawa, "Fujiwhara on five centuries of Lake Suwa in central Japan,"*Arch. Meteorol. Geophys. Bioklim.*, **B6,** 152-166 (1954).
15. B. M. Gray, "Early Japanese winter temperatures," *Weather,* **29,** 103-107 (1974).
16. H. E. Landsberg and R. E. Kaylor, "Statistical analysis of Tokyo winter temperature approximations, 1443-1970," *Geophys. Res. Lett.* **4,** 105-107 (1977).
17. A. White, *A Relation of Maryland Together with a Map of the Country, the Conditions of Plantation, His Majesties Charter to the Lord Baltimore, Translated into English*, William Peasley, London, 1635.
18. American Philosophical Society, *The Journal of Charles Mason and Jeremiah Dixon, 1763-1768*, Memoirs *76*, 1969.
19. E. Coues, *History of the Expedition Under the Command of Lewis and Clark*, Dover, New York, (first published 1893)
20. H. M. van den Dool, H. J. Krijnen, and C. J. E. Schuurmas, "Average winter temperatures at DeBilt (The Netherlands), 1634-1977," *Clim. Change,* **1,** 319-330 (1978).
21. D. W. Moodie and A. J. W. Catchpole, "Environmental data from historical documents by content analysis: Freeze-up and break-up of estuaries on Hudson Bay 1714-1871," *Manitoba Geogr. Stud.* **5,** Winnipeg, (1975).
22. A. J. W. Catchpole and D. W. Moodie, "Archives and the environmental scientists," *Archivaria*, No. 6, 113-136 (1978).
23. J. Hempstead, "Diary of Joshua Hempstead of New London, Connecticut, Covering a Period of Forty-Seven Years from September 1711 to November 1758," In *Collections of the New London County Historical Society,* Vol. 1, Providence, R.I., 1901.
24. W. Merle, *Consideraciones Temperici Pro 7 Annis*, Per Magistrum Willelmum Merle, Socium Domus de Merton (manuscript, Bodleian Library). Reproduced and translated by G. J. Symons, London, Edward Stanford, 1891.
25. E. N. Lawrence, "The earliest known journal of the weather," *Weather,* **27,** 494-501 (1972).
26. R. Mortimer, "William Merle's weather diary and the reliability of historical evidence for medieval climate," *Clim. Monit.*, **10,** 42-45 (1981).

27. L. Thorndike, "A daily weather record from the years 1399–1401," *ISIS*, **57**, 90–99 (1966).
28. R. H. Frederick, H. E. Landsberg, and W. Lenke, "Climatological analysis of the baseler weather manuscript: A daily weather record from the years 1399–1401," *ISIS*, **57** (1966).
29. F. Klemm, "Über die frage des beobachtungsortes des baseler wettermanuskriptes von 1399–1401," *Meteorolog. Rundschau*, **22**, 83–85 (1969).
30. C. Frisch, ed., *Joannis Kepleri Opera Omnia*, Vol. 7, Heyder & Zimmer Frankfurt a.M. 1858, pp. 618–653.
31. H. Flohn, "Zwei bayerische wetterkalender aus der reformationszeit; From: Prof. Dr. Albert Baumgartner zum 60. Geburtstag," *Wissensch. Mttlg. Meteorol. Inst. München*, **35**, 173–177 (1979).
32. J. M. Havens, "The 'first' systematic American weather observations," *Weatherwise*, **8**, 116–117 (1955).
33. B. Kington, "Searches for historical weather data: Appeals and responses," *Weather*, **35**, 124–134 (1980).
34. G. Hellmann, "Die Entwicklung der meteorologischen Beobachtungen in Deutschland von den ersten Anfängen bis zur Einrichtung staatlicher Beobachtungsnetze," *Abh. Preuss. Akad. Wissensch. Phys. Math. Klasse* (1), (1926).
35. F. Klemm, "Die Entwicklung der meteorologischen Beobachtungen in Franken und Bayern bis 1700," *Annual. Meteorol., Neue Folge*, (8) (1973).
36. F. Klemm, "Die Entwicklung der meteorologischen Beobachtungen in nord und Mitteldeutschland bis 1700," *Annal. Meteorol., Neue Folge* (10) (1976).
37. F. Klemm, "Die Entwicklung der meteorologischen Beobachtungen in Südwestduetschland bis 1700," *Annal. Meteorol., Neue Folge* (13) (1979).
38. P. K. Wang, "On the relationship between winter thunder and the climatic change in China in the past 2200 years," *Clim. Change*, **3**, 37–46 (1980).
39. E. Halley, "An historical account of the trade winds and monsoons, observable in the seas between the tropics, with an attempt to assign the physical cause of the said winds," *Philos. Transact.*, London (1686).
40. G. Hadley, "Concerning the cause of the general trade winds," *Philos. Transact.*, **39**, 58–62 (1735).
41. M. Maury, *Physical Geography of the Sea*, Sampson Low, London, 1855.
42. B. Kinsman, "Historical notes on the original Beaufort scale," *Marine Observer*, **39**, 116–124 (1969).
43. H. L. Crutcher, "Wind, numbers, and Beaufort," *Weatherwise*, **28**, 260–271 (1975).
44. H. H. Lamb, *Climate Present, Past and Future*, Vol, 2, Methuen London, 1977.
45. W. Reid, *The Progress of the Development of the Law of Storms and of the Variable Winds, with the Practical Application of the Subject to Navigation*, John Weale, London, 1849.
46. H. Landsberg, "The storm of Balaklava and the daily weather forecast," *Scientific Monthly*, **79**, 347–352 (1954).
47. W. E. Knowles Middleton, *Invention of the Meteorological Instruments*, Johns Hopkins University Press, Baltimore, 1969.
48. W. E. Knowles Middleton, *The History of the Barometer*, Johns Hopkins University Press, Baltimore, 1964.
49. W. E. Knowles Middleton, *A History of the Thermometer and its Use in Meteorology*, Johns Hopkins University Press, Baltimore, 1966.
50. N. Goodison, *English Barometers, 1680–1860*, Clarkson N. Potter, New York, 1968.
51. J. D'Alancé, *Traittez des Baromètres, Thermomètres et Notiomètres*, Henry Wettstein, Amsterdam, 1688.
52. W. Lenke, "Bestimmung der alten Temperaturwerte von Tübingen und Ulm mit Hilfe von Häufigkeitsverteilungen," *Ber. Dt. Wetterdienstes*, **10** (75), Offenbach a.M. (1961).

53. R. J. Camerarius, *Ephemerides Meteorologicae Tubingenses*, Kroniger & Goebel, Augsburg, 1696.

54. G. Manley, "Central England temperatures: monthly means 1659–1973," *Q. J. Roy. Meteorol. Soc.*, **100,** 389–405 (1974).

55. B. J. Mason, "Towards the understanding and prediction of climatic variations," *Q. J. Roy. Meteorol. Soc.*, **102,** 473–499 (1976).

56. C. D. Schönwiese, "Central England temperature and sunspot variability 1660–1975," *Arch. Meteorol. Geophys. Bioklimatol.*, **B26,** 1–16 (1978).

57. J. Dettwiller, "Les températures annuelles à Paris," *Bull. inf. Dir. Météor.*, Boulogne Billancourt, **53** (October 1981).

58. A. Labrijn, "Het klimat van Nederland gedurende de laatse twee een halve eeur," *Mededel. Verhandl. Kgl. Nederl. Meteorol. Inst.*, **49** (1945).

59. B. Brumme, "*Klimadaten 1683–1774 von Mitteldeutschland und Berlin nach einer Bearbeitung der Beobachtungs-Tagebücher der Familie Kirch*," Thesis, University of Bonn, 1978.

60. B. Brumme, "Neubearbeitung der Berliner Temperaturreihe am Beispiel der Periode 1700 bis 1710," *Annal. Meteorol.* (N.F.), **15,** 215–216 (1980).

61. P. Horrebow, *Tractatus Historico-Meteorologicus*, Copenhagen, 1780.

62. H. E. Hamberg, "Moyennes mensuelles et anuelles de la température et extrèmes de températures pendant les 150 années 1756–1903 à l'observatoire du Stockholm," *Kgl. Sv. Vetenskapsakad. Handl.*, **40** (1906).

63. C. Kirch, "Observationes quaedam meteorologicae in Pennsylvania habitae: quarum collatio cum berolinensibus instituta est," *Miscellanea Berolinensia,* **5,** 123–121 (1737). [A summary of the observations was published by J. M. Havens. A note on early meteorological observations in the United States with reference to the Germantown temperature record of 1731–32, *Bull. Am. Meteorol. Soc.*, **39,** 211–216 (1958). The whole paper is translated and the data are given in modern units in H. E. Landsberg, "Some early Philadelphia, Pennsylvania, temperature records," University of Maryland. Tech. Note BN-495 (1967).]

64. J. M. Havens, "A note on early meteorological observations in the United States with reference to the Germantown temperature records of 1731–1732," *Bull. Am. Meteorol. Soc.*, **39,** 211–216 (1958).

65. John Lining, "Extracts of two letters from Dr. John Lining, physician at Charles-Town in South Carolina to Dr. James Jurin, M.D., F.R.S.," *Philos. Transact.*, **42,** 491–509 (1743).

66. John Lining, "A letter from Dr. John Lining to C. Mortimer, M.D., Sec. R.S. concerning the weather in South Carolina, with abstracts of the tables of his meteorological observations in Charles-town," *Philos. Transact.*, **45,** 336–344 (1749).

67. Richard Brooke, "A thermometrical account of the weather for one year, beginning September 1752. Kept in Maryland, by Mr. Richard Brooke, Physician and Surgeon in that Province. Communicated by Mr. Henry Baker, F.R.S.," *Philos. Transact.*, **51,** 58–69 (1760a).

68. Richard Brooke, "A thermometrical account of the weather, for three years, beginning September 1754, as observed in Maryland. Communicated by Mr. H. Baker, F.R.S.," *Philos. Transact.*, **51,** 70–82 (1760b).

69. H. E. Landsberg, C. S. Yu, and Louise Huang, "Preliminary reconstruction of a long time series of climatic data for the eastern United States," University of Maryland. Tech. Note BN-571 (1968).

70. H. v. Rudloff, *Die Schwankungen und Pendelungen des Klimas in Europa seit dem Beginn der regelmässigen Instrumentenbeobachtungen (1670)*, Friedrich Vieweg, Braunschweig, 1967.

71. W. C. Derham, "A journal of meteorologicasl observations made at Petersburgh, by the Rev. Mr. Tho. Consett, from Nov. 24, 1724 to June 23, 1725, abstracted for the use of the royal society," *Philos. Transact.*, **338,** 101–104 (1733).

72. G. Martine, *Essays and Observations on the Construction and Graduation of Thermometers and on the Heating and Cooling of Bodies*, Alexander Donaldson, Edinburgh, 1744.

73. L. Cotte, *Traité de Météorologie*, Imprimérie Royale, Paris, 1774.

74. J. H. v. Swinden, "Dissertation sur la comparaison des thermomètres," Marc-Michel Rey, Amsterdam (1778).

75. J. F. Luz, *Vollständige und auf Erfahrung gegründete Anweisung die Thermometer zu verfertigen*, Christoph Weigel und Adam Gottlieb, Schneidersche Kunst-und Buchhandlung, Nürnberg, 1781.

76. A. Kh. Khrgian, *Ocherki Razvitiya Meteorologii*, Vol. 1, Gidro meteorologicheskoe Izdatel'stvo, Leningrad, 1959. [English Translation: *Meteorology, A Historical Survey*, Israel Program for Scientific Translation, Jerusalem, 1970.]

77. A. Celsius, "Observationer om tvenne bestandiga grader pao en thermometer," *Vetenskap. Akad. Handl.*, **4**, 197–205 (1742).

78. J. N. DeLisle, "Les thermometres de Mercure rendus Universels, en leur faisant en tout temp la quantité dont le volume du mercure est diminué, par la temperature presente de l'air, au dessous de l'entendu qú il a dans l'eau bouillante," *Mém. Serv. Hist. Prog. Astron. Géogr. Phys.* (St. Petersburg), 263–284 (1738). Cited after Middleton (1966) p. 87.

79. H. E. Landsberg, "*Interpretation of a manuscript weather record from St. Petersburg (now Leningrad), 1725-1732*," University of Maryland. Tech. Note BN-969 (1981).

80. V. V. Betin and Yi. V. Preobrajenskii, *Suróvost Zim v Europe in Ledovitost Baltiki* (Severe winters in Europe and ice cover of the Baltic), Gidrometeorologicheskoe Izdatelstvo, Leningrad, 1962.

81. G. E. Melin, "*Dissertatio Physico-Mechanica de Thermometris*," University of Uppsala, 1710.

82. R. A. F. de Réaumur, "Règles sur la construction des thermomètres," *Hist. et Mém. de l'Acad. R. Sci.*, Paris, 452–507 (1730).

83. A. J. v. Oettingen, *Abhandlungen über Thermometrie von Fahrenheit, Réaumur, Celsius*, Ostwald's Klassiker der Exacten Wissenschaften, No. 57, Wilhelm Engelmann, Leipzig, 1894.

84. K. Meyer, *Die Entwicklung des Temperaturbegriffs im Laufe der Zeiten*, Die Wissenschaft, No. 48, Friedrich Vieweg, Braunschweig, 1913.

85. J. Christin, "Thermomètre de Lyon, divisé selon la mésure de la dilatation du mercure trouvée en 1743 par M. Christin de l'Académie des Beaux Arts de Lyon," *Almanac de Lyon*, 42–45 (1745).

86. G. W. Krafft, *Wahrhaffte und umständliche Beschreibung und Abbildung des im Monath Januarius 1740 in St. Petersburg aufgerichteten merckwürdigen Hauses von Eiss*, Kaiserl. Akad. Wissensch (St. Petersburg), 1741.

87. E. Renou, "Historie du thermomètre," *Ann. Soc. Météorol. Fr.*, 18–72 (1876).

88. B G. Wales-Smith, "Monthly and annual totals of rainfall representative of Kew, Surrey, from 1697-1970," *Meteorol. Mag.*, **100,** 345-362.

89. J. M. Craddock, "Annual rainfall in England since 1725," *Q. J. Roy. Meteorol. Soc.*, **102,** 823-840 (1976).

90. C. Pfister, "Die älteste Niederschlagsreihe Mitteleuropas: Zürich 1708–1754," *Meteorol. Rundsch,* **31,** 56–62 (1978).

91. C. Pfister, "Zum Klima des Raumes Zürich im späten 17. and frühen 18. Jahrhundert," *Vierteljahrschr. Naturf. Ges. Zürich,* **122,** 447–471 (1977).

92. S. Polli, "Analisi periodale di una series pluviometrica bisecolare (Padova 1717-1940)," *Riv. Meteorol. Aeronaut.*, **7** (1), 19–23 (1943).

93. W. Wada, "Korean rain-gauges of the fifteenth century," *Q. J. Roy. Meteorol. Soc.*, **37,** 83–85 (1911).

94. H. Arakawa, "On the secular variation of annual totals of rainfall at Seoul from 1770–1944," *Arch. Meteorol. Geophys. Biokl,* **B 6,** 406–412 (1956).

95. H. H. Clayton, *World Weather Records*, Smithsonian Miscellaneous Collection, 1st reprint, Vol. 79, 275–276 Washington, D.C., 1944.

96. N. Rosenan, One hundred years of rainfall in Jerusalem, *Isr. Explor. J.*, **5,** 137–153 (1955).

97. J. Nicolas, "Bioclimatologie humaine de Saint-Louis du Sénégal," *Mém. Inst. Fr. Afr. Noire*, Ifan-Dakar, **57** (1959).

98. J. A. Eddy, "Climate and changing sun," *Clim. Change,* **1,** 173–190 (1977).

99. D. M. Willis, M. G. Easterbrook, and F. R. Stephenson, "Seasonal variation of oriental sunspot sightings," *Nature,* **287,** 617–619 (1980).

100. A. Pilgram, *Untersuchungen über das Wahrscheinliche der Wetterkunde durch vieljährige Beobachtungen*, Joseph Edler von Kurzbeck, Wien, 1788.

101. J. Jurin, "Invitatio ad Observationes Meteorologicas communi consilio instituendas," *Philos. Trans.*, **32,** 422–427 (1723).

102. Societas Meteorologica Palatina, *Ephemerides,* Vols. 1–12 (1781–1792 data), Mannheim (1783–1795).

103. V. Belohlavek, "The meteorological tradition of Prague, Czechoslovakia," *Bull. Am. Meteorol. Soc.*, **58,** 1056–1057 (1977).

104. W. Attmannspacher, 200 Jahre meteorologische Beobachtungen auf dem Hohenpeissenberg 1781–1980, *Ber. Dtsch. Wetterdienstes*, **155** (1981).

105. J. A. Kington, "The Societas Meteorologica Palantina: An eighteenth-century meteorological society," *Weather,* **29,** 416–426 (1974).

106. M. Rigby, "Ephemerides of the Meteorological Society of the Palatinate," Environmental Data Service, NOAA. 10–16, (February 1973).

107. A. Cappel, "Societas Meteorologica Palatina (1780–1795)," Ann. Meteorol. N.F., (16), 10–27 (1980a).

108. A. Cappel, "Societas Meteorologica Palatina (1780–1795)," *Inf. Fachdienst (Deutscher Wetterdienst)*, **11,** 1–15 (1980b).

109. H. E. Landsberg, "A bicentenary of international meteorological observations," WMO Bulletin, **29,** 235–238 (1980).

110. E. Lingelbach, "Vom Messnetz der Societas Meteorologica Palatina zu den weltweiten Messnetzen heute," Ann. Meteorol. N.F., (16), 1–9 (1980).

111. A. v. Humboldt, "Des lignes isothermes et de la distribution de la chaleut sur le globe," *Mém. Phys. Chim., Soc. Arceuil,* **3,** 462–602 (1817).

112. H. W. Brandes, *Beiträge zur Witterungskunde,* Johann Ambrosius Barth, Leipzig, 1820.

113. L. F. Kämtz, *Lehrbuch der Meteorologie*, Vol. 2, Gebauersche Buchhandlung, Halle, 1832.

114. J. L. Boeckmann, *Wünsche und Aussichten zur Erweiterung und Veruollkommnung der Witterungslehre*, Karlsruhe, 1778.

115. J. L. Boeckmann, *1779 Karlsruher Meteorologische Ephemeriden*, Karlsruhe, 1780.

116. H. E. Landsberg, "Early stages of climatology in the United States," *Bull. Am. Meteorol. Soc.*, **45,** 268–274 (1964).

117. A. McAdie, "A colonial weather service," *Pop. Sci. Mon.*, **45,** 331–337 (1894).

118. W. Hodge, "George Washington, farmer and weather observer," *EDIS,* **11,** (2), 20–22 (1980).

119. E. DeKrafft, "Meteorological register for the years 1822, 1823, 1824, 1825," Washington, D.C., 1826.

120. T. Lawson, *Meteorological Register for the years 1826, 1827, 1828, 1829, 1830,* J. & H. G. Langley, Philadelphia, 1840.

121. D. G. Baker, "Temperature Trends in Minnesota," *Bull. Am. Meteorol. Soc.*, **41,** 18–27 (1960).

122. S. Forry, *The Climate of the United States and Its Endemic Influences*, J. & H. G. Langley, New York, 1842.

123. L. Blodget, *Climatology of the United States and of the Temperate Latitudes of the North American Continent*, J. B. Lippincott, Philadelphia, 1857.

124. H. H. Lamb, "What can we find out about the trend of our climate?," *Weather,* **18,** 194–216 (1963).

125. H. H. Lamb, *The Changing Climate*, Methuen, London, 1966.

126. H. H. Lamb, "Climatic Fluctuations," In H. Flohn, ed., *World Survey of Climatology*, Vol. 2, Elsevier, Amsterdam, 1969, pp. 173–249.

127. P. -K. Wang, "Meteorological records from ancient chronicles of China," *Bull. Am. Meteorol. Soc.*, **60,** 313–318 (1979).

128. C. Weikinn, *Quellentexte zur Witterungsgeschichte Europas von der Zeitwende bis zum Jahr 1850*, Vol. 1, Hydrographie; issue 1: 0–1500 (1958), Akademie Verlag, Berlin.

129. D. Herlihy, "Climate and Documentary Sources: A Comment," In R. I. Rotberg and T. K. Rabb, eds., *Climate and History*, Princeton University Press, Princeton, N.J., 1981, pp. 133–137.

130. F. E. Mathes, "Report of the Committee on Glaciers," *Trans. Am. Geophys. Un.*, 518–523 (1939).

131. B. S. Groveman and H. E. Landsberg, "Simulated northern hemisphere temperature departures 1579–1880," *Geophys. Res. Lett.*, **6,** 767–769 (1979).

132. H. E. Landsberg, B. S. Groveman, and I. M. Hakkarinen, "A simple method for approximating the annual temperature of the northern hemisphere," *Geophys. Res. Lett.*, **5,** 505–506 (1978).

133. I. I. Borzenkova, K. Ya. Vinnikov, L. P. Spirina, and P. I. Stechnowskii, "Izmenenie temperatury vozduche severnago polushariya za period 1881–1975," *Meteorlogiya Gidrol.*, **7,** 27–35 (1976).

134. A. B. Pittock, L. A. Frakes, D. Jenssen, S. A. Peterson, and J. W. Eillman, eds., *Climatic Change and Variability A Southern Perspective*, Cambridge University Press, Cambridge, 1978.

135. C. D. Schönwiese, *Klimaschwankungen*, Springer-Verlag, Berlin, 1979.

136. A. Berger, ed., *Climatic Variations and Variability: Facts and Theories*, D. Reidel, Dordrecht, 1981.

137. H. Björnson, "Glaciers in Iceland," *Jökull*, **29,** 74–80 (1979).

138. V. W. Steinjans, "A stochastic point-process model for the occurrence of major freezes in Lake Constance," *Appl. Stat.*, **25,** 58–61 (1976).

139. T. Yamamoto, "On the climatic change along the current historical times in Japan and its surroundings," *J. Geogr.*, **76,** 115–141 (1967).

140. T. Yamamoto, "On the nature of the climatic change in Japan since the "Little Ice Age" around 1800 A.D.," *J. Meteorol. Soc. Jpn.*, **49,** 789–812 (1971).

141. J. A. Eddy, "The Maunder Minimum," *Science*, **192,** 1189–1202 (1976).

142. J. A. Eddy, "Climate and Role of the Sun," In R. I. Rotberg and T. K. Rabb, eds., *Climate and History*, Princeton University Press, Princeton, N.J., 145–167, 1981.

143. H. E. Landsberg, "Variable solar emissions, the Maunder Minimum and climatic temperature fluctuations," *Arch. Met. Geoph. Biokl.*, **B28,** 181–191 (1980).

144. H. Fritz, *Verzeichnis beobachter Polarlichter*, C. Gerold's, Vienna, 1873.

145. W. Schröder, "Auroral frequency in the 17th and 18th centuries and the Maunder Minimum," *J. Atm. Terr. Phys.*, **41,** 445–446 (1979).

146. N. Bion, *L'Usage des Globes Celeste et Terrestre et des Sphères Suivant les Différens Systèmes du Monde*, 6th ed., Jacques Buorin, Paris, p. 348, 1751.

147. X. Zhen-Tao and J. Yao-Tiao, "The solar activity of the 17th century viewed in the light of the sunspot records in the local topographics of China," *Nanjing Daxue Xuebao*, **2,** 31–38 (1979).

148. C. Cullen, "Was there a Maunder Minimum?," *Nature*, **283,** 427–428 (1980).

149. W. Gleissberg, "Betrachtungen zum Maunder-Minimum der Sonnentätigkeit," *Sterne und Weltraum*, **16,** 299–233 (1977).

150. W. Gleissberg and T. Damboldt, "Reflections on the Maunder Minimum of sunspots," *J. Brit. Astron. Assoc.*, **89,** 440–449 (1979).

151. C. M. Botley, "The Maunder Minimum in perspective," *1979 Yearb. Astron.*, 187–190.

176. E. M. Agee, "Terrestrial cooling and solar variability," Marshall Space Flight Center, Alabama, NASA Contractor Report, (1982).

3

CLIMATE AND TREE RINGS

Charles W. Stockton, William R. Boggess, and David M. Meko*

Laboratory of Tree-Ring Research, University of Arizona Tucson, Arizona

The frequent occurrence of droughts, floods, extreme heat, and extreme cold raises the question as to whether these conditions are anomalies or merely recurring phenomena. In most cases, weather records are not long enough to provide an adequate answer to this question. Hence it is necessary to use proxy data to provide information on long-term climatic variations and trends.

Variations in the widths of annual growth rings of many temperate and sub-polar species of trees can be successfully used as proxy data to reconstruct climatic records backward in time. Although ring widths are conditioned by both genetic and physiologic factors, they also reflect varying moisture and temperature conditions during and prior to individual growing seasons.

Tree rings are superior to other proxy data sources because they can be precisely dated; both high- (short-term) and low-frequency (long-term) variations are retained; and large samples can be obtained from many parts of the world with a relatively small investment in time and effort. A major disadvantage is the relatively short time span covered by tree-ring series in contrast with other sources such as ice cores or pollen profiles.

The discipline involving the development of tree-ring data for dating or studying past events is known as *Dendrochronology.* Two sub-disciplines, *Dendroclimatology* and *Dendrohydrology,* concerned respectively with climatic and hydrologic reconstructions, have developed rapidly since the late 1960s. This rapid development has been made possible by a) the capability of modern computers to process the large amounts of data inherent in tree-ring analyses; and b) the use of sophisticated techniques to interpret the complex aspects of ring formation and to develop the relationships between ring characteristics and climatic or hydrologic parameters.

*Several individuals have contributed to this chapter. Their names are listed following subchapter headings as appropriate.

Although the traditional use of ring widths as a basis for dendrochronological studies has worked quite well for trees on dry sites of the southwest, other methods appear to have considerable promise—especially for trees growing in more humid regions. These include the thickness of earlywood or latewood layers, ring density as determined by x-ray densitometry, and the unstable isotope content of carbon, hydrogen, and oxygen. (see Appendix.)

DEVELOPMENT OF DENDROCHRONOLOGY

Dendrochronology originated in the American Southwest through the efforts of an astronomer, A. E. Douglass, working with living conifers and archaeological wood on the Colorado Plateau. Douglass' demonstration in 1914[1] of the principle of crossdating, the technique of building composite tree-ring chronologies from many crossdated samples, and of a statistical relationship between climate and tree growth established dendrochronology as a legitimate scientific pursuit. He subsequently demonstrated that ring-width patterns in wood samples of unknown age could be matched (crossdated) with those from living tree-ring chronologies, enabling extension of information obtainable from tree rings far beyond that obtained from currently living trees. This also allowed samples of wood and charcoal from archaeological and geological contexts to be absolutely dated. In 1937, Douglass' new field of endeavor resulted in the establishment of the Laboratory of Tree-Ring Research at the University of Arizona, which conducts a broad program of dendrochronological research throughout the world.

From its modest beginning in the southwestern United States, dendrochronology has grown rapidly and now involves hundreds of individuals and numerous institutions and laboratories in many countries of the world. The western United States remains the most intensively sampled region in the world, resulting in the construction of hundreds of chronologies and the dating of thousands of archaeological samples.

Although Douglass had a deep interest in establishing the relationships between climate and ring widths, the foundations of present-day dendroclimatology were laid by his student and later co-worker, Edmund Schulman. Schulman was an indefatigable collector and his collections, although centered in the western United States, includes samples from the southern United States, Mexico, and South America.[2] Schulman also did the initial work on constructing a continuous Bristlecone Pine chronology, which has now been extended 8800 years into the past by his successors at the Laboratory of Tree-Ring Research.

Dendrochronology was introduced into the Great Plains by Harry Weakly in Nebraska[3,4] and George Will in North Dakota,[5] both of whom did both archaeological dating and chronology building. Subsequent work by Ward Weakly in South Dakota[6] produced chronologies dating back to the twelfth century.

At present, dendrochronological work in the United States is largely centered in the Universities of Arizona, Arkansas, Nebraska, and Washington; the Lamont Doherty Geological Observatory at Columbia University; Oak Ridge National Laboratory; and the U.S. Geological Survey in Reston, Virginia.

SITE SELECTION AND SAMPLING

The annual growth layers characteristic of trees growing in temperate and subpolar regions are related to both genetic and environmental factors. The former controls the tolerance and response of trees to environmental variations while, in contrast, the supply of nutrients, moisture, solar radiation, and atmospheric gases essential to growth are products of the environment. The amount of growth in a given year is related to the abundance or scarcity of one or more of these constituents. In most temperate regions precipitation is the most likely limiting factor, both because of its variability and its relationship to available soil moisture. Temperature usually becomes limiting near the arctic and alpine timberline. As a result of the complex physiological growth processes and the interplay between genetic and environmental factors, there is considerable variation in the kind and amount of information that can be extracted from tree rings. Thus, site selection and sampling are of utmost importance when data are to be used for climatic reconstructions as opposed to other uses such as archaeological dating.

Trees most likely to preserve climatic information in their ring patterns are those growing on stress sites and relatively free from the influence of factors such as fire, insect attacks, disease, competition, and human interference. In the American southwest, for instance, trees growing on shallow soils near the lower forest border, and free from the influence of ground water are most likely to record variations in precipitation patterns. Such trees are often stunted in appearance, with thin crowns and spike tops (figure 1). Although stressed trees in humid regions are likely to have a different appearance, the same general principles hold for the selection of climatically sensitive sites.

Ideally, complete cross-sections of tree trunks provide the best possible material for dendrochronological studies. Since cross-sections are rarely available, individual trees are sampled by removing cores with an increment borer.[7] It is important that at least two cores be extracted from each tree. This facilitates the identification of possible multiple or missing rings which, if undetected, would result in errors in the precise dating of each annual ring.

The number of trees that should be sampled on a given site depends upon the variation in ring widths from core to core and from tree to tree. Experience in the southwest has shown that 10 trees, with two cores per tree, will provide a satisfactory sample on the most climatically sensitive sites. This number should be considered the minimum and increased from 20 to 30 or more trees as variation within and between trees becomes greater.

DATING AND MEASURING

Dating may be considered a simple matter of counting rings by those unfamiliar with dendrochronology. Ring counting, however, can be a dangerous procedure because the growth response of trees to moisture availability or temperature extremes can cause locally absent or multiple rings. For instance, growth may not even be initiated if severe antecedent conditions have persisted for some time. In

Figure 1. Old aged pinyon pine showing characteristics typical of climatic stress.

contrast, two rings can form if growth is interrupted in mid-season by unfavorable conditions and resumed if conditions improve. Such anomalies in ring formation can be accounted for by crossdating, a procedure through which patterns of narrow and wide rings are matched on a year-to-year basis both within and between trees. Skeleton plots are extremely useful when dealing with large amounts of material or samples from unfamiliar areas.[7]

Once skeleton plots are developed for cores from a given site, they can be quickly compared to determine similar ring patterns. These established patterns can be used to identify possible locations of missing or multiple rings. After all such anomalies have been reconciled, each ring can be dated as to its year of formation.

The basic tools for measuring ring widths remain relatively unchanged, consisting mainly of a binocular microscope coupled with a calibrated device by which the mounted core can be moved across the stage and properly aligned with the crosshair for measurement. Continual improvements in the mechanics and design of measuring apparatus have been made through the years.

In contrast, the evolution of equipment and methods of recording and processing measurements has been dramatic. The state-of-the-art is probably exemplified by the system currently in use at the Laboratory of Tree-Ring Research, University of Arizona. This system uses precision screw stages fitted with optical incremental shaft encoders and interfaced with microcomputers.[8] Using appropriate computer software the ring-width data can be transmitted to an interactive mainframe computer for subsequent data manipulation with numerous statistical programs.[9]

STATISTICAL TREATMENT AND STANDARDIZATION

When measured ring widths are plotted as a time series, a distinct trend resembling a descending exponential curve and an associated decrease in variance is evident for many tree species.[10] This trend and changing variance reflects the aging process of trees and must be removed and the ring widths changed to indices before the data can be used in statistical analyses. This is done in a process called *standardization,* where a modified exponential, or some other appropriate curve, is fitted to the ring-width time series. Orthogonal polynomials and cubic spline functions[10,11] have been used to fit more complex curve functions. An index is derived by dividing each ring width by the corresponding curve value. The resulting time series, known as a *standardized ring-width chronology,* is at least weakly stationary and the indices have a mean value of 1.00. The standardized chronologies are then combined to form a mean value function or site chronology. As pointed out by Douglass,[12] standardization prevents one chronology developed from fast growing trees from dominating others with slow or average growth when the various series are combined.

Standardized chronologies are routinely characterized by several statistical parameters which also help determine their suitability for use in dendroclimatic reconstructions. Parameters determined include the mean, variance, standard deviation, standard error, serial correlation, and mean sensitivity.[10] Only the latter, mean sensitivity, is unique to dendrochronology. It was defined by Douglass[12] as the average ratio of the absolute difference between two successive ring measurements divided by their mean. The values range from zero when there is no difference between adjacent rings to two when one ring is zero and the other non-zero. Although mean sensitivity is a measure of high frequenty variations, its distribution, and hence its power, is unknown. In general, chronologies best suited for climatic reconstructions are those with high mean sensitivities, large standard deviations, and low first order autocorrelation.

RECONSTRUCTION METHODS AND STRATEGIES

A period of overlap between existing meteorological data and standardized chronologies is essential for climatic reconstructions. A part of the overlapping period is used to develop a mathematical relationship between the desired climatic parameter and ring-width indices; this step is known as *calibration.* In a second step, known as *verification,* the remaining part of the overlapping data are used to verify the calibration (figure 2).

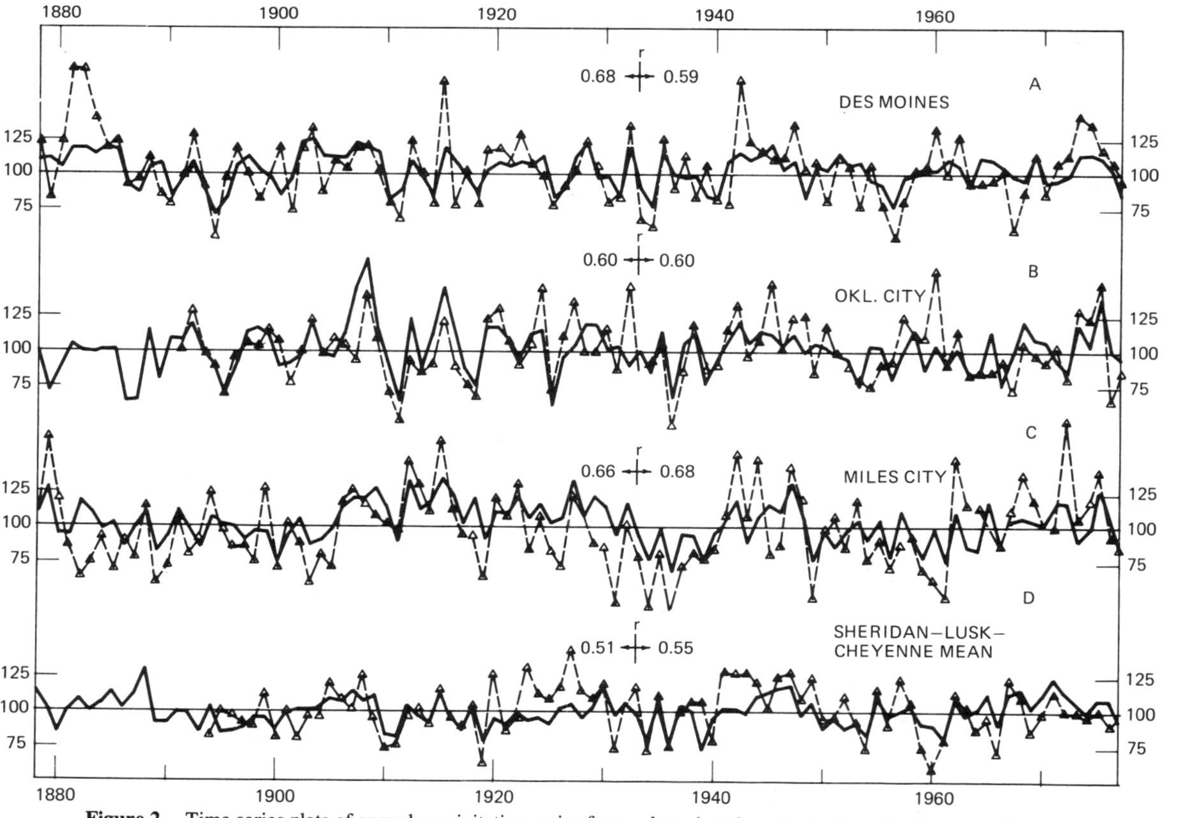

Figure 2. Time series plots of annual precipitation series from selected stations (dashed) within the central United States and regional precipitation reconstructions (solid). Series are plotted as percent of 1933–1977 normal. Correlation coefficients show relationship between actual series and reconstruction for both the 1933–1977 model calibration period and the earlier model verification period.

Earlier reconstructions used multiple step-wise regression methods, where a small set of independent variables were selected from a larger group. However, this method has inherent problems due to high intercorrelations among the independent variables.[13] To circumvent the problems caused by such intercorrelations, principal component analysis[14] is now used to create new sets of orthogonal (uncorrelated variables called empirical orthogonal functions (EOF's).[15,16] The EOF's are usually derived from the matrix of correlation coefficients among the independent variables. The first few EOF's usually represent most of the variability in dendroclimatic data. There is very little reason to include higher order EOF's in any analysis as they are probably mathematical artifacts of the technique and do not have any physical meaning.

It is important to remember that the tree-ring data series will reflect a climatic (or environmental) signal for the site from which the sample is selected. As with any atmospheric sensor, a large portion of the signal inherent in a series will be unique to that particular site. Additionally, the climatic signal within an appropriately sampled tree-ring data series may include inputs of various environmental factors varying throughout the year. Consequently, tne individual annual ring-width index represents a time integrated response to a complex interaction of many different atmospheric and site variables. In many instances, the variable signal dominant in the tree-ring time series can be predetermined by carefully defining the sampling population. For example, if precipitation information in the tree-ring series is to be maximized, samples should be selected from sites with minimal soil development, located on ridges or other well-drained sites and preferably near forest borders (if practicable). Similarly, if one is interested in fluctuating ground water tables, the sampling should be restricted to alluvial flood plains or other low lying places where the sampled trees are deriving their moisture from the capillary fringe area.

The part of the climatic signal related to a particular climatic factor can be identified and used to reconstruct, by appropriate calibration, that particular climatic variable incorporated within the tree-ring series. Calibration involves the development of a mathematical or statistical transfer function which is used to extract the climatic signal from the tree-ring data series. Transfer functions come in a multitude of forms. Some utilize single series, other sets of series. Additionally, the data sets of climatic series and tree-ring series may be transformed by again utilizing EOF's to isolate the signal in both series. This has the advantage of removing the noise (nonclimatic) portion of the signal from each series and allows the development of a transfer function by calibrating only the dominant signal of one series with the dominant signal of the other series. By using an unrotated principal component analysis, one can later re-transform the variables so the final results will be in terms of the original variable values.

The empirical nature of the transfer function development makes it necessary to test many different transfer functions. These functions must be proof-tested (verified) to evaluate the most appropriate one for the predetermined objectives of the analysis (figure 3). For this reason, a portion of the data utilized in the calibration procedure is held back for the purposes of evaluating (proof-testing) the transfer function on data not used in its development. Once the appropriate function is

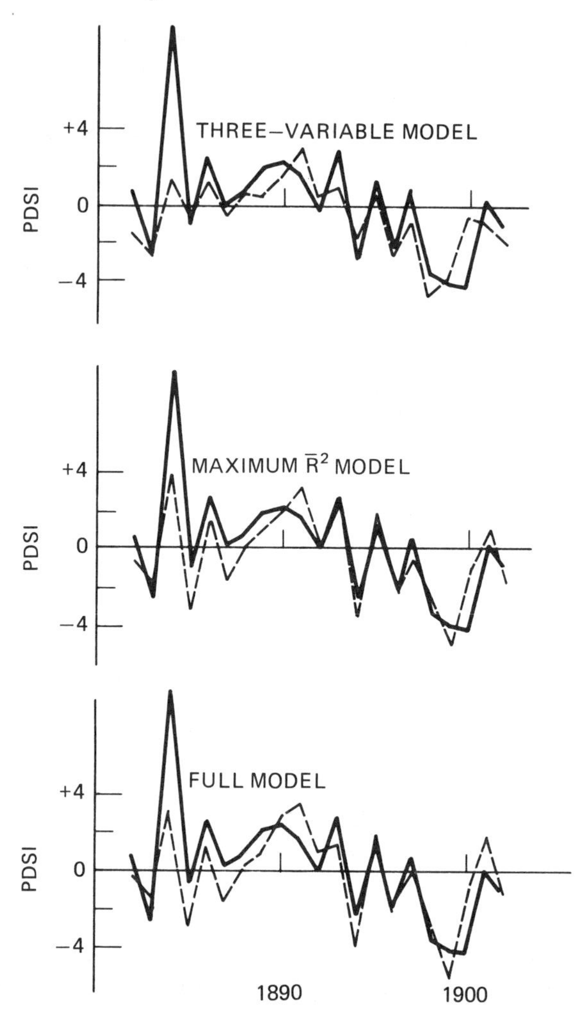

Figure 3. Examples of model verification for southern California drought reconstruction. The reconstructed (dashed) and actual (solid) July Palmer Drought Severity Index for the independent verification period (1882–1902) are shown. Model (a) is the simplest model incorporating 3 predictor tree-ring variables; (b) is the maximum-adjusted r^2 model based on 11 tree-ring predictor variables; and (c) is a model arrived at by letting the stepwise regression model run to an F-level of 0.01 for inclusion of tree-ring variables. Model (b) was selected as most appropriate.

selected, the entire data set is utilized in developing the final transfer function to obtain coefficients that are as stable as possible.

An additional complication arises in extending climatic signals from tree-ring data. It has been known for a long time that a certain amount of persistence exists in tree-ring data series that may represent a type of complicated biological memory of climate intermingled with a direct record of climate persistence. Early workers handled this problem by using moving averages with arbitrarily chosen coefficients or weights.[1,2] Although the results improved, they soon discovered unaccountable shifts in the time of critical events owing to the effects of the moving averages. More recent techniques include the use of lagged tree-ring variables and multivariate statistical techniques to determine the sign and magnitude of the coefficients used to transfer ring-width variance at different lags into an estimate of specific climatic variables.[17] An alternative approach involves the use of Box-Jenkins modeling[18] to determine by successive testing the appropriate persistence model. The models range from simple autoregressive (AR) or moving average (MA) to the complex universal autoregressive-moving average (APMA). Knowing the appropriate model allows one to proceed with the development of an appropriate transfer function utilizing the residuals from the Box-Jenkins modeling effort. The persistence which is well-defined by the modeling effort can be reincorporated into the final transfer function.

Meko[19] has recently compared the two approaches and finds that both have certain advantages and disadvantages under certain data conditions. For tree-ring series with appreciable autocorrelation, reconstructions by the Box-Jenkins method generally have a flatter frequency response than reconstructions by the lagged predictor model. This evenness of response is gained at the expense of reconstruction accuracy at longer wavelengths: the amplitudes of broad swings from the mean are underestimated by the Box-Jenkins approach. The Box-Jenkins method is preferable when the spectrum of the long-term reconstruction is of interest because the method attempts to minimize distortion of persistence. The lagged-predictor model, on the other hand, accepts frequency distortion as a trade off for higher total variance explained.

REQUIREMENTS FOR SPATIAL AND TEMPORAL COVERAGE

Reconstructing past large scale patterns of atmospheric circulation is of major interest in paleoclimatic research. Kutzbach and Guetter[20] have investigated the spatial density and location of climatic variables that are appropriate for estimating continental to hemispherical patterns of atmospheric circulation. Although they utilized existing meteorologial records (temperature and precipitation) of shorter duration to determine the optimal spatial sampling density for describing atmospheric circulation features, the results also appear to be applicable to tree-ring data. The reasoning is that a linkage exists in which physiological processes within the trees result in growth (as expressed by ring widths) that is directly related to the climatic variables of temperature and precipitation. These climatic variables are in turn closely related to the pattern of atmospheric circulation. For example, in parts of eastern United States, southerly winds are associated with warm, moist conditions as contrasted with northerly winds which are accompanied by cold and dry conditions. By evaluating these associations over long periods of time using tree rings, one is able to establish inconsistencies in present frequencies of recurrences and their relevance to present and future agricultural and hydrologic management practices.

In their study, Kutzbach and Guetter have shown that skill in reconstructing these large scale circulation patterns is closely tied to spatial sampling density of the network from which such large scale patterns are to be reconstructed. Using a rather large northern hemisphere database and techniques incorporating EOF's and linear regression, they arrived at five specific conclusions. They are: (1) for a fixed number of tree-ring sites, it is best to distribute the spacing of the sites uniformly over the area for which estimates of atmospheric circulation are desired. In certain cases, however, high density sites in restricted longitudinal sections may produce results of equivalent accuracy to those of low density sites spread over broader longitudinal sections. (2) For sampling densities—0.5 to 2.5 sites per 1,000,000 km^2—the ability to duplicate results on independent data (that is, data not used in the model development) increased with sample density. (3) Good sampling strategy appears to include the sampling of data sites beyond the boundaries of the region for which estimates of atmospheric circulation are desired. (4) The measures of adequacy obtained from independent data are usually substantially lower than that obtained by the model in the calibration operation. One would expect this difference to diminish as the length of records used becomes infinitely large. (5) Within the range of sampling densities they tested, Kutzbach and Guetter found *some* skill in the reconstruction of large scale circulation patterns from temperature and precipitation data from all spatial sample densities.

EXISTING WORLDWIDE TREE RING COVERAGE

The following evaluation of the worldwide tree-ring data base is not intended to be an exhaustive inventory. Rather it provides the reader with a concept of the spatial availability of tree-ring data as of 1980 for climatic reconstruction.

Western United States (Jeffrey S. Dean, University of Arizona)

Tree ring chronologies from the western United States suitable for dendroclimatic inferences cover an area of 4,532,320 square kilometers extending from 125° west longitude eastward to approximately 96° and north to south between 49° and 26° north latitude (see figure 4). Topography is varied and complex, but exhibits a strong north-south orientation. Climate is affected by both air mass movement and the large scale topography with precipitation varying from 80 mm to 2500 mm. Conifers dominate the forest vegetation.

Many hundreds of tree-ring chronologies are now available from the western United States. The Tree-Ring Laboratory alone has more than 700 such composite ring series in its computer data file. A compilation of 169 selected tree-ring chronologies with standardized ring-width indices (table 1, figure 4) has been selected from the much larger store of available series. These chronologies have been published and are stored in open computer data files. Most of the chronologies published before 1960, such as those in Schulman,[2] are not included because they have been augmented, updated, or superseded by more recently constructed ring series. In addition, most post-1960 chronologies were constructed using modern techniques of ring-width measurement, data collection, and standardization, and by merging

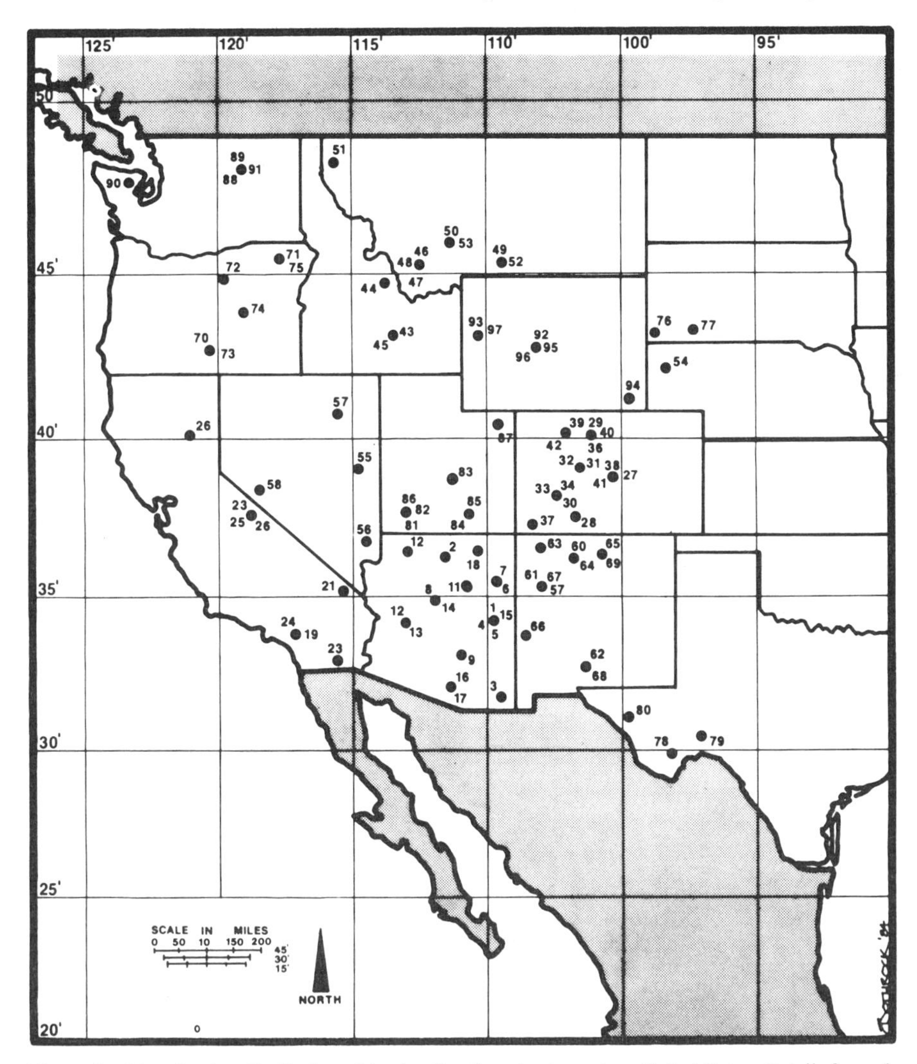

Figure 4. Map showing distribution of dendroclimatic series in western United States. Details for each series are listed in table 1.

the data from many samples into composite series. Chronologies thus compiled are both mechanically and statistically comparable to one another.

Eastern United States (Edward R. Cook, Lamont-Doherty Geological Observatory of Columbia University)

This sector of North America extends west from the Atlantic coastline to roughly 96° west longitude and south from about 50° north latitude to approximately 36° north latitude. The northern boundary approximates the boreal forest-deciduous

Table 1. Dendroclimatically Sensitive Series of Western United States[a]

Map No.	Site Name	State	Species	Location		Record Length
1	Alpine	Arizona	Pseudotsuga menziesii	33°54′	109°06′	1666–1967
2	Bright Angel	Arizona	Pseudotsuga menziesii	36°15′	112°08′	1695–1967
2	Bright Angel	Arizona	Pseudotsuga menziesii	36°17′	115°58′	1617–1967
2	Bright Angel	Arizona	Pseudotsuga menziesii	36°20′	112°04′	1613–1967
2	Bright Angel	Arizona	Pseudotsuga menziesii	36°16′	112°02′	1708–1967
3	Chiricahua Mtns.	Arizona	Pseudotsuga menziesii	31°57′	109°18′	1647–1969
4	Cienega	Arizona	Pinus ponderosa	33°18′	109°44′	1661–1966
5	Clark Peak Saddle	Arizona	Pseudotsuga menziesii	32°42′	109°59′	1630–1967
6	Defiance East	Arizona	Pinus ponderosa	33°50′	109°07′	1554–1965
7	Defiance West	Arizona	Pseudotsuga menziesii	35°52′	109°26′	1474–1965
8	Flagstaff	Arizona	Pinus ponderosa	35°15′	111°30′	570–1972
9	Galiuros A	Arizona	Pseudotsuga menziesii	32°35′	110°17′	1650–1965
9	Galiuros C	Arizona	Pseudotsuga menziesii	32°31′	110°15′	1780–1967
10	Granite Mtn.	Arizona	Pinus ponderosa	34°39′	112°32′	1605–1965
11	Hopi Mesa	Arizona	Pinus edulis	35°50′	110°26′	500–1971
12	June Tanks	Arizona	Pinus edulis	36°32′	112°55′	1700–1964
5	Lady Bug Pk.	Arizona	Pseudotsuga menziesii	32°37′	109°49′	1625–1967
13	Mayer	Arizona	Pinus ponderosa	34°23′	112°21′	1620–1965
14	Mogollon Rim	Arizona	Pseudotsuga menziesii	34°23′	111°00′	1593–1965
12	Mt. Trumbull	Arizona	Pinus edulis	36°25′	113°10′	1620–1964
4	Nantuck Gap	Arizona	Pseudotsuga menziesii	33°18′	109°45′	1650–1966
1	Nutrioso	Arizona	Pinus edulis	34°01′	109°07′	1587–1965
15	Rose Peak	Arizona	Pinus ponderosa	33°23′	109°22′	1660–1965
16	Santa Catalina High	Arizona	Pseudotsuga menziesii	32°25′	110°46′	1526–1968
17	Santa Rita High	Arizona	Pseudotsuga menziesii	31°43′	110°51′	1645–1966
17	Santa Rita Low	Arizona	Pseudotsuga menziesii	31°44′	110°50′	1646–1965

18	Tsegi Canyon	Arizona	Pseudotsuga menziesii	36°45′	110°30′	381–1970
4	Turkey Creek	Arizona	Pinus edulis	33°23′	109°47′	1700–1966
12	Tuweep	Arizona	Pinus edulis	36°28′	113°08′	1730–1964
8	Walnut Canyon A	Arizona	Pseudotsuga menziesii	35°11′	111°30′	1646–1966
8	Walnut Canyon	Arizona	Pinus ponderosa	35°12′	113°31′	1447–1966
19	Baldwin Lake South	California	Pinus ponderosa	34°15′	116°48′	1513–1966
20	Campito Mtn.	California	Pinus longaeva	37°30′	118°13′	−170–1970
21	Clark Mtn.	California	Abies concolor	35°32′	115°35′	1596–1968
20	Cottonwood	California	Pinus longaeva	37°34′	118°12′	590–1967
20	Crooked Creek	California	Pinus longaeva	37°31′	118°09′	1038–1953
22	Keen Camp Summit	California	Pseudotsuga macrocarpa	33°43′	116°05′	1458–1966
23	Log Cabin Mine Rd.	California	Pinus jeffreyi	37°57′	119°09′	1485–1963
23	Log Cabin Mine Rd.	California	Juniperus occidentalis	37°57′	119°09′	1500–1964
23	Log Cabin Mine Rd.	California	Abies concolor	37°57′	119°09′	1685–1964
24	Mill Creek Summit	California	Pseudotsuga macrocarpa	34°23′	118°04′	1599–1966
23	Mono Craters	California	Pinus jeffreyi	37°55′	119°01′	1699–1966
20	Sheep Mtn.	California	Pinus longaeva	37°32′	118°13′	470–1970
25	Sherwin Summit	California	Pinus jeffreyi	37°31′	118°38′	1635–1964
25	Sherwin Summit	California	Pinus monophylla	37°32′	118°37′	1536–1964
19	Southern California	California	Pseudotsuga macrocarpa	34°03′	117°05′	1458–1966
26	Susanville	California	Pinus ponderosa	40°29′	120°33′	1485–1963
20	White Mtns, Master 2	California	Pinus longaeva	37°25′	118°10′	800–1963
20	White Mtns.	California	Pinus longaeva	37°25′	118°10′	−5141–1962[c]
27	Almagre Mtns.	California	Pinus aristata	38°47′	104°58′	560–1968
28	Antonito Merged	California	Pseudotsuga menziesii	37°04′	106°11′	1298–1965
29	Big Thompson	Colorado	Pseudotsuga menziesii	40°25′	105°17′	1700–1964
30	Black Canyon, Gunnison	Colorado	Pseudotsuga menziesii	38°34′	107°42′	1478–1964
30	Black Canyon	Colorado	Pinus edulis	38°33′	107°45′	1457–1964
31	Chicago Creek	Colorado	Pseudotsuga menziesii	39°41′	105°38′	1441–1964
32	Eagle East	Colorado	Pinus edulis	39°40′	106°43′	1314–1964
32	Eagle	Colorado	Pseudotsuga menziesii	39°39′	106°52′	1107–1964

Table 1. *(Continued)*

Map No.	Site Name	State	Species	Location		Record Length
29	El Dorado Canyon	Colorado	Pseudotsuga menziesii	39°56′	105°17′	1610–1964
29	El Dorado Springs	Colorado	Pinus ponderosa	39°56′	105°17′	1750–1964
33	Escalante Forks	Colorado	Pinus edulis	38°40′	108°20′	1640–1964
34	Gunnison Upper	Colorado	Pseudotsuga menziesii	38°41′	106°52′	1322–1964
35	Hermit Lake	Colorado	Pinus aristata	38°06′	105°37′	1259–1968
29	Horsetooth Reservoir	Colorado	Pinus ponderosa	40°32′	105°09′	1650–1964
29	Horsetooth Summit	Colorado	Pseudotsuga menziesii	40°34′	105°12′	1500–1964
36	Idaho Springs East	Colorado	Pseudotsuga menziesii	39°45′	105°28′	1710–1964
29	Kassler	Colorado	Pseudotsuga menziesii	39°27′	105°07′	1810–1964
29	Kassler	Colorado	Pinus ponderosa	39°02′	105°07′	1690–1964
29	Lyons	Colorado	Pinus ponderosa	40°10′	105°17′	1730–1964
37	Mesa Verde	Colorado	Pseudotsuga menziesii	37°10′	108°30′	480–1971
38	Mount Evans	Colorado	Pinus aristata	39°38′	105°36′	977–1968
39	New North Park	Colorado	Pseudotsuga menziesii	40°55′	106°20′	1354–1964
40	Niwot Ridge	Colorado	Picea engelmannii	40°03′	105°33′	1528–1968
29	Owl Canyon	Colorado	Pinus edulis	40°48′	105°11′	1530–1964
41	Salida A	Colorado	Pseudotsuga menziesii	38°28′	105°55′	1440–1964
41	Salida B	Colorado	Pinus edulis	38°28′	105°54′	1700–1964
41	Salida Merged	Colorado	Pseudotsuga menziesii	38°29′	105°56′	1328–1964
37	Schulman Old Trees	Colorado	Pseudotsuga menziesii	37°12′	108°30′	1450–1963
42	Steamboat Springs	Colorado	Pseudotsuga menziesii	40°45′	106°51′	1736–1964
29	Van Bibber	Colorado	Pinus ponderosa	39°48′	106°15′	1720–1964
29	Waterdale	Colorado	Pinus ponderosa	40°25′	105°12′	1680–1964
43	Ketchum East	Idaho	Pseudotsuga menziesii	43°46′	115°16′	1777–1965
43	Ketchum	Idaho	Pseudotsuga menziesii	43°48′	114°16′	1521–1965
44	Salmon River South	Idaho	Pseudotsuga menziesii	44°58′	113°57′	1569–1965

45	Warm Springs	Idaho	Pseudotsuga menziesii	43°41′	114°22′	1709–1965
46	Butte	Montana	Pseudotsuga menziesii	45°50′	112°21′	1688–1965
47	Dell, Admin.	Montana	Pinus flexilis	44°39′	112°46′	1311–1965
47	Dell, Sheep Creek	Montana	Pseudotsuga menziesii	44°33′	112°48′	1368–1965
48	Divide, Big Hole R.	Montana	Pseudotsuga menziesii	45°46′	112°47′	1628–1965
49	Gardiner 1	Montana	Pseudotsuga menziesii	45°00′	110°42′	1445–1965
49	Gardiner	Montana	Pinus flexilis	45°00′	110°42′	1721–1965
50	Helena	Montana	Pinus flexilis	46°43′	111°48′	1565–1965
51	Libby	Montana	Pinus flexilis	48°23′	115°21′	1738–1965
52	Livingston	Montana	Pinus flexilis	45°36′	110°33′	1532–1965
52	Springdale	Montana	Pinus flexilis	45°43′	110°14′	1422–1965
53	Townsend Ridge	Montana	Pseudotsuga menziesii	46°20′	111°14′	1750–1965
54	Ft. Robinson Timber Res.	Nebraska	Pinus ponderosa	42°41′	103°36′	1610–1964
55	Big Wash East	Nevada	Pinus monophylla	38°53′	114°10′	1612–1964
55	Blue Lake	Nevada	Pinus flexilis	39°00′	114°18′	1200–1968
56	Charleston Peak	Nevada	Pinus longaeva	36°17′	115°38′	966–1964
55	Hill 10842, Snake Range	Nevada	Pinus longaeva	38°57′	114°13′	1100–1967
57	Lamoille Canyon	Nevada	Pinus edulis	40°40′	115°27′	1578–1965
58	Little Whiskey Flat	Nevada	Pinus monophylla	38°13′	118°40′	1614–1964
55	Mt. Washington Upper	Nevada	Pinus longaeva	38°53′	114°20′	737–1965
55	Wheeler Pk. Moraine	Nevada	Pinus flexilis	39°00′	114°18′	820–1967
55	Wheeler Pk. Switchback	Nevada	Pinus flexilis	39°02′	114°17′	1620–1967
55	Wheeler Pk. Viewpoint	Nevada	Pseudotsuga menziesii	39°01′	114°16′	1620–1967
59	Cebolleta Mesa	New Mexico	Pinus edulis[b]	34°50′	107°45′	680–1972
60	Chama Valley	New Mexico	Pseudotsuga menziesii[b]	36°17′	106°32′	759–1972
61	Cibola	New Mexico	Pinus ponderosa[b]	35°20′	108°30′	439–1972
62	Cloudcroft Low	New Mexico	Pinus edulis	32°57′	105°49′	1670–1965
62	Cloudcroft	New Mexico	Pinus edulis	32°57′	105°40′	1515–1960
63	Gobernador	New Mexico	Pinus edulis	36°45′	107°30′	623–1971
64	Jemez Mtns.	New Mexico	Pinus edulis[b]	35°40′	106°30′	598–1972
65	Jicarita	New Mexico	Pinus aristata	36°03′	105°32′	1436–1968

Table 1. (*Continued*)

Map No.	Site Name	State	Species	Location		Record Length
66	Luna	New Mexico	Pseudotsuga menziesii	33°50′	109°00′	1666–1967
66	Luna	New Mexico	Pinus ponderosa	33°51′	109°01′	1693–1965
67	Mt. Taylor	New Mexico	Pinus edulis	35°14′	107°41′	1611–1972
68	Organ Mtns.	New Mexico	Pseudotsuga menziesii	32°21′	106°33′	1597–1970
69	Pecos	New Mexico	Pseudotsuga menziesii	35°35′	105°46′	1750–1965
63	Pueblito Canyon	New Mexico	Pinus edulis	36°42′	107°20′	1594–1971
66	Rainy Mesa	New Mexico	Pseudotsuga menziesii	33°35′	108°35′	1520–1967
65	Red Dome	New Mexico	Pinus aristata	36°33′	105°23′	1535–1968
65	Rio Grande, North	New Mexico	Pinus ponderosa[b]	36°15′	105°40′	1104–1972
69	Santa Fe	New Mexico	Pinus edulis[b]	35°42′	105°50′	878–1972
62	Silver Springs	New Mexico	Pseudotsuga menziesii	33°00′	105°40′	1542–1965
66	Tularosa Divide	New Mexico	Pseudotsuga menziesii	33°45′	108°30′	1749–1967
62	Wofford Tower	New Mexico	Pseudotsuga menziesii	32°59′	105°42′	1663–1965
62	Wofford & Cloudcroft	New Mexico	Pseudotsuga menziesii	32°59′	105°42′	1515–1965
70	Abert Rim Lookout	Oregon	Pinus ponderosa	42°23′	120°14′	1511–1964
71	Chief Joseph Mtn.	Oregon	Pinus albicaulis	45°17′	117°17′	1538–1964
72	Dufer-Happy Ridge	Oregon	Pinus ponderosa	45°16′	121°19′	1600–1964
73	Lakeview	Oregon	Pinus ponderosa	42°06′	120°34′	1421–1964
74	Paulina	Oregon	Pinus ponderosa	44°16′	119°53′	1600–1965
71	Slickrock	Oregon	Pinus flexilis	45°17′	117°19′	715–1965
75	Union	Oregon	Pinus ponderosa	45°09′	117°37′	1565–1964
76	Parker Peak Lookout	South Dakota	Pinus ponderosa	43°24′	103°40′	1620–1964
76	Pilger Mtn. Lookout	South Dakota	Pinus ponderosa	43°28′	103°54′	1520–1964
77	Rosebud Sioux Res.	South Dakota	Pinus ponderosa	43°14′	100°58′	1810–1964
78	Big Bend, Boot Spring	Texas	Pseudotsuga menziesii	29°15′	103°18′	1631–1965
78	Big Bend, Pine Canyon	Texas	Pinus ponderosa	29°16′	103°14′	1752–1965

79	Frio, Kent Creek Ranch	Texas	Pinus edulis	29°53′	99°52′	1819–1966
80	McDonald Observatory	Texas	Pinus edulis	30°41′	104°01′	1748–1965
81	Alton	Utah	Pinus ponderosa	37°25′	112°30′	1600–1964
82	Bryce Canyon	Utah	Pseudotsuga menziesii	37°30′	112°10′	1270–1964
83	Emery	Utah	Pinus ponderosa	38°59′	111°19′	1537–1964
84	Karparowitz, East	Utah	Pinus edulis	37°24′	111°12′	1605–1964
85	Natural Bridges	Utah	Pinus edulis[b]	37°37′	110°00′	94–1971
86	Navajo Point	Utah	Pinus longaeva	37°41′	112°53′	1800–1965
87	Nine Mile Canyon	Utah	Pseudotsuga menziesii	39°47′	110°18′	1194–1964
81,82	Water Canyon-Alton	Utah	Pinus ponderosa	37°32′	112°18′	1340–1964
82	Water Canyon	Utah	Pinus ponderosa	37°40′	112°06′	1336–1964
88	Boulder Creek	Washington	Pinus ponderosa	48°35′	120°10′	1468–1965
89	Conconully	Washington	Pinus ponderosa	48°41′	119°41′	1690–1965
90	Dungeness River	Washington	Pseudotsuga menziesii	47°54′	123°08′	1780–1965
88	Pipestone Canyon	Washington	Pseudotsuga menziesii	48°25′	120°03′	1700–1966
91	Quartz Mtn. Lookout	Washington	Pseudotsuga menziesii	48°39′	118°40′	1690–1965
91	Quartz Mtn. Junction	Washington	Pseudotsuga menziesii	48°39′	118°44′	1656–1965
91	Quartz Mtn. Lookout	Washington	Pinus ponderosa	48°39′	118°40′	1562–1976
91	Quartz Mtn. Lookout	Washington	Larix occidentalis	48°39′	118°40′	1598–1965
92	Alcova-A	Wyoming	Pinus ponderosa	42°31′	106°44′	1544–1964
92	Alcova-B	Wyoming	Pinus ponderosa	42°32′	106°44′	1762–1964
93	Elbow Campground	Wyoming	Pseudotsuga menziesii	43°13′	110°47′	1470–1965
94	Laramie A	Wyoming	Pseudotsuga menziesii	41°06′	106°05′	1444–1964
94	Laramie B	Wyoming	Pinus ponderosa	41°06′	106°05′	1667–1964
95	Pedro Mtns. B	Wyoming	Pinus flexilis	42°21′	106°51′	1508–1964
95	Pedro Mtns. A	Wyoming	Pinus ponderosa	42°22′	106°51′	1610–1964
96	Seminoe Reservoir	Wyoming	Pinus ponderosa	42°07′	106°55′	1536–1964
97	Wind River Mtns. B	Wyoming	Pseudotsuga menziesii	42°58′	109°46′	1562–1964
97	Wind River Mtns. A	Wyoming	Pinus flexilis	42°56′	109°46′	1682–1964

[a]Material from the files of the Laboratory of Tree-Ring Research, University of Arizona (see figure 4).
[b]Species listed is for living tree samples; older portion of this record contains samples from other species.
[c]This series consists of filtered values only.

forest ecotone in southern Canada and the western limit extends to the prairie-deciduous forest ecotone. Climate is generally temperate with an overall surplus of annual precipitation. The region is part of the Eastern Deciduous Forest Biome[21] and is characterized by an abundance of deciduous and coniferous species occurring in closed canopy forests.

Tree-ring chronology development in the eastern United States and southeastern Canada has accelerated greatly since 1970 as data processing techniques have improved, and from the realization that useful quantitative climatic information could be obtained from mesic forest trees. Although a complete network does not yet exist for the entire sector, the present rapid rate of chronology development should produce such a network within a few years. The collection described here includes only those chronologies which extend back beyond 1700 and end no sooner than 1966.

The general locations of the tree-ring sites are shown in figure 5 and table 2 provides pertinent information about each chronology. Although some of the collections have not yet been fully analyzed, all will extend beyond 1700. The total now stands at 50 chronologies. Although major gaps still exist in the network, recent communication with natural resources officials indicate that well over 100 potential sites exist in the areas not presently covered. Thus, in this sector of North America alone, a network comprising as many as 200 chronologies dating back to 1700 is highly feasible. Such a network will be more compatible to that in western North America than exists at the present.

Southeastern United States (David W. Stahle, University of Arkansas)

The southeastern United States lies within latitudes 26° and 38° north and longitudes 76° and 100° west (figure 6). Climate is classified as humid subtropical with

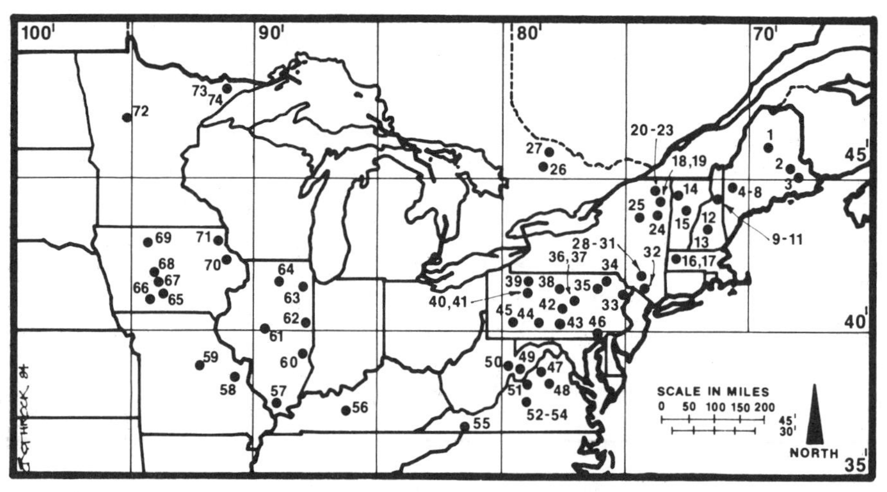

Figure 5. Map showing distribution of dendroclimatic series in northeastern United States. Details are listed in table 2.

Table 2. Tree Ring Sites in the Northeastern United States[a].

Map Ref. No.[b]	Site Location	Species	Years[c]	No. Cores A.D. 1700 or Beyond
1	Hamlin Ridge, ME	Picea rubens	1610–1981	9
2	Mattawamkeag, ME	Tsuga canadensis	1680–1981	2
3	No. 3 Pond, ME	Tsuga canadensis	1684–1981	14
4	Elephant Mt., ME	Picea rubens	1667–1976	—
5	Elephant Mt., ME	Picea rubens	1667–1976	—
6	Elephant Mt., ME	Picea rubens	1667–1976	—
7	Elephant Mt., ME	Picea rubens	1667–1976	—
8	Elephant Mt., ME	Picea rubens	1667–1976	—
9	Nancy Brook, NH	Picea rubens	1561–1981	18
10	Mt. Washington, NH	Picea rubens	1678–1976	6
11	Gibbs Brook, NH	Tsuga canadensis	1509–1981	19
12	Fox Forest, NH	Tsuga canadensis	1600–1981	—
13	Paradise Point, NH	Tsuga canadensis	1622–1981	2
14	Camel's Hump, VT	Picea rubens	1635–1971	6
15	Granville Gulf, VT	Tsuga canadensis	1670–1981	8
16	Livingston, MA	Picea rubens	1697–1971	1
17	Mohawk Trail, MA	Tsuga canadensis	1698–1980	1
18	Winch Pond, NY	Pinus strobus	1690–1978	2
19	Wilmington Notch, NY	Pinus strobus	1670–1981	4
20	Roaring Brook, NY	Tsuga canadensis	1599–1978	12
21	Roaring Brook, NY	Picea rubens	1618–1978	13
22	Roaring Brook, NY	Pinus strobus	1632–1981	4
23	Adirondack Mt. Reserve, NY	Tsuga canadensis	1608–1981	13
24	Pack Forest, NY	Tsuga canadensis	1595–1976	9
25	W. Branch Sacandaga River, NY	Tsuga canadensis	1665–1981	3
26	Pot Lake-Northwest Lake, Ont.	Tsuga canadensis	1641–1982	24
27	Dickson Lake, Ont.	Pinus resinosa	1550–1982	27
28	Mohonk Lake, NY	Tsuga canadensis	1636–1973	10
29	Mohonk Lake, NY	Pinus strobus	1626–1973	12
30	Mohonk Lake, NY	Pinus rigida	1622–1973	5
31	Mohonk Lake, NY	Quercus prinus	1690–1973	2
32	Dark Hollow Trail, NY	Quercus alba	1648–1977	5
33	Dingmans Falls, PA	Tsuga canadensis	1620–1981	5
34	Salt Springs State Park, PA	Tsuga canadensis	1619–1981	5
35	Rickett's Glen State Park, PA	Tsuga canadensis	1637–1981	9
36	Bear Run, PA	Tsuga canadensis	1641–1981	6
37	Woodward Gap, PA	Quercus prinus	1600–1981	—
38	East Branch Swamp, PA	Tsuga canadensis	1540–1981	20
39	Tionesta, PA	Tsuga canadensis	1426–1978	25
40	Cook Forest, PA	Quercus alba	1660–1981	10
41	Longfellow Trail, PA	Pinus strobus	1679–1981	14
42	Alan Seegar Natural Area, PA	Tsuga canadensis	1609–1981	20
43	Hemlocks Natural Area, PA	Tsuga canadensis	1535–1981	11

Table 2. (*Continued*)

Map Ref. No.	Site Location	Species	Years	No. Cores A.D. 1700 or Beyond
44	Sweet Root Natural Area, PA	Tsuga canadensis	1612–1981	12
45	Ohiopyle State Park, PA	Tsuga canadensis	1622–1981	11
46	Otter Creek Natural Area, PA	Quercus prinus	1631–1981	6
47	Watch Dog, VA	Quercus prinus	1640–1981	7
48	Pinnacle Point-Hawksbill Gap, VA	Quercus alba	1612–1981	15
49	Cedar Knob, WV	Juniperus		
50	Gardineer, WV	Picea rubens	1652–1976	9
51	Ramsey's Draft, VA	Tsuga canadensis	1595–1981	7
52	Blue Ridge, VA	Quercus prinus	1594–1982	18
53	Patty's Oaks, VA	Quercus alba	1569–1982	10
54	Sunset Field, VA	Tsuga canadensis	1531–1982	10
55	Mt. Rogers Hemlocks, VA	Tsuga canadensis	1645–1982	6
56	Mammoth Cave, KY	Quercus alba	1648–1966	2
57	Ferne Clyffe, IL	Quercus alba	1669–1972	7
58	Babler State Park, MO	Quercus alba	1641–1980	—
59	Boone County, MO	Juniperus virginiana	1650–1978	—
60	Fox Ridge State Park, IL	Quercus alba	1674–1980	5
61	Lincoln's New Salem State Park, IL	Quercus alba	1671–1980	6
62	Kickapoo State Park, IL	Quercus alba	1670–1980	2
63	Kankakee River State Park, IL	Quercus alba	1686–1980	2
64	Starved Rock State Park, IL	Quercus alba	1633–1980	12
65	Lake Ahquabi and N.E. Warren County, IA	Quercus alba	1574–1980	—
66	Pammel State Park, IA	Quercus alba	1635–1980	—
67	Duvick Back Woods and Saylorsville, IA	Quercus alba	1654–1980	—
68	Ledges State Park, IA	Quercus alba	1663–1980	—
69	Woodman Hollow State Preserve, IA	Quercus alba	1698–1979	—
70	White Pine Hollow State Preserve, IA	Quercus alba	1631–1980	5
71	Yellow River State Forest, IA	Quercus alba	1651–1980	7
72	Itasca State Park, MN	Pinus resinosa	1672–1971	2
73	Saganaga Lake, MN	Pinus resinosa	1620–1972	17
74	Seagull Lake, MN	Pinus resinosa	1625–1971	12

[a]From files of Lamont Doherty Geological Observatory.

[b]Map reference number refers to figure 5.

[c]An incomplete starting year indicates that the site is being developed.

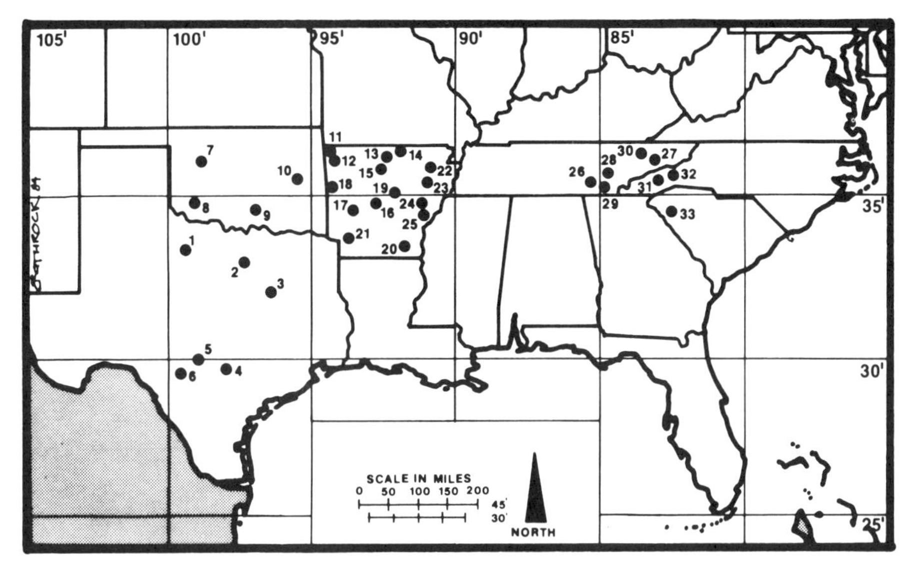

Figure 6. Map showing distribution of dendroclimatic series in southeastern United States. Details are listed in table 3.

precipitation varying from near 40 inches in eastern Oklahoma and Texas to 60 inches or more along the Gulf of Mexico. Temperatures are generally mild, resulting in long growing seasons.[22] Vegetation is diverse with southern pines dominating the coastal plains. The interior highlands support hardwood forests of oak-hickory which is mixed with shortleaf pine in the Ozark mountains. Mixed mesophytic forests cover the Cumberland Plateau.[21]

Tree-ring studies in the Southeast began early[23-25] although sustained efforts to specifically develop a network of modern chronologies started only recently.[26,27] At least 46 chronologies have been developed (table 3, figure 6) but these vary in length and quality, and do not provide uniform coverage of the Southeast.

Shortleaf (*Pinus echinata*), loblolly (*P. taeda*), and longleaf pine (*P. palustris*) have all proven suitable for dendrochronology and individuals of each species may exceed 300 years in age. Other species with considerable potential in the Southeast include white oak (*Quercus alba*), post oak (*Q. stellata*), Eastern red cedar (*Juniperus virginiana*), Eastern hemlock (*Tsuga canadensis*), red spruce (*Picea rubens*), and bald cypress (*Taxodium distichum*).

White oak and post oak have provided excellent, climatically sensitive chronologies in the central United States.[26-28] Small, old-growth trees of both species may be found in the Southeast on noncommercial sites that escaped cutting because of their low productivity. Hemlock and red spruce both reach their southern limits at higher elevations of the southern Appalachians. Both species have been previously dated and may provide relatively sensitive chronologies. Eastern red cedar (*Juniperus virginiana L.*) is widely distributed in the Southeast and old-growth stands with small, stressed trees are still remarkably prevalent along limestone bluffs. Red

Table 3. Tree Ring Chronologies of the Southeastern United States[a]

Map No.	Site Name	Collectors	Reference	Species	Dating	Comments[b]
1	D. Nichols Ranch, Texas	Stahle, Hawks		Quercus stellata	1695–1980	
2	Ft. Worth Nature Center, Texas	Stahle, Hawks	26	Quercus stellata	1737–1980	
3	Oak Park, Texas	Cook, Harlan	26	Quercus stellata	1699–1974	
4	Walls Ranch, Texas	Harlan		Quercus stellata	1790–1978	A
5	Rabke Ranch, Texas	Stokes, Harlan		Quercus stellata	1734–1980	A
6	Frio, Kent Creek Ranch, Texas	Harlan	142	Pinus edulis	1819–1966	
7	Oakwood, east, Oklahoma	Stahle, Hawks		Quercus stellata	1772–1980	
8	Quanah Mountain, Oklahoma	Stahle, Hawks		Quercus stellata	1686–1980	
9	Lake of the Arbuckles, Oklahoma	Stahle, Hawks, Cook, Harlan		Quercus stellata	1696–1980	
10	Lake Eufaula, Oklahoma	Stahle, Hawks		Quercus stellata	1745–1980	
11	Wedington Mountain, Arkansas	Stahle, Jurney, Cande		Quercus stellata	1725–1977	
12	Dutch Mills Ruin, Arkansas	Stahle, Jurney	31	Quercus sp.	1624–1842	B
	Ridge House, Arkansas	Stahle, Jurney	31	Quercus sp.	1674–1838	B
13	Magness Barns, Arkansas	Stahle	31	Quercus sp.	1598–1858	B
	Jackson Cabin, Arkansas	Stahle	31	Quercus sp.	1654–1849	B
	Horton House, Arkansas	Stahle	31	Pinus echinata	1750–1881	B
	Drury House, Arkansas	Stahle	31	Pinus echinata	1649–1850	B
14	Wolf House, Arkansas	Stahle, Jurney, Wolfman	31	Pinus echinata	1672–1827	B
	Lancaster Barns, Arkansas	Stahle	31	Juniperus virginiana	1733–1866	B
	Mt. Olive Group, Arkansas	Stahle	31	Juniperus virginiana	1710–1859	B
15	Pope County, Arkansas	Bell	26	Quercus alba	1642–1939	
	Russellville, Arkansas	Stockton et al.	26	Quercus alba	1713–1972	
	Russellville-North, Arkansas	Estes		Pinus echinata	1726–1977	
16	Montgomery County, Arkansas	Bell	26	Pinus echinata	1666–1939	

	Big Brushy Mountain, Arkansas	Stockton et al.	26	Pinus echinata	1760–1972	Update of 16
17	Polk County, Arkansas	Bell	26	Quercus alba	1677–1939	
	Brush Heap Mountain, Arkansas	Stockton et al.	26	Quercus alba	1720–1972	Update of 17
18	Blackfork Mountain, Arkansas	Stahle, Hawks		Quercus alba	1650–1980	
19	Lake Winona Natural Area, Arkansas	Stahle, Hawks		Pinus echinata	1667–1980	
20	Levi-Wilcoxon Forest, Arkansas	Stahle, Hawks		Pinus echinata	1779–1980	
21	Hempstead County, Arkansas	Stahle, Hawks		Taxodium distichum	1766–1980	
	Lafayette County Jail, Arkansas	Stahle	28	Taxodium distichum	1727–1828	B
22	Egypt, Arkansas	Stahle, Hawks		Taxodium distichum	1417–1980	
23	Black Swamp, Arkansas	Stahle, Hawks		Taxodium distichum	1464–1980	
24	Bayou Meto, Arkansas	Stahle, Hawks		Taxodium distichum	1522–1980	
25	Sugarberry Natural Area, Arkansas	Stahle, Hawks		Quercus lyrata	1764–1980	
26	Warren County, Tennessee	Bell	26	Quercus alba	1669–1940	
27	Steiner's Woods, Tennessee	Stockton and TVA	26	Quercus alba	1625–1972	
28	Fall Creek Falls, Tennessee	Stockton et al.	26	Quercus alba	1767–1972	
29	Savage Gulf, Tennessee	Stockton et al.	26	Pinus echinata	1700–1972	
30	Norris Watershed Boundary, Tennessee	Stockton and TVA	26	Pinus echinata	1681–1972	
31	Newfound Gap, North Carolina	Ames, Harsha	26	Picea rubens	1686–1972	
32	Linville Gorge, North Carolina	Cook		Quercus alba	1616–1978	
33	Clemson Forest, South Carolina	Cleaveland	26	Pinus echinata	1685–1973	C

[a]Unreferenced material from files at Tree-Ring Laboratory, University of Arizona.

[b](A) Dating completed, analysis underway; (B) historic chronology; (C) includes three chronologies for earlywood, latewood, and total ring width.

cedar may attain great age and individuals 400 to 600 years old have been located.[29,30]

Baldcypress, in spite of its riverine habitat, may have the best potential of any southeastern species for long-term chronologies. Chronologies in excess of 500 years old, recently developed in eastern Arkansas, show strong internal and regional crossdating that suggests a macroclimatic control of cypress growth. Pronounced seasonal and annual variations in stream levels, quite common in the Southeast, appear to be an important factor in determining cypress ring widths. Cypress chronologies may crossdate regionally with those from upland oak, pine, and cedar which are largely limited by precipitation.[31]

The extension of southeastern chronologies beyond A.D. 1600 should be possible with long-lived species such as baldcypress and red cedar, and through the recovery of historic, archaeological, and subfossil wood. Early historic structures often include old-growth timbers that were cut locally from the virgin forests and can help improve and extend modern chronologies.[31] Although wood and charcoal suitable for dendrochronology are only occasionally preserved in prehistoric archaeological sites in the Southeast, enough has been recovered to remain optimistic about the potential for archaeological tree-ring dating.[32,33]

Subfossil or bog wood a few hundred to many thousand years old has been recovered from buried deposits throughout the Southeast.[34] A significant amount of durable baldcypress, in particular, has been uncovered throughout its present and former range.[35–37]

Northern North America (Gordon C. Jacoby, Jr., Lamont-Doherty Geological Observatory, Columbia University)

The boreal forest and forest-tundra ecotone cover most of northern North America except for the relatively small area of mountain and coastal forest in western Canada and southeast Alaska. These forest regions are bordered on the north by treeless tundra and on the south by the eastern deciduous forests, the prairies, or the montane forests of British Columbia. Climate is generally characterized by short, cool summers and cold to frigid winters. Annual precipitation varies from under 25 cm to more than 200 cm in limited areas along the northwest Pacific coast.

Tree-ring series from the boreal forests have been successfully related to temperature,[38] moisture,[39] atmospheric pressure,[40] and ice conditions.[41] Although some of these relationships are statistically simple, the underlying causal mechanisms are complex and involve soil temperatures, permafrost, edaphic conditions, air mass movements, and lag effects of previous years on the growth year.

The northern region of North America is less densely sampled than some other areas and much of the sampling was done several decades ago. Because many of the meteorological records are short in northern areas, the abbreviated calibration period for the earlier chronologies is a serious problem.

The relatively low ring-width variability indicates that dendroclimatology in the more northern regions would benefit greatly from improved data analysis and new techniques such as densitometry.

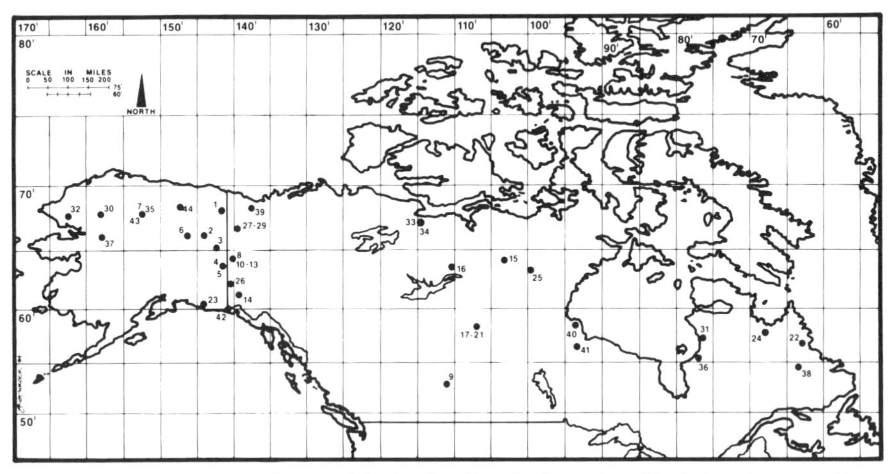

Figure 7. Map showing distribution of dendroclimatic series in northern North America. See table 4.

Locations of existing chronologies are shown on figure 7 and in table 4. This listing only includes chronologies developed after 1960 and extending back to at least 1800. More complete listings are published in references 41–43.

Due to low diversity and lack of longevity in the northern areas, there are only a limited number of useful species. White spruce (*Picea glauca* [Moench] Voss) is the most useful, can live to over 500 years, and is distributed throughout the boreal zone. Western hemlock (*Tsuga heterophylla* [Rat.] Sarg.), mountain hemlock (*T. mertensiana* [Bong.] Carr.), and Sitka spruce (*P. sitchensis* [Bong.] Carr.) all can live over 500 years. Larch or tamarack (*Larix laricina* [DuRoi] Koch) and black spruce (*P. mariana* [Mill.] B.S.P.) are also widely distributed and achieve considerable age. Other species most likely to be useful are subalpine fir (*Abies lasiocarpa* [Hook] Nutt.), Alaska cedar (*Chamaecyparis nootkatensis* [D. Don] Spach), lodgepole pine (*Pinus contorta,* Dougl.), northern white cedar (*Thuja occidentalis* L.), and western red cedar (*T. plicata* Donn).

Mexico and Central America

All available sources have been searched for the existence of tree-ring data in Mexico and Central America suitable for climatic interpretation. In general, it appears that data for south of about 23°N latitude have little to no value for climatic reconstruction purposes. Some data have been collected as far south as Oaxaca, Mexico, and even into El Salvador; however the records are typically short and do not possess properties deemed desirable for mot dendroclimatic work. Therefore, they have not been included in table 5 or shown on figure 8.

Some of the higher elevations in southern Mexico and central America could possibly yield dendrochronological samples suitable for climatic reconstruction. Although some work has been done in this area (for example, Hastenrath[44]), consid-

Table 4. Recent Tree Ring Chronologies or Collections in the Boreal Forest Regions of North America[a]

Map Reference[b]	Site Name	State/ Province	Species	Time Span	Latitude (°N)	Longitude (°W)	Reference
1	Procrastination Creek	AKA	Picea glauca	1633–1962	67°40′	142°30′	40
2	Twelve Mile Summit	AKA	Picea glauca	1650–1962	65°20′	146°00′	40
3	Salcha River Headwaters	AKA	Picea glauca	1650–1962	64°55′	144°00′	40
4	Dawson Junction	AKA	Picea glauca	1443–1962	64°10′	141°30′	40
5	Mount Fairplay	AKA	Picea glauca	1680–1962	63°50′	142°00′	40
6	Hermann's Cabin	AKA	Picea glauca	1750–1962	65°20′	147°30′	40
7	Chandalar Lake	AKA	Picea glauca	1785–1962	67°30′	148°30′	40
8	Terasmae	YT	Picea glauca	1624–1964	64°51′	138°19′	42
9	Peyto Lake	ALB	Picea engelmannii	1680–1965	51°30′	116°30′	143
10	Chapman Lake	YT	Picea glauca	1710–1966	64°51′	138°19′	144
11	Swede Creek	YT	Picea glauca	1800–1966	64°08′	139°43′	40
12	Sixty Mile	YT	Picea glauca	1790–1966	64°08′	140°35′	40
13	Gold Creek	YT	Picea glauca	1750–1966	64°06′	140°49′	40
14	Bullion Creek	YT	Picea glauca	1690–1966	61°01′	138°37′	40
15	Thelon Game Sanctuary	NWT	Picea glauca	1574–1969	63°50′	104°12′	144
16	Lake Beniah	NWT	Picea glauca	1747–1970	63°29′	112°17′	40
17	Athabasca River	ALB	Picea glauca	1708–1970	58°22′	111°32′	39
18	Quatre Fourches	ALB	Picea glauca	1765–1970	58°47′	111°27′	39
19	Revillon Coupe	ALB	Picea glauca	1783–1970	58°52′	111°18′	39
20	Peace River II	ALB	Picea glauca	1698–1970	58°59′	111°26′	39
21	Claire River	ALB	Picea glauca	1760–1970	58°53′	111°53′	39
22	Nain Forest (A+B)	LAB	Picea glauca	1769–1973	56°33′	62°00′	146
23	Herring-Alpine	AKA	Tsuga heterophylla	1422–1972	60°26′	147°45′	146
24	Ft. Chimo (comb.)	QUE	Larix laricina	1650–1974	58°22′	68°23′	146
25	Dubwant River	NWT	Picea mariana	1740–1974	62°37′	101°17′	147
26	Wolverine Plateau	YT	Picea glauca	1690–1975	61°30′	140°43′	42

27	Twisted Tree-Heartrot Hill	YT	Picea glauca	1459–1975	65°00′	138°20′	L-DGO[c]
28	River Crag	YT	Picea glauca	1635–1975	65°40′	138°00′	L-DGO[c]
29	Cat Track	YT	Picea glauca	1696–1975	65°57′	137°15′	L-DGO[c]
30	Arrigetch	AKA	Picea glauca	1586–1975	67°27′	154°03′	L-DGO[c]
31	Gulf Hazard	QUE	Picea glauca	1681–1976	56°10′	76°34′	L-DGO[c]
32	412-Noatak	AKA	Picea glauca	1515–1977	67°56′	162°18′	L-DGO[c]
33	Coppermine Mts.	NWT	Picea glauca	1428–1977	67°14′	115°55′	L-DGO[c]
34	September Mts.	NWT	Picea glauca	1340–1977	67°11′	116°08′	L-DGO[c]
35	Lake Sylva (comb.)	AKA	Picea glauca	1634–1977	67°00′	148°00′	42
36	Cri Lake	QUE	Picea glauca	1700–1977	55°20′	77°40′	150
37	Walker Lake (comb.)	AKA	Picea glauca	1627–1977	67°00′	154°00′	149
38	Border Beacon	LAB	Picea glauca	1660–1976	55°20′	63°15′	146
39	Spruce Creek	YT	Picea glauca	1570–1977	68°31′	138°40′	42
40	Churchill	MAN	Picea glauca	1650–1978	58°43′	94°04′	L-DGO
41	Sky Pilot Creek	MAN	Picea glauca	1725–1978	56°24′	94°22′	L-DGO
42	Dwarf Trees	AKA	Tsuga mertensiana T. heterophylla	1266–1979	60°01′	141°49′	L-DGO
43	Sukakpak Mt.	AKA	Picea glauca	1250–1979	67°36′	149°48′	L-DGO
44	Sheenjek River	AKA	Picea glauca	1506–1979	68°37′	143°40′	L-DGO

[a]Compiled by Linda D. Ulan and Gordon C. Jacoby, 1980.

[b]Map reference is for figure 7.

[c]Lamont-Doherty Geological Observatory

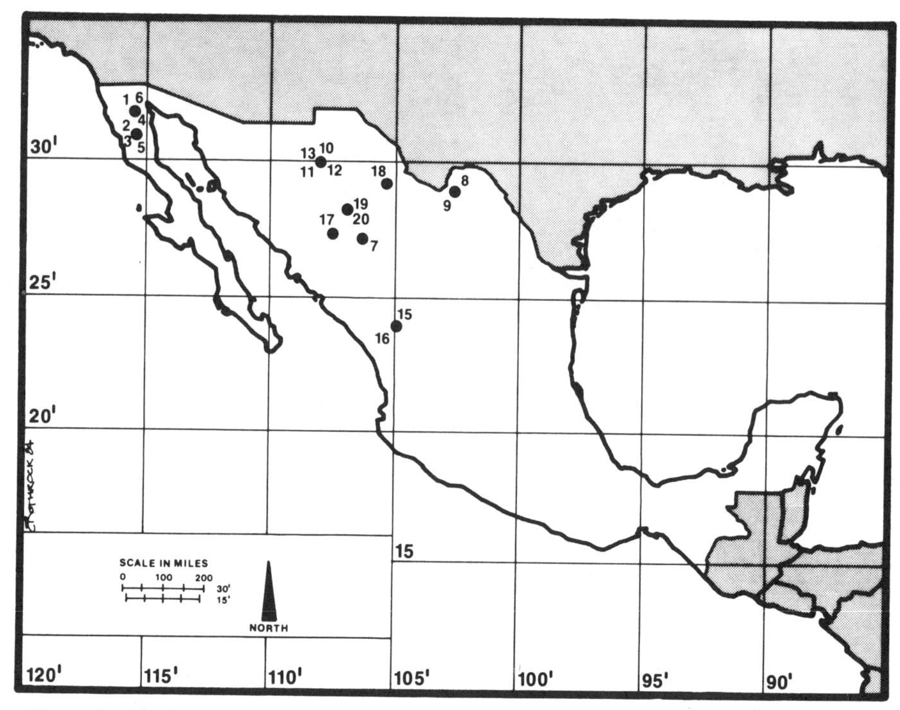

Figure 8. Distribution of tree-ring series suitable for climatic interpretation in Mexico. See table 5.

erably more in-depth study will be necessary before these data can be considered acceptable for climatic inference.

South America

The climatic regions of the South American continent range from areas of large tropical rainy conditions between 10°N latitude to 20°S latitude to humid subtropical to Mediterranean-type climate south of 20°S latitude. A rather narrow strip of dry climate occurs along the lee side of the Andes from about 20°S to 50°S and a long narrow strip of desert occurs along the western coast from the equator to about 32°S latitude. The potential for high-quality dendrochronological materials is greatest in the drier climatic regions. For this reason, most of the dendrochronological research in South America has been restricted to the drier areas of Argentina and Chile.

The South American continent is of special interest to dendroclimatologists because it extends about 900 km further south (55°S latitude) than any other forested continent. However, because of the tropical and subtropical conditions in Equador, Peru, Brazil, and southward along the eastern flank of the Andes in northern Argentina—north of about 23°S latitude—there appears to be little potential for den-

Table 5. Mexico and Central America Tree Ring Sites[a]

Map No.	Site Name	Species	Location: Latitude (°N)	Location: Longitude (°W)	State	Record Length
1	Baja N. Pond	Pinus monophylla	32°15′	115°50′	Baja	1569–1961
2	Baja C. Tasajera	Libocedrus decurrens	31°00′	115°25′	Baja	1473–1973
3	Baja C. San Pedro Martir, Low	Pinus jeffreyi	31°00′	115°24′	Baja	1463–1971
4	Vallecitos, San Pedro Martir	Pinus jeffreyi	31°05′	115°25′	Baja	1564–1971
5	Baja C. Tasajera	Abies concolor	31°00′	115°25′	Baja	1664–1971
6	Baja Topo	Pinus jeffreyi Pinus monophylla	32°00′	115°50′	Baja	1617–1971
7	Sierra Madre Rio Verde	Pseudotsuga menziesii	26°18′	106°30′	Chihuahua	1634–1973
8	Sierra Del Carmen	Pseudotsuga menziesii	28°56′	102°37′	Coahuila	1675–1971
9	Sierra Del Carmen	Pinus cembroides	28°58′	102°37′	Coahuila	1829–1971
10	Headwaters; Cañon Grande	Pinus cembroides	30°32′	108°35′	Chihuahua	1678–1969
11	Sierra Madre; Tres Rios	Pinus ponderosa	30°20′	108°30′	Chihuahua	1636–1965
12	Rancho Escondido (old road)	Pinus ponderosa	30°10′	108°15′	Chihuahua	1720–1965
13	Rancho Escondido (new road)	Pinus (several var.)	30°08′	108°15′	Chihuahua	1630–1965
14	El Vergel	Pinus cembroides	26°45′	106°06′	Durango	1763–1965
15	El Salto; West	Pseudotsuga menziesii	23°20′	105°36′	Durango	1592–1965
16	El Salto; Rita Tunnel	Pseudotsuga menziesii	23°45′	105°31′	Durango	1668–1965
17	Creel Airport	Pseudotsuga menziesii	27°42′	107°36′	Chihuahua	1643–1972
18	Sierra Del Nido	Pseudotsuga menziesii	29°31′	106°49′	Chihuahua	1569–1971
19	Teporachic MWP	Pinus ayacahuite	27°48′	106°55′	Chihuahua	1755–1973
20	Nonoavae	Pinus cembroides	27°24′	106°50′	Chihuahua	1750–1973

[a]Material from files at the Laboratory of Tree-Ring Research, University of Arizona.

droclimatological research. Problems include but are not limited to poor ring definitions, circuit irregularity, short maximum ages, and even multiple annual rings.

Schulman[2] pioneered the development of climatic sensitive chronologies from the southern Andes region using the coniferous species (*Austrocedrus chilensis, Araucaria araucana*) and *Fitzroya cupressoides*. Later work by LaMarche et al.,[45] concentrated on conifers of the coastal ranges and the Andean foothills between about 32°S and 40°S in Chile and as far as 54° south in Argentina. Exploratory samples were also taken of several nonconiferous species. However, no attempt was made to sample along the southern Chilean coast.

The concentrated research effort by LaMarche and associates from 1973 to 1979 resulted in extensive collections within Argentina and Chile. These data are available in Volumes 1 and 2 of the *Southern Hemisphere Chronology Series V* published by the Laboratory of Tree-Ring Research, University of Arizona.[45] In Argentina there are 21 chronologies available suitable for climatic inferences. Some of these dated series extend back to the mid-1100s but sample density in most of the series is probably not of sufficient magnitude beyond the 1600s. Site locations are shown in figure 9. Additional information for each series is shown in table 6. The spatial distribution of the dendroclimatic series that presently exist in Chile is shown in figure 10 and listed in table 7.

Africa

Continental Africa represents a large unknown to dendrochronology although there are places within the continent where the established dendrochronological principles have been successfully applied (for example, the work of Munaut et al.[46]) on *Cedrus Atlantica* in Morocco; Aloui[47] on *Pinus pinaster* in Tunisia; and LaMarche et al.[48] on *Widdringtonia cedarbergensis* in South Africa). Little success has been achieved in dendrochronological work for the vast area between these examples in extreme northern and southern Africa. Dyer[49] associates the difficulties to the climate of the continent, suggesting that in temperate climates, the climatic signal inherent in tree rings is likely to be weak and current technology does not allow analysis under such conditions. Furthermore, he suggests that areas where summer rainfall predominates, creates difficulty for dendrochronological applications because the seasonal timing of the rainfall weakens the clarity of the rings themselves. Except for extreme northern and southern Africa, summer is the predominant rainfall season. It appears the wide variability of success in dendrochronology in Africa may be attributed to the variation in climatic controls and the resulting spatially changing climate.

Table 8 shows the extent of known dendroclimatic series available in Africa. Hughes et al. (ref. 50, p. 54) indicate additional collections of *Pinus pinaster* and *Quercus foginea* in Tunisia, but these series are all less than 100 years in length.

Other collections have been made in Africa but they are not of the quality necessary for climatic interpretative work. Examples include additional samples of *Widdringtonia*, *Podocarpus* collected by LaMarche et al.[48] in South Africa. Stewart[51]

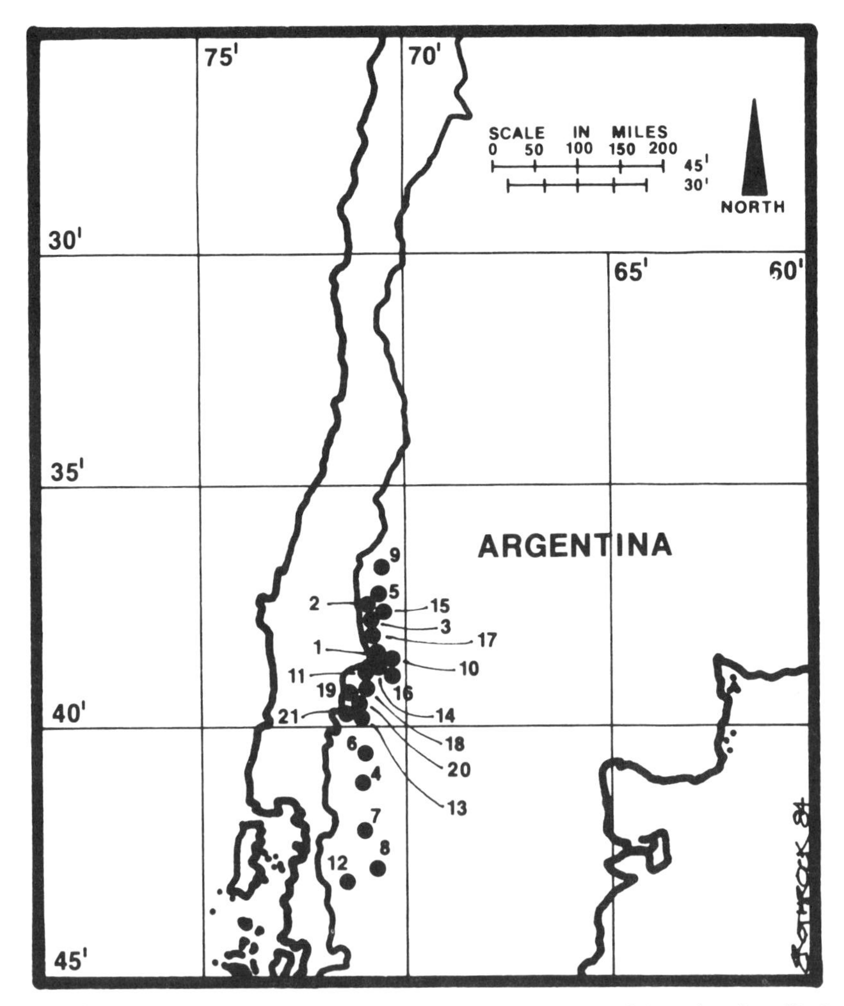

Figure 9. Locations of climatically sensitive tree-ring series in central Argentina. See table 6.

reported that samples of *Cupressus dupreziana* at least 2000 years of age have been collected from the Tassili Plateau in Algeria. Additional stands of *Cupressus sempervirens* are known to exist in Tunisia and Libya but have not been collected for dendroclimatic interpretation.

Australia and New Zealand

Tree-ring research has had a relatively short history within and around Australia. As pointed out by Ogden[52] the native populations used very little wood in their structures, thus attracting little attention from archaeologists as was the case in the

Table 6. Existing Dendroclimatic Series in Argentina[a]

Name	Map Code	Species	Location	Record Length
Angostura Lago	1	Araucaria araucana	38°53′S 71°10′W	1717–1974
Caviahue	2	Araucaria araucana	37°52′S 71°01′W	1444–1974
Chenque Pehuén	3	Araucaria araucana	38°06′S 70°51′W	1246–1974
Cerro los Leones	4	Austrocedrus chilensis	41°05′S 71°09′W	1539–1974
Copahue	5	Araucaria araucana	37°48′S 71°04′W	1640–1974
Cuýin Manzano	6	Austrocedrus chilensis	40°43′S 71°08′W	1543–1974
El Maitén	7	Austrocedrus chilensis	41°59′S 71°15′W	1690–1974
Estancia Teresa	8	Austrocedrus chilensis	42°57′S 71°26′W	1540–1974
Huinganco	9	Austrocedrus chilensis	37°04′S 70°36′W	1418–1975
Rio Kilca	10	Araucaria araucana	38°55′S 70°46′W	1700–1974
Lonco Luam	11	Araucaria araucana	38°59′S 71°03′W	1306–1974
Laguna-Terraplen	12	Austrocedrus chilensis	43°01′S 71°34′W	1700–1974
Estancia Manuil-Malal	13	Araucaria araucana	39°41′S 71°13′W	1690–1974
Lago Moquehue	14	Araucaria araucana	38°52′S 71°15′W	1601–1974
Puente del Agrio	15	Araucaria araucana	37°49′S 70°57′W	1486–1974
Primeros Pinos de Aluminé	16	Araucaria araucana	38°53′S 70°37′W	1140–1974
Pino Hachado	17	Araucaria araucana	38°38′S 70°45′W	1459–1974
Rahue	18	Araucaria araucana	39°24′S 70°48′W	1483–1974
Lago Racachorói	19	Araucaria araucana	39°13′S 71°10′W	1392–1976
Lago Racacharói	20	Austrocedrus chilensis	39°13′S 71°10′W	1572–1976
Lago Tromen	21	Araucaria araucana	39°36′S 71°22′W	1617–1976

[a]See reference 45.

Figure 10. Location of climatically sensitive tree-ring series in central Chile. See table 7.

southwestern United States. In addition, the two main genera (*Eucalyptus* and *Acacia*) are relatively short-lived and normally do not produce clearly defined annual rings. As a result, little attention was paid to tree rings until their use as proxy data sources in climatic reconstructions was widely recognized during the 1970s.

From the dendroclimatic viewpoint, the most productive and available series suitable for climatic interpretive work comes from the efforts of LaMarche et al.[53] Using background work concentrating on identification and location of suitable genera, LaMarche and his co-workers have collected and processed a total of 17 site chronologies from Australia which are suitable for climatic interpretation. Of the 17, 15 are from the island of Tasmania and two from western Australia. The western Australian series are both less than 100 years in length and for this reason

Table 7. Existing Dendroclimatic Series in Chile[a]

Name	Map Code	Species	Location	Record Length
Abanico	1	Austrocedrus chilensis	37°21′S 71°36′W	1733–1975
Piedra del Aguila	2	Araucaria araucana	37°50′S 73°02′W	1242–1975
Caramávida	3	Araucaria araucana	37°41′S 73°10′W	1440–1975
El Chacay	4	Austrocedrus chilensis	37°21′S 71°30′W	1641–1975
El Asiento	5	Austrocedrus chilensis	32°40′S 70°49′W	1011–1972
San Gabriel	6	Austrocedrus chilensis	33°46′S 70°13′W	1131–1975
Hueicolla	7	Pilgerodendron uviferum	40°08′S 73°31′W	1869–1975
Santa Isabelde las Cruces	8	Austrocedrus chilensis	34°52′S 70°45′W	1568–1975
Volcan Lonquimay	9	Araucaria araucana	38°23′S 71°34′W	1664–1975
Alto de las Mesas	10	Austrocedrus chilensis	34°55′S 70°42′W	1796–1975
Nalcas	11	Araucaria araucana	38°20′S 71°29′W	1386–1975

[a]See reference 45.

Table 8. African Tree Ring Sites

Site Name	Country	Species	Record Length	Reference
Die Bos[a]	South Africa	Widdringtonia cedarbergensis	1564–1976	48
Tleta de Ketama[b]	Morocco	Cedrus Atlantica	1790–1975	46
Djebel Tidighin	Morocco	Cedrus Atlantica	1754–1975	46
304[b,c]			1700–1975	145
307			1537–1975	145
310			1604–1977	145
321			1840–1977	145

[a]Latitude 32°24′S, longitude 19°13′E.

[b]Latitude 35°30′S, longitude 4°30′W.

[c]Sites 304, 307, 310, and 321 from Morocco are from reference 145.

are not included in table 9. Although additional collections have been made in Australia (see LaMarche et al.[53]) it appears that only those from Tasmania can be considered suitable for dendroclimatology.

Dendrochronology has received considerably more detailed attention in New Zealand than in Australia and includes applications to archaeology, geology, geomorphology, and forest mensuration. However, attempts to develop long, climatically sensitive series have not been generally successful. Prior to the works of LaMarche and his associates in 1977 and 1978, only one other attempt had been successful in developing a meaningful dendroclimatic series (Carter).[54] LaMarche et al.[53] list 13 climatically sensitive series from North Island and eight from South Island. However, no actual climatic reconstructions have yet been attempted from these data. A list of the dated series is shown in table 10 and on figure 11.

United Kingdom (John P. Cropper, University of Arizona)

Consisting of four separate countries (England, Scotland, Wales, and northern Ireland) the United Kingdom is located between latitudes 49° 57′N to 60° 40′N and longitudes 1° 46′E to 8° 10′W. The island location prevents temperature extremes; average daily means vary from 7.8°C in the north to 11.1°C in the south. Precipitation is more variable, ranging from 400 cm in the western Scottish highlands to less than 50 cm in the southeast, outside London. United Kingdom flora has developed since the last Ice Age with deciduous trees the dominant woody species.

After much hesitation as to the possibility of developing tree-ring chronologies in the United Kingdom, work began in earnest in the early 1970s. In 1971, Fletcher, at Oxford, developed chronologies from oak.[55] These, however, were developed for use as a dating tool rather than for climatic interpretation. In a review of tree ring research in Europe, Eckstein[56] listed only two chronologies for the United King-

Table 9. Dendroclimatic Series from Australia[a]

Site Name	Province	Species	Location		Record Length
Mt. Arrowsmith	Tasmania	Phyllocladus aspleniifolius	42°12′S	146°06′E	1548–1974
Beyond Burn	Tasmania	Athrotaxis cupressoides	42°39′S	146°34′E	1028–1975
Bruny Island (a)	Tasmania	Phyllocladus aspleniifolius	43°22′S	147°16′E	1542–1975
Clear Hill	Tasmania	Phyllocladus aspleniifolius	42°40′S	146°16′E	1554–1974
Cradle Mountain	Tasmania	Athrotaxis selaginoides	41°38′S	145°56′E	1198–1975
Coolangatta Road	Tasmania	Phyllocladus aspleniifolius	43°22′S	147°16′E	1711–1975
Dove River Road	Tasmania	Phyllocladus aspleniifolius	41°35′S	146°00′E	1776–1975
Franklin River Valley	Tasmania	Phyllocladus aspleniifolius	42°12′S	145°59′E	1673–1975
Holley Range Road	Tasmania	Phyllocladus aspleniifolius	42°46′S	146°04′E	1675–1974
Lockley Spur Road	Tasmania	Phyllocladus aspleniifolius	43°25′S	143°16′E	1579–1975
Lyell Highway (b)	Tasmania	Phyllocladus aspleniifolius	42°12′S	146°02′E	1548–1975
Mount Field (c)	Tasmania	Phyllocladus aspleniifolius	42°41′S	146°33′E	1028–1975
Lake Newdegate	Tasmania	Athrotaxis cupressoides	42°40′S	146°33′E	1286–1974
Pieman River	Tasmania	Phyllocladus aspleniifolius	41°47′S	145°25′E	1310–1975
Pine Lake	Tasmania	Athrotaxis cupressoides	41°45′S	146°42′E	1514–1974
Statlelens	Tasmania	Phyllocladus aspleniifolius	41°11′S	148°00′E	1507–1974
Sheepwash Creek	Tasmania	Phyllocladus aspleniifolius	43°28′S	154°17′E	1542–1975
Weindorfer Forest	Tasmania	Nothofagus gunnii	41°38′S	145°56′E	1732–1975

[a]See reference 53.

[b]Mean value function of Coolongatta Road, Lockley Spur Road, and Sheepwash Creek.

[c]Mean value function of Lyell Highway and Mt. Arrowsmith.

[d]Mean value function of Beyond Burn and Lake Newdegate.

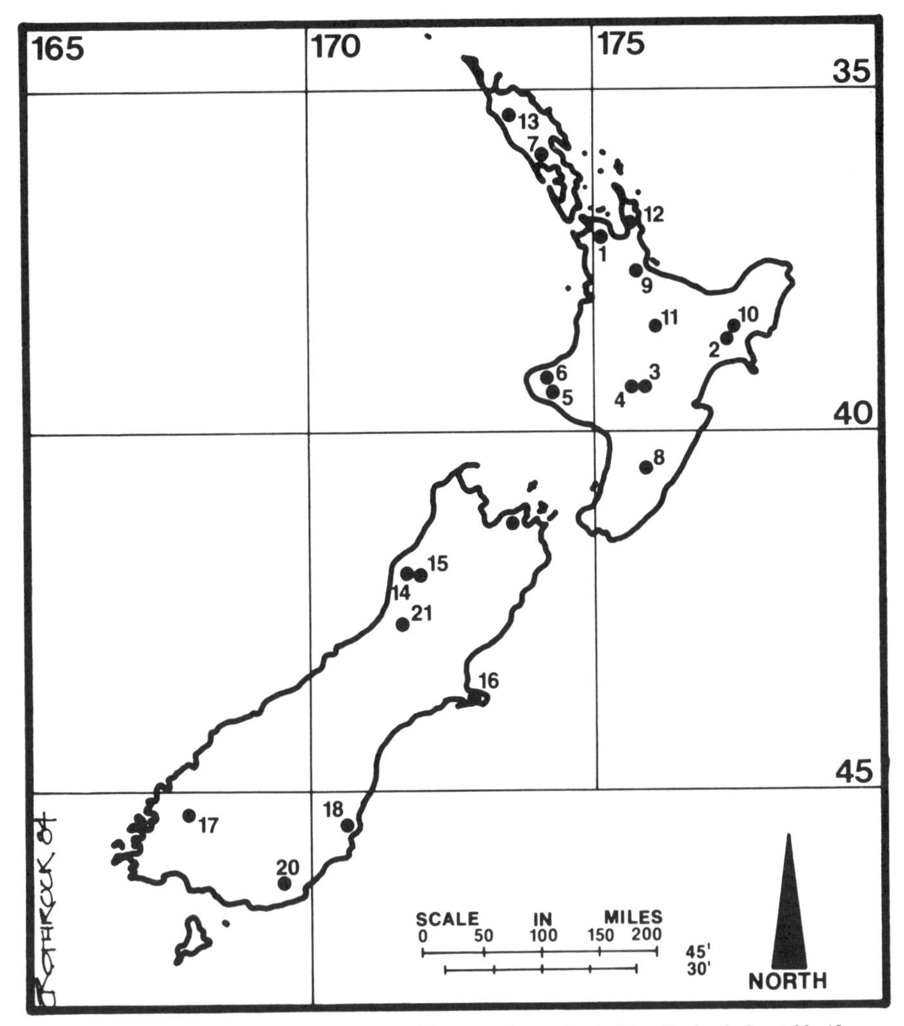

Figure 11. Location of climatically sensitive tree-ring series in New Zealand. See table 10.

dom. One of these was placed in time only by radiocarbon analysis with no calendar years assigned and the other was from northern Ireland.[57]

The production of tree ring chronologies in the United Kingdom has presented many problems primarily due to the relatively short (300 year) life span of native oaks and the fact that few ever reach that age before being felled. Since the earlier works, provisional studies and chronology development have progressed with a high dependence on dating material using correlation techniques.[58,59]

The general location of existing chronologies is shown on figure 12. The institutions and investigators primarily responsible for the development of the chronologies are listed, along with other pertinent information in table 11.

Table 10. Dendroclimatically Sensitive Series of New Zealand[a]

Map No.	Site Name	Province	Species	Location		Record Length
1	Konini Forks	North Island	Agathis australis	37°04′S	175°08′E	1712–1976
2	Lake Wai Kareiti	North Island	Phyllocladus glaucus	38°42′S	177°12′E	1535–1976
3	Mangawhero River Br. A	North Island	Dacrydium colensoi	39°21′S	175°29′E	1464–1976
4	Mangawhero River Br. B	North Island	Libocedrus bidwillii	39°21′S	175°29′E	1662–1976
5	Mt. Egmont	North Island	Libocedrus bidwillii	39°15′S	174°05′E	1616–1976
6	North Egmont	North Island	Libocedrus bidwillii	39°17′S	174°06′E	1625–1976
7	Paparoa	North Island	Phyllocladus trichomanoides	36°07′S	174°15′E	1779–1975
8	Takapari	North Island	Libocedrus bidwillii	40°05′S	176°00′E	1256–1976
9	Te Aroha	North Island	Phyllocladus glaucus	37°30′S	175°50′E	1779–1975
10	Urewera	North Island	Libocedrus bidwillii	38°41′S	177°12′E	1346–1976
11	Waimanoa	North Island	Phyllocladus glaucus	38°34′S	175°42′E	1745–1976
12	Waiomu	North Island	Phyllocladus trichomanoides	37°02′S	175°32′E	1664–1976
13	Waipoua	North Island	Phyllocladus glaucus	35°41′S	173°33′E	1585–1976
14	Ahaura A	South Island	Dacrydium colensoi	42°23′S	171°48′E	1403–1976
15	Ahaura B	South Island	Libocedrus bidwillii	42°23′S	171°48′E	1525–1976
16	Armstrong Reserve	South Island	Libocedrus bidwillii	43°50′S	173°00′E	1450–1978
17	Manapouri	South Island	Dacrydium biforme	45°32′S	167°18′E	1567–1976
18	Mt. Cargill	South Island	Libocedrus bidwillii	45°50′S	170°32′E	1492–1975
19	Okiwi	South Island	Phyllocladus trichomanoides	41°07′S	173°40′E	1724–1976
20	Owaka	South Island	Libocedrus bidwillii	46°23′S	169°27′E	1732–1976
21	Pegleg Creek	South Island	Phyllocladus alpinus	42°54′S	171°34′E	1717–1976

[a]See reference 53.

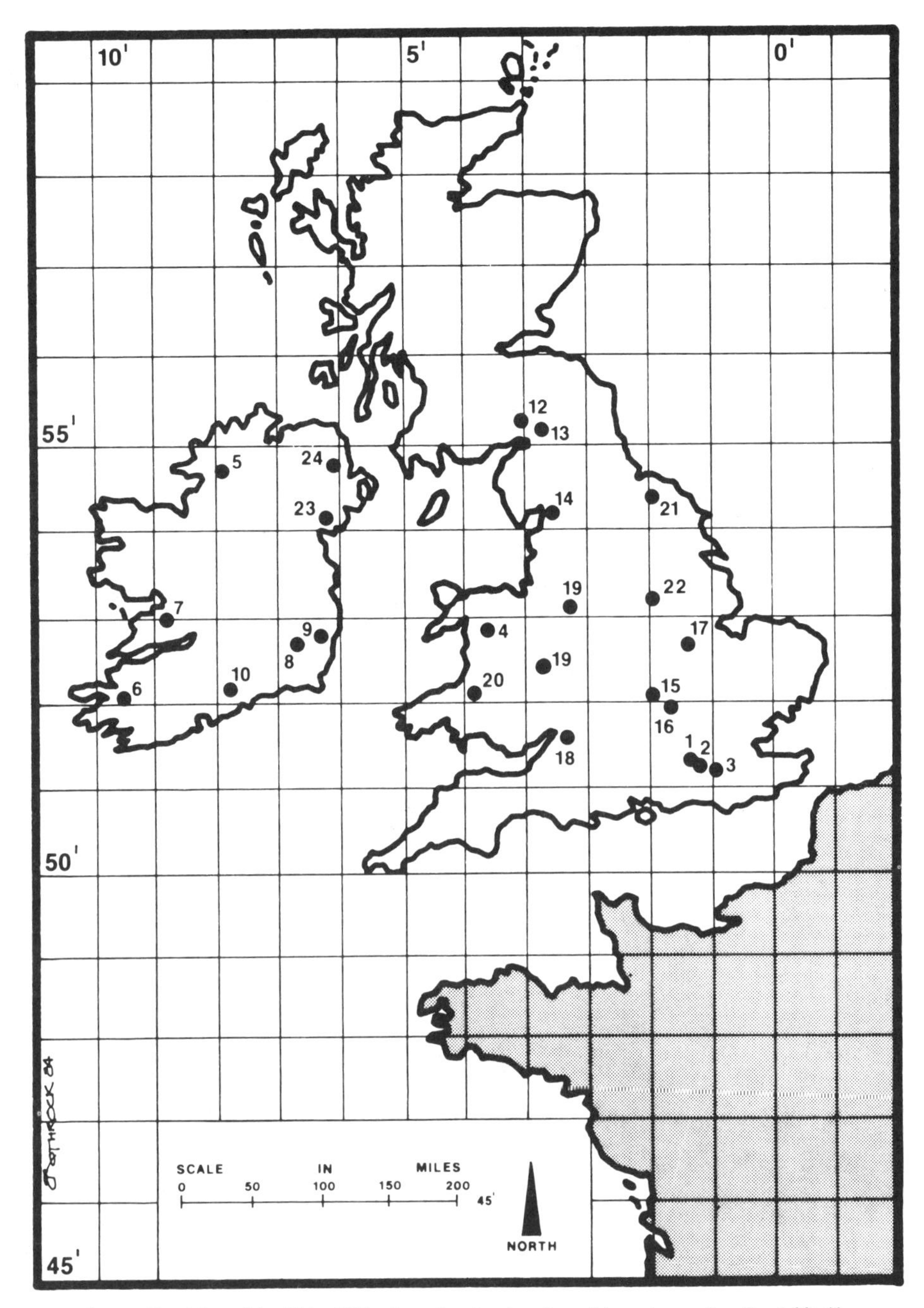

Figure 12. Map of the United Kingdom showing location of tree-ring series. See table 11.

Table 11. Information on Tree Ring Chronologies in the United Kingdom

Name	Institute	Site	Map No.	Species	Year		Trees
					First	Last	
D. Brett	Bedford College University of London	Regent's Park	1	Ulmus sp.	1840	1970	5
		Brompton Cemetery	2	Ulmus sp.	1900	1970	3
		London Group	3	Ulmus sp.	1900	1970	10
M. K. Hughes, et al.	Liverpool Polytechnic	Maentwrog	4	Quercus petraea (Mattuschka) Liebl.	1710	1974	35
Pilcher and Baillie	Queens University Belfast	Ardara	5	Quercus petraea (Mattuschka) Liebl.	1803	1978	
		Killarney	6	Quercus petraea (Mattuschka) Liebl.	1809	1978	
		Lough Doon	7	Quercus petraea (Mattuschka) Liebl.	1850	1978	
		Eniscorthy	8	Quercus petraea (Mattuschka) Liebl.	1811	1978	
		Glen of Downs	9	Quercus petraea (Mattuschka) Liebl.	1809	1978	
		Cappoquin	10	Quercus petraea (Mattuschka) Liebl.	1813	1978	

		Glenluce	11	Quercus sp.	1798	1978	
		Raehills	12	Quercus sp.	1824	1975	
		Lockwood	13	Quercus sp.	1571	1975	
		Scorton	14	Quercus sp.	1813	1978	
		Oxford (1)	15	Quercus sp.			
		Oxford (2)	16	Quercus sp.			
		Blickling	17	Quercus sp.	1717	1979	
		Bath	18	Quercus sp.	1754	1979	
		Ludlow	19	Quercus sp.	1825	1978	
R. Morgan	Queens University Belfast	Towy Valley	20	Quercus sp.			
		Castle Howard	21	Quercus sp.			
		Padley Wood	22	Quercus sp.			
Pilcher	Queens University Belfast	Rostrevor	23	Quercus petraea	1750	1975	18
Baillie	Queens University Belfast	Belfast Modern	24	Quercus sp.			

Scandinavia (Bengt Jonnson, Royal College of Forestry, Umea, Sweden)

Scandinavia is composed of Denmark, Norway, and Sweden; Finland is also included in this paper. The climate is unexpectedly mild, considering Scandinavia's northern latitude. Except for the Atlantic coastline, precipitation generally occurs during the summer, averaging from 500–700 mm annually. Conifers dominate the forest vegetation although deciduous trees are present in the south and west.

Only three chronologies have been specifically collected for the International Tree-Ring Data Bank at the University of Arizona. Their location is shown on figure 13 and characteristics in table 12. The small number of available chronologies is related to the fact that, due to the importance of forest products to the Scandinavian economy, most studies with dendrochronological applications have been highly oriented toward forestry. Many of these studies have focused on the effects of temperature and precipitation on diameter increment and thus provide valuable information on the relationship between these climatic factors and annual ring widths.

Published studies, beginning with the work of Hesselman[60] have covered a wide range of conditions from Denmark to north of the Arctic Circle. In general most of the work has been with species of pine, spruce, and fir. Various relationships have been established between diameter increment and climatic factors both during the year of ring formation and the preceding year. Other important studies relating climate and growth include those of Wallen,[61] Laitakari,[62] Kolmodin,[63,64] Eide,[65] Erlandsson,[66] Ording,[67] Ruden,[68] Hustich,[69] Mikola,[70] Holmsgaard,[71] Eklund,[72,73] Jonsson,[74] Tveite,[75] Vestjordet,[76] and Strand.[77]

A large amount of material is available that may be potentially useful for dendroclimatic studies. For instance, during 1941–1965, approximately 48,000 cores from pine, 47,000 from spruce, and 12,900 from birch were collected on 2,075 yield sample plots; 983 plots were in virgin forests and 1,092 in thinned stands.[78] Annual rings were measured to .01 mm and computer recorded. This material is currently at the Swedish University of Agricultural Science. Similar material is available in Norway and Denmark but on a smaller scale.

In addition, some 45,000 cores are collected annually by the Swedish National Forest Survey from randomly chosen trees evenly distributed throughout the country's forests. Annual rings are measured to 0.1 mm and the data are computerized.

Europe, excluding U.S.S.R. (Dieter Eckstein, University of Hamburg, Hamburg, Germany)

The forest vegetation in Europe is rather manifold due to the prevailing climatic and other site factors. Thus, the conditions for dendrochronological work are rather complex and can only be described along general lines.

Dendrochronology in Europe can be traced back to the beginning of this century[79] but it was not before the early 1940s that Hüber[80] gave an essential impact to further development and systematic application. Until recently, the primary concern has been with the dating of timber. In this regard several longterm tree-ring chronologies

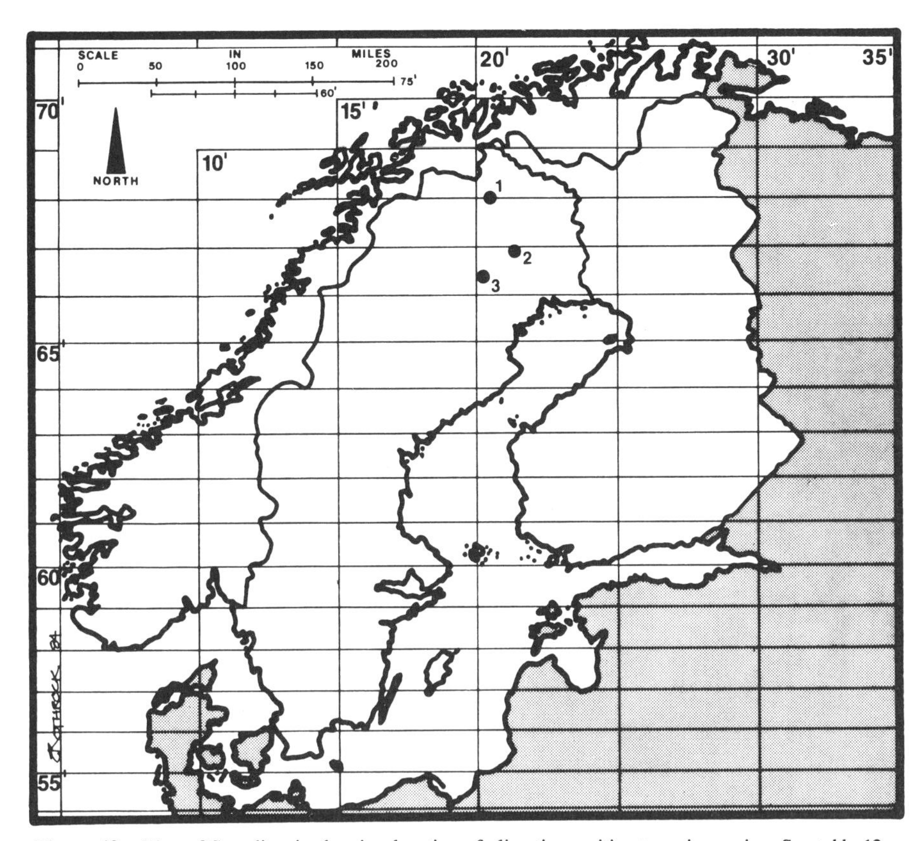

Figure 13. Map of Scandinavia showing location of climatic sensitive tree-ring series. See table 12.

Table 12. Information on Tree Ring Chronologies from Scandinavia[a]

Collector	Institute	Site	Species	Year First	Year Last	Trees	Cores
Bengt Jonsson	Swedish Univ. of Agricultural Sciences, Umea	Årosjokk	Pinus sylvestris	1638	1971	15	16
Bengt Jonsson	Swedish Univ. of Agricultural Sciences, Umea	Muddus	Pinus sylvestris	1532	1972	21	41
Bengt Jonsson	Swedish Univ. of Agricultural Sciences, Umea	Arjeplog	Pinus sylvestris	1552	1974	20	40

[a]Material available in the International Data Bank, University of Arizona, Tucson.

are being constructed, ranging from the present to prehistoric periods and based on samples of considerable geographical range. The longest tree-ring series is called the Central European oak chronology and goes back consistently to about 3968 B.C., covering most of the Holocene as far back as 8500 years.[81,82] Its validity for dating has been proven for the range between Czechoslovakia in the east, Normandy and even southeast England in the west, and Switzerland and northern Italy in the south. Towards and along the coastal region of the European mainland, chronologies exhibit a smaller range of validity, presumably because of maritime climatic influences on tree growth.

The chronologies mentioned above have been very useful for dating art-historical objects like wooden panels of Rembrandt paintings,[83] architectural monuments such as the Romanesque Cathedral in Trier,[82] and archaeological excavations like the Swiss lake dwellings from Neolithic to Bronze Age.[84] It has also been useful for dating geomorphological processes such as the alterations in the valleys of some European rivers.[85] Furthermore, the chronologies help to calibrate radiocarbon dating.[86] In contrast, dendroclimatology does not have a long tradition in Europe and must be considered as an emerging discipline. Nevertheless, dendroclimatological results will be of great interest for an area like Europe with such a long cultural history; not only with respect to global climatology, but also for the reconstruction of past local environments. The climatic database in Europe, necessary as a starting point for such analyses, is better than elsewhere in the world.

There are some valuable older studies and observations on the relationship between tree growth and climate.[79,87,88] These studies were completed without the help of electronic data processing and therefore could not make use of large amounts of complex data and multivariate statistical procedures. Many studies suggest that the overall climatic information obtainable from tree-ring analyses is low.[89–94] Oak is one of the most promising tree species for dendrochronological purposes in the European deciduous forest belt. However, the amount of climatic information stored in the annual rings of northwest German oaks, for example, varies around 30 percent; a great deal of ring-width variation is also caused by the growth conditions of the prior year. In addition, the main climatic influence factors change rapidly from one site to the other.[95] More climatic information is apparently extractable from coniferous species in the Alpine region. In this case, wood density has proven a better recorder of climate, specifically of summer temperature, than ring width.[90] If rainfall is to be considered more precisely, the vessel system of deciduous trees may contribute a suitable parameter, as indicated by preliminary studies.[96]

In general, there are a number of basic points to be considered for evaluating dendroclimatology in Europe.[97,98] They are:(1) The largest part of the area under consideration has relatively few tree species (although the Mediterranean Region is relatively rich in tree species, only a few data have been derived so far, especially concerning growth periodicity); (2) the trees are mostly short-lived; and (3) the forests have been influenced by man since prehistoric times. Moreover, many of the current chronologies used for dating are unsuitable for climatic studies. This is because only their youngest parts (back to about A.D. 1800) have been based on living trees growing under well-known site conditions. The chronologies ranging further back are based on nonliving trees inhabiting a large geographic area and, to a certain extent, obtained from sites which today are treeless or covered with tree species that are insignificant for dendrochronology. Thus, chances of evaluating the climate-growth relationship for these forests are low. If the problems associated with composite chronologies can be solved, a reasonably dense grid of tree-ring data already exists, or should become available in the near future, covering the period back to about A.D. 1000. It is, however, also necessary to start sampling in specific regions of south and southeast Europe, where no systematic collection of tree-ring data has been done so far.

Table 13 summarizes the existing tree ring chronologies, as far as information has been obtainable. Locations of the available tree-ring series are shown in figure 14. Since the data have been supplied in a rather different and unbalanced way, they will be presented on their smallest common basis only, as indicated in the headings of the table. As far as it is possible to judge, all chronologies listed are well replicated. Even if the data have not been stored in the International Tree-Ring Data Bank, it can be assumed that most of them will be available upon request.

Soviet Union

The area of the Soviet Union is enormous and covers a wide range of physiographic provinces. Consequently, the longer available dendrochronological records represent

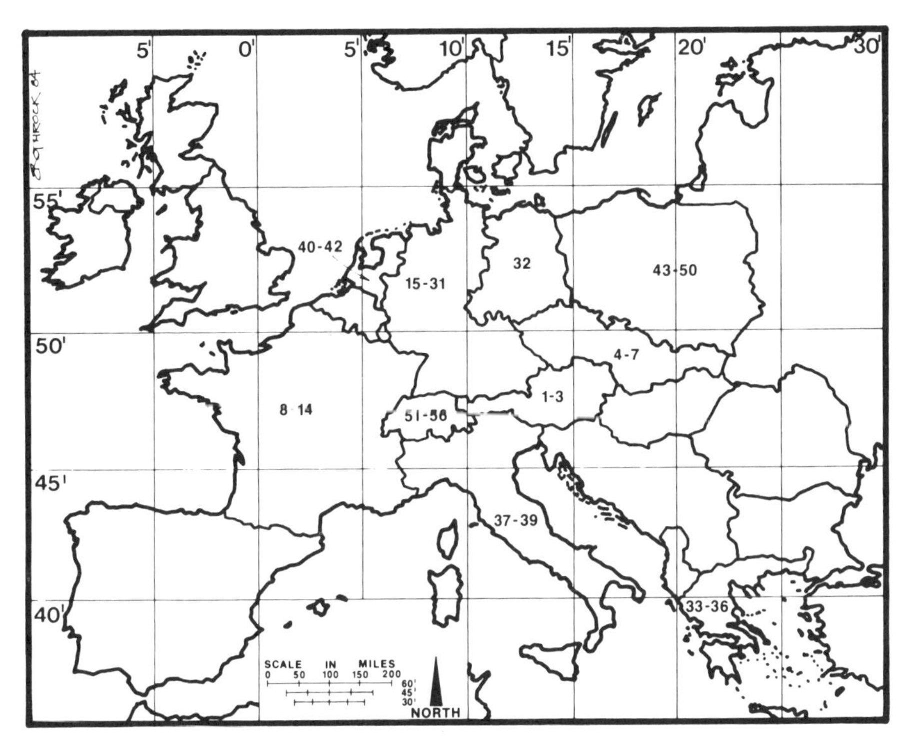

Figure 14. Map of a portion of Europe showing approximate location of existing tree-ring series believed to contain climatic information. See table 13.

Table 13. Existing Tree Ring Data in Europe.

Site Designation	Country	Species	Location[a]		Period	Source of Data[a]
			Latitude (°N)	Longitude (°E)		
1	Austria	Pinus cembra	47	13	1466–1971	12
2		Picea abies	47	13	1276–1974	12
3		Larix decidua	47	13	1333–1972	12
4	Czechoslovakia	Abies alba	49	17	1701–1943	11
5		Picea excelsa	49	17	1725–1943	11
6		Fagus silvatica	49	17	1660–1943	11
7		Picea excelsa	49	17	1488–1963	10
8	France	Larix decidua	45	05	933–1974	16
9		Pinus halepensis	45	06	1807–1973	16
10		Quercus pubescens	45	05	1850–1976	16
11		Picea excelsa	45	05	1741–1973	16
12		Abies alba	45	05	1653–1975	16
13		Quercus sp.	55	03	1280–1610	12
14		Quercus sp.	53	01	1798–1979	18
15	Germany	Quercus sp.	52	07	−717–1981	7
16		Quercus sp.	52	07	−3968–1981	5,7,11
17		Quercus sp.	46	12	−4050– −1720(b)	11
18		Quercus sp.	46	12	−8550– −6400(b)	11
19		Abies alba	46	12	820–1981	11, 12
20		Picea excelsa	46	12	1250–1981	11
21		Abies alba	46	12	1541–1961	11
22		Picea excelsa	46	12	1573–1961	11

23		Larix decidua	46	12	1340–1947	11
24		Pinus sylvestris	46	12	1178–1971	11
25		Fagus sylvatica	46	12	1684–1962	11
26		Quercus robur	46	12	1776–1972	5
27		Quercus sp.	53	08	1080–1967	1
28		Quercus sp.	54	10	436–1981	1
29		Quercus sp.	54	10	1087–1981	1
30		Picea excelsa	54	10	1592–1981	1
31		Quercus sp.	54	10	1004–1981	3
32	Germany (GDR)	Quercus sp.	53	13	1424–1981	2
33	Greece	Pinus nigra	38	22	1705–1979	19
34		Pinus leucodermis	38	22	1255–1979	19
35		Abies cephallonica	38	22	1676–1978	19
36		Quercus sp.	38	22	1116–1978	19
37	Italy	Abies alba	44	10	1539–1972	11
38		Picea excelsa	44	10	1530–1700	17
39		Abies alba	44	10	1334–1561	17
40	Netherlands	Quercus sp.	53	05	1036–1981	1, 20
41		Quercus sp.	53	05	1109–1637	1, 20
42		Quercus sp.	53	05	420– 817	1, 20
43	Poland	Picea excelsa	50	20	1766–1965	9
44		Pinus cembra	50	20	1732–1969	9
45		Fagus sylvatica	50	20	1820–1979	9
46		Larix decidua	50	20	1806–1977	9
47		Pinus carpatica	50	20	1890–1977	9
48		Pinus sylvestris	50	20	1820–1975	9
49		Acer pseudoplatanus	50	20	1841–1977	9
50		Pinus montana	50	20	1870–1976	9
51	Switzerland	Pinus cembra	47	07	1765–1975	14
52		Larix decidua	47	07	1792–1975	14

Table 13. (*Continued*)

Site Designation	Country	Species	Location[a] Latitude °N	Longitude °E	Period	Source of Data[c]
53		Picea abies	47	07	1688–1975	14
54		Pinus montana	47	07	1832–1975	14
55		Pinus cembra	47	07	1589–1972	4
56		Larix decidua	47	07	1529–1972	4

[a]Approximate only, precise location is unavailable.

[b]Radiocarbon dated.

[c](1) Ordinariat fürHolzbiologie, Universität Hamburg, FRG. (2) Zentralinstitut für Alte Geschichte, Berlin-Ost, GDR. (3) Institut für Forstbenutzung, Universität Göttingen, FRG. (4) Institut de Palynologie et Phytosociologie, Université de Louvain, Belgium. (5) Institut für Ur - und Fruhgeschichte, Universität Koln, FRG. (6) Centre d'Archéologie médiévale, Université de Caen, France. (7) Rheinisches Landesmuseum, Trier, FRG. (8) Centre Technique du Bois, Paris, France. (9) Instytut Hodowlilasu, Kraków, Polska. 910) Societas Botanica Cechoslovaca, Prague, CSSR. (11) Institut für Botanik, Universitat Nohenheim, FRG. (12) Dendrochronologisches Labor, Asenham, FRG. (13) Büro für Archäologie, Zürich, Schweiz. (14) Eidgenössische Forstliche Versuchxanstalt, Birmensdorf, Schweiz. (15) Musée Cantonal d'Archéologie, Neuchâtel, Schweiz. (16) Laboratoire de Botanique Historique et Palynologie Université d'Aix-Marseille, France. (17) Tree Ring Laboratory, E. Corona, Rome, Italia. (18) Palaeoecology Laoboratory, University of Belfast, N. Ireland. (19) Department of Classics, Cornell University, Ithaca, N.Y. USA. (20) State Service for Archaeological Investigations, Amersfoort, Netherlands.

a variety of species extending over a broad geographic range from the Karelia and Carpathian Region in the west to Lake Baikal Region in the east (see figure 15). Climate within the Soviet Union ranges from cold Arctic conditions to warm Mediterranean conditions in the south. Vegetation is reflective of the diverse climate.

Apparently several groups are working in dendrochronology within the Soviet Union. In Soviet Central Asia, Mukhamedshin and Sartbaev[99] and Mukhamedshin[100] show data from dendrochronological analysis of 1580 specimens of junipers. They report a 1214-year chronology consisting of 18 trees ranging in age from 452 to 1214 years and growing at an altitude of 2900 to 3500 meters on the north slope of the Altai Mountains in western Mongolia. It is unclear as to the variety of species incorporated into this juniper chronology. It is reported that in the lower, more arid sites, climatic response of *Juniperus seravachania* and *Juniperus semiglobosa* is mainly to moisture but that high growing season temperature may produce a negative growth response. On colder and more moist sites, *Juniperus turkestania* responds to June and July temperature. Additional dendroclimatic work is reported to have been accomplished by Molchanov[101] but the availability of these data are unknown. Apparently, considerable work has been done in the Lake Baikal region of Siberia. Galazii[102] reports on a lake level fluctuation study involving tree rings. Data obtained by the Laboratory of Tree-Ring Research from the Dendrochronological Laboratory of the Institute of Botany of Lithuania[104,105] contain fairly long series of Dayurski Larch and Siberian Cedar from the same area. It is uncertain if these data are the same referred to by Galazii.[102] Tarankov[103] made a survey of dendroclimatology in the eastern Soviet Union and lists 138 references to past investigations. Together with reviewing techniques, he presents data for sites in the Soviet far east including series up to 400 years in length. Species include *Taxus cuspidata*, *Pinus konaiensis*, *Picea jezoensis*, *Abies holophylla*, *Abies nephnolepips* and *Larix genelinii*. The availability of these data are unknown.

Much data relating to the dendrochronology in the USSR has been accumulated in recent publications of the Laboratory of Dendroclimatochronology, Institute of Botany, Academy of Sciences of the Lithuanian S.S.R.[104,105] These sources list dendrochronological data collected by and described by contributing investigators.

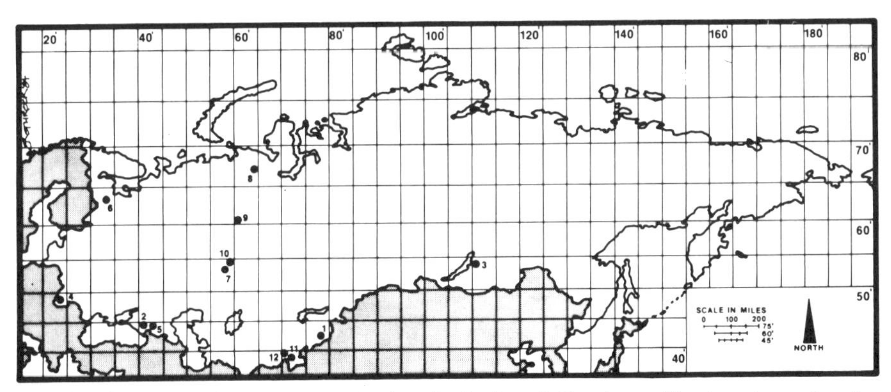

Figure 15. Map of the Soviet Union showing distribution of climatic sensitive tree-ring series. See table 14.

Table 14 Dendroclimatic Chronologies from USSR[a]

Map Reference	Location	Species	Time Span	Latitude (°N)	Longitude (°E)	Number of Trees	Investigator
1	Northeast Tien Shan Mtns.	Picea shrenkiana	1590–1978	43	78	13	N.M. Borsheva
2	Western Caucasus Mtns.	Picea orientalis	1660–1977	43	41	64	V.I. Brukshtus
3	Lake Baikal	Dayurski Larch	1431–1944			11	
		Siberian Cedar	1415–1944	54	109	9	G.I. Galazii
		Siberian Cedar	1858–1959			20	
4	Carpathian Mtns.	Pinus cembra	1594–1978	48	41	10	V.G. Kolishuk
5	Northern Caucasus Mtns.	Picea orientalis	1370–1972	43	31	60	T.T. Bitvinskas
		Pinus sylvestris	1507–1977				
6	Karelian SSR	Pinus sylvestris	1554–1967				T.T. Bitvinskas
		Pinus sylvestris	1447–1958	62	33		
		Pinus sylvestris	1578–1979				
7	Southern Ural Mtns.	Larix sp.	1570–1976	54	57		T.T. Bitvinskas
	Ural Mtns.	Larix sp.	1575–1977	53	57		T.T. Bitvinskas
8	Sub-arctic Ural Mtns.	Larix sibirica	1691–1969			20	
		Larix sibirica	1541–1968	67	65	20	S.G. Shiyatov
9	Northern Ural Mtns.	Larix sibirica	1590–1969	60	59	25	S.G. Shiyatov
		Pinus sylvestris	1557–1969	60	59	11	S.G. Shiyatov
10	Southern Ural Mtns.	Pinus sylvestris	1469–1973	54	59	22	S.G. Shiyatov
11	Southern Tien Shan Mtns.	Juniperus turkestanica	750–1973	40	73	18	K.D. Mukhamedshin
12	Western Tien Shan Mtns.	Juniperus turkestanica	1163–1970	39	70	—	E.V. Maximova

[a]See references 104 and 105.

Some of the longer tree-ring series from these publications have been selected for closer scrutiny. They are listed in table 14 and their locations shown on figure 15. Close examination and cross comparison of these records shows a wide variety in quality and apparent suitability for paleoclimatic purposes. This is evidenced by the rather abrupt changes in the time series properties of some of the series. Rather poor agreement of chronologies from the same species on neighboring sites seems to pose potential problems with calendrical accuracy and suggests rather limited usefulness for climatic studies.

China (Jiacheng Zhang, State Meteorological Administration, Beijing, China)

Situated in the southeastern part of the Eurasian continent, China has a total land area of 9.6 million square kilometers. It stretches from 4° to 53°N latitude and from 73° to 135°E longitude. Topography is quite varied ranging from low-lying areas in the east to high plateaus and mountains in the west. Although China's climate is dominated by monsoonal winds, great differences are found due to its extensive land mass and complex topography. Vegetation is equally varied, ranging from rain forests to temperate coniferous and deciduous forests through extensive areas of grasslands and deserts.

Dendroclimatological work began in China in the 1930s for the purpose of clarifying the relations between tree growth and climate and reconstructing past climate.

Meteorologist Zheng Zizheng in 1935 collected dozens of samples from trees aged over 350 years at West Hills, Jingshan, Ziannontan, Tanzesi and Beijing, and attempted to compare the tree growth with rainfall.[106] Biologist Deng Shuqun took three samples of *Picea asperata* from the Heihe valley in Oilianshan and a sample of *Abies chensiensis* from Beilongjiang valley near Xigu (both in Gansu Province).[107] According to his analysis, not only do the minima in the growth cycles of trees seem to bear a relation to the maxima of the sunspot cycle and, in several cases, to the occurrence of known droughts in the history of Gansu, but the maxima in tree rings also match quite closely with the sunspot minima. Attempting to assess the streamflow variation in the upper reaches of the Huanghe, Han Shoutang et al.[108] collected tree samples from Beijing, Xian, Lanzhou, Yuzhong, and Minxian. One result of this work was that rainfall variability seemed to be closely correlated with ring-width variability in these arid regions.

Since the 1970s, the Institute of Geography (Academic Sinica), Academy of Meteorological Sciences (Central Meteorological Bureau), Lanzhou Institute of Glaciology, and others, in cooperation with local meteorological organizations, have engaged in a great deal of tree-ring research, but few results have been published.

Lin Zhengyao and Wu Xiangding[109] have established a tree-ring series longer than 600 years, reflecting the climatic change on the Tibet Plateau in historical times. Ding Shicheng [110] studied several ring-width index series at Maijiang, Jilin Province, in relation to climate, and found that these series correlated well with temperature. This work is of great significance, as the study of low temperatures

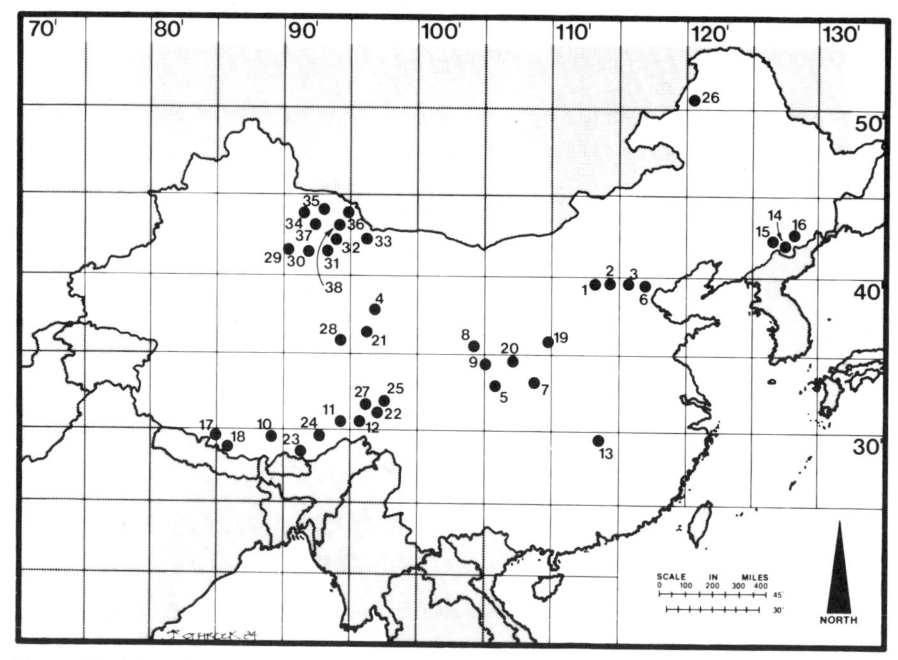

Figure 16. Map of China showing location of existing tree-ring series. Table 15 gives additional information about each site.

and cold damage is of importance to agricultural production in northeastern China. Liu Guangyuan et al. (unpublished) developed a tree-ring chronology of about 1000 years in length. The samples were taken near the upper treeline of the cold alpine belt in the neighborhood of the Quilanshan glaciers. The series is the longest one in China at present, and could be used as an indicator of past temperature fluctuation and in interpreting the advances and retreats of glaciers on Quilianshan. Temperature and rainfall series several hundred years in length were reconstructed for some regions of the Tibet Autonomous Region on the basis of dendroclimatological research by Wu Xiangding and Lin Zhenguao.[109]

Dendroclimatological work is advancing in China. Large numbers of samples have been collected from the eastern part of Tianshan, the Hengduan Mountains, and other areas. New collections in the arid and semi-arid regions of northwestern China are being planned and chronology development is under way. Table 15 gives information on tree-ring series which have been or are to be published in China. The general location of existing chronologies is shown on figure 16.

DENDROCLIMATOLOGY: APPLICATIONS

Regional and Local Studies

Colorado River Streamflow Reconstruction

The importance of the Colorado River as a source of water for agriculture, hydroelectric power generation, and human consumption in the southwestern United

Table 15. Information on Tree Ring Chronologies in China

Name	Institute	Site	Map No.	Species	Year		Trees	Reference[a]
					First	Last		
Zheng Zizheng		Xiannontam	1	Thuja orientalis	1650	1933	3	106
		Jingshan	2	Pinus bungeana	1800	1933	2	106
		West Hills	3	Pinus bungeana	1730	1933	1	106
Deng Shuqun	Institute of Botany, Academia Sinica	Heihe valley	4	Picea asperata	1802	1945	3	107
		Xiguxian	5	Abies chensiensis	1744	1945	1	107
Han Shoutang et al.	Institute of Forestry Committee of Huanghe Administration	Beijing	6	Pinus tabulaeformis	1881	1954	2	108
		Changanxian	7	Pinus tabulaeformis	1936	1954	2	108
		Yuzhong	8	Picea Sp.	1872	1954	2	108
		Minxian	9	Picea Sp.	1805	1954	2	108
Lin Zhengyao et al.	Institute of Geography, Academia, Sinica	Langkazi	10	Sabina Sp.[b]	1747	1974	1	109
		Lingzi	11	Picea Sp.	1785	1974	1	109
		Shejinashan	12	Sabina Sp.	1513	1974	1	109
Zen Shengjiang et al.	Meteorological Bureau of Hunan	Lingxiang	13	Pinus massoniana	1769	1974	1	109
Ding Shicheng	Meteorological Bureau of Jilin	Manjiang	14	Fraxinus mandshurica	1820	1974	4	109
		Manjiang	15	Pinus koraiensis	1835	1974	5	109
		Manjiang	16	Larix gmelini	1793	1974	3	109
Wu Xiangding et al.	Institute of Geography, Academia, Sinica	Jielong	17	Sabina recurva	1649	1974	2	109
		Zhangmu	18	Tsuga dumosa	1728	1974	1	109
Li Zhaoyuan et al.	Meteorological Bureau of Shaanxi, Central Meteorological Bureau	Teibaixian	19	Larix Sp.	1810	1974	1	109
		Huanglin	20	Cupressus funebris	1650	1974	1	109
Zhuo Zhenda et al.	Lanzhou University	Tiagjunxian	21	Sabina prezewalskii Kom.	1059	1975	1	109

Table 15. (*Continued*)

Name	Institute	Site	Map No.	Species	Year First	Year Last	Trees	Reference[a]
Wu Xiangding et al.	Institute of Geography, Academia Sinica	Basu	22	Sabina Sp.	1263	1974	1	109
		Luozha	23	Sabina Sp.	1761	1974	4	109
		Langxian	24	Picea Sp.	1797	1974	10	109
		Madalashan	25	Sabina Sp.	1851	1975	8	109
Gong Gaofa et al.		Genhe	26	Larix Sibirica	1851	1975	2	109
Wu Xiangding et al.		Changdu	27	Picea Sp.	1624	1975	10	109
Liu Guang yuan et al.		Delingha	28	Sabina prezewalskii Kom.	1043	1977	3	[c]
Li Jiangfeng, et al.	Meteorological Bureau of Xinjing Uygur Aut. Rgn., Meteorological Bureau of Hami, Institute of Weather and Climate, Central Meteorological Bureau	Hamixian	29	Larix Sibirica	1678	1979	25	[c]
		Hamixian	30	Larix Sibirica	1653	1979	8	[c]
		Hamixian	31	Larix Sibirica	1756	1979	19	[c]
		Yiwuxian	32	Larix Sibirica	1456	1979	9	[c]
		Yiwuxian	33	Larix Sibirica	1630	1979	10	[c]
		Balikunxian	34	Larix Sibirica	1814	1979	5	[c]
		Balikunxian	35	Larix Sibirica	1710	1979	8	[c]
		Balikunxian	36	Larix Sibirica	1463	1979	10	[c]
		Balikunxian	37	Larix Sibirica	1760	1979	5	[c]
		Balikunxian	38	Larix Sibirica	1511	1979	10	[c]

[a]All but two references are in Chinese.

[b]*Sabina* is a synonym for *Juniperus*

[c]To be published.

States cannot be overstated. This 1440-mile long river, with a drainage area of 244,000 square miles, flows through some of the most arid land and spectacular scenery in the Southwest. Importantly most of the flow originates from the 109,800 square mile Upper Basin and from less than 20 percent of the total drainage area. The average annual flow from the Upper Basin is estimated to be about 13.5 million acre feet (maf). About 5 maf is diverted annually from the basin, more than for any other river basin in the United States.

Water from the river is over-allocated as a result of the Colorado River Compact of 1922, which divided the annual flow equally between the Upper and Lower Basins — 7.5 maf to each. A later treaty allocated 1.5 maf to Mexico. The allocations were based on an anticipated average annual flow of 16.2 maf. Ironically, the river has not equaled this figure since the compact was signed. For this reason tree ring data were used to reconstruct river flow back in time well beyond the period of instrumented record for comparison purposes.[111,112]

Tree-ring data from 30 different sites (figure 17), representing as many of the major runoff-producing areas as possible, were used in the reconstructions. Based on several different models and tree-ring grids, the best estimate of the average annual flow at Lee Ferry, Arizona (the dividing point on the river between the Upper and Lower Basins), is 13.5 $\pm$ 0.5 maf. The reconstructed hydrograph (figure 18) shows the river flow back to A.D. 1510. Note that the period from 1906 through 1930 was the longest period of sustained high flows during the past 450 years. It was from these high flow years that the 16.2 maf figure used for the compact was obtained. Although this anticipated flow was based on the best data available at the time, the period of record simply was not typical of the long-term flow characteristics of the river.

Based on gaged data during the past 50 years, the 13.5 $\pm$ 0.5 maf determined from the reconstructions appears to be an improved estimate of the average annual runoff and thus may provide valuable information for management decisions. Another important point often overlooked in hydrologic forecasting is emphasized by this study. That is, there is no assurance that a period of gaged records for a river represents a random sample from the infinite population of past events. As in the case of the Colorado River, statistical estimates from such data may contain a considerable degree of bias and produce erroneous results.

Salt-Verde River Streamflow Reconstruction

Draining some 13,000 square miles in central Arizona the Salt and Verde Rivers provide the primary source of water for the metropolitan Phoenix area and a considerable acreage of irrigated agriculture (figure 19). Hydroelectricity is generated at four of the six storage dams on the two rivers and all impoundments are widely used for water-based recreational activities.

Both rivers originate at higher elevations with the extreme altitude approaching or exceeding 12,000 feet above sea level. Precipitation is quite variable, ranging from 8 inches in the more arid portions to more than 32 inches in the mountains. Winter storms from large scale cyclonic systems moving in from the Pacific Ocean are the main source of storm runoff.

Rapid urbanization has caused increasing concern over future water supplies and

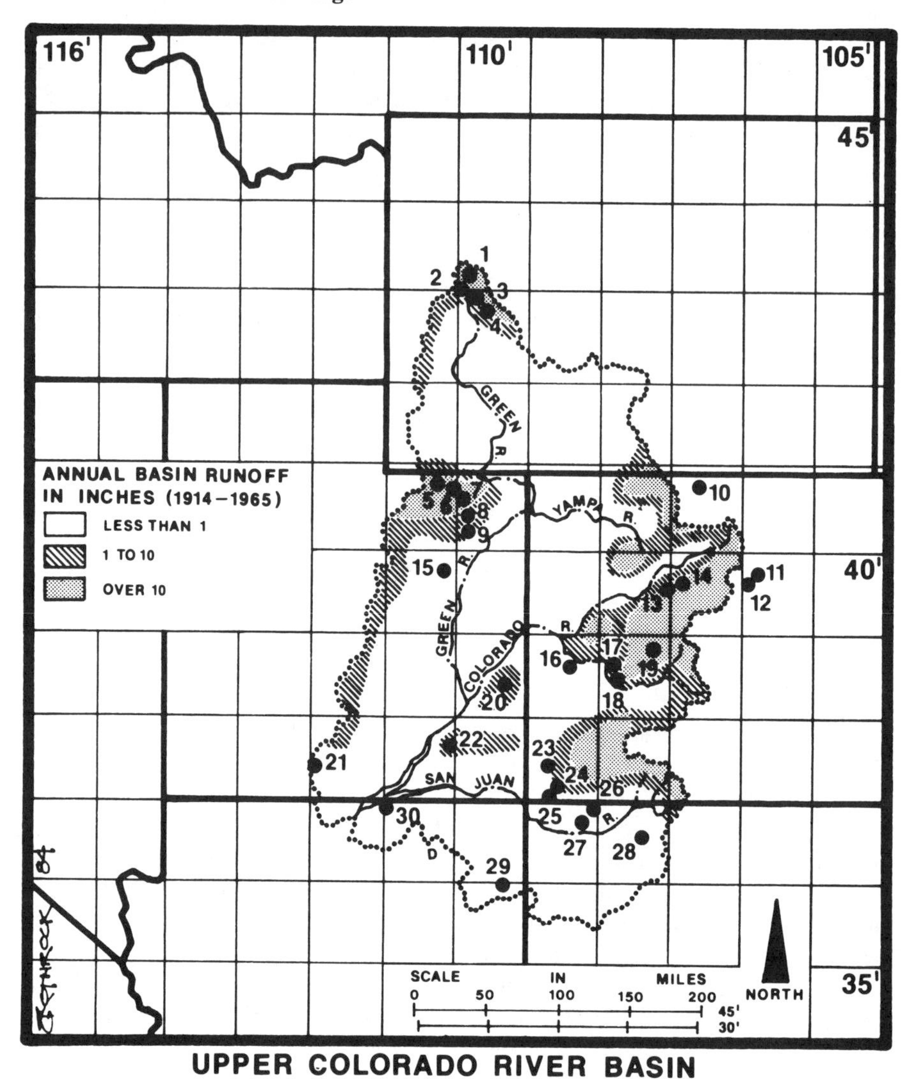

Figure 17. Map of Upper Colorado River Basin showing annual runoff in inches (shaded), and location of tree-ring series (dots).

aggravated flooding in the already flood-prone Salt River Valley. A basic tenet in both of these issues is whether the period of gaged records, although relatively long, adequately represents the long-term flow characteristics of the rivers. Some insight into this problem has been gained by using tree-ring data as a proxy source to reconstruct streamflow back in time and thus supplement gaged records.[113]

Data from 13 tree-ring sites and 9 climatic stations within the basins were used in conjunction with some 80 years of streamflow data to develop and verify relationships between ring-width indices and stream discharge. These relationships

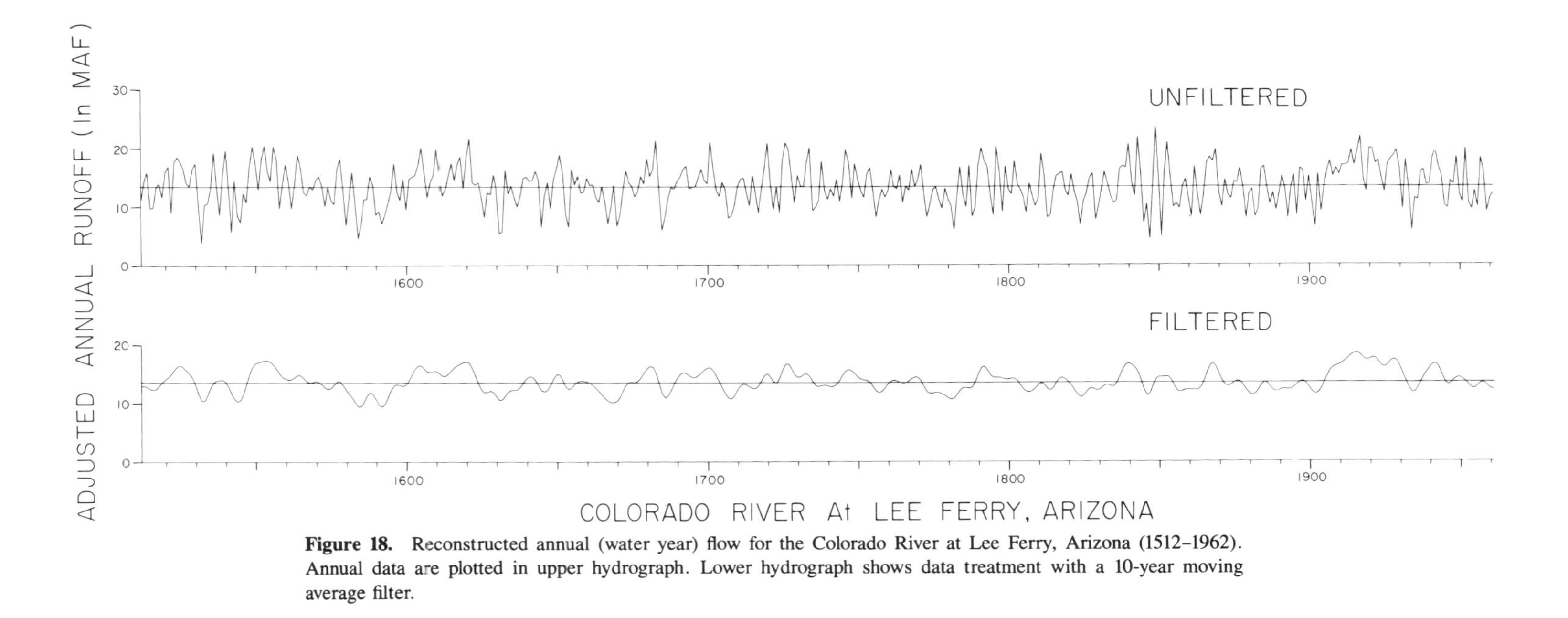

Figure 18. Reconstructed annual (water year) flow for the Colorado River at Lee Ferry, Arizona (1512–1962). Annual data are plotted in upper hydrograph. Lower hydrograph shows data treatment with a 10-year moving average filter.

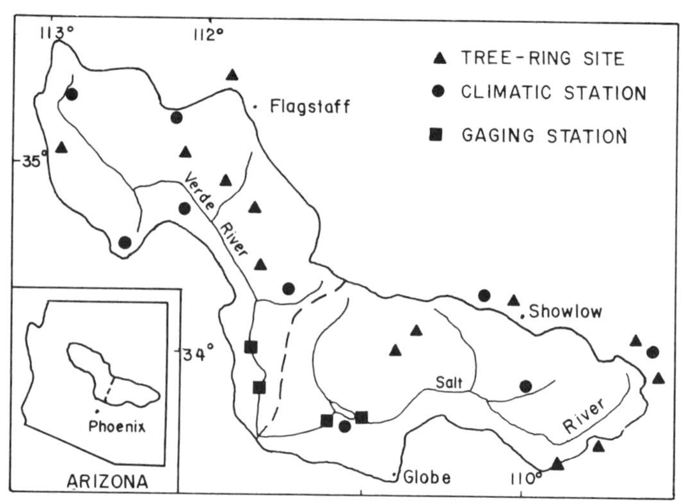

Figure 19. Location map of Salt and Verde River Basins, Arizona. Tree-ring sites, climatic stations, and gaging stations used in reconstruction (figure 20) are shown).

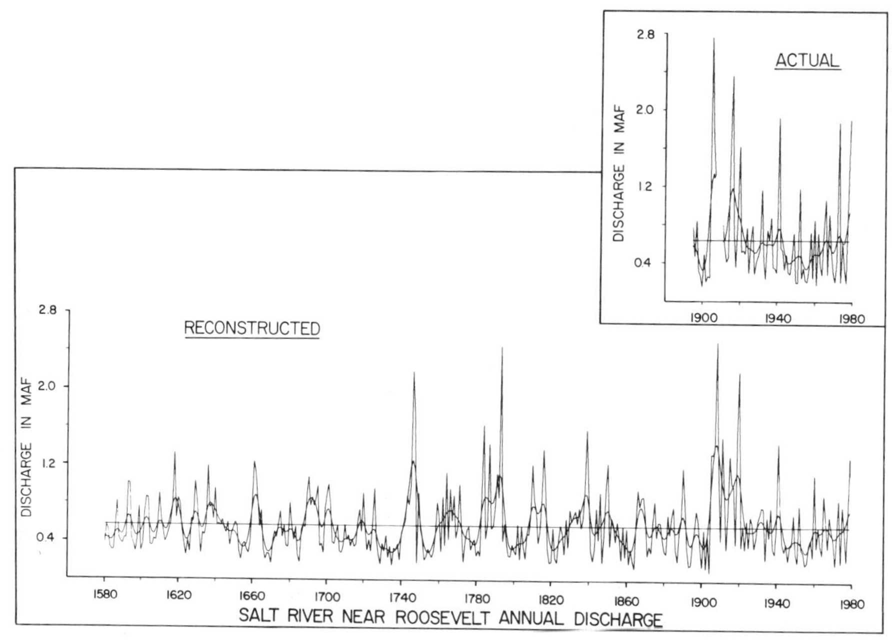

Figure 20. Reconstructed water year hydrograph for the Salt River near Roosevelt, Arizona. Period is 1580–1980. The gaged discharge normal for the same station for 1889–1980 is also shown for comparison purposes.

provided a basis for reconstructing seasonal and annual flows back to A.D. 1580 (figure 20).

The reconstructions show that several periods of extended low flows have occurred during the past 400 years, many of which were more severe than for any comparable period since 1890. The low flow periods appear to have a recurrence interval of about 22 years. Also, the gaged records contain an above average number of high seasonal and annual flows when compared to the entire 400 years. The reconstructions contain important implications for future water supply and flood potential in the Salt River Valley.

Southern California Drought Reconstruction

The mid-1970s drought in the western United States dramatically emphasized the impact of prolonged dry weather on water supplies for municipal, industrial, and agricultural uses. California was particularly hard hit in 1976 through 1978. By the spring of 1977, water stored in many reservoirs was nearly exhausted and major centers of population were forced to implement severe rationing of available supplies. The impact of the drought was not as severe in southern California largely because of water imported from reservoirs on the Colorado River. This region, however, is particularly vulnerable to severe droughts because of its large population and industrial development. For this reason, an attempt was made to place the 1970s drought in perspective over time by reconstructing the drought history of the region from tree-ring data.[114]

Techniques developed in a study of large scale drought in the western United States (discussed elsewhere in this chapter) were applied to southern California. Data from eight tree-ring sites and three climatic stations were used to develop a relationship between tree-ring width indices and the Palmer Drought Severity Index, PDSI.[115] A regional PDSI developed for July of each year between 1882 and 1977 served as the predictand and the ring-width indices as predictors in multiple regression analyses to develop the prediction (reconstruction) equations. From these, the drought history of the region was developed for the period A.D. 1700–1963 (figure 21), the time span common to all tree-ring sites.

Based on the reconstructions, extreme droughts (PDSI $\leq$ 3.5) were relatively rare in the first half of the twentieth century, but have been more frequent since 1950. The two driest years were in 1857 and 1961. Droughts that last more than

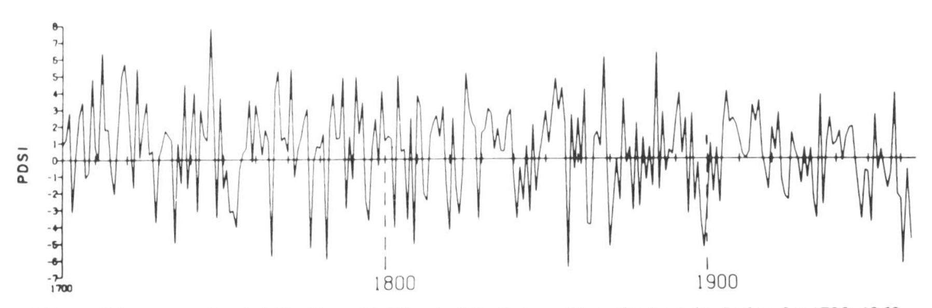

Figure 21. Reconstructed Southern California July Palmer Drought Severity Index for 1700–1963.

one year are extremely important in water resources planning. Three consecutive years of drought were rare over the study period, with none at the extreme level. Since 1963, the actual PDSI records show a severe two-year drought in 1963–64 and three-year drought in 1970–72. Evidence so far suggests that the current half-century may well become the driest 50-year period since A.D. 1700.

California Precipitation Reconstruction (Geoffrey A. Gordon, University of Maine)

Recent experience with water supplies in California presents an interesting example of the increasing impact of naturally occurring variations in a resource as the ratio of demand to supply grows. It is generally accepted that significant variations in precipitation and hence water supply over California are related to the anomalies in a large-scale controlling mechanism, the Pacific subtropical high pressure cell.[116–118] This fact provides an opportunity to calibrate a dendroclimatic network that has been demonstrated to be reflective of large-scale climatic conditions[119] in terms of California statewide precipitation.

A network of 52 chronologies for the period 1599 to 1963 was assembled over an area suitable for characterizing large-scale variations in circulation pattern. Corresponding to those data, a group of monthly precipitation records for the period 1872 to 1963 were collected over the state of California and were converted to monthly statewide precipitation averages. The tree-ring data were calibrated with the precipitation data using a modification of a model discussed by Fritts (ref. 10, pp. 428–429) and described in detail by Fritts and Gordon.[120] Calibration was performed over the period 1901 to 1963 and data from 1872 to 1900 was used for independent testing. The reconstruction consisted of the average statewide total precipitation for the water years (October through September) 1600–1963 (figure 22).

The reconstruction was dominated by low-frequency variations that may be artifacts, resulting from the combination of data from stations that were selected from a highly complex temporal and spatial field of variation. This factor makes comparison with more regional reconstructions difficult and may distort the pattern of variation in the reconstruction from what might be expected. The most striking

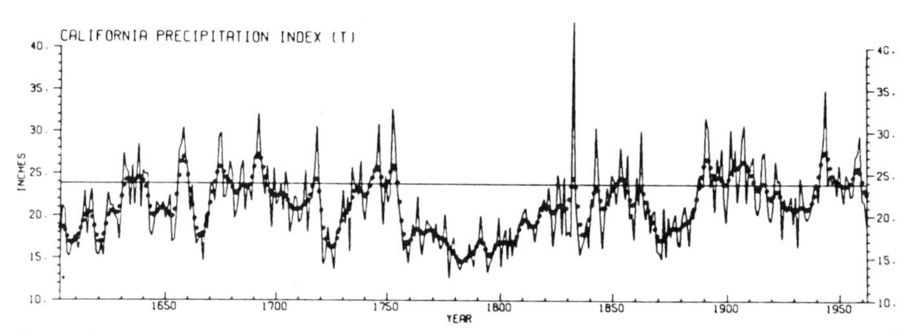

Figure 22. Reconstructed statewide precipitation index in inches for California. Mean line drawn for 1901–1963 water year averages of 23.82 inches (605 mm).

feature of the reconstruction was a lengthy dry period extending from about 1760 to 1820, characterized by a dramatic threefold increase in the frequency of years with deficient rainfall compared to actual experience in the period 1901 to 1960. This feature and other reconstructed dry periods are in qualitative agreement with other sources of information.[121,122]

From this reconstruction it appears that variations of a sufficient magnitude and duration to have a significant impact have probably occurred six times since 1600. From the perspective of a 360-year record, the period since 1890 may be viewed as one of precipitation surplus on a statewide level. This reconstruction should in no way be considered a final product. Further experimentation with different model structures and with the dendroclimatological database are essential.

Drought Reconstruction in the Hudson Valley (Edward R. Cook, Lamont-Doherty Geological Observatory, Columbia University)

Although the New York Metropolitan Area is normally well endowed with abundant precipitation, the area population and the per capita consumption of water have increased to the point where even short-term precipitation deficits can have a major impact on water supplies. During the unprecedented multi-year drought of the 1960s, reservoir supplies for New York City declined to below 30 percent of capacity before the drought ended in the autumn of 1966. A more recent drought that began in the spring of 1980 reduced the same reservoir supplies to about 25 percent of capacity in less than nine months. Before heavy rains fell in February 1981, some municipalities were down to a 14 day water supply and strict water-use restrictions were imposed. These two examples emphasize the increasing vulnerability of the New York Metropolitan Area to drought.

If the 1980s drought grew in severity and persisted as long as that of the 1960s, the impact on water supplies would be catastrophic. Thus, it is important to determine the likelihood of another drought equal to the severity and duration of the 1960s event. Probability estimates based on measured meteorological variables are not always sufficient to answer questions about extreme events because the period of record is too short. Thus, long, drought-sensitive tree-ring series were used to reconstruct drought periods back to 1694.[123,124]

Six tree-ring chronologies were developed from sites in the Hudson Valley climatic division of New York, all with the common period of 1694–1973. A principal components regression procedure was used to calibrate the tree-ring data in terms of the July Palmer Drought Severity Index, PDSI,[115] computed from divisional average climatic data. The calibration period was 1931–1970. The multiple regression calibration equation was then applied to the tree-ring data prior to 1931 to reconstruct July PDSI. A test of the reconstruction against independent July PDSI data (1896–1930) revealed that it was valid.

The reconstruction indicated that the 1960s drought has been the most severe such event in the past 287 years. Thus, the probability is very low that another drought of similar magnitude will happen very soon. However, the reconstruction indicated that less severe droughts were more frequent prior to 1900 and, in general, more persistent. Considering the sensitivity of the current water supply system in the New York Metropolitan Area to even short-term precipitation deficits, a return to a climate regime like the one responsible for the pre-1900 droughts could seriously strain the water resources of the region.

Streamflow and Temperature Reconstruction in the Southern Hemisphere

Chronologies developed by LaMarche et al.[53] were used by Campbell[125,126] to reconstruct warm season streamflow and by LaMarche and Pittock[127] to reconstruct warm season temperatures in Tasmania, Australia.

Based on data from 11 Tasmanian chronologies (table 9) Campbell[126] demonstrated that ring widths and cumulative November through March streamflow for eight streams were similarly responsive to temperature and precipitation variations. An empirical relationship was derived in which the primary variance components of the tree-ring data set were calibrated against those of the streamflow for the period 1958–1973 (16 years). The resulting transfer function was then utilized to reconstruct November–March streamflow at each of the eight stations for the period 1775–1973 (199 years). Although, as Campbell[125] points out, the tree-ring sites were widely spaced and not ideally located to provide information about runoff from the watersheds to which they were compared, the results do show evidence that tree-ring chronologies selected from temperate rain forest species may provide limited information about past hydrologic variations.

In the first known attempt to reconstruct long-term, southern hemisphere temperatures, LaMarche and Pittock[127] used the same tree-ring data sets as Campbell, along with separate spatial grids of 15 temperature and 78 precipitation stations selected for their location and length, continuity, and homogeneity. Climate response functions were derived from monthly total precipitation and average monthly temperatures for each of the climatic grids using principal component analysis. The results indicated that the dominant response of the tree-ring data was to warm season temperatures. The 15-station grid was then utilized to develop transfer functions and subsequent verification and long-term reconstruction of station temperatures spanning the period 1776–1972. Because of the similarity between the individual reconstructions only one, along with the appropriate observed station record, is shown on figure 23.

The reconstructions show slow variations on the order of 1° to 2°C. Over periods of several decades and the means of different decades appeared to differ significantly. The warmest periods were from the early 1830s through the late 1850s and the 1890s to about 1915. Cooler periods centered around 1815, the late 1850s, 1870s, and during the 1920s.

Although the authors feel that their results indicate considerable potential for dendroclimatic studies in Australia and other parts of the Southern Hemisphere, they carefully point out some important qualifications in the database and methodology employed in the reconstructions. In particular they suggest that different techniques, such as independent one-to-one comparisons of individual observational

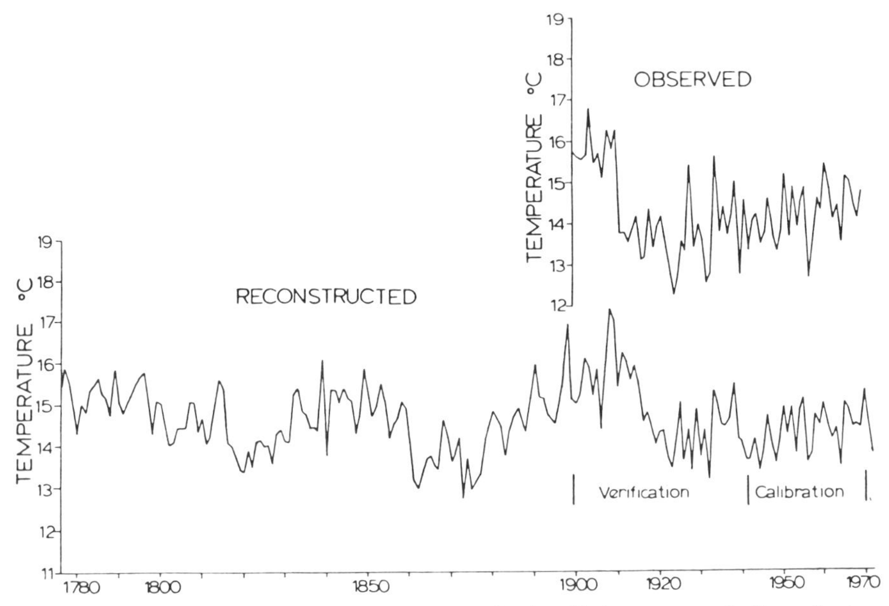

Figure 23. Observed and reconstructed warm season (October–May) temperature in Tasmania, Australia.

station records with single tree-ring data series, may overcome the tendency toward duplicity invoked by the principal component analysis.

In the first study of its kind in South America, seven tree ring chronologies from *Auraucaria araucaria* (Mol.) C. Koch and *Austrocedrus chilensis* ([D. Don] Endl.) were used to reconstruct the flow of the Rio Neuquen and Rio Limay rivers in Argentina back to A.D. 1601.[128,129] The two rivers are located in the Patagonia Andes of Argentina and drain an area from latitude 36 South to about 42 South.

Correlation between the reconstructed and gauged values for annual runoff was 0.73 for the period of overlapping records, about 60 years. Although the measured streamflow data appears to show more extreme high and low values than the reconstruction, particularly for the years of greatest flows in the early part of the records, the covariance in the low frequency range appears to be high.

Large Scale Climate, Drought, and Atmospheric Studies

Drought Area Index and Possible Relation to Solar Variation

A year rarely passes that some part of the world is not affected by severe drought. Although some regions are more drought-prone than others, few are immune. Severe or prolonged droughts affect two basic human needs: food and water. Thus their severity, duration, and recurrence interval are of critical concern to all segments of society.

Drought is difficult to define in specific terms because it does not mean the same things to all people. Basically it is a meteorological phenomenon related to specific patterns of atmospheric circulation and pressure and can be characterized as an anomaly created by prolonged moisture deficiencies.

A number of investigators have suggested an approximate 22-year periodicity for major droughts affecting the Northern Hemisphere, especially the North American continent.[130-133] However, the meteorological database is not long enough to verify this cycle much beyond the beginning of the present century.

Some insight has been gained into the cyclic aspects of drought by studies at the University of Arizona where tree-ring data have been used to reconstruct more than 300 years of drought history for the western United States (essentially the area west of the Mississippi River.[134-136] These studies have confirmed the previously suggested cycle in that the maximum area covered by moderate to severe drought occurs approximately every 22 years. In addition, this periodicity bears a strong statistical relationship to the Hale Double Sunspot Cycle.

The drought reconstructions were accomplished by developing causal relationships between ring-width indices from a grid of 40 to 65 sites and the Palmer Drought Severity Index.[115] This index (PDSI) is based on an empirical water balance approach and a given value (table 16) indicates the same degree of drought or nondrought from one location to another.

Monthly PDSI values from 1931–1970 for each of the 204 climatic divisions in the western United States were obtained from the National Climatic Data Center In Asheville, North Carolina. These 204 divisions were grouped into 40 regions as interest lay in developing large scale drought patterns. The regional PDSI was calculated as the arithmetic mean of values for the divisions comprising the region.

The PDSI for July was selected as an annual indicator of drought; a single value was essential for compatability with the annual ring-width indices. The July PDSI was chosen because diameter growth of the species involved is essentially complete by July. Also, droughts during the present century tended to peak in July when moisture demands are greatest, both for crops and human consumption.

A Drought Area Index (DAI), indicative of large scale drought patterns, was

Table 16. Drought Classification by Palmer Drought Severity Index (PDSI)

Palmer Index	Degree of Drought
$\text{PDSI} \leq -4.0$	Extremely dry
$-4.0 < \text{PDSI} \leq -3.0$	Severely dry
$-3.0 < \text{PDSI} \leq -2.0$	Moderately dry
$-2.0 < \text{PDSI} \leq -1.0$	Mildly dry
$-1.0 < \text{PDSI} < +1.0$	Near normal
$+1.0 \leq \text{PDSI} < +2.0$	Mildly wet
$+2.0 \leq \text{PDSI} < +3.0$	Moderately wet
$+3.0 \leq \text{PDSI} < +4.0$	Severely wet
$+4.0 \leq \text{PDSI}$	Extremely wet

developed through multivariate analysis. Initially, eigenvector analysis was applied to the PDSI values for the 40 climatic regions and the tree-ring series from the appropriate grid. A few of the eigenvectors were chosen from each data field that represented a reasonably high percentage of the total variance. In this procedure fields of highly correlated variables can be represented by a smaller number of orthogonal functions (eigenvectors) and their corresponding orthogonal amplitudes.

The eigenvector amplitudes, as discussed above, for the period of overlap between the PDSI and tree-ring series (1931–1960) were then related to each other by canonical analysis. The resulting transfer function was then used to translate tree-ring data into PDSI values for each year back to A.D. 1700. An annual DAI was developed for PDSI values of −1, −2, −3, and −4. For example, the DAI for a PDSI of −2 would be the number of regions, out of the 40 total, where the PDSI equalled or exceeded a value of −2.

When the annual DAI series is plotted against the time a distinct periodicity in the occurrence of dry periods is evident (figure 24). This suggested periodicity was confirmed by a spectral analysis which shows a peak at 22 years which is significant at the 99 percent level. The periodicity diminishes at more intense drought categories; significance is above 95 percent for PDSI < -2 and less than 95 percent for PDSI values of −3 and −4.

The statistically significant periodicity of about 22 years in the reconstructed DAI values suggested a possible relationship with the Hale solar cycle which has a similar cyclic occurrence. This was investigated by using both cross-spectral and harmonic dial analyses.

The squared coherency is analogous to a correlation coefficient between two series as a function of frequency. Squared coherency between the DAI and Hale sunspot series peaked significantly (95 percent level) near 22 years; the relationship was strongest for the most severe drought (PDSI < -3) category tested.

Harmonic dial analysis was used to test the degree of consistency with which the DAI series marched in step with the Hale sunspot cycle. Both the DAI and sunspot series were first bandpassed filtered to emphasize variance near 22 years. Maxima from the filtered DAI series were then plotted on a dial where the plotted position of each peak was determined by (1) the number of years since the preceding sunspot minimum, and (2) the amplitude of the DAI peak. Strong phase locking is indicated by the clustering of points along one radial direction far from the center

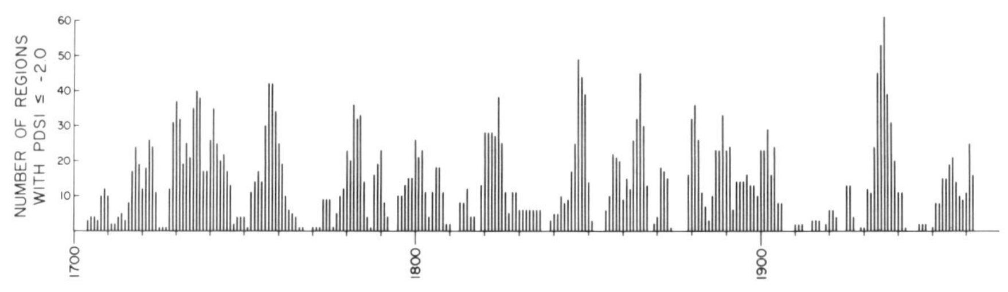

Figure 24. Year area in drought (Palmer Drought Severity Index less than or equal to −2) in western United States for period 1700–1962. The data have been treated with a three-year moving average filter.

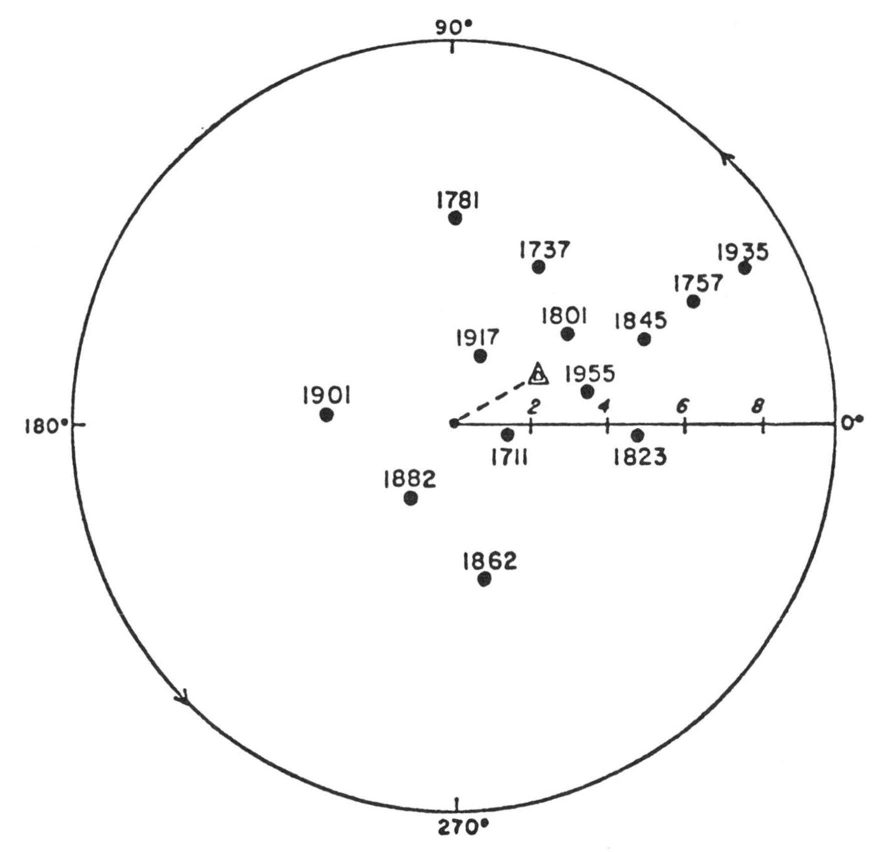

Figure 25. Harmonic dial showing phase relationship between the drought area index (DAI where PDSI ≤ -1) and the Hale Solar Cycle. Significance level is 0.011.[134]

of the dial; significance is tested by an F-test on the centroid of points. The dial for a particular DAI series is in figure 25. Drought maxima tend to lag sunspot numbers by about two years, with a significance level of 99 percent. Testing of various DAI series from other tree-ring grids and levels of drought severity yielded similar results.

Temperature and Precipitation Reconstructions (H. C. Fritts, University of Arizona)

Fritts, Lofgren and Gordon[119] used 65 ring-width chronologies from western North America which span 1601–1963[137] to reconstruct spatial variations in climate. The climatic data included (1) grids of temperature observed at 77 stations and precipitation observed at 96 stations covering the United States and southwestern Canada, (2) grids of temperature observations at 46 stations and precipitation observations at 52 stations covering the western half of the United States and southwestern Canada, and (3) a grid of sea level pressure observations at 96 evenly spaced points spanning North America, the North Pacific and eastern Asia.[119]

The 65 chronologies were each filtered for first-order autocorrelation with the following algorithm:

$$X_t = x_t - \rho(s_{t-1} - \overline{m})$$

where x_t is the standardized index for year t, ρ is the first-order autocorrelation, $\overline{m}$ is the mean index, and X_t is the filtered value for the respective chronology. The tree-ring data both filtered and unfiltered for autocorrelation were subjected to principal component analysis. The largest 15 principal components of each set, making a total of 30, were included as multivariate statistical predictors in a transfer function to reconstruct the seasonal variations in climate.

A variety of predictor models were used, including components from years $t - 1$, t, $t + 1$, and $t + 2$, where t is the spring and summer season current with growth and autumn and winter for the season occurring immediately before growth. Each model differed in the kinds and numbers of principal components used as predictors as well as the lags and leads considered in the structure.[119] These were applied to varying numbers of predictand principal components for each climatic variable and season.

The strategy was to calibrate each model with the twentieth century instrumented record (dependent data from 1899–1963) and then verify the reconstructions against available independent instrumented data. Seasonal averages or totals, rather than annual climatic data, were calibrated so that differences in response to each season could be accommodated.

A stepwise canonical analysis program modified from Blasing[138] was used to select those canonical variates that contributed significantly to the analysis. Estimated principal components of climate were in turn converted to spatial anomalies of seasonal climate for each season from 1602 to 1962.

Independent verification data for temperature and precipitation were the nineteenth century observational records available from stations used for calibration. The observational data available at a station were compared to the reconstruction and six verification statistics were calculated. Since little nineteenth century data were available for verification of pressure, a calibration procedure which left out different sub-samples of data one at a time was developed by Gordon,[139] to obtain independent pressure estimates. These independent estimates were then compared to the twentieth century instrumented record used for calibration to obtain the independent verification statistics.

Following the ideas of Bates and Granger[140] which showed that the reliability of a statistical forecast may be improved by averaging statistical estimates from several different models, the results from two or three of the best reconstructions for each variable and season were averaged. The calibration and verification statistics were recalculated, and the results were examined for indications of improvement. On the average there was an 8.5 percent improvement in variance calibrated (adjusted for loss of degrees of freedom) when the results from two or three different models were averaged.

It was hypothesized that the multivariate methods using principal components emphasized large scale features of seasonal climate at the expense of the details of year-to-year variations at each calibrated site. To test this possibility, the estimates at individual reporting stations were averaged to obtain mean values for specific climatic regions, and these were compared to the seasonal values of the instrumented data averaged for the same region. There was on the average a 10.4 percent improvement over the calibration at the individual grid points or stations. This confirmed that the large scale patterns in climate were best reconstructed.

It was also noted that the decade-long averages appeared to be better reconstructed than the individual seasonal estimates. This feature was tested by smoothing the regional averages with a low-pass digital filter passing 50 percent at eight-year periods[10] and recalculating the calibration and verification statistics after correcting for losses in degrees of freedom due to the filtering. This resulted in a 7.7 percent further improvement in the percent variance in common as compared to the unfiltered regional averages. The overall calibration of the filtered seasonal data over all regions and adjusted for loss of degrees of freedom due to regression was 38.4 percent.

Even though seasonal climatic data were calibrated to accommodate the seasonal differences in tree responses, the statistics for the seasonal climatic estimates seemed low. It appeared that estimates for a given season may be perturbated by climatic conditions occurring in other seasons. This arises because the ring width is known to be an integrated measure of climate for all four seasons.[10] While the multivariate calibration was used to resolve some of the seasonal differences, it apparently accommodated only a portion of the inter-season variation. If the seasonal reconstructions derived from annual tree growth are perturbated by conditions in other seasons, it was hypothesized that underestimates for one season should be accompanied by overestimates in other seasons. When seasonal estimates of a particular variable are averaged over the year, these perturbations should cancel one another and a more precise annual estimate should result. To test this possibility, the seasonal estimates were combined into annual averages for temperature and pressure and annual totals for precipitation. The annual instrumented record was used to obtain measurements equivalent to calibration and verification statistics. The average adjusted percent variance reduced for all annual variables was 30.9 percent as compared to 20.3 percent for the average of the season, which is a 10.6 percent improvement. When these annual values were averaged into climatic regions and then smoothed to emphasize low-frequency variation, there was an additional 12.0 percent and 15.6 percent improvement. The average adjusted variance agreement between the filtered annual estimates and the filtered instrumented data for the regions was 58.5 percent compared to 38.4 percent for the seasons. These values amounted to 70.6 percent for annual pressure, 62.1 percent for annual temperature, and 48.8 percent for annual precipitation. The verification statistics also improved when the reconstructions were averaged.

Figure 26 shows maps of the uncorrected percent variance calibrated at each station over the annual temperature and precipitation grids, along with a summary of the corresponding verification statistics calculated from independent data.

Temperature was reconstructed and verified for large areas eastward from the

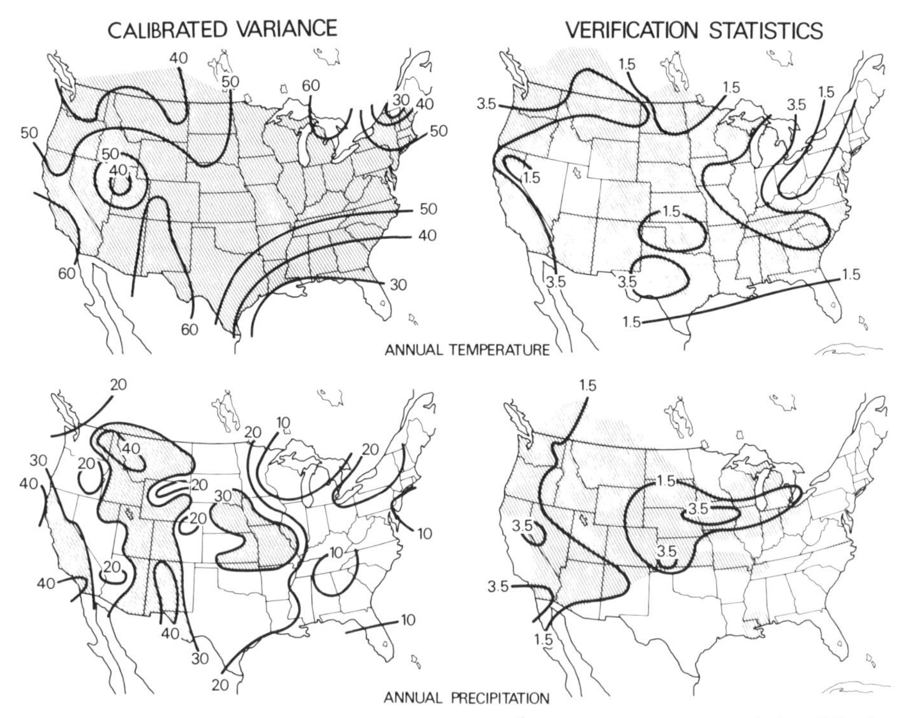

Figure 26. Isopleths of the percent calibrated variance (r^2) (left) and verification statistics (right) for annual temperature and precipitation. Shading on left side indicates significance at the 95 percent level. Shading on the right designates those areas where the reduction of error[9] is positive. The isopleths indicate the number of other verification tests out of a total of five that passed the significance test. The 1.5 isopleth distinguishes those stations that did not verify (with one or zero tests significant) from those that did verify (with two or more tests significant).

tree-ring data grid into eastern United States. The poorest calibration and verification was in Florida and the Gulf Coast, the New England states, and over the western portion of the tree grid in the Great Basin of Nevada and Utah where tree-ring data were sparse or lacking.

Precipitation did not calibrate as well as temperature and did not verify as far eastward from the tree-ring database into eastern North America. Yet there was some verification through the Central Plains and Great Lakes. Calibration was poor in the lee of the Sierra Nevada and Rocky Mountains. There was little success in verifying precipitation in the northern Great Lakes, Atlantic coastal states, Appalachian Mountains and from southern New Mexico to the southeast Gulf Coast. Some difficulties were encountered in verifying precipitation along the coast of the Pacific Ocean.

Atmospheric Pressure Variation (H. C. Fritts, University of Arizona)

Surface pressure (figure 27) appears to calibrate well, especially over south-central North America, the North American Arctic, the subtropics, the far western North

CALIBRATED VARIANCE

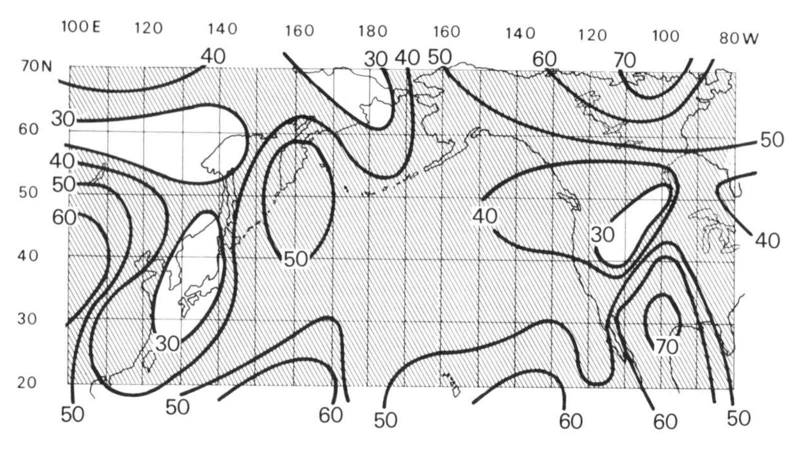

VERIFICATION STATISTICS

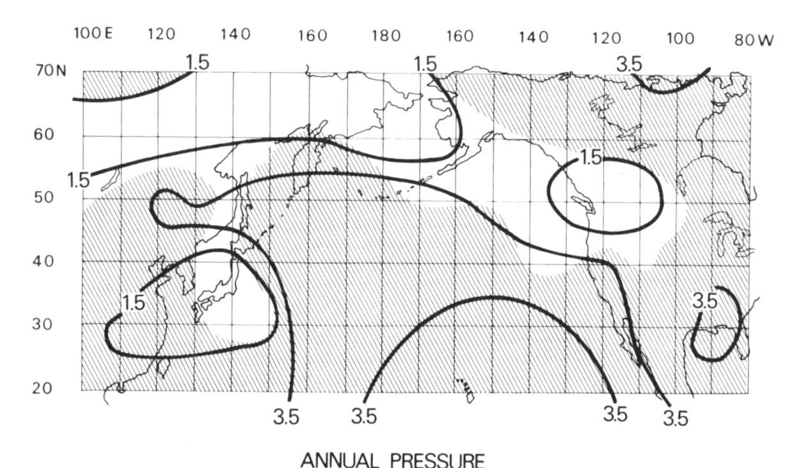

ANNUAL PRESSURE

Figure 27. Same as 26 except for annual pressure.

Pacific south of Kamchatka, and central Asia. Verification is best, however, through the central North Pacific, along the California coast, and over the Gulf of Mexico and the southern states of the United States. Verification is poorest in Siberia, the far western North Pacific, and from the Aleutians to the Pacific west-central regions of North America.

Figure 28 includes examples of reconstructed anomalies in pressure, temperature, and precipitation for each decade in the first half of the nineteenth century expressed as departures from the twentieth century average values. In 1801 to 1810, a strong southerly displaced Aleutian Low was reconstructed along with a weak blocking high positioned over central North America. The anomalously southerly

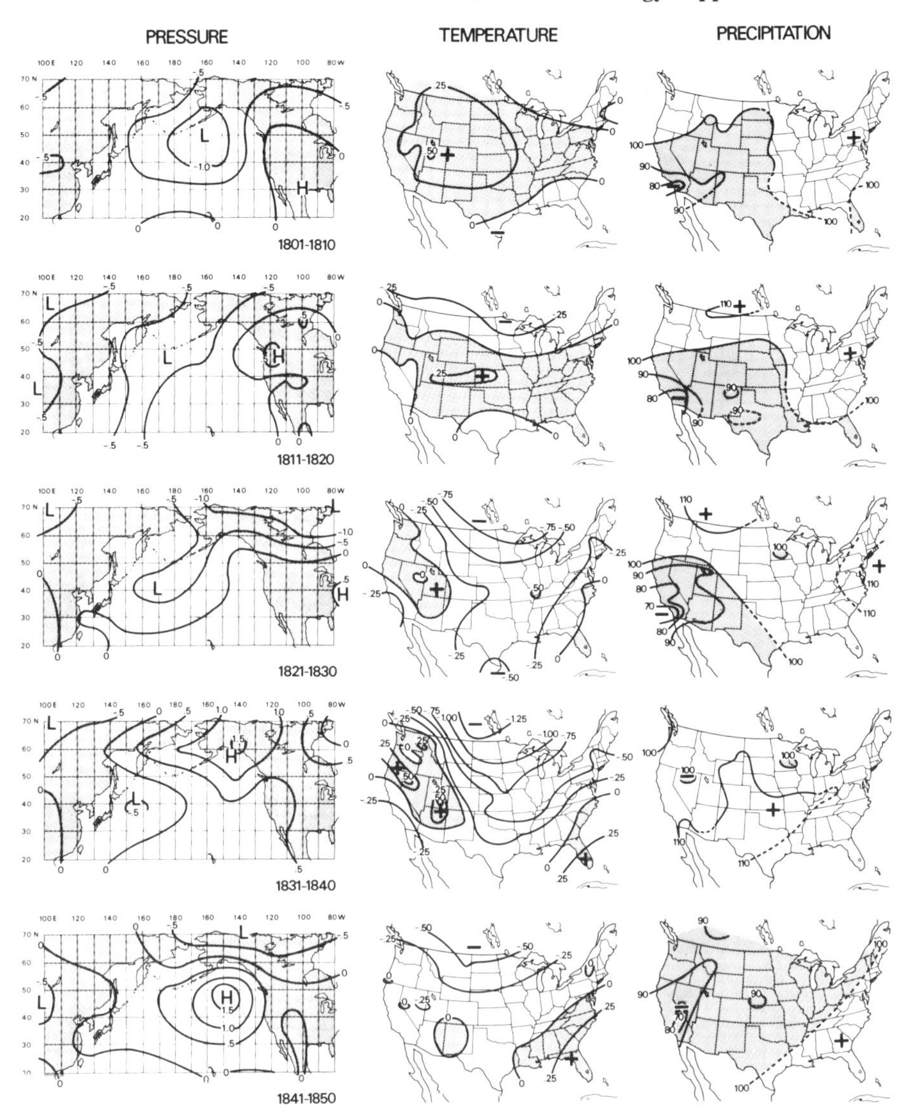

Figure 28. The reconstruction of mean annual surface pressure, temperature, and precipitation for decades in the first half of the seventeenth century plotted as departures or percentages of the 1901–1970 averages. Shaded areas are warm, dry anomalies, and dashed line indicates areas with poor verification.

flow between the low and high anomaly is consistent with the warmer temperatures that were reconstructed there and the below average precipitation reconstructed in the western and southwestern United States. The consistency between these three independent calibrations helps to support their validity. Storms from the North Pacific probably moved somewhat north of their twentieth century position into western Canada and then traveled east-southeastward into the eastern United States where moisture conditions were reconstructed to have been above the twentieth century average.

An anomalous blocking high was reconstructed to persist in 1811 to 1820, but the Aleutian Low anomaly weakened. Temperatures were reconstructed not to be as warm, but droughts persisted in the southwestern United States. Temperatures along the northern borders of the United States were reconstructed to have been lower than the twentieth century average.

In the 1821–1830 reconstructions, the blocking high disappeared from western North America. A vigorous zonal flow was established over North America with colder temperatures and more precipitation in central North America. Anomalous drought (compared to twentieth century data) persisted in the far western area as the storms were reconstructed to move most commonly through northwestern United States, somewhat north of the twentieth century mean position.

The decade 1831–1840 was characterized by increasing anomalous pressure over Alaska, indicating southward displacement of storms over the North Pacific. A strengthened Icelandic Low was indicated by a weak low anomaly over Hudson Bay. Storms entering the west coast of the United States appeared to bring 10 percent higher than average moisture elsewhere throughout the country. Temperatures were reconstructed to have been moderate to cool in the west and markedly colder than the twentieth century east of the Rocky Mountains.

In 1841 to 1850 a blocking high was reconstructed over the eastern North Pacific. A high in this position would bring southward advection of cool, polar air into the West, as indicated by the lower temperature reconstruction. However, mild conditions were reconstructed along the Atlantic coast, where more northward air flow would be expected. Precipitation was reconstructed to have been somewhat below average, perhaps indicating that the storms were displaced north of their twentieth century position. The Southeast was reconstructed to have been anomalously warm and moist, but one must keep in mind that the verification of precipitation in this area was too poor to justify any confidence that these moisture conditions actually existed.

An example of the filtered time series for the annual temperature (1602–1962) for regions in the United States are presented in figure 29. The dots on the right correspond to the filtered values of the instrumented data with which the tree rings were calibrated. The similarity between the instrumented record and the reconstructions attests to the high variance reduced in the filtered sets by the calibration. Variations in reconstructed temperature over North America since 1602 are evident, but large regional differences are also indicated.

While the reliability of the reconstructions diminishes to the east as the distance increases between the climate stations and the sites of the trees, one can conclude that considerable information concerning the large scale spatial variations can be reconstructed for large segments of the continent, sometimes for areas well beyond the site locations of the chronologies.

When more trees have been collected and their chronologies are available for a large portion of North America, the revised reconstructions will undoubtedly improve along the eastern margins of the continent.[141]

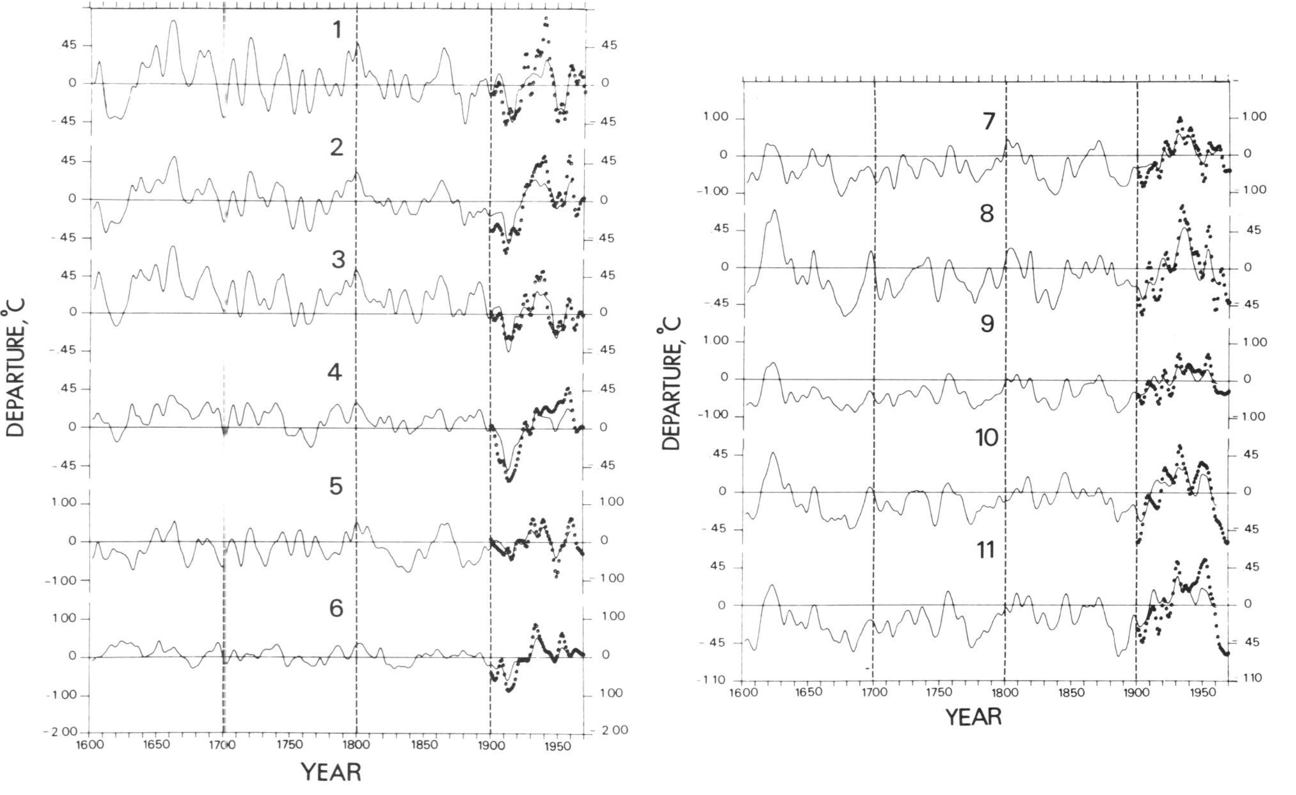

Figure 29. Reconstruction of the regionalized annual temperature for eleven regions in the United States. The data have been treated with a low-pass digital filter with a frequency response of 0.5 at periods of eight years and plotted as departures from the 1901–1970 averages. Dots on the right are the filtered, regionalized instrumented data used for calibration. The regions are (1) Columbia Basin, (2) California Valleys, (3) Intermountain Basins, (4) Southwest Deserts, (5) Northern High Plains, (6) Southern High Plains, (7) Northern Prairie, (8) Southern Prairie, (9) Great Lakes and Midwest, (10) Southeast, and (11) East.

This study as well as those of Kutzbach and Guetter[20] and Fritts and Shatz[137] provides an appropriate strategy for choosing tree-ring chronologies and other comparable proxy data to reconstruct hemisphere-wide climatic patterns.[141] Some accuracy can be obtained for large areas of the globe whenever the climatic system is sufficiently intercorrelated.

FUTURE DEVELOPMENT OF DENDROCLIMATOLOGY

The science of dendroclimatology is moving forward on many fronts. Foremost among these efforts is to increase the worldwide data coverage. A concentrated effort was begun in 1980 to obtain a worldwide tree-ring network consisting of high quality properly replicated tree-ring samples. An International Tree-Ring Data Bank has been formed and is housed at the Laboratory of Tree-Ring Research, University of Arizona, Tucson. The data bank has been designated as the most logical repository and clearing house for data to be used in this international effort.[50] With a well designed worldwide tree-ring database, methods like those described earlier in this chapter can be used to reconstruct and study past hemispherical and perhaps worldwide climatic variations.

Another area of considerable potential is in the extension of climatic information from the isotopic concentrations within tree-ring data series.

REFERENCES

1. A. E. Douglass, "A Method of Estimating Rainfall by the growth of Trees." In E. Huntington, ed., *The Climatic Factor*, Carnegie Institute, Washington, D.C., 1914, Pub 192, pp. 101–122.
2. E. Schulman, *Dendroclimatic Changes in Semiarid America*, University of Arizona Press, Tucson, 1956.
3. H. E. Weakly, "A tree-ring record of precipitation in Western Nebraska," *J. For.*, **41**(11), 816–819 (1943).
4. H. E. Weakly, "History of drought in Nebraska. *J. Soil Water Conserv.*, **17,** 271–275 (1962).
5. G. F. Will, "Tree-ring studies in North Dakota," *N. Dakota Agr. Expt. Sta. Bull.*, **338,** 1–24 (1946).
6. W. F. Weakly, "Tree-Ring Dating and Archaeology in South Dakota," Ph.D. diss. The University of Arizona, Tucson, 1968.
7. M. A. Stokes and T. L. Smiley, *An Introduction to Tree Ring Dating*, University of Chicago Press, 1968.
8. W. J. Robinson and R. Evans, "A microcomputer-based tree-ring measuring system," *Tree-Ring Bull.* **40,** 59–64 (1980).
9. D. A. Graybill, "Revised computer programs for tree-ring research," *Tree-Ring Bull.* **39,** 77–82 (1979).
10. H. C. Fritts, *Tree Rings and Climate*, Academic, London, 1976.
11. E. R. Cook and K. Peters, "The smoothing spline: A new approach to standardizing forest interior tree-ring width series for dendroclimatic studies," *Tree-Ring Bull.* **41,** 45–58 (1981).
12. A. E. Douglass, *Climatic cycles and tree growth, Vol. 3 A study of cycles*, Carnegie Institute Washington, D.C., 1936, Publ. 289.

13. M. G. Kendall, *A course in Multivariate Analysis*, Charles Griffin, London, 1957.
14. N. W. Cooley and P. R. Lohnes, *Multivariate Data Analysis*, Wiley, New York, 1971.
15. E. N. Lorenz, "Empirical orthogonal functions and statistical weather prediction," *Scientific Report No. 1*, M.I.T., Department of Meteorology, Cambridge, Mass., 1956.
16. J. E. Kutzbach, "Empirical eigenvectors of sea-level pressure, surface temperature and precipitation complexes over North America," *J. Appl. Meterol.*, **6,** 791–802 (1967).
17. C. W. Stockton, "The feasibility of augmenting hydrologic records using tree-ring data," Ph.D diss. University of Arizona, Tucson, 1971.
18. G. E. P. Box and G. M. Jenkins, *Time Series Analysis: Forecasting and Control*, Holden-Day, San Francisco, 1976.
19. D. M. Meko, "Applications of Box-Jenkins methods of time series analysis to the reconstruction of drought from tree rings," Ph.D diss. University of Arizona, Tucson, 1981.
20. J. E. Kutzbach and P. J. Guetter, "On the design of paleoenvironmental data networks for estimating large-scale patterns of climate," *Quat. Res.* **14**(2), 169–187 (1980).
21. E. L. Braun, *Deciduous Forest of Eastern North America*, Hafner, New York, 1950.
22. G. T. Trewartha and L. H. Horn, *An Introduction to Climate*, McGraw-Hill, New York, 1980.
23. G. R. Willey, "Notes on central Georgia dendrochronology," *Tree-Ring Bull.* **4**(2), 6–8 (1937)
24. F. Hawley, *Tree-Ring Analysis and Dating in the Mississippi Drainage*, University of Chicago Press, 1941.
25. E. Schulman, "Dendrochronology in the pines of Arkansas," *Ecology* **23**(3), 308–318 (1942).
26. E. DeWitt and M. Ames, *Tree-ring chronologies of eastern North America*, Laboratory of Tree-Ring Research, University of Arizona. Chronology Series 4, 1978.
27. D. W. Stahle, J. G. Hehr, and G. C. Hawks, "Tree-ring network collections in the central United States (abstract)," *Assoc. Southeast. Biol. Bull.*, **28**(2), 45 (1981).
28. D. N. Duvick and T. J. Blasing, "A dendroclimatic reconstruction of annual precipitation amounts in Iowa since 1680," *Water Resour. Res.* **17**(4), 1183–1189 (1981).
29. R. E. Bell, "Dendrochronology at the Kincaid Site." In F. C. Cole, ed., *Kincaid: A Prehistoric Illinois Metropolis*, University of Chicago Press, 1951, pp. 223–292.
30. R. Guyette, E. A. McGinnes, G. E. Probasco, and K. E. Evans, "A climate history of Boone County, Missouri from tree-ring analysis of eastern red cedar," *Wood Fiber*, **12**(1), 17–28, 1980.
31. D. W. Stahle, "Tree-ring dating of historic buildings in Arkansas," *Tree-Ring Bull.* **39,** 1–29 (1979).
32. W. M. Walker, "The Troyville Mounds, Catahoula Parish, Louisiana," *Bur. Ethnol. Bull.*, **113** (1936).
33. H. W. Hamilton, "The Spiro Mound," *Missouri Archeol.*, **14,** 1–26, 1952.
34. G. B. Mills, "Of men and rivers," U.S. Army Corps of Engineers, Vicksburg, Mississippi, 1978.
35. A. Bibbins, "The buried cypress forests of the Upper Chesapeake." *Rec. Past,* **4**(2), 47–53. Records of the Past Exploration Society, Washington, D.C. (1905).
36. P. B. Sears and G. Couch, "Microfossils in an Arkansas peat and their significance," *Ohio J. Sci.*, **32**(1), 63–68 (1932).
37. H. H. Moore, *Andrew Brown and Cypress Lumbering in the Old Southwest.* Louisiana State University Press, Baton Rouge, 1967.
38. G. C. Jacoby and E. R. Cook, "Past temperature variations inferred from a 400-year tree-ring chronology from Yukon Territory, Canada," *Arctic Alp. Res.*, **13**(4), 409–418 (1981).
39. C. W. Stockton and H. C. Fritts, *Long-term Reconstruction of Water Levels for Lake Athabasca (1810-1967) by Analysis of Tree Rings,* Water Resources Bull., **9**(5), 1006–1027 (1973).
40. T. J. Blasing and H. C. Fritts, "Reconstructing past climatic anomalies in the North Pacific and western North America from tree-ring data," *Quat. Res.*, **6,** 563–579 (1976).
41. G. C. Jacoby and L. D. Ulan, "Reconstruction of past ice conditions in a Hudson Bay estuary using tree rings, Nature, **298,** 637–639 (1982).

42. J. P. Cropper and H. C. Fritts, "Tree-ring width chronologies from the North American Arctic." *Arctic Alp. Res.*, **13**(3), 245–260 (1981).

43. G. C. Jacoby and L. D. Ulan, "Review of Dendroclimatology in the Forest-Tundra Ecotone of Alaska and Canada." In C. R. Harington, ed., *Syllogeus: Climatic Change in Canada, vol. 2,* No. 33, National Museums of Canada, Ottawa, 1980, pp. 97–128.

44. S. Hastenrath, "Recent Climatic Fluctuations in the Central American Area and Some Geoecological Effects." In Carl Troll, ed., *Geoecology of the Mountainous Regions of the Tropical Americas,* Proceedings UNESCO Mexico Symposium, New York, pp. 131–138, 1966.

45. V. C. LaMarche, Jr., R. L. Holmes, P. W. Dunwiddie, and L. G. Drew, *Tree-Ring Chronologies of the Southern Hemisphere, No. 1, Argentina and No. 2, Chile: Chronology Series V,* Laboratory of Tree-Ring Research, University of Arizona, Tucson, 1979.

46. A. V. Munaut, A. J. Berger, J. Guiot, and L. Mathieu, "Dendroclimatological studies on cedars in Morocco," unpublished paper, Laboratoire de Palynologie et Phytosociologie, Université Catholique de Louvain-la-Neuve, 1978.

47. A. Aloui, "Quelques aspects de la dendroclimatologie en Khroumiries (Tunisie). Q.E.A. de'Ecologie méditerraneeanne," Université Aix-Marseille, 1978.

48. V. C. LaMarche, Jr., R. L. Holmes, P. W. Dunwiddie, and L. G. Drew, *Tree-ring Chronologies of the Southern Hemisphere, No. 5, South Africa: Chronology series V,* Laboratory of Tree-Ring Research, University of Arizona, Tucson, 1979.

49. T. G. J. Dyer, "Southern Africa." In M. K. Hughes, P. M. Kelly, J. R. Pilcher, and V. C. LaMarche, eds., *Climate from Tree Rings,* Cambridge University Press, London, 1982, pp. 82–83.

50. M. K. Hughes, P. M. Kelly, J. Pilcher, and V. C. LaMarche, Jr., eds., *Second International Workshop on Dendroclimatology, July, 1980: Report and Recommendations,* Edmund Norvic Press, Belfast, Northern Ireland, 1980.

51. P. Stewart, "*Cupressus dupreziana,* threatened conifer of the Sahara." *Biol. Conservator,* **2**(1), 10–12 (1969)

52. J. Ogden, "Australasia." In M. K. Hughes, P. M. Kelly, J. R. Pilcher, and V. C. LaMarche, eds., *Climate from Tree Rings,* Cambridge University Press, 1982, pp. 90–104.

53. V. C. LaMarche, Jr., R. L. Holmes, P. W. Dunwiddie, and L. G. Drew, *Tree-Ring Chronologies of the Southern Hemisphere Volume 4, Australia and 5 New Zealand: Chronology Series V,* Laboratory of Tree-Ring Research, University of Arizona, Tucson, 1979.

54. C. N. Carter, "Studies in dendrochronology," North Island, New Zealand. B.Sc. (Hans.) project. Victor University, Wellington, 1971.

55. J. Fletcher, "Oak chronologies for eastern and southern England: Principles for their construction and application: their comparison with others in northwest Europe." In J. Fletcher, ed., *Dendrochronology in Europe,* National Maritime Museum, Greenwich Arch. Series No. 4, 1978. See pp. 139–156.

56. D. Eckstein, "Tree-ring research in Europe," *Tree-Ring Bull.*, 1–19 (1972).

57. M. G. L. Baillie, "*A dendrochronological study in Ireland with reference to the dating of Medieval and post-Medieval Timbers,*" Ph.D Diss. Queen's University, Belfast, 1973.

58. H. C. Fritts, "Computer programs for tree-ring research," *Tree-Ring Bull.*, **25**(3/4), 2–7 (1963).

59. M. G. L. Baillie and J. R. Pilcher, "A simple crossdating program for tree-ring research," *Tree-Ring Bull.*, **33,** 7–14 (1973).

60. H. Hesselman, "Om tallens höjdtillväxt och skottbildning somrarne 1900–1903," *Sv. Skogsvåardsfören. Tidsskr.*, **B2,** Stockholm (1904 a). "Om tallens diametertillväxt under de sista tio åren," *Sv. Skogsvaardsfören. Tidsskr.*, **B2,** Stockholm (1904 b).

61. A. Wallén, "Om temperaturens och nederbördens inverkan pà granens och tallens höjd—och radietillväxt á Stamnäs kronopark 1890–1914," Skogshögskolans festskrift 1917, pp. 413–427, 1917.

62. E. Laitakari, "Untersuchungen über die einwirkung der witterungsverhältnisse auf den längen—und dickenwachstum der kiefer," *Acta For. Fenn.*, **17,** Helsingfors (1920).

63. G. Kolmodin, "Tillvàxtundersokningar i norra dalarna," *Sv. Skogsvàardsfören. Tidsskr.*, **21,** 1-35, Stockholm (1923).

64. G. Kolmodin, "Väderlekens inflytande på tallens diametertillväxt," *Sv. Skogsvàardsfören Tidsskr.*, **33,** 321-379, Stockholm (1935).

65. E. Eide, "Über Sommertemperatur und Dickenwachstum im Fichtenwald," *Medd. Fr. Det Norske Skogforsøksves.*, **B2,** Oslo, (1926). (In Norwegian, German summary).

66. S. Erlandsson, "Dendrochronological studies," *Data Fr. Stockholms Högskolas Geokronol. Inst.*, **23,** Stockholm, (1936).

67. A. Ording, "Arringanalyser pà gran och furu," *Medd. Fra. Det Norske Skogforsøksves.*, **7**(2), 1-354, Oslo (1941). (Annual ring analyses on spruce and pine. In Norwegian, English summary.)

68. T. Ruden, "A valuation of the methods employed in dendrochronology and annual ring analyses," *Medd. Fra. Det Norske Skogsorsoksves.*, **9**(2), 181-267, Oslo (1945). (In Norwegian, English summary.)

69. I. Hustich, "On the correlation between growth and the recent climatic fluctuation," *Geogr. Ann.*, **31**(1-4), 90-105, Stockholm (1949).

70. P. Mikola, "On variations in tree growth and their significance to growth studies," *Commun. Inst. For. Fenn.*, **38**(5), 1-133, Helsingfors (1950).

71. E. Holmsgaard, "Tree-ring analysis of Danish forest trees," *Forstl. Forsøgsv, Dan.*, **22**(1), 1-246 (1955). (In Danish, English summary.)

72. B. Eklund, "Variation in the widths of the annual rings in pine and spruce due to climate conditions in northern Sweden during the years 1900-1944," *Medd. Fr. Statens Skogsforskningsinst.*, **44**(8), 1-150 (1954). (In Swedish, English summary.)

73. B. Eklund, "The annual ring variations in spruce in the centre of northern Sweden and their relation to the climatic conditions," *Medd. Fr. Statens Skogsforskningsinst* **47**(1), 1-63 (1957). (In Swedish, English summary.)

74. B. Jonsson, "Studies of variations in the widths of annual rings in Scots pine and Norway spruce due to weather conditions in Sweden," Department of Forest Yield Research, Royal College of Forestry. Research Notes No. 16, Stockholm, 1969. (In Swedish, English summary.)

75. B. Tveite, *Sur nedbør—Skogproduksjon,* SNSF—prosjektet, TN 11/75, Oslo, 1975.

76. E. Vestjordet, *Acid Precipitation—Forest Yield,* SNSF-prosjektet, IR 12/75, Oslo, 1975. (In Norwegian, English summary.)

77. L. Strand, *Acid Precipitation and Regional Tree Ring Analyses,* SNSF-prosjektet, IR 73/80, Oslo, 1980.

78. M. Näslund, "New material for forest yield research: Pine, spruce and birch," *Stud. For. Suec.*, **89,** Stockholm, (1971). (In Sweden, English summary.)

79. J. C. Kapteyn, "Tree growth and meteorological factors," *Rec. Bot. Néerl.*, **11,** 70-93. (1914).

80. B. Hüber, "Aufbau einer mitteleuropäischen Jahrring-Chronologie," *Mitt. Akad. Dtsch. Forst wiss.*, **1,** 110-125 (1941).

81. B. Becker, "Dendrochronology and Calibration of the Radiocarbon Timescale." In D. Eckstein, S. Wrobel, R. W. Aniol, eds., *Dendrochronology and Archaeology in Europe.* Mitteilungen Bundesforschungsanstalt Forst-u.Holzwirtschaft, Hamburg No. 141, 1983.

82. E. Hollstein, "Mitteleuropäische Eichenchronologie." In *Trierer Grabungen und Forschungen,* Vol. 11, 1980.

83. J. Bauch, D. Eckstein, "Woodbiological investigations on panels of Rembrandt paintings," *Wood Sci. Technol.*, **15,** 251-263 (1981).

84. U. Ruoff, "Zürcher Seeufersiedlungen, Altersbestimmung mit Hilfe der Dendrochronologie," *Helv. Archaeol.*, **12,** 89-97 (1981).

85. B. Becker, "Dendrochronological observations on the postglacial river aggradation in the southern part of Central Europe," *Bull. Geol.* **19,** 127-136 (1975).

86. A. F. M. de Jong, W. G. Mook, and B. Becker, "Confirmation of the Suess wiggles: 3200-3700 B.C.," *Nature,* **280,** 48-49 (1979).

87. B. Hüber, "Die jahresringe der bäume als hilfsmittel der klimatologie und chronologie, *Die Naturwiss,* **35,** 151–154 (1948).

88. O. Fürst, "Vergleichende untersuchungen über räumliche und zeitliche unterschiede interannueller jahrringbreitenschwankungen und ihre klimatologische auswertung," *Flora,* **153,** 469–508 (1963).

89. D. Eckstein and B. Schmidt, "Dendroklimatologische untersuchungen an stieleichen aus dem maritimen klimagebiet schleswig-holsteins, *Angew. Bot.*, **48,** 371–383 (1974).

90. F. H. Schweingruber, H. C. Fritts, O. U. Bräker, L. G. Drew, and E. Schär, "The x-ray technique as applied to dendroclimatology," *Tree-Ring Bull.*, **38,** 61–91 (1978).

91. F. Serre, "The dendroclimatological value of the European larch (Larix decidua Mill.) in the French Maritime Alps," *Tree-Ring Bull.*, **38,** 25–34 (1978).

92. Z. Bednarz, "Tatra Mts. Dendroclimatological Data Base." In M. K. Hughes, P. M. Kelly, J. R. Pilcher, and V. C. LaMarche, eds., *Climate from Tree-Rings,* Cambridge University Press, 1982.

93. D. Eckstein, R. W. Aniol, "Dendroclimatological reconstruction of the summer temperature for an alpine region," *Mitt. Forstl. Bundesvers. Anst. Wien.*, (142)391–398 (1981).

94. J. Guiot, A. Berger, A. Munaut, "Response Function in Dendroclimatology: Comparison of Different Methods and Recommendations." In M. K. Hughes, P. M. Kelly, J. R. Pilcher, and V. C. LaMarche, eds., *Climate from Tree-rings,* Cambridge University Press, 1982.

95. D. Eckstein, W. Liese, B. Schmidt, "Dendroklimatologie und Dendroökologie," *Allgem. Forstzeitschr.*, **34,** 1364–1368 (1979).

96. D. Eckstein and E. Frisse, "Investigations on the Influence of Temperature and Precipitation on Vessel Area and Ring Width of Oak and Beech." In M. K. Hughes, P. M. Kelly, J. R. Pilcher, and V. C. LaMarche, eds., *Climate from Tree-Rings,* Cambridge University Press, 1982.

97. D. Eckstein, "Global Tree-Ring Data Base: Europe." In M. K. Hughes, P. M. Kelly, J. R. Pilcher, and V. C. LaMarche, eds., *Climate from Tree-Rings,* Cambridge University Press, 1982.

98. J. R. Pilcher, "Comment to: Global Tree-Ring Data Base: Europe." In M. K. Hughes, P. M. Kelly, J. R. Pilcher, and V. C. LaMarche, eds., *Climate from Tree-Rings,* Cambridge University Press, 1982.

99. K. D. Mukhamedshin and S. K. Sartbaev, "The Increment Cycle of Juniper in the High Mountain Conditions of the Tien-Shan." In *Investia Adedemia Nanka Kirgiz SSR,* 1972, pp. 55–60.

100. K. D. Mukhamedshin, "Variation in the Increments of Juniper in the High Mountains of the Tien-Shan Over the Last Millenium of the Holocene." In *Sovesheh po Probl. As trofiz Yav lemga i Radiouglerod, Tb ilisi ISSR,* 1974, pp. 149–161.

101. A. A. Molchanov, *Dendrochronological Principles of Long Range Forecasting,* Gid. rot. cometizdat, Moscow, 1976.

102. G. I. Galazii, "The Annual Increment of Trees in Relation to Changes in Climate, Water Level and Relief on the N.W. Shore of Lake Baikal." In *ta Sib. Otd. An SSR,* 1972, pp. 71–214.

103. V. I. Tarankov, "Introduction to Dendroclimatologyin the Far East." In *ta Dal 'nevost novchtsenter AN SSR,* 1973, pp. 7–23.

104. T. T. Bitvinskas, ed., *Dendroclimatochronological Scales of the Soviet Union,* Lab. of Dendroclimatochronology Inst. of Botany, Acad. of Sci. of the Lithuaniun S.S.R., Kaunas, Lithuania, 1979.

105. T. T. Bitvinskas, ed., *Dendroclimatogical Scales of the Soviet Union, Part II,* Lab. of Dendroclimatochronology Inst. of Botany, Acad. of Sci. of the Lithuaniun S.S.R., Kaunas, Lithuania, 1981.

106. Z. Zheng, *Monthly Local Records,* Vol. 8, No. 6, State Met. Admin., Beijing 1935, (In Chinese).

107. S. Deng, *Botanical Bulletin of Academica Senica,* Vol. 2, No. 3, Academica Senica, Beijing, 1948, (in Chinese).

108. S. Han, *Collection of Experiences on Hydrologic Computations,* Water Conservancy Press, Beijing, 1958.

109. Z. Lin and X. Wu, *Proc. Clim. change ultra-long range forecasting,* Science Press, Beijing, China, 1977.

110. Zhang Jiacheng, personal communication, Academy of Meteorological Science, State Meteorological Admin., Beijing, 1982.

111. C. W. Stockton, "Long-Term Streamflow Reconstruction in the Upper Colorado River Basin Using Tree Rings." In C. G. Clyde, D. H. Falkenberg, and J. P. Riley, eds., *Colorado River Basin Modeling Studies,* Utah Water Research Lab., Utah State University, Logan, 1976, pp. 410–441.

112. C. W. Stockton and W. R. Boggess, "Tree Rings: A Proxy Data Source for Hydrologic Forecasting." In R. M. North, L. B. Dworsky, and D. J. Allee, eds., *Proc., Unified River Basin Manage. Symp.,* American Water Resources Association, Minneapolis, Minn., Tech. Publ. No. TPS81-3, 609-624, (1980).

113. L. P. Smith and C. W. Stockton, "Reconstructed streamflow for the Salt and Verde Rivers from tree-ring data," *Water Resour. Bull.,* **17**(6), 939–947 (1981).

114. D. M. Meko, C. W. Stockton, and W. R. Boggess, "A tree-ring reconstruction of drought in southern California," *Water Resour. Bull.,* **16**(4), 544–600 (1980).

115. W. C. Palmer, "*Meteorological drought,*" U.S. Weather Bureau Research Paper 45, U.S. Department of Commerce, (1965).

116. P. Williams, Jr., "The variation of the time of maximum precipitation along the west coast of North America," *Bull. Am. Meteorol. Soc.,* **29**(4), 143–145 (1948).

117. C. B. Pyke, "Some meteorological aspects of the seasonal distribution of precipitation in the western United States and Baja, California," Contribution No. 139, Water Resources Center, University of California, Los Angeles, 1972.

118. O. E. Granger, "Secular fluctuations of seasonal precipitation in lowland California," *Mon. Weather Rev.,* **105**(4), 386–397 (1977).

119. H. C. Fritts, G. R. Lofgren, and G. A. Gordon, "Variations in climate since 1602 as reconstructed from tree rings," *Quat. Res.,* **12**(1), 18–46 (1979).

120. H. C. Fritts and G. A. Gordon, "Annual Precipitation for California, U.S.A., Since 1600 Reconstructed from Western North American Tree Rings." In M. K. Hughes, et al., eds., *Climate from Tree Rings,* Cambridge University Press, London, 1982.

121. H. B. Lynch, *Rainfall and Stream Run-off in Southern California since 1769,* Metropolitan Water District of Southern California, Los Angeles, 1931.

122. S. T. Harding, "Recent Variations in the Water Supply of the Western Great Basin," Water Resources Center Archives. Archive Series Report No. 16. University of California, 1965.

123. E. R. Cook and G. C. Jacoby, "Tree-ring-drought relationship in the Hudson Valley, New York," *Science,* **198**, 399–401 (1977).

124. E. R. Cook and G. C. Jacoby, "Evidence for quasi-periodic July drought in the Hudson Valley, New York," *Nature,* **282**(5737), 390–392 (1979).

125. D. A. Campbell, "The feasibility of Using Tree-Ring Chronologies to Augment Hydrologic Records in Tasmania, Australia," M.S. thesis, University of Arizona, Tucson, 1980.

126. D. A. Campbell, "Preliminary estimates of summer streamflow for Tasmania." In M. K. Hughes, P. M. Kelly, J. R. Pilcher, and V. C. LaMarche, eds., *Climate from Tree Rings,* Cambridge University Press, London, 1982, pp. 170–177.

127. V. C. LaMarche, Jr. and A. B. Pittock, "Preliminary temperature reconstructions for Tasmania." In M. K. Hughes, P. M. Kelly, J. R. Pilcher, and V. C. LaMarche, eds., *Climate from Tree Rings,* Cambridge University Press, London, 1982, pp. 177–185.

128. R. L. Holmes, C. W. Stockton, and V. C. LaMarche, Jr., "Extension of river-flow records in Argentina. In M. K. Hughes, P. M. Kelly, J. R. Pilcher, V. C. LaMarche, eds., *Climate from Tree Rings,* Cambridge University Press, London, 1982, pp. 168–170.

129. R. L. Holmes, C. W. Stockton, and V. C. LaMarche, Jr., "Extension of river flow records in Argentina from long tree-ring chronologies," *Water Resour. Bull.* **15**(4), 1081–1085 (1979).

130. W. C. Palmer, "Climatic variability and crop production," *Proc. Semin. Weather Food Supply,* Center for Agricultural and Economic Development, Iowa State University, Ames, Iowa. CAED Report 20: 173–187, 1964.

131. H. C. Willett, "Solar-climatic relationships in the light of standardized climatic data," *J. Atmos. Sci.*, **22,** 120 (1965).

132. W. O. Roberts, "Relation between solar activity and climate change," Goddard Space Flight Center. Special Report SP-366, NASA. W. R. Baudeen and S. O. Maran, eds., p. 13, 1975.

133. L. M. Thompson, "Cyclic weather patterns in the middle latitudes," *J. Soil Water Conserv.*, **28,** 87–89 (1973).

134. C. W. Stockton and D. M. Meko, "A long-term history of drought occurrence in western United States as inferred from tree rings," *Weatherwise,* **28**(6), 244–249 (1975).

135. J. M. Mitchell, Jr., C. W. Stockton, and D. M. Meko, "Evidence of a 22-Year Rhythm of Drought in the Western United States Related to the Hale Solar Cycle Since the 17th Century." In B. M. McCormac and T. A. Seliga, eds., *Solar-Terrestrial Influence on Weather and Climate,* D. Reidel, Dordrecht, 1979, pp. 125–143.

136. C. W. Stockton, J. M. Mitchell, Jr. and D. M. Meko, "Tree-ring evidence of a relationship between drought occurrence in the western United States and the Hale sunspot cycle," *Conf. Proc. Great Plains: Perspect. Prospects,* University of Nebraska Press, Lincoln, 1981.

137. H. C. Fritts and D. J. Shatz, "Selecting and characterizing tree-ring chronologies for dendroclimatic analysis," *Tree-Ring Bull.* **35,** 31–40 (1975).

138. T. J. Blasing, "Time Series and Multivariate Analysis in Paleoclimatology." In H. H. Shugart, Jr., ed., *Time Series and Ecological Processes,* Society for Industrial and Applied Mathematics, Philadelphia, 1978, pp. 213–228.

139. G. A. Gordon, "Verification of Dendroclimatic Reconstructions." In M. K. Hughes, P. M. Kelly, J. Pilcher, and V. C. LaMarche, eds., *Climate from Tree Rings,* Cambridge University Press, London, 1982.

140. J. M. Bates and C. W. J. Granger, "The combination of forecasts," *Oper. Res. Q.*, **20**(4), 451–468 (1969).

141. H. C. Fritts, "An Overview of Dendroclimatic Techniques, Procedures, and Prespects." In M. K. Hughes, P. M. Kelly, J. Pilcher, and V. C. LaMarche, eds., *Climate from Tree Rings,* Cambridge University Press, London, 1982.

142. L. G. Drew, *Tree-Ring Chronologies of Western North America II: Arizona, New Mexico, Texas: Chronologies Series 1,* Laboratory of Tree-Ring Research, University of Arizona, 1972.

143. M. L. Parker and W. E. S. Henoch, "The use of engelmann spruce latewood density for dendrochronological purposes," *Can. J. For. Res.*, **1**(2), 90–98 (1971).

144. L. G. Drew, *Tree-Ring Chronologies of Western America VI: Western Canada and Mexico: Chronology Series 1,* Laboratory of Tree-Ring Research, University of Arizona, 1975.

145. M. K. Hughes, P. M. Kelly, J. Pilcher, and V. C. LaMarche, eds., *Climate from Tree Rings,* Cambridge University Press, London, 1982.

146. H. C. Fritts, personal communication, Laboratory of Tree-Ring Research, University of Arizona, Tucson, 1978.

147. P. A. Kay, "Dendroecology in Canada's forest-tundra transition zone," *Arctic Alp. Res.*, **10**(1), 133–138 (1978).

148. M. L. Parker, personal communication, Western Forest Products Laboratory, Vancouver, B.C., Canada, 1981.

149. L. B. Brubaker, personal communication, College of Forest Resources, University of Washington, Seattle, 1978.

150. M. L. Parker, L. A. Josa, S. G. Johnson and P. A. Bramhall, "Dendrochronological Studies of the Coasts of James Bay and Hudson Bay (Parts I and II)." In C. R. Harrington, ed., *Syllogeus: Climatic Change in Canada No. 2,* National Museums of Canada, No. 33, Ottawa, 1981, pp. 129–188.

APPENDIX: PALEOCLIMATIC STUDIES USING ISOTOPES IN TREES

M. Stuiver and R. L. Burk, Quaternary Isotope Laboratory and Department of Geological Sciences, University of Washington, Seattle, Washington

Tree-ring isotope chemistry is a new area of research that has already significantly contributed to paleoclimatic studies. The "tools" for these investigations are the isotope ratios derived from the carbon (^{12}C, ^{13}C, ^{14}C), oxygen (^{16}O, ^{17}O, ^{18}O), and hydrogen (^{1}H, ^{2}H or D) incorporated in cellulose. Direct correlations have been made between stable isotope ratios and climatic data whereas ^{14}C research of tree-ring cellulose has provided new information on solar variability. Because of the widespread distribution of trees in mid-latitudes these methods are particularly attractive and may prove to be crucial to global paleoclimatic reconstructions.

$^{13}C/^{12}C$ and Climate

The link between $^{13}C/^{12}C$ ratios in tree-rings and temperature is uncertain and there appears to be conflicting evidence even on the sign of the temperature coefficient. Urey[1] first suggested that the ^{13}C content of plant material may be related to temperature. In addition to such a direct photosynthetically controlled change in the trees' isotope ratios the changes in atmospheric carbon isotopic ratios are also reflected in the cellulose isotopic composition. The $^{13}C/^{12}C$ ratio of atmospheric CO_2, in turn, depends on a temperature dependent exchange reaction with the oceans.[2]

There are various estimates of the $^{13}C/^{12}C$ temperature coefficient of cellulose. Theoretical calculations by Libby[3] show a predicted 0.36‰ C^{-1}, whereas Troughton and Card[4] using controlled environment chambers determined a temperature coefficient for C−3 plants of −0.0135‰ C^{-1}.

Using whole wood analyses Pearman et al.[5] show temperature correlations significant at the 1% level of 0.35 ± 0.11‰ and 0.48 ± 0.12‰ C^{-1} for the two different King Billy Pines (*Anthrotaxis selaginoides* D. Don) which grew within 200 meters of each other in northern Tasmania.

Wilson and Grinsted[6] derive a temperature coefficient for both cellulose and lignin of 0.3‰ C^{-1} for *Pinus radiata* from New Zealand and show annual variations in $\delta^{13}C$* values. Keeling[7] has measured a Northern Hemispheric annual fluctuation of 0.3‰ in atmospheric CO_2 with the atmosphere heavier in autumn; this effect according to Keeling is due to the respiration of land plants. Lerman and Long[8] have also suggested the importance of surrounding plants in determining the fluctuations of $^{13}C/^{12}C$ observed in tree rings. However, the variations shown by Wilson and Grinsted,[6] are out of phase with those expected from changes in the activity of terrestrial plants and hence are interpreted directly in terms of temperature. *Pinus radiata* lays down wood year-round so that full seasonal effects should be present.

Seasonal variations have been described for $^{13}C/^{12}C$ and $^{18}O/^{16}O$ ratios.[6,9] This is in apparent contradiction to work using tree-ring widths which demonstrate a lag in the growth response of a tree with respect to the occurrence of climate.[10]

Farmer[11] gives a $-0.7‰\ C^{-1}$ coefficient using whole wood analyses. This factor has an opposite sign from the factors given by Wilson and Grinsted, and Pearman et al.[5,6] Grinsted et al.[12] conclude that other factors such as precipitation and humidity changes are also important. Freyer[13] summarizes the various conflicting results regarding a temperature coefficient for ^{13}C. Tans[14,15] suggests that Oak $\delta^{13}C$ values can be used as a paleo-rain gauge and a paleothermometer; however, both climatic variables combined account for only 50 percent of the $\delta^{13}C$ variance and it is also not possible to distinguish between the two types of information without help from other methods.

Factors such as temperature, precipitation and humidity may be integrated into the correlations noted by Mazany.[16] They find a correlation between the regional tree-ring width indices and $\delta^{13}C$ values from wood at an archaeological site in Chaco Canyon, New Mexico. Mazany et al.[16] ascribe the relationship to a connection between vegetation activity and the $\delta^{13}C$ in the local atmospheric CO_2 and consequently more negative $\delta^{13}C$ in plant cellulose.

Farquhar[17] suggests that carbon isotope discrimination is a function of the ratio of intercellular and atmospheric CO_2 concentrations. Temperature, irradiance, and nutrition are among the factors which influence intercellular CO_2 concentrations. No uncomplicated relationship between $\delta^{13}C$ in tree rings and climate seems likely based on this work.

Bender and Berge[18] showed that $^{13}C/^{12}C$ ratios in timothy grass vary with both temperature and nutrient level and that the highest isotope discrimination occurs at optimum growth conditions. This result agrees with the equations presented by Farquhar.[17]

Because the global carbon reservoirs (ocean, atmosphere, and biosphere) differ in isotope abundance ratios, any redistribution of carbon between these reservoirs results in a perturbation of the reservoir $^{13}C/^{12}C$ isotope ratios. A net transfer of isotopically light biospheric carbon to the atmosphere and oceans results in a $^{13}C/$

*Delta values give the relative deviation of the sample isotope ratio relative to a standard isotope ratio, that is:

$$\delta^{13}C = \left[\frac{^{13}C/^{12}C\ \text{sample} - {}^{13}C/^{12}C\ \text{std}}{^{13}C/^{12}C\ \text{std}}\right] \times 1000‰$$

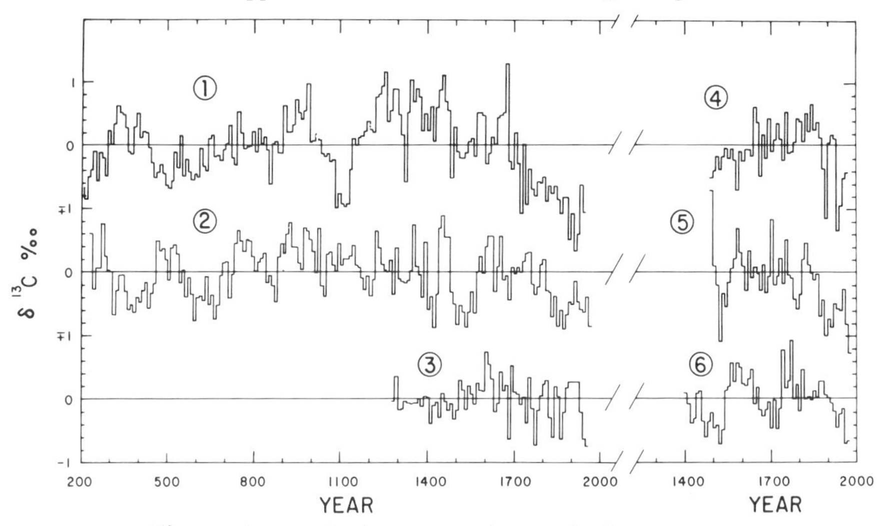

Figure 1. The $\delta^{13}C$ records, normalized on constant ring area for those trees where a relationship between ring area and $\delta^{13}C$ is apparent, of six trees: (1) Inyo Bristlecone Pine, (2) National Park Sequoia, (3) Whittaker Sequoia, (4) Coos Bay Fir, (5) Bjorka Spruce, and (6) Valdivia Alerce. Specific details on collection sites are found in reference 19.

^{12}C lowering of the latter reservoirs. The record of $^{13}C/^{12}C$ time variations of the atmosphere, as derived from tree-ring records (figure 1) can be used to study natural and anthropogenic changes in global carbon reservoirs. Model calculated pre-industrial CO_2 levels of 276 ± 16 ppm were derived from the figure 1 data by Stuiver et al.[19]

Analytical Methods —$^{13}C/^{12}C$

Analysis of tree rings for $^{13}C/^{12}C$ ratios is accomplished by oxidation of carbon to carbon dioxide, which is subsequently run on a mass spectrometer. Some authors[5,11] used whole wood analyses; however, the majority use a cellulose fraction for combustion because a portion of the resin component in whole wood may be deposited after the ring is formed.

There are three types of combustions systems: (1) a flow-through system where the sample is oxidized in a stream of oxygen gas that passes through a combustion oven[14,20], (2) a circulation system similar to flow-through except that a Toepler pump is used to circulate the sample gas through the oxidation furnace[21–23], and (3) a closed combustion method where organic matter is oxidized in a Vycor tube using cupric oxide.[19,24]

O^{18}/O^{16} Studies

Much of the interest in $^{18}O/^{16}O$ in tree rings has been generated by published correlations with temperature.[25–30] However, results have been conflicting and it is now apparent that the method is more complex than originally assumed. There are also conflicting assessments of the mechanisms of isotopic fractionation in plants.[31–33]

Libby and Pandolfi[25,26] showed significant correlations between whole wood $\delta^{18}O$ values and temperature. For *Quercus petraea* Libby et al. quoted a temperature fractionation factor of approximately 18‰ per degree Celsius mean annual temperature, whereas a *Cryptomeria japonica* had a factor of approximately 1‰ C^{-1}. Epstein and Yapp[32] pointed out that these variations in $^{18}O/^{16}O$ ratios and also published deuterium/hydrogen (D/H) ratios are much larger than would be expected from variations in meteoric water which Libby et al.[26] assumed was the major factor controlling the isotopic compositions of wood. Libby et al.[26] graphed D/H versus $^{18}O/^{16}O$ in whole wood and showed a slope of 8 which suggests that these isotopes are incorporated without any of the evapotranspiration effects observed in references 30 and 32.

Gray and Thompson[28] showed a correlation coefficient of 0.97 for $\delta^{18}O$ in cellulose versus September to August mean annual temperature for a white spruce (*Picea glauca*) grown in Edmonton, Alberta, Canada. Their temperature coefficient is 1.3 $\pm$ 0.1‰ $C.^{-1}$. More recent work at other localities does not show as high a correlation and suggests that meteoric water is dominant in determining the $\delta^{18}O$ value of cellulose. Over a range of latitudes from the Yukon to California for non-coastal sites, the temperature coefficient was 0.54‰ C^{-1}.[34]

Epstein and Yapp[32] demonstrated that $\delta^{18}O$ in cellulose relates to humidity and the isotopic composition of meteoric water and questioned the basis for the correlation between temperature and $\delta^{18}O$ observed in the Edmonton spruce. Perry[35] questioned the influence of winter net photosynthesis on oxygen isotope composition as suggested by Gray and Thompson but did expect temperature dependent isotopic fractionation between starch that may be hydrolyzed to glucose during the winter. Gray and Thompson[29] showed that $\delta^{18}O$ in whole wood is less sensitive to mean annual temperature variations than $\delta^{18}O$ in cellulose.

Wilson and Grinsted [31] showed a seasonal variation in $\delta^{18}O$ within individual rings of *Pinus radiata* grown in New Zealand. They recognized the effects of evapotranspiration and the possible isotopic equilibration between CO_2 and H_2O in the leaf. Part of the variation in $\delta^{18}O$ is tentatively attributed to the temperature coefficient for the CO_2-H_2O isotope equilibrium. Temperature is not an important factor in determining the oxygen isotope composition of the bean plant tissue (*Phaseolus vulgaris*) according to experiments using controlled environment chambers.[36,37] Humidity and the isotopic composition of the irrigation water appeared to be the major factors which influence the $\delta^{18}O$ values of these plants.Long et al.[38] suggests this combined humidity and irrigation water effect as an explanation of the variations found in samples collected at different latitudes. Burk and Stuiver[30] show that the latitude and altitude effects for precipitation $\delta^{18}O$ values demonstrated by Dansgaard[39,40] can also be delineated for $\delta^{18}O$ in tree-ring cellulose at selected locations in North America. $\delta^{18}O$ in precipitation over a range of latitudes correlates with mean annual temperature and $\delta^{18}O$ in tree-ring cellulose correlates with $\delta^{18}O$ in precipitation and hence temperature. This temperature information is available only if $\delta^{18}O$ in precipitation correlates with mean annual temperature, and humidity is fairly constant. For Pacific coastal sites, Burk and Stuiver[30] calculate a change in the $\delta^{18}O$ value of cellulose of 0.41‰ C^{-1}. When this same value is applied to Mt. Rainier, Washington, a lapse rate of 5.2° ±0.5°C per 100 meters is calculated, which compares with the accepted mean annual lapse rate of 5°C per 1000 meters elevation.

Epstein et al.[32] developed two models for the fixation of CO_2 based on measurements on deuterium and oxygen isotopes in cellulose, water distilled from plants, and environmental water. One model (A) assumed isotopic equilibration of CO_2 with plant water and their other model (B) assumes no isotopic equilibration of CO_2 with plant water. Significantly, both models assumed that during fixation of CO_2, one molecule of CO_2 is fixed along with one molecule of H_2O, which was conceptually written as

$$\text{ribulose 1,5-diphosphate} + CO_2 + H_2O \rightarrow 2(3 - \text{phosphoglocerate})$$

thus two-thirds of the oxygen in cellulose comes from CO_2 and one-third from water. This assumption contradicts the classic model of photosynthesis. Epstein et al. also favor model B, which contradicts work quoted by Carrier[41] and certain assumptions made by Wilson and Grinsted,[6] Fehri and Letolle[36] and Long et al.[38] In a later paper, DeNiro and Epstein[33] showed that for wheat plants, the oxygen in carbon dioxide completely exchanged with the oxygen of plant water during synthesis of cellulose. On the basis of these later experiments, model A was favored. A more detailed study of biochemistry involved in photosynthesis has led Epstein and DeNiro (personal communication) to view the oxygen in CO_2 as virtually the sole source of oxygen in cellulose. The two-thirds CO_2 and one-third H_2O proportions are now viewed simply as an empirical relationship.

Fehri et al.[42] and Fehri and Letolle[36,37] suggest that 62 percent of the oxygen in cellulose comes from CO_2 and 38 percent from leaf water. These results do not appear to be in agreement with others[9,30,33]. DeNiro and Epstein demonstrate equilibration of plant water with CO_2. Based on this work and experiments by Burk[9,30], Burk and Stuiver (1981) assume that CO_2 equilibrates with leaf water and can thereby use equations for the $\delta^{18}O$ values of leaf water[43] along with the equation for CO_2-H_2O equilibration to develop an equation for calculating cellulose values.

Using only $\delta^{18}O$ values it does not appear that precise temperature information can be isolated without making assumptions about humidity and soil water $\delta^{18}O$. Apparently D/H values are only mildly affected by changes in humidity[44] and if this is the case it may be possible to use both to separate out the temperature signal.

Analytical Methods - $^{18}O/^{16}O$

There are three principal methods used to oxidize the carbon in wood or cellulose to carbon dioxide. One depends on oxidation by $HgCl_2$ and the other two on pyrolysis.

The mercuric chloride method was introduced by Rittenburg and Ponticorvo[45] and was used in modified form.[9,21,26,46,47] Reported analytical precision for this method ranges from 0.2 to 0.3‰. Gray and Thompson[29] and Epstein et al.[32] use the nickel pyrolysis method. A sample is placed in an evacuated closed nickel reaction vessel and heated to 1100°C. At that temperature, hydrogen gas diffuses through the walls of the reaction vessel and the oxygen from cellulose is converted to CO or CO_2. The CO is converted to CO_2 using high voltage discharge. As with the mercuric chloride method, the reported analytical precision is 0.2 to 0.3‰.

Hardcastle and Friedman[18] and Ferhi et al.[49] pyrolize organic matter at 1250°C using diamonds as a reducing agent. The reported analytical precision of this method is ± 0.35‰.

D/H and climate

Schiegl[50] showed a correlation between D/H ratios in tree rings and temperature. The correlation was assumed to be based on the correlation between D/H in precipitation and temperature. Schiegl notes the importance of leaf water and various factors such as humidity and leaf boundary layer conditions which may affect the isotopic composition of leaf water.

Epstein et al.[47] suggest that evapotranspiration affects D/H values in leaf water but does not find any connection with changes in humidity. Wilson and Grinsted[51] showed a direct correlation between temperature and D/H ratios in *Pinus radiata*. Because the expected fractionation effects of the evapotranspiration would be in the opposite direction from their results, Wilson and Grinsted conclude that the changes are due to temperature alone affecting one or more of the biochemical reactions that lead to the fixation of cellulose.

Epstein and Yapp[52] analyzed D/H values for a Scots Pine from Loch Affric, Scotland, and a Bristlecone Pine from the White Mountains of California. They found a good correlation between the two records and interpreted this correlation as reflecting a sensitivity to large scale climatic trends. For the Bristlecone Pine they note a periodicity of 22 years which coincides with the occurrence of droughts in the great plains of North America.

Yapp and Epstein[53] applied some of their earlier work by analyzing ancient wood samples (9500-22000 years B.P.) that had been dated by carbon–14 dating techniques. On the basis of D/H in meteoric water inferred from D/H ratios in cellulose nitrate from wood, they suggest that the average temperature for ice-free areas of North America was about the same as at present although the summers were probably cooler to maintain the ice and winters warmer to allow greater snowfall.

Studies conducted on wheat and maize plants in a controlled environment chamber give a temperature fractionation factor of $-1.39 \pm 0.35‰\ C^{-1}$ (wheat) and $-1.45 \pm 0.72‰\ C^{-1}$(maize).[54] This experimental work contradicts Libby's[96] theoretical estimate of $+2.0‰\ C^{-1}$. Schiegl[50] suggests that humidity is an important factor in determining leaf water D/H isotopic composition yet White[44] and Epstein et al.[47] do not find this relationship. If both D/H and $\delta^{18}O$ are subject to humidity effects, then it will not be possible to resolve temperature information using both isotopes together.

White (personal communication) finds linear relationship between the D/H ratio of atmospheric water vapor and relative humidity. When this relationship is combined with the leaf water composition model used by Schiegl,[50] D/H values in leaf water do not show much of a humidity effect whereas $\delta^{18}O$ still has a considerable variation with humidity (see $^{18}O/^{16}O$ section).

It has been assumed that D/H in cellulose nitrate prepared from wood reflects the D/H value of precipitation[47] and any temperature connection was due to temperature fractionation of precipitation as described by Dansgaard.[40] However, DeNiro and Epstein[33] and Wilson and Grinsted[51] suggest that there is also a temperature dependent biochemical fractionation factor in the various steps leading to the formation of cellulose.

White (personal communication) shows a factor of -3.0 to $-3.5‰\ C^{-1}$ based on some preliminary controlled environment work at the Duke University Phytotron.

D/H Analytical Methods

The appropriate method for determination of D/H values in cellulose has been a subject of considerable debate. The O-H hydrogens in cellulose are exchangeable and hence may not have the same isotope ratio as the cellulose originally synthesized by the tree. One approach to this problem has been to exchange these exchangeable hydrogens with hydrogens from water of known D/H composition.[21] Another approach is to replace the exchangeable OH hydrogens with nitrates during the formation of cellulose nitrate[55] from ordinary cellulose.[56] The cellulose nitrate is then combusted in excess O_2 to produce H_2O and CO_2. The CO_2 can be run for ^{13}C analysis (White, personal communication) and the water is passed over hot uranium metal and converted to hydrogen to be run on a mass spectrometer. Certain methods of preparing cellulose nitrate may have caused significant errors in published D/H ratios.[57]

Thompson et al.[58] report progress on a modification of the nickel pyrolysis method which allows determination of both $^{18}O/^{16}O$ and D/H simultaneously. Friedman and Gleason[59] present evidence that exchanged extracted wood, exchanged cellulose, and cellulose nitrate can all be utilized to derive the D/H value of water assimilated by the tree.

^{14}C and Climate

Neutrons, produced by cosmic radiation in the upper atmosphere, are responsible for the global ^{14}C production. The ^{14}C, after oxidation to $^{14}CO_2$, mixes with atmospheric CO_2 and labels this compound with a $^{14}C/^{12}C$ ratio of approximately 10^{-12}.

The $^{14}C/^{12}C$ history of atmospheric carbon relates to (1) the variable ^{14}C production rate in the atmosphere and (2) the changes in the size of, and exchange rates between, the various terrestrial carbon reservoirs. The ^{14}C production rate in the upper atmosphere is influenced by long-term ($\sim 10^3$ yr) geomagnetic field intensity changes, and by shorter-term (~ 10–10^2 yr) variations in magnetic properties of the solar wind.

Climatic change may cause variations in the size, and exchange rate, of the various carbon reservoirs. The influence of such climate-related changes is most obvious on atmospheric CO_2 content. Holocene levels are in the 260 to 270 ppm range, whereas during the last glacial atmospheric CO_2 levels may have been as low as 200 ppm.[60]

The atmospheric ^{14}C changes accompanying these changes have not yet been determined because tree-ring ^{14}C records are not available for the most recent glacial-interglacial transition.

Variations in the $^{14}C/^{12}C$ ratio are derived from tree rings by measuring the

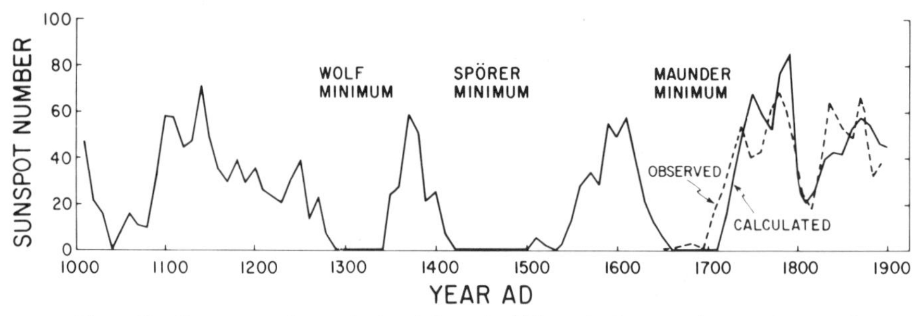

Figure 2. Sunspot numbers calculated from the ^{14}C production record, see reference 61.

current ^{14}C activity and correcting for decay since the time of formation. Because changes in solar wind properties influence atmospheric ^{14}C levels, the record of ^{14}C change can be used to gain information on solar variability.[61] For this type of study, the influence of the long-term geomagnetic field change is first removed from the ^{14}C record, and the remaining variability is attributed to solar change (which is expressed in yearly numbers of sunspots). The historically observed sunspot number record, and the record calculated from the ^{14}C data, agree quite well (figure 2).

The detailed history of solar change obtained from the atmospheric ^{14}C variations can be used to evaluate claims of a sun-weather relationship. Stuiver[62] compared the record of solar change with (1) oxygen isotope records from Devon Island and Greenland (Camp Century and Crête), (2) tree-ring width records of Lapland and Nevada, (3) tree-ring density records of Switzerland, and (4) historical winter severity indexes of England and Russia. The conclusion of this study is that a statistically significant relationship between combined regional climate and ^{14}C time series is absent for the current millennium.

Because many factors influence local climatic change, a solar influence on climate can perhaps be expected in a few areas only. Of the above records, only the Camp Century and Russian climate records possibly demonstrate a climate-solar relationship. However, the basic properties of the Russian climate record and the record of ^{14}C change differ substantially[63] and the statistically significant correlation found for this record may be only coincidence.

ACKNOWLEDGMENTS

The isotope studies at the Quaternary Isotope Laboratory are supported by NSF Grants ATM-8318665 and EAR-8115994, and DOE contract 19X-43303-C.

REFERENCES

1. H. C. Urey, "The thermodynamic properties of isotopic substances," *J. Chem. Soc.*, 562–581 (1947).
2. H. Craig, "Carbon-13 in plants and the relationships between carbon 13 and carbon 14 variations in nature," *J. Geology*, **62,** 115–149 (1954).

3. L. M. Libby, "Multiple thermometry in paleoclimate and historic climate," *J. Geophys. Res.*, **77,** 4310–4317 (1972).
4. J. D. Troughton and K. A. Card, "Temperature effects on the carbon-isotope ratio of C_3, C_4 and Crassulacean-acid-metabolism," *(CAM) Planta*, **123,** 185–190 (1975).
5. G. I. Pearman, R. J. Francey, and P. J. B. Fraser, "Climatic implications of stable carbon isotopes in tree rings," *Nature*, **260,** 771–773 (1976).
6. A. T. Wilson and M. J. Grinsted, "$^{12}C/^{13}C$ in cellulose and lignin as palaeothermometers," *Nature*, **265,** 133–135 (1977).
7. C. D. Keeling, "The concentration and isotopic abundance of carbon dioxide in the atmosphere," *Tellus*, **12,** 200–203 (1960).
8. J. C. Lerman and A. Long, "Carbon-13 in tree rings: local or canopy effect?" *Proc. Inter. Meet. Stable Isot. Tree-Ring Res., New Paltz, New York, May 22–25, 1979*, U.S. Department of Energy, Washington, D.C., 22–34 (1980).
9. R. L. Burk, "Factors affecting $^{18}O/^{16}O$ ratios in cellulose," Ph.D. diss., University of Washington, Seattle, 1979.
10. H. C. Fritts, *Tree Rings and Climate*. Academic, New York, 1976.
11. J. G Farmer, "Problems in interpreting tree-ring ^{13}C records," *Nature*, **279,** 229–231 (1979).
12. M. J. Grinsted, A. T. Wilson, and C. W. Ferguson, "$^{13}C/^{12}C$ ratio variations in *Pinus Longaeva* (Bristlecone Pine) cellulose during the last millennium," *Earth Plan. Sci. Lett.*, **42,** 251–253 (1979).
13. H. D. Freyer, "Record of environmental variables by ^{13}C measurements in tree-rings," *Proc. Inter. Meet. Stable Isot. Tree-Ring Res., New Paltz, New York, May 22–25, 1979*, U.S. Department of Energy, Washington, D.C., 13–21 (1980).
14. P. O. Tans, "Carbon 13 and carbon 14 in trees and the atmospheric CO_2 increase," Ph.D. diss. University of Groningen, Holland, 1978.
15. P. O. Tans, "Past atmospheric CO_2 levels and the $^{13}C/^{12}C$ ratios in tree rings," *Tellus*, **32,** 268–283 (1980).
16. T. Mazany, J. C. Lerman, and A. Long, "Carbon-13 in tree-ring cellulose as an indicator of past climates," *Nature*, **287,** 432–435 (1980).
17. G. D. Farquhar, "Carbon Isotope Discrimination by Plants: Effects of Carbon Dioxide Concentration and Temperature via the Ratio of Intercellular and Atmospheric CO_2 Concentrations." In G. I. Pearman, ed., In *Carbon Dioxide and Climate: Australian Research*, Australian Academy of Science, Canberra, (1980), pp. 105–110.
18. M. M. Bender, and A. J. Berge, "Influence of N and K fertilization and growth temperature on $^{13}C/^{12}C$ ratios of timothy (*Phleum pratense* L.)," *Oecologia*, **44,** 117–118 (1979).
19. M. Stuiver, R. L. Burk and P. D. Quay, "$^{13}C/^{12}C$ ratios and the transfer of biospheric carbon to the atmosphere," *Journ. Geophys. Res.*, in press, (1984).
20. W. G. Mook, "Geochemistry of the stable carbon and oxygen isotopes of natural waters in the Netherlands," Ph.D. diss. University of Groningen, Holland, 1968.
21. M. J. Grinsted, "A study of the relationships between climate and stable isotope ratios in tree rings," Ph.D. diss. University of Waikato, Hamilton, New Zealand, 1977.
22. J. M. Hayes, D. J. DesMarais, D. W. Peterson, D. A. Schaeller, and S. P. Taylor, "High precision stable isotope ratios from microgram samples," *Adv. Mass Spectrom. 7A, Proc. 7th Int. Conf.*, Florence Heyden, London, 475–480 (1976).
23. H. D. Freyer, "On the ^{13}C record in tree rings: Part I. ^{13}C variations in northern hemispheric trees during the last 150 years," *Tellus*, **32,** 124–137 (1979).
24. D. L. Buchanan and B. J. Corcoran, "Sealed tube combustions for the determination of carbon-14 and total carbon," *Analytical Chemistry*, **31,** 1635–1638 (1959).
25. L. M. Libby and L. J. Pandolfi, "Calibration of isotope thermometers in an oak tree using official weather records," *Proc. Colloq. Int. C.N.R.S.*, **219,** 299–310 (1973).

26. L. M. Libby, L. J. Pandolfi, P. H. Payton, J. Marshall III, B. Becker, and V. Giertz-Sienbenlist, "Isotope tree thermometers," *Nature*, **261,** 284–288 (1976).
27. J. Gray and P. Thompson, "Climatic significance of the oxygen isotopic composition of cellulose from tree rings," *Adv. Mass Spectrom. 7A, Proc. 7th Int. Conf.*, Florence Heyden, London, 509–513 (1976a).
28. J. Gray and P. Thompson, "Climatic information from $^{18}O/^{16}O$ ratios of cellulose in tree rings," *Nature*, **262,** 481–482 (1967a).
29. J. Gray and P. Thompson, "Climatic information from $^{18}O/^{16}O$ analysis of cellulose, lignin and whole wood from tree rings," *Nature*, **270,** 708–709 (1977).
30. R. L. Burk and M. Stuiver, "Oxygen isotope ratios in trees reflect mean annual temperature and humidity," *Science*, **211,** 1417–1419 (1981).
31. A. T. Wilson and M. J. Grinsted, "The Possibilities of Deriving Past Climate Information from Stable Isotope Studies on Tree Rings." In B. W. Robinson, ed., *Stable Isotopes in the Earth Sciences*, New Zealand Dep. Science Industrial Research Bull. *220*: 61–66 (International stable isotope conf).
32. S. Epstein and C. Yapp, "Isotope tree thermometers," *Science*, **266,** 477–478 (1977).
33. M. DeNiro and S. Epstein, "$^{18}O/^{16}O$ ratios of terrestrial plant cellulose are independent of the $^{18}O/^{16}O$ ratio of atmospheric carbon dioxide," *Science*, **204,** 51–54 (1979).
34. J. Gray and P. Thompson, "Natural variations in the ^{18}O content of cellulose," *Proc. Int. Meet. Stable Isot. Tree-Ring Res., New Paltz, New York, May 22–25, 1979*, U.S. Department of Energy, Washington, D.C., (1980).
35. D. A. Perry, "Oxygen isotope ratios in spruce cellulose," *Science*, **266,** 476–477 (1977).
36. A. Fehri and R. Letolle, "Transpiration and evaporation as the principal factors in oxygen isotope variations of organic matter in land plants," *Physiol. Veg.*, **15,** 363–370 (1977).
37. A. Fehri and R. Letolle, "Relation entre de milieu climatique et les teneurs en oxygene-18 de la cellulose des plantes terrestres." *Physiol. Veg.*, **17,** 107–117 (1979).
38. A. Long, J. C. Lerman, and A. Fehri, "Oxygen-18 in tree rings, paleothermometers or paleohygrometers?," U.S. Geological Survey. Open-file Report 78-701, 253 (1978).
39. W. Dansgaard, "The abundance of ^{18}O in atmospheric water and water vapor," *Tellus*, **5,** 461–469 (1953).
40. W. Dansgaard, "Stable isotopes in precipitation," *Tellus*, **16,** 436–468 (1964).
41. I. M. Carrier, "Isotopic composition of the Phosphoglyceric acid carboxl group in *Euglena gracilis* after steady-state photosynthesis with $^{13}C^{18}O_2$," *2nd Int. Congr. Photosynth., Stress, 1971*, **3,** 1963–1970 (1971).
42. A. Fehri, R. Letolle, A. Long, and J. C. Lerman, "Factors controlling the variations of oxygen-18 in plant cellulose," *Proc. Int. Meet. Stable Isot. Tree-Ring Res., New Paltz, New York, May 22–25, 1979*, U.S. Department of Energy, Washington, D.C. 71–83, (1980).
43. F. Farris and B. Strain, "The effects of water-stress on leaf H_2O enrichment," *Radiat. Environ. Biophys.*, **15,** 167–202 (1978).
44. J. W. C. White, "The relationship between the non-exchangeable hydrogens of tree-ring cellulose and the source waters for tree sap," *Proc. Int. Meet. Stable Isot. Tree-Ring Res., New Paltz, New York, May 22–25, 1979*, U. S. Department of Energy, Washington, D.C. 58–65 (1980).
45. D. Rittenburg, and L. Pontecorvo, "A method for determination of the 18-0 concentration of the oxygen of organic compounds," *Int. J. Appl. Rad. Isot.*, **1,** 208–214 (1956).
46. A. Long, T. Mazany, C. Steelink, and J. C. Lerman, "Oxygen isotope exchange in cellulose during isolation from wood," manuscript (1977).
47. S. Epstein, P. Thompson, and C. J. Yapp, "Oxygen and hydrogen isotopic ratios in plant cellulose," *Science*, **198,** 1209–1215 (1977).
48. K. G. Hardcastle and I. Friedman, "A method for oxygen isotope analyses of organic material," *Geophys. Res. Lett.*, **1,** 165–167 (1974).

49. A.M. Fehri, R. R. Letolle, and J. C. Lerman, "Oxygen isotope ratios of organic matter: Analyses of natural compositions," *Proc. 2nd Int. Conf. Stable Isot. October 20–23, 1975, Oak Brook, Illinois. Conf.*, 751027 ERDA, 716–724 (1975).
50. W. E. Schiegl, "Climatic significance of deuterium abundance in growth rings of *Picea*," *Nature*, **251,** 582–584 (1974).
51. A. T. Wilson and M. J. Grinsted, "Paleotemperatures from tree-rings and the D/H ratio of cellulose as a biochemical thermometer," *Nature*, **257,** 387–388 (1975).
52. S. Epstein and C. J. Yapp, "Climatic implications of the D/H ratio of hydrogen in C-H groups in tree cellulose," *Earth Plan. Sci. Lett.*, **30,** 252–261 (1976).
53. C. J. Yapp and S. Epstein, "Climatic implications of D/H ratios of meteoric water over North America (9500–2200 years B.P.) as inferred from ancient wood cellulose C-H hydrogen," *Earth Plan. Sci. Lett.*, **34,** 333–350 (1977).
54. C. M. Van der Straaten, "Deuterium in organic matter," Ph.D. diss., University of Groningen, Holland. (1981)
55. D. A. I. Goring and T. E. Timmell, "Molecular properties of a native wood cellulose," *Evensk Papperstidn.*, **6,** 524–527 (1960).
56. S. Epstein, C. J. Yapp, and J. H. Hall, "The determination of the D/H ratio of non-exchangeable hydrogen in cellulose extracted from aquatic and land plants," *Earth Plan. Sci. Lett.* **30**: 241–251, (1976).
57. M. S. DeNiro, "The effects of different methods of preparing cellulose nitrate on the determination of the D/H ratios of non-exchangeable hydrogen of cellulose," *Earth Plan. Sci. Lett.*, **54,** 177–185 (1981).
58. P. Thompson, J. Gray, and S. J. Long, "A preliminary report on the simultaneous extraction of hydrogen and oxygen from organic materials," *Proc. Int. Meet. Stable Isot. Tree-Ring Res., New Paltz, New York, May 22–25, 1979*, U. S. Department of Energy, Washington, D.C., 142–146 (1980).
59. I. Friedman and J. Gleason, "Deuterium content of lignin and of the methyl and OH hydrogen of cellulose," *Proc. Int. Meet. Stable Isot. Tree-Ring Res., New Paltz, New York, May 22–25, 1979*, U. S. Department of Energy, Washington, D.C., 50–55 (1980).
60. A. Neftel, H. Oeschger, J. Schwander, B. Stauffer, and R. Zumbrum, "Ice core sample measurements give atmospheric CO_2 content during the last 40,000 years," *Nature*, **295,** 220–223 (1982).
61. M. Stuiver and P. D. Quay, "Changes in atmospheric carbon-14 attributed to a variable sun," *Science*, **207,** 11–19 (1980).
62. M. Stuiver, "Solar variability and climatic change during the current millennium," *Nature*, **286,** 868–871 (1980).
63. M. Stuiver, "Statistics of the A.D. Record of Climatic and Carbon Isotopic Change," *Radiocarbon*, **25,** 219–228 (1983).

4

HOLOCENE PALYNOLOGY AND CLIMATE

Thompson Webb III

Department of Geological Sciences, Brown University
Providence, Rhode Island

Former vegetational patterns are one of the main sources of quantitative climatic information for the period of the past 15,000 years. Pollen data preserved in radiocarbon-dated sediments record these vegetational patterns, and pollen diagrams reveal the temporal changes at specific sites. Geographic networks of the pollen diagrams permit mapping of the changing vegetational and associated climatic patterns.

Several studies have documented how closely the modern vegetation reflects the climate today, especially at a continental and global scale. Köppen[1] in 1900 highlighted this relationship because he used vegetational patterns to guide his classification of the globe into a system of climatic regions. Borchert,[2] Bryson,[3] and Krebs and Barry[4] have all mapped seasonal patterns in the atmospheric circulation and shown that they match certain broad scale vegetational patterns. For example, Bryson[3] illustrated that the wintertime position of Arctic air masses lies along the southern border of the boreal forest. Burke et al.[5] have recently demonstrated a physiological basis for this association, because the freezing-stress mechanism in certain deciduous trees will not work for temperatures below −40°C, and these temperatures only occur in wintertime Arctic air masses.

Just as modern climatic patterns are evident in the contemporary distribution of vegetation, so too are past climatic patterns evident in former distributions of the broad scale vegetation. The task is to map the patterns of former vegetation and to calibrate them in climatic terms. The pollen content of lake and bog sediments provides the main data used in this mapping and calibration work.

In North America, Eurasia, and South America, lakes and bogs are abundant in the area covered by ice sheets during the last glaciation, and they contain sediments dating back to 15,000 years ago. Beyond the ice border, lakes and bogs are less common, but among them are a rare few with records of 100,000 years or longer.[6,7]

Pollen-bearing sediments in moderate-sized basins (10 to 1000 ha) accumulate at rates of 10 to 200 cm per 1000 years, and palynologists generally sample these cores in 200- to 500-year intervals. Samples from lakes with annually laminated sediments, however, can be taken in intervals as small as two years.[8] Time resolution of vegetational events within an individual core can be as low as 5 to 50 years, but the resolution of synchronous events within a network of radiocarbon-dated cores is 300 to 500 years.[9] Pollen data thus fill the gap in paleoclimatic coverage between the detailed short-term records of tree rings and the general long-term records from marine micro-fossils (figure 1).

Multivariate statistical procedures exist for calibrating pollen data in climatic terms[10,11] and are similar to those used for calibrating tree rings[12] (chapter 3) and marine planktonic data[13] (chapter 5). These procedures use data sets with paired samples of modern climate and pollen data. Multiple regression analyses of these data permit calculation of calibration functions that transform fossil pollen data into estimates of precipitation, temperature, and other climatic variables such as the duration of specific air masses.[14] Paleoclimatic maps can then be produced in regions that contain geographic networks of radiocarbon-dated pollen data.

Research is now in progress to produce maps of temperature for 12,000, 9000, 6000, 3000, and 500 years B.P. in eastern North America, and in parts of north-

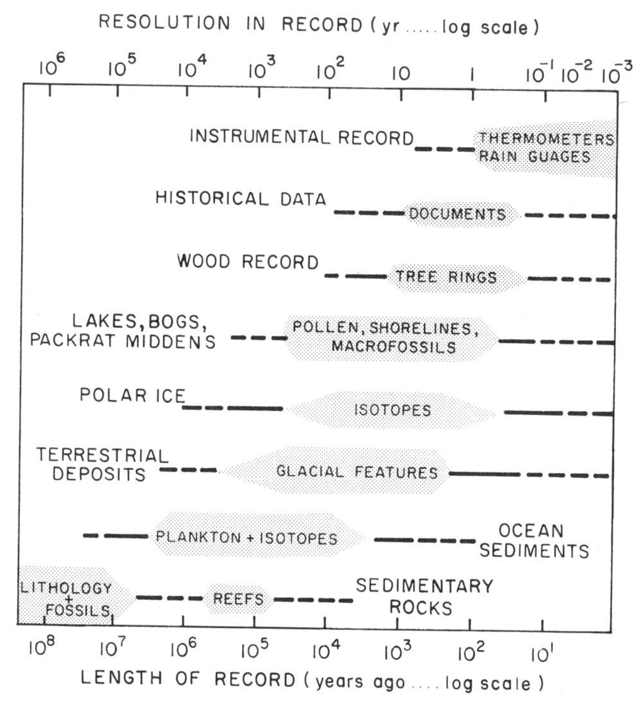

Figure 1. The time span and resolution of various climatic records from decades (thermometers) to hundreds of millions of years (fossils). The thickness of each box reflects the relative contribution for each time scale in the past. Adapted from ref. 91.

western North America, Europe, and the western Soviet Union. Each of these regions contains networks of radiocarbon-dated pollen diagrams from which maps of the pollen data have been produced. These maps show broad scale vegetational changes indicative of past changes in climate. Networks and maps of the modern pollen data also exist, and these show patterns within the modern data that reflect climatic gradients, thus indicating that the modern data can be used to calibrate the changes in the Holocene data. Systems of computer programs are available for managing and displaying the data, and these systems include the series of statistical programs needed to calibrate the pollen data.[15,16]

The procedures followed in producing paleoclimatic maps from pollen data are best illustrated by an example, which is described in the next section. This section demonstrates the production of temperature and precipitation maps from the northern Midwest. I then review the status of mapping and calibration work for pollen data from eastern North America, the Alaska-Yukon region, Europe, the Soviet Union, New Zealand, Chile, and northwest India. In a final section, I have described the COHMAP (Cooperative Holocene Mapping Project) plan to produce global climatic maps for the Holocene using not only pollen data but also data and maps of glacial moraines,[17] plant macrofossils,[18] lake-levels[19] (chapter 7), and marine micro-fossils[20] (chapter 5). These other data sets enlarge the coverage of Holocene data to a global network of paleoclimatic information by providing coverage in polar regions, arid lands, and oceans where pollen data are rare. This COHMAP effort recognizes the need for global maps in order (1) to understand past climates,[21] (2) to illustrate the broad spatial patterns in climate,[19] and (3) to verify the ability of general circulation models to simulate certain features of past climates.[22]

PALEOCLIMATIC MAPS FROM THE MIDWEST

The production of paleoclimatic maps from pollen data involves two sets of procedures that each require an ordered series of computer programs. The first series of programs is used to manage and map fossil pollen data, and the second series is used to calculate calibration functions and to derive the climatic estimates.

The Midwest is a good region in which to illustrate these two sets of procedures because its topography is relatively flat, and broad networks of both modern and fossil pollen samples exist there.[23] The data from these samples are now available in the Pollen Data Bank at Brown University to which many analysts have contributed their published data. These analysts used standard procedures in collecting the samples and analyzing the fossil data.[24] With hand-driven piston corers, they obtained 2 to 15 m of sediments from a network of 43 lakes and bogs. One to several radiocarbon dates were obtained from each core, and 1 ml subsamples were taken for pollen analysis at intervals of 5 to 25 cm. These subsamples were then treated with a sequence of caustic chemicals including sulphuric and hydrofluoric acids that dissolved the sediment matrix and left a residue rich in pollen. Single drops of the residue, which contained hundreds of pollen grains, were spread onto microscope slides. Two hundred to 1000 grains from each sample were then identified and counted when viewed at 400 to 1000 magnifications (figure 2).

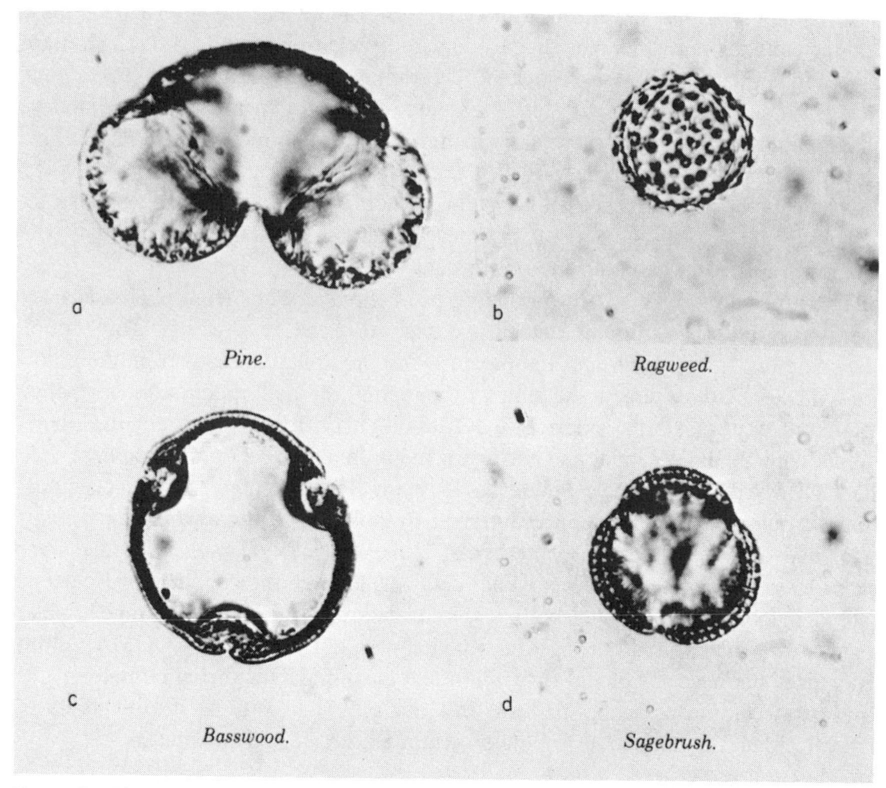

Figure 2. Photographs of pollen grains for (*a*) pine (*Pinus*—70 mμ), (*b*) ragweed (*Ambrosia*—30 mμ), (*c*) basswood (*Tilia*—60 mμ), and (*d*) sagebrush (*Artemisia*—30 mμ) taken at 1000× under an oil emersion lens (from ref. 91). Taxonomic identifications are based on the differences in shape and surface texture.

The pollen counts from each core are standardly published in a pollen diagram in which the data are presented as pollen percentages (figure 3). Because the percentages are multinomially distributed, the precision of each percentage value can be estimated and depends upon how large the percentage is and how many grains of all types were counted.[25]

Maps of Holocene Pollen Data

Contour maps of the modern and fossil pollen data are necessary prerequisites to the production of paleoclimatic maps. They show the patterns of pollen changes in time and space and thus help illustrate the existence (1) of patterns in the fossil data that reflect past changes in climate[23,26] and (2) of modern pollen data capable of calibrating these patterns in the fossil data.[27-29] These maps illustrate the patterns in pollen percentages, and hence in the vegetation, at selected dates such as today, 1000, 3000, and 10,000 years B.P. Difference maps between two dates show the

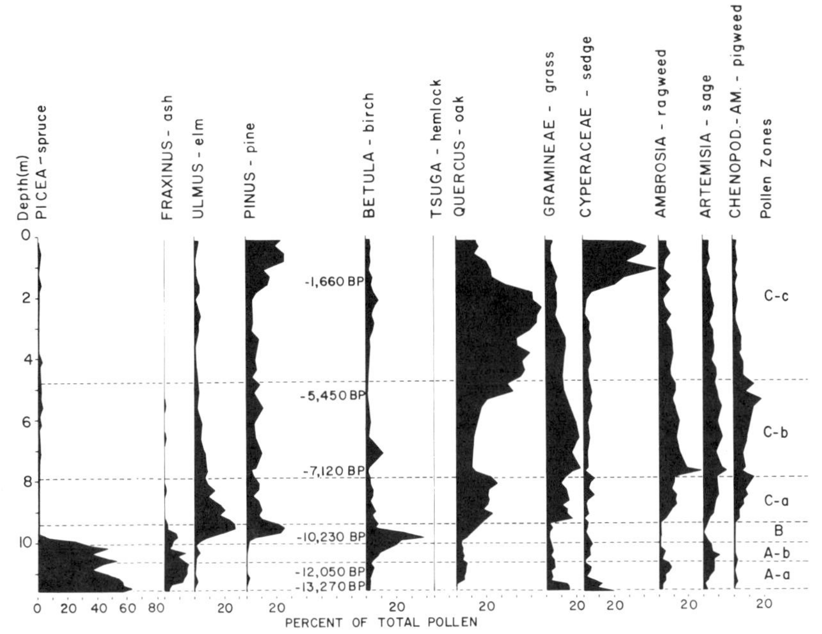

Figure 3. Pollen diagram for Kirchner Marsh is east-central Minnesota.[92] It shows the percentages of the different pollen types plotted against depth in meters from the sediment surface. A time scale can be estimated from the 6 radiocarbon dates. High percentages of spruce (*Picea*) pollen indicate conifer parkland conditions about 12,000 years B.P. High percentages of herb pollen types (Gramineae—grass, Chenopodiaceae/Amaranthaceae—pigweed, *Artemisia*—sagebrush, and *Ambrosia*—ragweed) indicate prairie conditions between 5000 and 7000 years B.P. and high percentages of oak (*Quercus*) pollen indicate that deciduous forest grew near the site from 5000 to 1600 years B.P. Reprinted from the *Journal of Interdisciplinary History* (1980), by permission of the *Journal of Interdisciplinary History* and the MIT Press, Cambridge.[30]

changes in these patterns, and isochrone maps of selected isopolls (isofrequency contours) illustrate the long-term trends in changing plant abundances.

When similar changes for one or several pollen types occur from New England to Minnesota or when the long-term increases (or decreases) in plant abundance parallel the directions of major climatic gradients, then climate can be postulated as a likely cause of these changes.[30] A reversal in the long-term trends is another feature in the pollen diagrams and maps that can help in identifying past variations that probably reflect climatic changes.[31] Von Post[32] used the term *revertence* to describe such a reversal in a pollen diagram and proposed that this type of change was induced by climate. For the various pollen types, isochrone maps are an excellent format for illustrating the long-term trends and the reversals in these trends.

Production of the maps of the Holocene pollen data require the following seven step[33]:

1. Use of the same pollen sum at all sites for calculating the pollen percentages.
2. Plotting scatter diagrams of radiocarbon or biostratigraphic dates versus depth for each site and checking these diagrams for anomalous dates.

3. Choosing the set of dates and the interpolation procedure to use at each site and estimating the date for each depth with a pollen sample.
4. Producing tables listing the depths, dates, and pollen percentages for each sample and checking for errors.
5. Interpolation of pollen percentages for the dates to be mapped at each site.
6. Plotting the interpolated values on maps, contouring the values, and checking for any anomalies that may result from errors or dating biases (when erroneous dates are discovered and corrected, then steps 3 to 6 must be repeated).
7. Selection of illustrative isopolls for the major pollen types and mapping of these as isochrones.

Webb et al.[23] followed this series of steps and gained maps indicating continuous climatic change in the Midwest during the past 11,000 years. The isochrone maps show broad scale, long-term changes and revertence in the distribution patterns for the major pollen types (figure 4). The consistency in the changes suggests that climatic variations may be the main, ultimate cause for the mapped changes in pollen distributions. The 5 percent contour for spruce (*Picea*) pollen shows that spruce populations moved northward into Ontario from 10,000 to 8000 years B.P. (figure 4*a*). Later, after 4000 years B.P., the abundance of spruce trees increased southward into northern Minnesota and Michigan. These trends parallel the major gradient in modern temperatures across the Midwest (figure 5*a*) and are also similar to the changes in spruce populations in New England and southern Quebec.[26,33,34]

The isochrones for oak (*Quercus*) pollen show a similar northward movement in the early Holocene and southward movement after 6000 years B.P. (figure 4*b*,*d*). North-south trends are also evident in the isochrone maps of pine (*Pinus*) and birch (*Betula*) pollen.[23] For prairie-forb pollen, which is the sum of ragweed (*Ambrosia*) + sage (*Artemisia*) + daisy (Compositae) + pigweed (Chenopodiaceae/Amaranthaceae) pollen, however, the primary movement of the 20 percent isochrone is east-west (figure 4*c*). The trends in its isochrones show that the prairie/forest border moved eastward from 10,000 to 7000 years B.P. and then retreated westward after 6000 years B.P..[26,35] These east-west trends parallel the precipitation gradient in the Midwest (figure 5*b*).

The mapped changes in tree pollen suggest that temperatures in the northern Midwest increased from 10,000 to 6000 years B.P. and then decreased after that time, particularly after 4000 years B.P. The map of prairie-forb pollen suggests that precipitation decreased from 10,000 to 6000 years B.P. with most of the decrease in Minnesota and the Dakotas. After 7000 to 6000 years B.P., precipitation again increased. In support of this interpretation, independent evidence from cores from many Iowa and Minnesota lakes show that water levels were lower at 7000 years B.P. than they are today.[36] These qualitative interpretations of climatic change motivate the calculation of calibration functions in order to gain quantitative estimates for the temperature and precipitation variations.

Maps of Modern Pollen Data

Maps of the modern data show that the main gradients in several pollen types parallel those of certain climatic variables (figure 5). The north-south gradient of

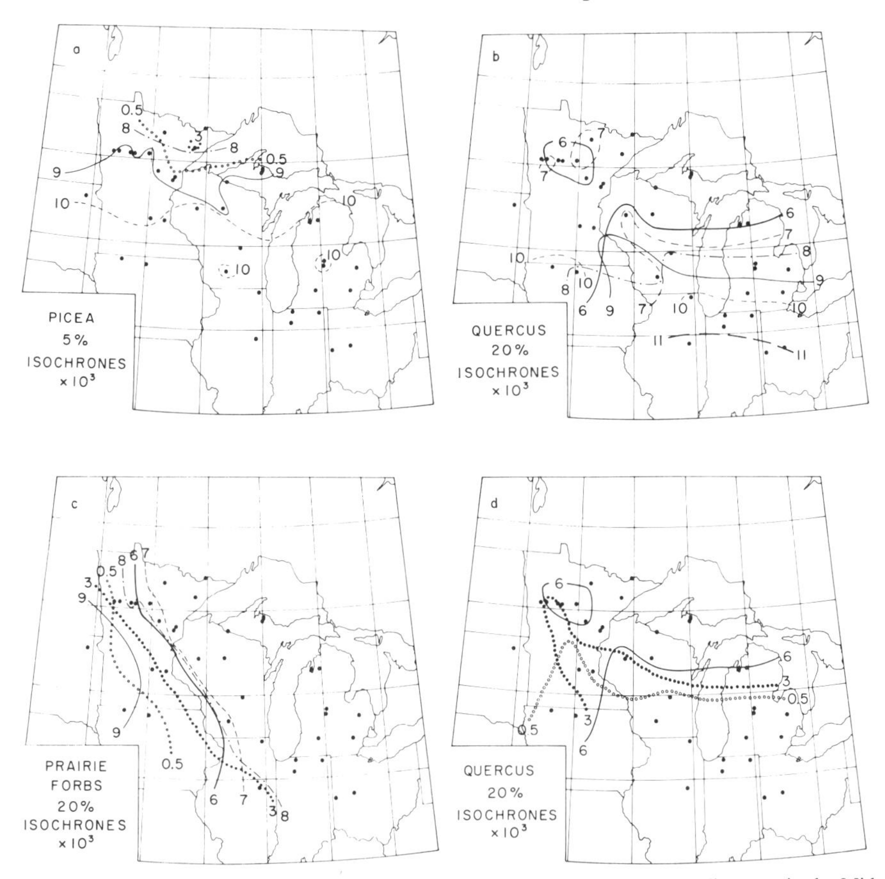

Figure 4. Isochrone maps for 10,000 to 500 B.P. from radiocarbon-dated pollen diagrams in the Midwest: (*a*) map of *Picea* pollen in which 5 percent or more spruce pollen is to the north of the isochrones, (*b*) and (*d*) maps of *Quercus* pollen in which oak pollen is to the south of the isochrones, and (*c*) map of prairie forb (*Artemisia* + *Ambrosia* + Compositae + Chenopodiaceae/Amaranthaceae) pollen in which 20 percent or more of the pollen is to the west of the isochrones.[23] Reprinted with permission of the University of Minnesota Press.[23]

annual temperature east of 95°W is inversely related to the distribution of spruce pollen (figure 5*a*,*c*) and is directly related to the distribution of oak pollen (figure 5*d*). The general east-west gradient in annual precipitation between 40 and 50°N (figure 5*b*) is directly related to the distribution of oak pollen (figure 5*d*) and inversely related to the distribution of prairie-forb pollen (figure 5*e*). The mapped patterns of temperature and precipitation show that these climatic variables are orthogonal over the Midwest (Figure 5*a*,*b*), thus enabling the reconstruction of two essentially independent components of the midwestern climate.

The Problems of Human Disturbance and Vegetational Lags

Before calculation of the calibration functions, one further question remains concerning whether the modern and Holocene pollen data are in equilibrium with

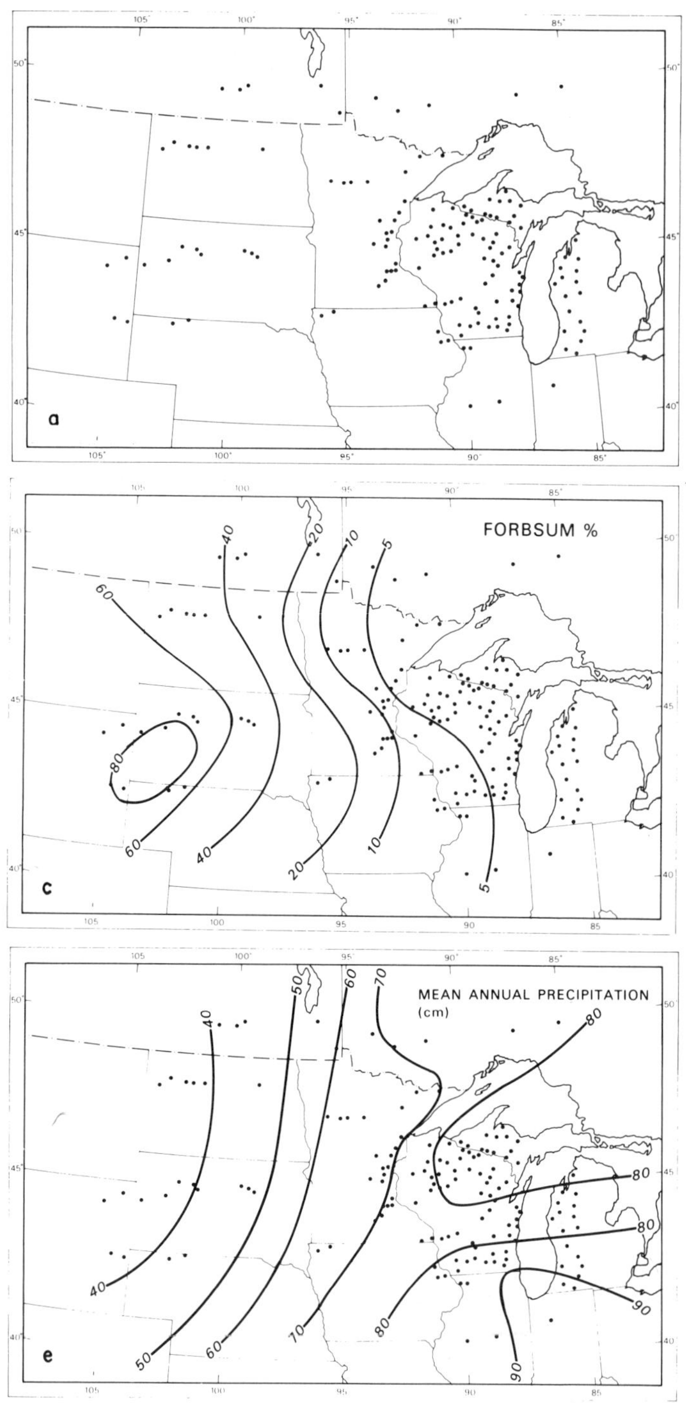

Figure 5. Maps of (*a*) the sites with modern pollen data, (*b*) mean annual temperature today, (*c*) prairie forb pollen today, (*d*) spruce pollen today, (*e*) mean annual precipitation today, and (*f*) oak pollen today.

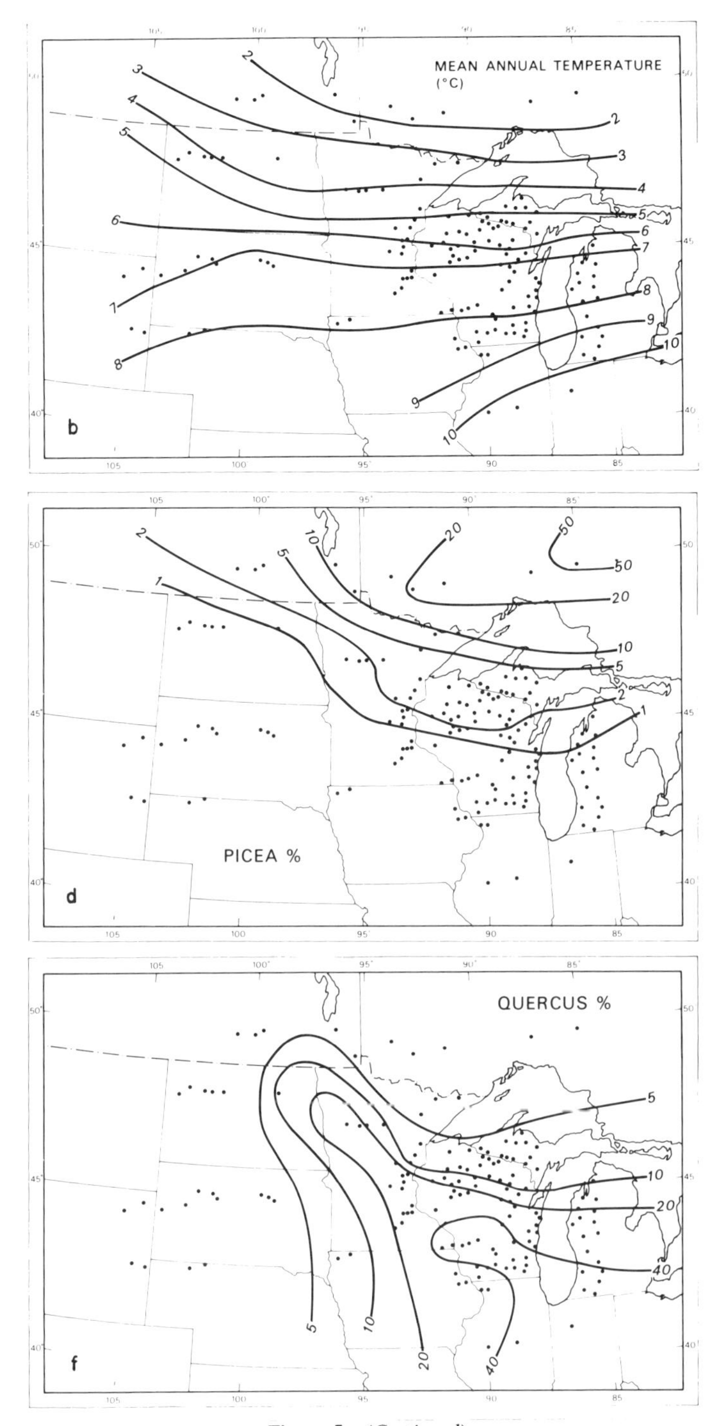

Figure 5. *(Continued)*

climate.[31,37] This question is best broken into two questions, and the answer to each question will depend on the time and space scales being studied. These questions are: (1) Has human disturbance of modern vegetation affected how the percentages for certain pollen types change along major climatic gradients? and (2) Has the Holocene vegetation lagged in its response to long-term climatic changes?

The answer to both questions is *yes;* but, for the time and space scales under study, the affects of human disturbance and of vegetational lags are generally too small and too short-term to influence the major pollen/climate relationships represented in the calibration functions. Webb[38] compared the modern and presettlement pollen records in lower Michigan and found that once ragweed (*Ambrosia*) pollen and other weedy types were removed from the pollen sum, the modern and presettlement pollen recorded the same major broad scale vegetational gradients. Because the broad scale gradients are those that dominate in the calculation of the calibration functions, human disturbance can be discounted as a major determinant in the relationships represented in the calibration functions.[11]

In most Holocene pollen records, the sample interval of 200 to 500 years acts to dampen or eliminate the affects of vegetational lags of a few decades to centuries. When the Holocene records are mapped in 1000- to 3000-year intervals, the main concern must focus on situations when the vegetation lags the climate by 1000 years or more. Bernabo and Webb[26] and Webb et al.[23] have examined the Holocene pollen record in central eastern North America and the Midwest for evidence of such lags. They found several instances where the range extension of one pollen type lags that of another by over 1000 years,[39] but these interspecific migration-lags reflect the individualistic behavior of the different taxa and probably are long-term equilibrium responses within the vegetation to climatic differences in seasonality between the early and late Holocene. Further discussion of this issue appears in Webb et al.[23] and Howe and Webb.[11]

Climatic Calibration of the Pollen Data

Several numerical techniques have been used in calibrating pollen data in climatic terms. These include multiple regression analysis,[11] principal components analysis,[40,41] and canonical correlation analysis.[14] Webb and Clark[10] described the many similarities among these techniques and concluded that multiple regression was the simplest and most straightforward procedure to use. Multiple regression yields a model of the following form:

$$ {}_n\mathrm{c}_1 = {}_nP_m b_1 + {}_n\epsilon_1 $$

where n = number of samples
m = number of pollen types
${}_n\mathrm{c}_1$ = an $n \times 1$ observation vector for a given climatic variable
${}_nP_m$ = an $n \times m$ matrix of pollen percentages
${}_mb_1$ = an $m \times 1$ vector of regression coefficients
${}_n\epsilon_1$ = an $n \times 1$ vector of deviation from the model results.

Howe and Webb[11] and Bartlein and Webb[42] have described the several steps required in the multiple regression analysis of pollen data. Arigo et al.[15] and Bartlein and Webb[16] document the ordered series of seven or more computer programs used in implementing these steps and calculating calibration functions.

The aim when calculating a regression model is to minimize any violations of the statistical assumptions that ordinary least-squares estimation requires.[42] Satisfying these assumptions means that standard procedures can be used to calculate confidence intervals and that the regression coefficients will be unbiased. When a calibration function is calculated, model-specification errors are the main source of assumption violations. These errors arise from several sources including (1) selection of too large or too small a geographic region containing the modern data used in the regression analysis, (2) neglect of nonlinear relationships between the dependent and predictor variables, and (3) omission of potentially useful predictors from the model. When specification errors are minimized, other sources of assumption violations are generally minimized as well. The particular sequence of steps used in model building includes:

1. Selection of the pollen types with mean values greater than 1 percent and recalculation of the pollen percentages using a sum of just these types.
2. Examination of bivariate scatter diagrams (a) to help choose transformations for linearizing the pollen/climate relationship and (b) to help screen for outliers and bad data values.
3. Estimation of a regression model with all candidate predictors included, and the calculation of various diagnostic statistics to help screen for further unrepresentative or overly influential data points and to assess the homogeneity of the calibration data set.
4. Estimation of a regression model with a subset of the candidate predictors chosen by a criterion that minimizes bias within the coefficients.
5. A formal analysis of the regression diagnostics for the chosen equation and modification of the equation, if necessary.

These procedures were used to calculate calibration equations for annual temperature and precipitation for data sets between 40 and 50°N. Modern data from 85 to 95°W were used for the temperature equation, and modern data from 85 to 105°W were used for the precipitation equation (figure 5). Scatter diagrams revealed pronounced nonlinearities in the relationships between the climate variables and selected pollen types (figure 6*a–d*). These bivariate relationships may be linearized by transforming one or both variables. Raising the percentages of oak pollen to the 0.25 power linearized its relationship with mean annual temperature (figure 6*e*), and raising the percentage of prairie-forb pollen to the 0.50 power linearized its relationship with mean-annual precipitation (figure 6*f*). Similar transformations also linearize the climatic relationships for several other pollen types (table 1).

The temperature equation, which accounts for 83 percent of the total variance and has a standard error of 0.7°C, includes 10 terms. The negative coefficient for spruce pollen and the positive coefficient for oak pollen (table 1) reflect the relationships evident on the scatter diagrams (figure 6). The precipitation equation has

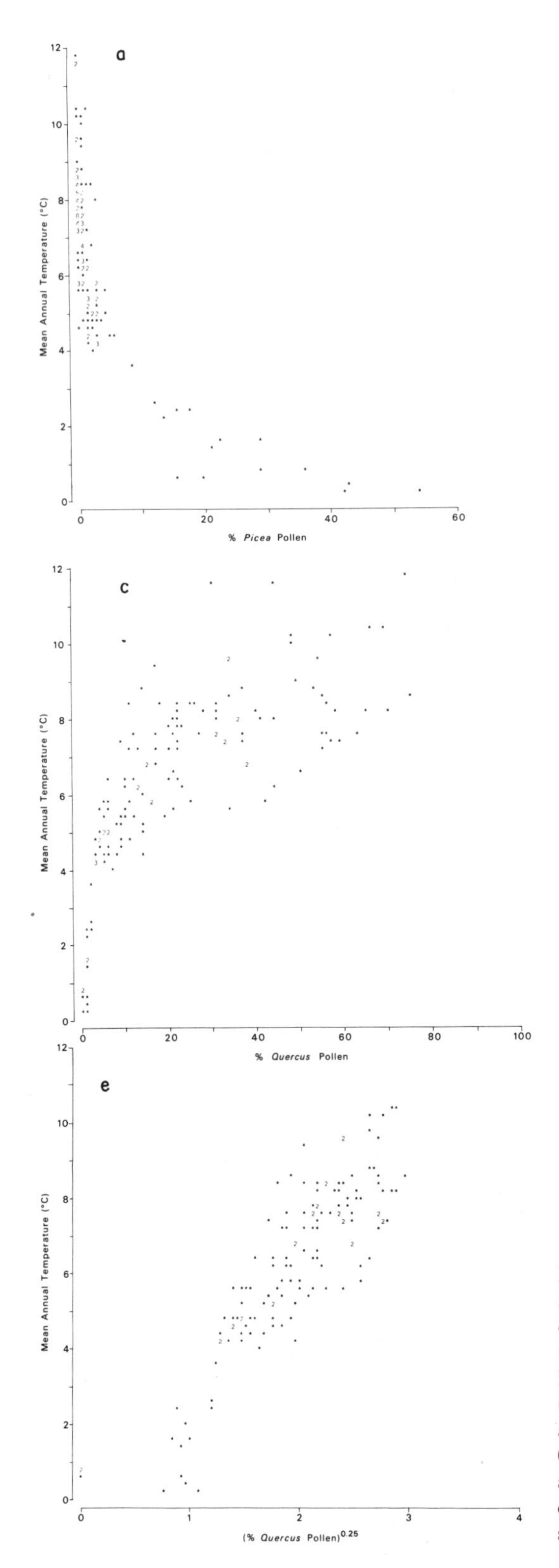

Figure 6. Scatter diagrams for mean annual temperature vs. the percentages of (*a*) Spruce (*Picea*) pollen, (*b*) oak (*Quercus*) pollen, (*c*) oak (*Quercus*) pollen, (*d*) prairie forb pollen, (*e*) oak (*Quercus*) pollen raised to the 0.25 power, and (*f*) prairie forb pollen raised to the 0.50 power. Ragweed (*Ambrosia*) pollen, an indicator of human disturbance, was deleted from prairie forb pollen for these scatter plots and for climatic calibration.

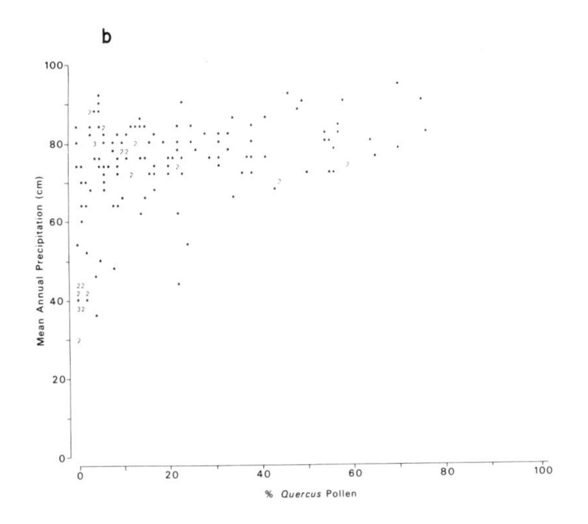

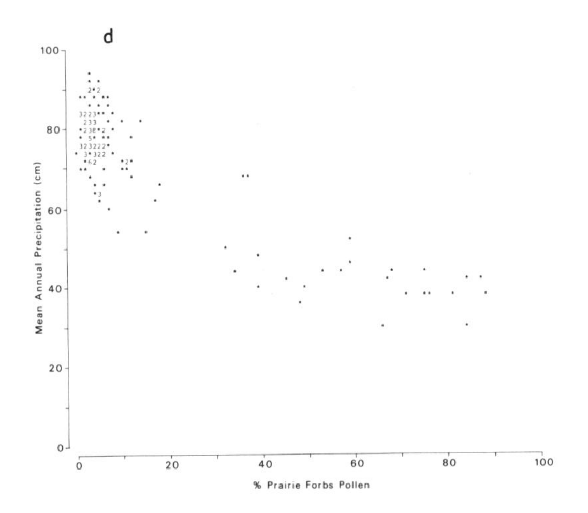

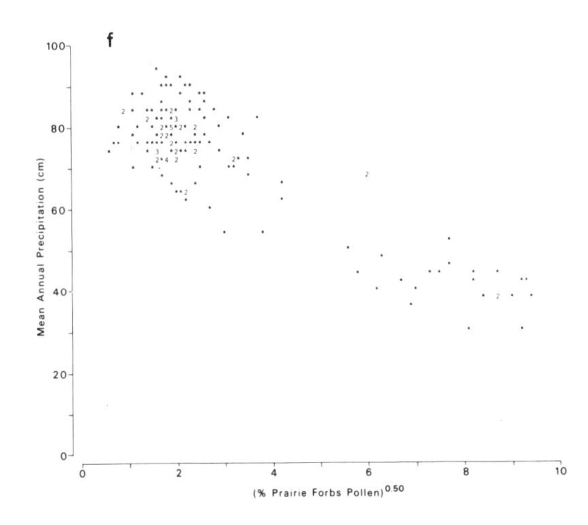

Figure 6. *(Continued)*

Table 1. Calibration Functions and Their Summary Statistics for Mean Annual Temperature and Precipitation

Calibration Functions

(a) Mean Annual Temperature (°C)

$$\begin{aligned}\text{TMEANYR} = {} & \underset{(.81)}{5.55} - \underset{(.17)}{.46}\ \text{PICEA}^{.25} - \underset{(.06)}{.17}\ \text{PINUS}^{.50} - \underset{(.07)}{.29}\ \text{BETULA}^{.50} \\ & + \underset{(.05)}{.20}\ \text{FRAXINUS} + \underset{(.27)}{1.43}\ \text{QUERCUS}^{.25} - \underset{(.02)}{.05}\ \text{ULMUS} \\ & + \underset{(.06)}{.14}\ \text{ACER} + \underset{(.09)}{.24}\ \text{TSUGA}^{.50} - \underset{(.13)}{.29}\ \text{ALNUS}^{.50}\end{aligned}$$

(b) Mean Annual Precipitation (cm)

$$\begin{aligned}\text{PRECPYR} = {} & \underset{(4.86)}{92.75} + \underset{(1.49)}{3.91}\ \text{ABIES}^{.50} + \underset{(.81)}{2.08}\ \text{JUNIPERUS}^{.50} - \underset{(.41)}{1.67}\ \text{PINUS}^{.50} \\ = {} & \underset{(1.35)}{2.48}\ \text{FRAXINUS}^{.25} + \underset{(1.42)}{6.21}\ \text{QUERCUS}^{.25} - \underset{(1.54)}{4.17}\ \text{ULMUS}^{.25} \\ & + \underset{(.93)}{3.68}\ \text{TSUGA}^{.25} - \underset{(1.28)}{5.12}\ \text{ALNUS}^{.25} - \underset{(.43)}{5.87}\ \text{FORB SUM}^{.50}\end{aligned}$$

(Values in parentheses are the standard errors of the regression coefficients.)

Summary Statistics

	(a) Temperature	(b) Precipitation
n (number of observations)	108	135
p (number of predictors)	9	9
R^2 (squared multiple correlation coefficient)	84.6	88.2
Adjusted R^2 (adjusted for *p*)	83.2	87.3
Standard error	.687	5.39
Statistics for test of		
Inhomogeneity of residual variance	1.3478[a]	.7743[a]
Non-normality of residuals	.997[a]	.994[a]
Spatial autocorrelation of residuals	.16[b]	.09[b]

[a]Indicates that the statistic is not significant.
[b]Indicates that the test statistic is significant.

10 terms with a negative coefficient for the forb sum of sage, pigweed, and daisy pollen and a positive coefficient for oak pollen. This equation accounts for 87 percent of the total variance and has a standard error of 5.4 cm (table 1). These two equations represent empirical relationships between the available pollen and climate data in the calibration regions and should not be used on data from outside of the area where these relationships obtain.

Before the climatic maps for 9000, 6000, and 3000 years B.P. were produced, Bartlein et al.[43] checked the performance of the calibration functions at each Hol-

Table 2. Regression Summaries and *t*-Tests for Modern Observed Temperature and Precipitation Values Versus the Estimates of 0 years B.P.

Regression Equation	Climate Variable	Dependent Variable	Independent Variable	b_0 Intercept	*t*-test $b_0 = 0$	b_1 Slope	*t*-test $b_1 = 1.0$	r^2	n
1	Mean annual temperature	Modern observed	Estimate for 0 years B.P.	−0.0938	−0.205	1.0294	0.420	88.6	30
2	Annual precipitation	Modern observed	Estimate for 0 years B.P.	−8.6410	−1.170	1.1303	1.298	77.9	38

ocene site by comparing the climatic estimates for 0 years B.P. with the modern observed values. They looked for any large deviations at individual sites, and they also tested for any systematic under- or over-estimation of the climatic values (i.e., for bias) by regressing the climatic estimates from the 0 years B.P. data against the modern observed values. For both the temperature and precipitation estimates, these regressions yielded intercepts insignificantly different from 0 and slopes insignificantly different from 1 (table 2). These results revealed no evidence of bias among the modern samples from the fossil sites.

Paleoclimatic Maps

The maps in the Midwest show that from 9000 to 6000 years B.P., annual precipitation declined almost everywhere (figure 7*a*). Precipitation was 20 percent higher at 9000 years B.P. than at 6000 years B.P. in parts of western Minnesota and Iowa and was at least 10 percent higher in western Wisconsin. At 6000 years B.P., precipitation was less than 80 percent of its modern values over parts of western Wisconsin, southern Minnesota, and Iowa (figure 7*b*). These precipitation variations provide estimates for the magnitude of climatic changes that most likely caused the prairie/forest border first to move east from 9000 to 6000 years B.P. and later to retreat westward (figure 4*c*). The precipitation variations were probably associated with a regional increase in the duration of mild, dry Pacific air masses that resulted from an increase in the frequency of zonal-type circulation patterns.[14] Calibration of the pollen data in terms of air mass durations supports this interpretation.

Geomorphic activity in the Driftless Area of southwestern Wisconsin is broadly correlated with these climatic variations.[44,45] During the interval of maximum dryness just before 6000 years B.P., little or no alluvial sediments were deposited in the valley of Brush Creek, Wisconsin.[46] As precipitation increased after 6000 years B.P., significant changes in the nature of geomorphic activity occurred that resulted in widespread alluviation within Driftless Area valleys.

From 6000 to 3000 years B.P., annual precipitation increased about 10 percent nearly everywhere in the Midwest except from Iowa to northern Indiana (figure 7*e*). The reconstructed decrease in precipitation there is related to an eastward increase in the abundance of prairie-forb pollen (figure 4*c*). This increase probably resulted from a southward shift in the axis of strongest westerly flow. After 3000

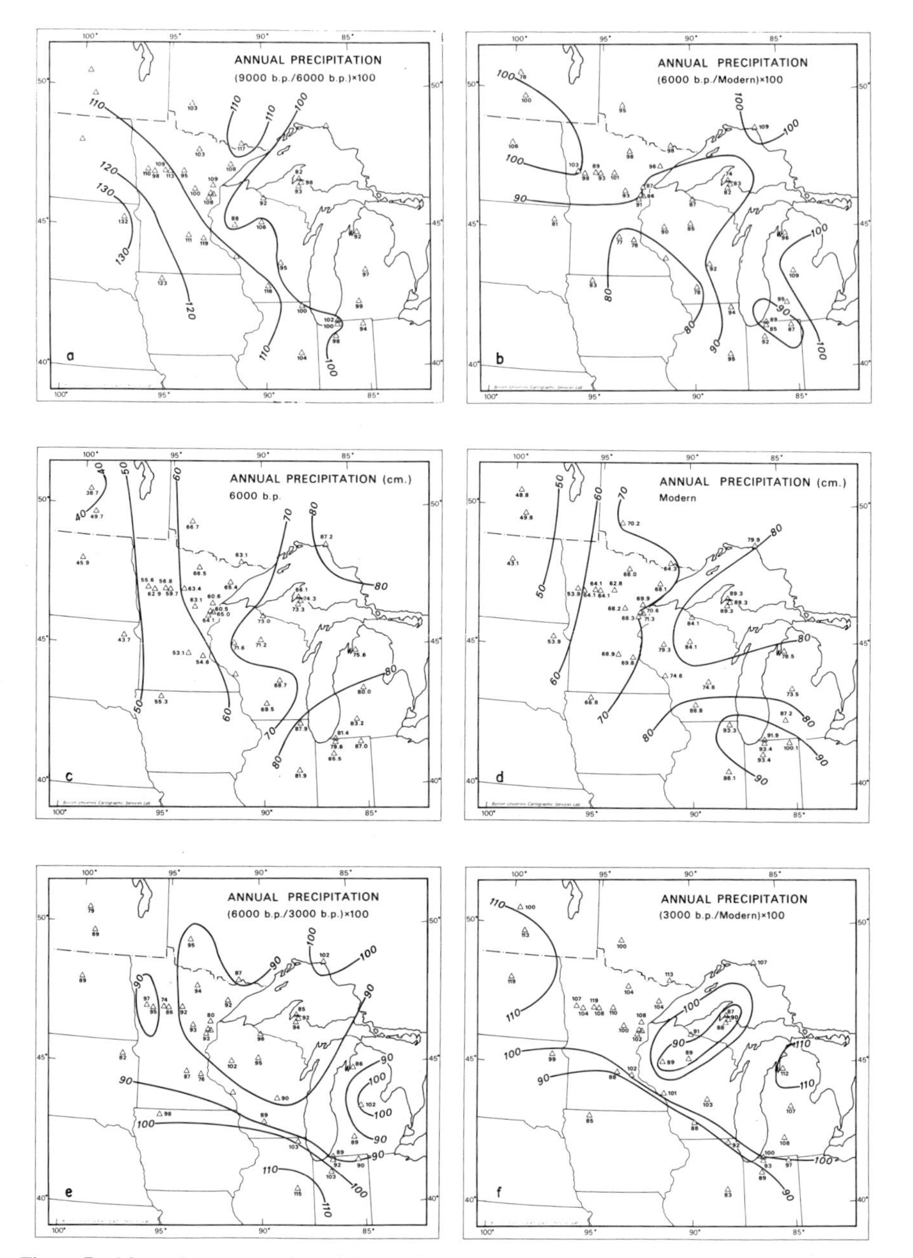

Figure 7. Maps of mean annual precipitation for (*a*) the ratio of values between 9000 and 6000 years B.P., (*b*) the ratio between 6000 years B.P. and today, (*c*) 6000 years B.P., (*d*) today, (*e*) the ratio between 6000 and 3000 years B.P., and (*f*) the ratio between 3000 years B.P. and today.

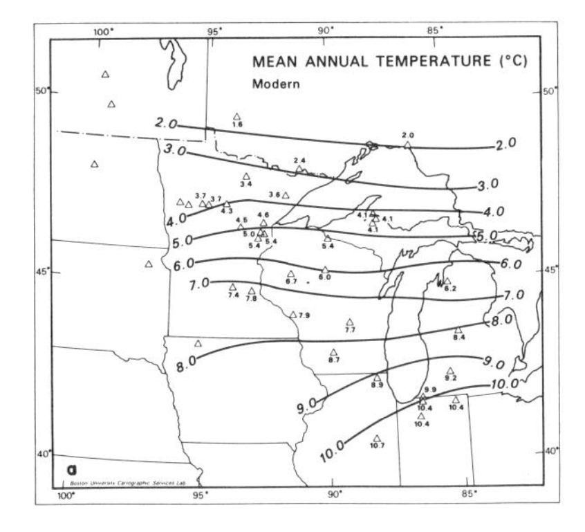

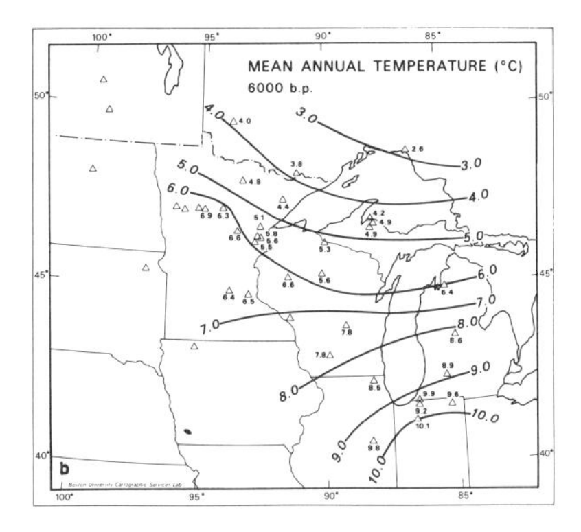

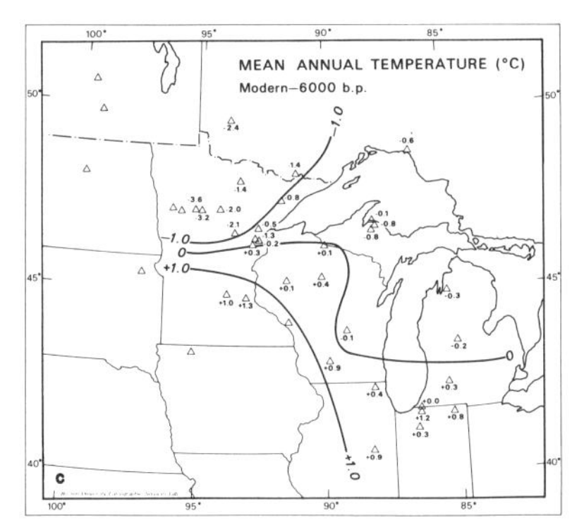

Figure 8. Maps of mean annual temperature (°C) for today and 6000 years B.P.: (*a*) observed temperatures today, (*b*) estimated temperatures for 6000 years B.P. from pollen data, and (*c*) differences in °C between today and 6000 years B.P. Negative values indicate higher temperatures at 6000 years B.P.

years B.P., annual precipitation increased across the southern edge of the region as the duration of moist Atlantic (maritime tropical) air masses increased there (figure 7*f*).

Annual temperatures in the Midwest were up to 0.5°C lower at 9000 years B.P. than at present and increased between 1°C and 2°C nearly everywhere between 9000 and 6000 years B.P. (figure 8). This temperature calibration provides an estimate for the magnitude of regional climatic warming associated with the northward retreat of the boreal forest into Canada (figure 4*a*). In parts of northern Minnesota, annual temperatures were over 2.0°C higher at 6000 years B.P. than at present, whereas the temperatures were about 1°C lower than at present across the southern part of the study area. This pattern of temperature change probably reflects the replacement of the eastward extending wedge of Pacific air masses by both Arctic air masses in the north and Atlantic air masses in the south. Subsequent research by Bartlein et al.[43] has shown that the midwestern vegetation at 6000 years B.P. probably reflects a change in mean July temperature more than mean annual or January temperature.

One significant feature of the paleoclimatic maps is that the regional patterns appear in both the magnitude and the direction of change from one period to the next (figures 7 and 8). When the data set is expanded to New England and Quebec, "synoptic" scale maps can be reconstructed. The regional patterns evident on these maps should reveal the changing pattern of atmospheric circulation across northeastern North America.

DEVELOPMENT OF A GLOBAL DATABASE FOR HOLOCENE CLIMATES

Climatic changes are global in character and therefore require global scale monitoring systems to map their patterns. The midwestern pollen maps and paleoclimatic maps serve as a useful standard for what is needed in other regions. For many mid- and high-latitude areas, such as eastern North America, Alaska/Yukon, Europe, the USSR, and New Zealand, networks of pollen data exist and provide the main source of Holocene paleoclimatic information. These data sets are described in the next five sections. For arid lands and the oceans, the main data sources include former water levels in lakes, plant macro-fossils in packrat middens, and marine plankton data in deep-sea cores. I briefly describe these data in the final section after reviewing the global data set for 18,000 years B.P. I then present a global map of the available data that can illustrate the global scale patterns in Holocene climates. The compilation of this data set is currently in progress and a major focus of COHMAP research.

Regional Networks of Holocene Pollen Data

Major regional networks of Holocene pollen sites exist in eastern North America, the Alaska-Yukon region, Europe, the USSR, and New Zealand. Australia, South America, Africa, and southern Asia contain several radiocarbon-dated diagrams, but the coverage in these areas is too sparse for isopoll mapping. Summaries of the published pollen data exist or are being produced from all of the areas with dense networks of sites. Climatic calibration of Holocene pollen data from terrestrial sites, however, has been restriced to the midwestern United States,[10,11,14,30,43] the Canadian Arctic,[41,47–50] southwestern British Columbia,[51] Chile,[40] and northwest India.[52,53]

Eastern North America

A broad network of pollen sites exists within eastern North America (figure 9). Bernabo and Webb[26] mapped the data from 40 to 48°N, where the coverage is densest, and illustrated the major broad scale vegetational changes during the Holocene. Other mapped summaries of Holocene changes within this region include maps of either the pollen data or inferred vegetation from central Canada,[54] from south of 50°N,[39,55] from the Northeast for 14,000 to 9000 years B.P.,[56] and from southern Quebec.[33] These studies describe many changes in the vegetation across

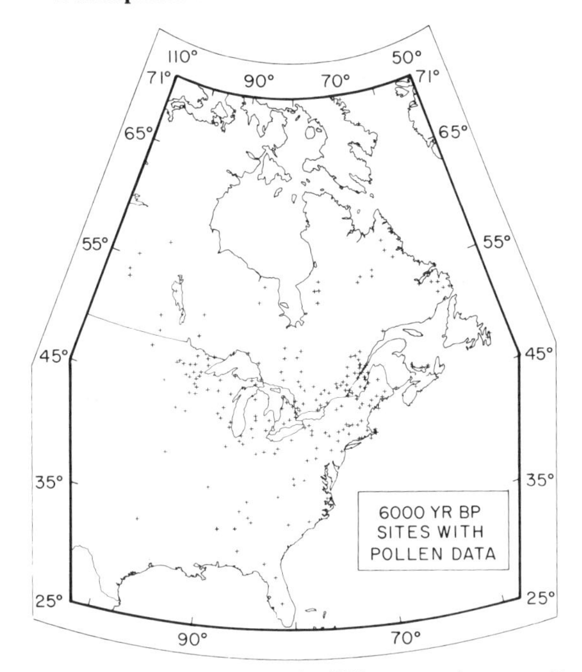

Figure 9. Location of sites with pollen data for 6000 years B.P. in eastern North America.

eastern North America, several of which suggest major climatic changes of the same magnitude as those illustrated in the Midwest.

A large network of over 3000 samples with modern data are available for calibrating the major changes in the Holocene data. Several analysts have described and mapped the patterns in the modern pollen data.[27–29,57–60] Most of these pollen data along with contemporary climate data are stored in a computer file at Brown University and are available for climatic calibration work. Calibration functions like those from the Midwest are being produced for the Southeast, Northeast, southern Canada, central Canada, and the Arctic.[93] Kay,[47] Andrews et al.,[41] and Andrews and Diaz[48] have already completed initial calibration of the pollen data from several sites in the Canadian Arctic. These studies indicated a 1 to 2°C decrease in July mean temperature in the region between 55 and 65°N since 3000 years B.P.

Alaska and the Yukon

In the topographically diverse region of Alaska and the Yukon, the network of sites with radiocarbon-dated pollen data[61–64] is a bare minimum for mapping the Holocene vegetational patterns (figure 10). Preliminary isopoll, difference, and isochrone maps for grass (Gramineae), sedge (Cyperaceae), birch (*Betula*), alder (*Alnus*), and spruce (*Picea*) pollen show both south-north and east-west trends.[61] The network of modern data is still too sparse for calculating calibration functions, but work is in progress to fill in several of the gaps.

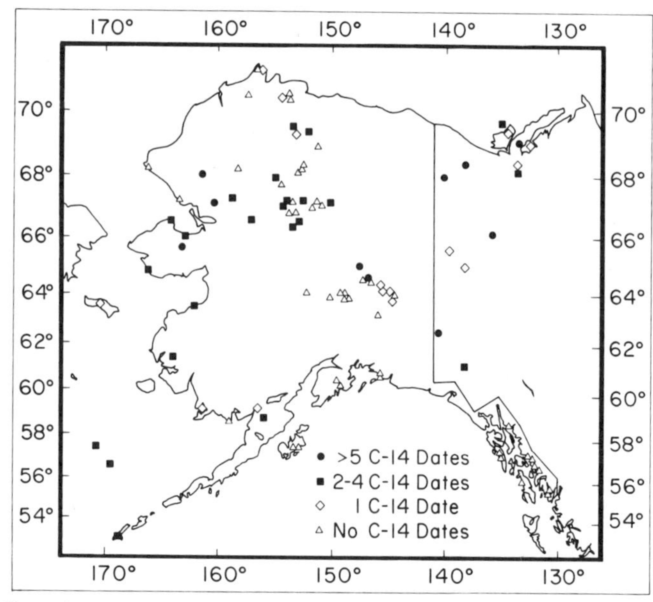

Figure 10. Sites with Holocene pollen data in Alaska and the Yukon (from ref. 61). Symbols show the number of radiocarbon dates at each site.

Europe

Huntley and Birks[65] recently completed a major compilation and mapping of the published radiocarbon-dated diagrams from Europe (figure 11). Their study included maps of the modern distribution of 53 pollen types and maps in 1000- to 2000-year intervals for most of these pollen types from 13,000 years B.P. to present. These maps update the pioneering isopoll maps for Poland[66] and for Germany,[67] which were completed before any radiocarbon dates were available. New isopoll maps are now available for Poland.[68]

The maps by Huntley and Birks[65] show several patterns of change in the Holocene pollen record. These include south-to-north trends in the pollen of major tree taxa during the early Holocene and east-to-west trends for spruce (*Picea*), hornbeam (*Carpinus*), and beech (*Fagus*) pollen in the mid- to late-Holocene. No evidence exists for steppe vegetation in Hungary and Romania at 7000 years B.P. The pollen maps show that this vegetation type became established there since this time. Work is in progress to use the modern data from 450 sites in order to calculate calibration functions for reconstructing maps of Holocene climates.

The Soviet Union

Neustadt[69] was the first to compile and map the pollen data from the Soviet Union; but, like the studies by Szafer[66] and Firbas,[67] radiocarbon dates were not used in producing his maps. Khotinskii[70] updated Neustadt's study by presenting data from 43 radiocarbon-dated pollen diagrams, and Peterson[71,94] has used these sites plus others in producing isopoll and isochrone maps for the major pollen types in the western Soviet Union (figure 12). His maps show broad scale patterns with both

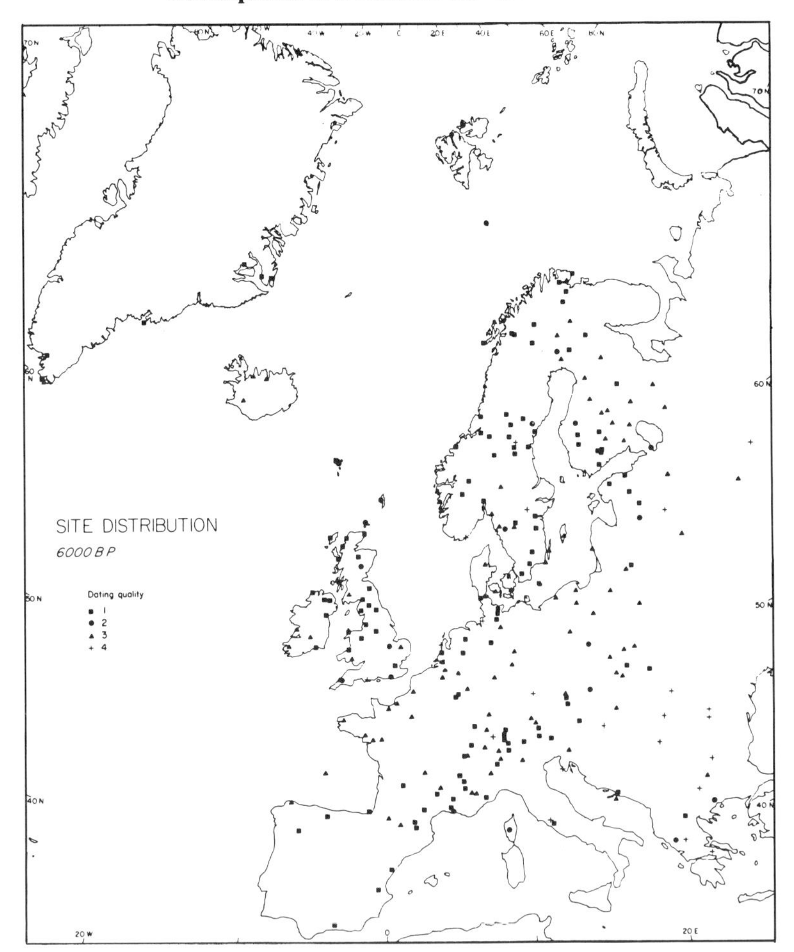

Figure 11. Sites with pollen data for 6000 years B.P. in Europe. Dating quality varies from good (1) to poor (4). See discussion in Huntley and Birks.[65] Reprinted by permission of Cambridge University Press.

north-south and east-west trends. Khotinskii[72] identified three different pollen stratigraphies in the western, central, and eastern Soviet Union and interpreted these differences in terms of distinct climatic histories in each of the three regions.

Peterson[94] has also compiled a data set of over 544 samples with modern pollen data from the Soviet Union west of 100°E. These maps show high abundances of herb pollen types in steppe and desert regions and a narrow band of pollen from thermophilous deciduous trees in the west. Climatic calibration of the fossil data indicates a temperature decrease of about 2°C since 6000 years B.P. at four sites near 57°N, 35°E.

New Zealand

Salinger[73] has recently compiled the data from radiocarbon-dated pollen diagrams in New Zealand (figure 13). In this initial study, sites were chosen whose pollen

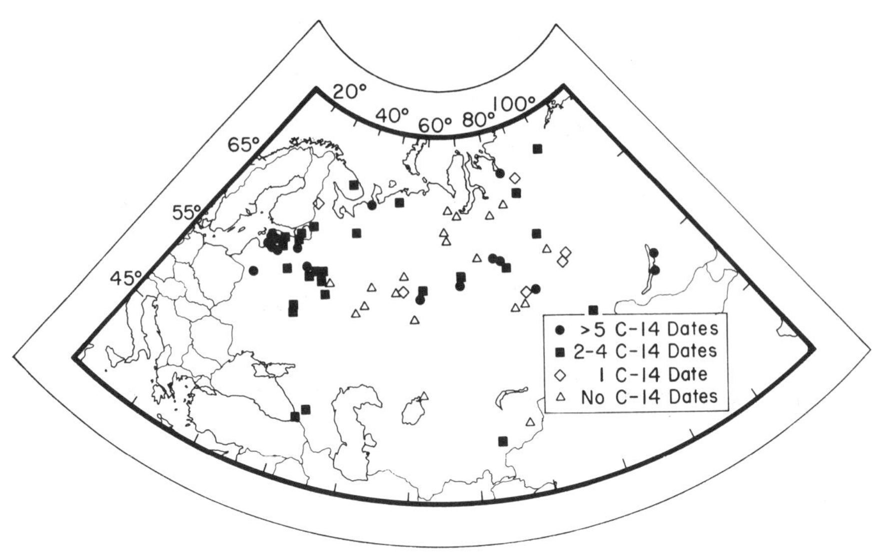

Figure 12. Location of sites with Holocene pollen data in the western USSR. Symbols and numbers show the number of radio-carbon dates at each site (from ref. 71).

data indicated wetter or drier conditions than today. The map for 6000 years B.P. shows regional patterns, and a set of modern data is being collected and analyzed for use in the calibration of the fossil data.

Climatic Calibration of Pollen Data in Other Regions

Besides the work in the Midwest and Arctic, calibration functions have been constructed for pollen data collected from terrestrial sites in northwest India,[52,53] Chile,[40] and southwest British Columbia[51] and from a deep-sea core off the Oregon coast.[74] The calibration work in the first of these areas features the use of independently derived climatic estimates to check the estimates derived from pollen data. The work in all four areas illustrates certain limitations to the method.

Singh's modern and fossil pollen data from the Rajasthan Desert[75-77] made it possible for Bryson and Swain[52] and for Swain et al.[53] to calculate precipitation estimates from the data collected at Lake Lunkaransar. A proposed lake level for Sambhar Lake at 6000 years B.P. also allowed Swain et al.[53] to use the energy budget model of Kutzbach[78] to gain an estimate for a 200 mm decrease in summer precipitation since 6000 years B.P. This estimate agrees with their estimates derived from the pollen data and is an independent quantitative verification of the calibration function result.

The general trends in the precipitation estimates are similar in the Bryson and Swain[52] and Swain et al.[53] studies, but the results of Bryson and Swain[52] have a much larger variance. Swain et al.[53] used a more conservative set of calibration equations than Bryson and Swain[52] and deleted certain pollen types whose past values were not matched in the modern data. This "no-analogue" condition is a limitation for all the types of data (e.g., marine plankton) to which calibration

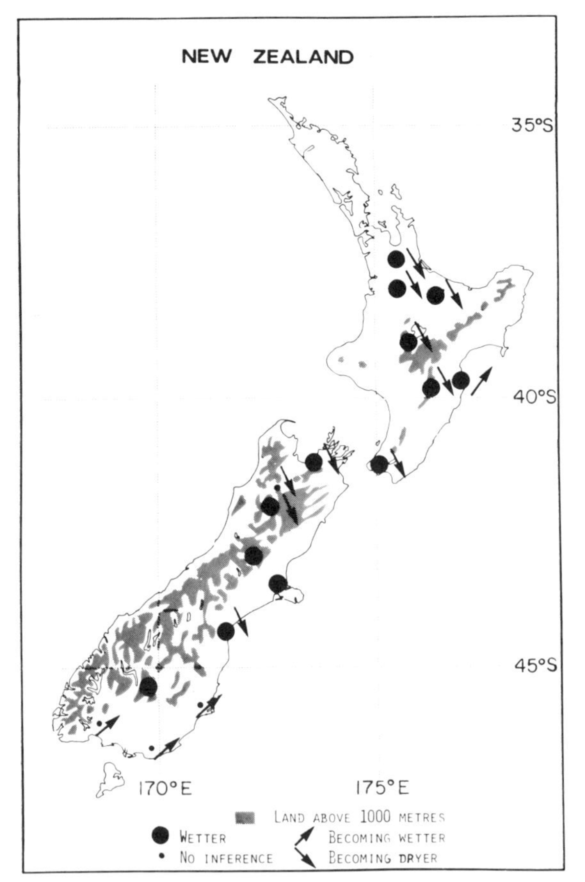

Figure 13. Location of sites with Holocene pollen data for 6000 years B.P. in New Zealand. The map symbols indicate the moisture conditions inferred for each site by Salinger.[73] Reprinted with permission of A. A. Bakema Publ.

functions are applied.[14,79] The decision to delete these pollen types eliminated much of the high-frequency variation evident in the time series presented by Bryson and Swain.[52]

The methods of Imbrie and Kipp[13] were used in calibrating a 16,000-year pollen record from a lake in Chile[40] and a 125,000-year pollen record in a deep-sea core off Oregon.[74] Problems exist with both of these studies. The estimates from the topmost sample in the Chilean lake underestimate the temperature today by 3°C and overestimate the precipitation today by 1 m. The Holocene temperature estimates have a range of 12°C, and the range in precipitation is over 4 m. These highly varying estimates probably result from a mismatch between the fossil and modern data similar to but more severe than that found in the Bryson and Swain[52] study. The calibration functions erred when the anomalously high fossil values forced these functions to extrapolate beyond the limits of the relationship that they model.

Increasing the geographical and ecological range of the set of modern samples in Chile may help eliminate this problem. Webb and Bryson[14] faced a similar problem for three pollen types in the Midwest and deleted these types from their analysis. Birks[31] criticized this practice. Subsequent research with different calibration methods and an increased set of modern data, however, has allowed these types to be included in the sets of modern and Holocene data.[43]

Heusser et al.[74] used a calibration function derived from modern pollen data from terrestrial sites in order to calibrate the pollen data in a deep-sea core off the Oregon coast. This practice is nonstandard and ignores the differences in pollen representivity between deep-sea and terrestrial sites. The data of Balsam and Heusser[80] and Heusser and Balsam[81] illustrate these differences when the pollen data from deep-sea cores are compared with pollen data from terrestrial sites.[27,82] Caution is therefore required in interpreting the climatic estimates gained from the deep-sea core. Before climatic calibration, studies are required to show how best to transform pollen percentages from deep-sea cores into percentages from terrestrial sites.

The Global Database for 18,000 years B.P.

Studies of the climatic conditions at 18,000 years B.P. show both the possibility and the value of assembling global data sets of paleoclimatic information.[21,22,83] Because global ice volume was at a maximum, the conditions at 18,000 years B.P. contrast markedly with the period of minimum global ice volume at 6000 years B.P. (see chapter 5). CLIMAP[21,83] defined the major boundary conditions required by general-circulation climate models, and Gates[84] and Manabe and Hahn[85] used these boundary conditions in models to simulate the global atmospheric patterns that may have existed at 18,000 years B.P. (chapter 8).

Peterson et al.[22] studied the terrestrial data from 18,000 years B.P. in order to gain independent paleoclimatic information that could be used to check the model simulations (figure 14). The broad geographic coverage needed in their study required compilation of many different types of geological information (table 3). They also stratified their data by the dating control at each site and used different map symbols to differentiate well-dated from poorly dated data (figure 14). Gaining accurate and precise dates for geological data is a major prerequisite for their use in paleoclimatic mapping studies. Imprecisely dated samples may result in a mixture of data from two time periods that differed markedly in their climatic patterns.

Peterson et al.[22] revealed the need both for improvements in the paleoclimatic data and for additional results from the global circulation models before full validation of the models' results could be completed. In 1979, simulation results were only available for July and August conditions at 18,000 years B.P.[84,85] The paleoclimatic data provided evidence for lower temperatures on most continents in good agreement with the model results,[84] but the temperature estimates from the geological evidence were too imprecise to check the geographic patterns in the model-simulated temperatures. For moisture conditions, former lake levels record the an-

Table 3. Continental Distribution of Conventional Weather-Observing Sites for Today and Paleoclimatic Sites for 18,000 years B.P.[a]

Region	Area (10^6 km^2)	Weather Observing Sites[b]				Paleoclimate Sites			
		Radiosonde/ rawinsonde sites	Sonde sites/ 10^6 km^2	Surface sites	Surface sites/10^6 km^2	Total sites	Total sites/10^6 km^2	Well- and moderately well-dated sites	Sites/10^6 km^2
North America	24.3	141	(5.8)	274	(11.3)	107	(4.4)[c]	66	(2.7)[c]
South America	17.8	17	(1.0)	84	(4.7)	20	(1.1)	7	(0.4)
Europe (includes European USSR)	10.5	142	(13.5)	815	(77.6)	71	(6.8)	26	(2.5)
Africa	30.3	45	(1.5)	277	(9.1)	65	(2.1)	35	(1.2)
Australia, Oceania	8.5	74	(8.7)	101	(11.9)	52	(6.1)	30	(3.5)
Asia (includes Asian USSR)	44.4	179	(4.0)	751	(16.9)	15	(0.3)	9	(0.2)
Antarctica	13.4	8	(0.6)	18	(1.3)	0	(0)	0	(0)
Total or (Average)	149.2	606	(4.1)	2320	(15.5)	330	(2.2)	173	(1.2)

[a]Reprinted with permission of *Quaternary Research.*

[b]Weather observing site information taken from WMO (1971): 606 radiosonde/rawinsonde sites report twice or more daily. 2320 surface sites report eight times daily.

[c]Site density would be approximately doubled if one takes into account that about one-half of North America was under an ice sheet.

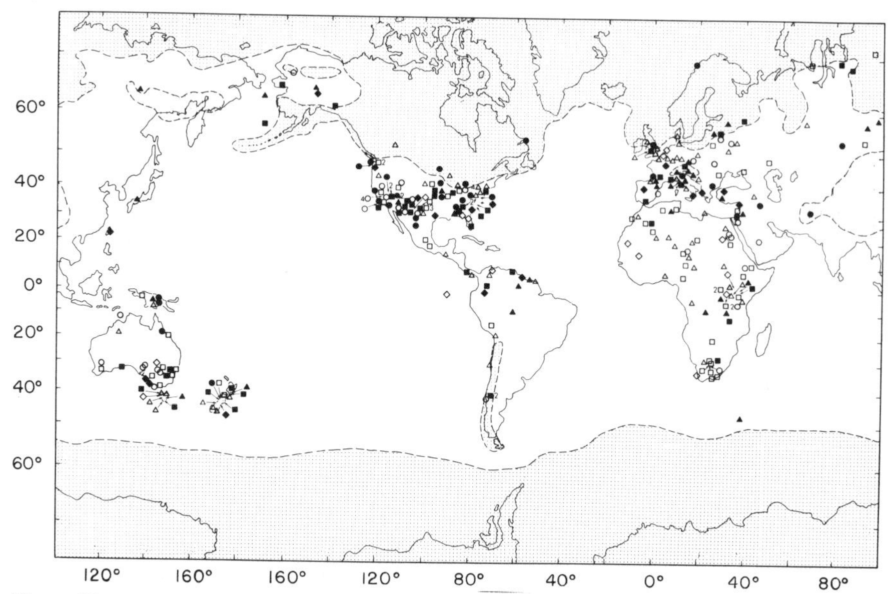

Figure 14. Location of all sites with paleobotanical (solid figures) or other geological (outline figures) evidence for 18,000 years B.P. Numerals indicate multiple sites or multiple sources of evidence at one location. Circles are well-dated sites with bracketing dates for the interval 23,000 to 13,000 years B.P. Squares are sites with one date in the interval 23,000 to 13,000 years B.P. and with the 18,000 years B.P. level located. Diamonds are sites with dates outside the interval 23,000 to 13,000 years B.P., but with the 18,000 years B.P. level located. Reprinted with permission of *Quaternary Research.*

nual average water budget and therefore can only be used to check simulation results that are representative of the whole annual cycle not just of July or August.

Global Databases for Holocene Climates

Compilation of global paleoclimatic data sets for selected dates during the Holocene reveal climatic conditions that contrast both with the climate of today and with that of 18,000 years B.P. Two dates were of initial interest, 9000 and 6000 years B.P. Data from 6000 years B.P. are broadly representative of the so-called Hypsithermal period that dates from 8000 to 4000 years B.P., and data from 9000 years B.P. represent the time when the earth's orbital position differed more from 18,000 years B.P. and today than it did at 6000 years B.P., when the global ice volume reached a minimum. A global data set for 9000 years B.P. can also be used to check the climate model simulations that Kutzbach[86] and Kutzbach and Otto-Bliesner[87] have recently produced.

The current COHMAP research effort follows the lead set by CLIMAP (Climate, Long-range Investigation, Mapping, and Prediction) which was an international interdisciplinary paleoclimatic research group. The major goals of the COHMAP research are (1) to provide global maps of the available data and the climatic patterns during the various phases of an interglacial period, (2) to estimate the global mean temperature at selected dates during the various phases of an interglacial period, (3) to obtain estimates of the rates and magnitudes of temperature and moisture changes in various regions, and (4) to use the data to check the results of climate model simulations for 6000 and 9000 years B.P. The paleoclimatic maps for 6000 years B.P. will show how the climatic patterns at the time of minimum ice volume contrast with the climatic reconstructions for 18,000 years B.P.[21,83–85] The warm period around 6000 years B.P. may also provide information about the climatic patterns of future climates that are warmer than today.[88] Kellogg and Schware[88] discussed this use of the data for understanding the possible climatic effects of increased levels of carbon dioxide in the atmosphere, but such use requires caution. The climatic patterns at 6000 and 9000 years B.P. will also help in testing the importance of the earth's orbital variations in influencing the climates of the past 12,000 years.[86]

The global data sets for 6000 and 9000 years B.P. were assembled by combining the dense networks of pollen data with other data sets; and, except for southeastern Asia, the central Pacific Ocean and the Arctic Ocean, few major gaps exist in the resulting global coverage (figure 15). When lake-level data were added to the pollen data, the density of sites compared favorably with the number of radiosonde stations in North America, Europe, Africa, and South America (table 4).

The global data sets for 6000 and 9000 years B.P. include pollen, lake level, glacial-moraine, marine plankton, and packrat-midden data. Radiocarbon dates provide the time control for each of these data sets; and the number, quality, and distribution of the dates at each site was used to assess the dating accuracy and precision of the data. Denton and Hughes[17] have reviewed the glacial data and the available radiocarbon dates. They used this information to produce isochrones for the margins of the continental ice sheets during the Holocene. Lake-level data provide most of the information from the arid lands in Africa, Australia, and western North America (chapter 7). Some sparsely distributed pollen data supplement the lake-level data in these continental regions as well as in South America. The coverage of marine plankton data is best in the Atlantic Ocean,[20] but radiocarbon-dated cores exist in the Indian[89] and Southern Oceans as well as the marginal regions of the Pacific Ocean. Slow sedimentation rates precluded use of any information for the Central Pacific and Arctic Oceans. Botanical data from packrat middens[18,90] provide an additional paleoclimatic database for western North America.

Estimates of climatic variables can be obtained from the pollen, marine plankton (chapter 5), and lake-level data (chapter 7). These estimates will be combined with maps of the main patterns in the vegetational and oceanographic data. These maps will provide a basis for inferring the major patterns in atmospheric circulation at selected dates during the Holocene. Comparison of the maps for different dates will show the dynamic relationships among the cryosphere, hydrosphere, atmosphere, and biosphere.

The compilation, mapping, and climatic calibration of pollen data are only one part of the major effort to map, model, and understand Holocene climatic change. The broad coverage and accurate dating of the pollen data make them an essential part of this paleoclimatic research.

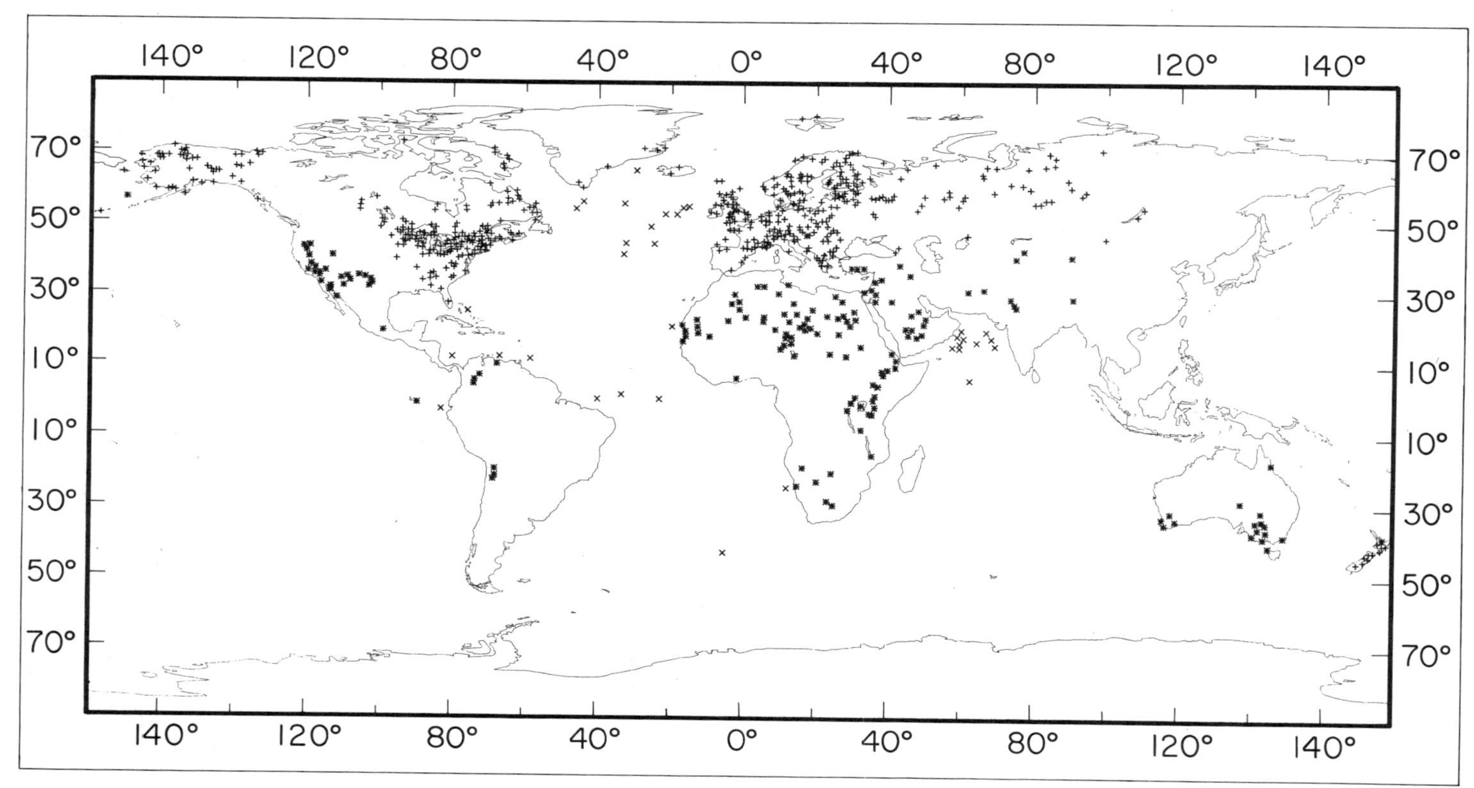

Figure 15. Location of sites in the global data base of Holocene data. Pluses locate sites with pollen, asterisks locate sites with lake-level data, and x's locate sites with marine plankton data.

Table 4. Continental Distribution of Sites with Holocene Data and with Radiosonde Stations Today

Region	Global Holocene Data Set		Radiosonde Stations
	Pollen	Lakes	
North America	400	42	141
Europe + Eur.	250	1	142
USSR	35	5	
Africa	10	90	45
Asia	25	30	179
Australia, Oceania	20	15	74
	Plankton		
Atlantic	25		
Pacific	10		
Indian	10		
Southern	4		
Marginal seas	6		

ACKNOWLEDGMENTS

A grant from the National Science Foundation Program of Climate Dynamics (ATM81-11897) and a contract (DE-AC02-79EV10079) from the U.S. Department of Energy Carbon Dioxide Research Division supported the writing of this chapter. P. J. Bartlein did the climatic calibrations in the Midwest and critically read an early draft. I thank R. Arigo, J. Avizinis, R. M. Mellor, and S. Suter for technical assistance and P. M. Anderson, L. B. Brubaker, B. Huntley, F. A. Perrott, G. M. Peterson, W. F. Ruddiman, and W. G. Spaulding for comments on a final draft.

REFERENCES

1. W. Köppen, "Versuch einer Klassifikation der Klimate, vorsugsweise nach ihren Bezeihungen zur Pflanzenwelt," *Geogr. Z.*, **6,** 593–611; 657–679 (1900).
2. J. R. Borchert, "The climate of the central North America grassland," *Ann. Assoc. Am. Geog.*, **40,** 1–29 (1950).
3. R. A. Bryson, "Air masses, streamlines, and the boreal forest," *Geogr. Bull.*, **8,** 228–269 (1966).
4. J. S. Krebs and R. G. Barry, "The Arctic front and the tundra-taiga boundary in Eurasia," *Geogr. Rev.*, **40,** 548–554 (1970).
5. M. J. Burke, L. V. Gusta, H. A. Quamme, C. J. Weiser, and P. H. Li, "Freezing injury in plants," *Ann. Rev. Plant Physiol.*, **27,** 505–528 (1976).
6. T. A. Wijmstra, "Palynology of the first 30 metres of a 120 m deep section in northern Greece," *Acta Bot. Neerl.*, **18,** 511–527 (1969).
7. G. M. Woillard and W. G. Mook, "Carbon-14 dates at Grande Pile: Correlation of land and sea chronologies," *Science*, **215,** 159–161 (1982).

8. A. M. Swain, "A history of fire and vegetation in northeastern Minnesota as recorded in lake sediments," Ph.D. diss. University of Minnesota, Minneapolis, 1974.
9. T. Webb III, "Temporal resolution in Holocene pollen data," *Proc. 3rd North Am. Paleontol. Conv.*, **2,** 569–572 (1982).
10. T. Webb III and D. R. Clark, "Calibrating micropaleontological data in climatic terms: A critical review," *N.Y. Acad. Sci., Ann.*, **288,** 93–118 (1977).
11. S. E. Howe and T. Webb III, "Calibrating pollen data in climatic terms: Improving the methods," *Quat. Sci. Rev.*, **2,** 17–51 (1983).
12. H. C. Fritts, *Tree Rings and Climate,* Academic, London, 1976.
13. J. Imbrie and N. G. Kipp, "A New Micropaleontological Method for Paleoclimatology: Application to a Late Pleistocene Caribbean Core." In K. K. Turekian, ed., *The Late Cenozoic Glacial Ages,* Yale University Press, New Haven, 1971, pp. 71–181.
14. T. Webb III and R. A. Bryson, "Late- and postglacial climatic change in the northern Midwest, U.S.A.: Quantitative estimates derived from fossil pollen spectra by multivariate statistical analysis," *Quat. Res.*, **2,** 70–115 (1972).
15. R. Arigo, S. E. Howe, and T. Webb III, "Computer programs for climatic calibration of pollen data." In B. E. Berglund, ed., *Palaeohydrological Changes in the Temperate Zone in the Last 15,000 Years, IGCP 158B. Lake and Mire Environments. Project Guide Vol. 3. Specific Methods,* Lund, Sweden, 1982, pp. 79–109.
16. P. J. Bartlein and T. Webb III, "Annotated Computer Programs for Climatic Calibration of Pollen Data: A User's Guide." *American Association of Stratigraphic Palynologists Contribution Series,* in press.
17. G. H. Denton and T. J. Hughes, eds., *The Last Great Ice Sheets,* Wiley, New York, 1981.
18. W. G. Spaulding, E. B. Leopold, and T. R. Van Devender, "Late Wisconsin Paleoecology of the American Southwest." In S. C. Porter, ed., *Late-Quaternary Environments of the United States,* Vol. 1, *The Late Pleistocene,* University of Minnesota Press, Minneapolis, 1983, pp. 259–293.
19. F. A. Street and A. T. Grove, "Global maps of lake-level fluctuations since 30,000 B.P." *Quat. Res.*, **10,** 83–118 (1979).
20. W. F. Ruddiman and A. McIntyre, "The North Atlantic Ocean during the last deglaciation," *Palaeogeogr., Palaeoclimatol., Palaeoecol.*, **35,** 145–214 (1981).
21. CLIMAP Project Members, "The surface of the ice-age earth," *Science,* **191,** 1131–1136 (1976).
22. G. M. Peterson, T. Webb III, J. E. Kutzbach, T. van der Hammen, T. A. Wijmstra, and F. A. Street, "The continental record of environmental conditions at 18,000 yr B.P.: An initial evaluation," *Quat. Res.*, **12,** 47–82 (1979).
23. T. Webb III, E. J. Cushing, and H. E. Wright, Jr., "Holocene Changes in the Vegetation of the Midwest." In H. E. Wright, Jr., ed., *Late-Quaternary Environments of the United States,* Vol. 2, *The Holocene,* University of Minnesota Press, 1983, pp. 142–165.
24. H. J. B. Birks and H. H. Birks, *Quaternary Palaeoecology,* Edward Arnold, London, 1980.
25. L. J. Maher, "Nomograms for computing 0.95 confidence limits of pollen data," *Rev. Palaeobot. Palynol.*, **13,** 85–93 (1972).
26. J. C. Bernabo and T. Webb III, "Changing patterns in the Holocene pollen record of northwestern North America: A mapped summary," *Quat. Res.*, **8,** 64–96 (1977).
27. R. B. Davis and T. Webb III, "The contemporary distribution of pollen in eastern North America: A comparison with the vegetation," *Quat. Res.*, **5,** 395–434 (1975).
28. T. Webb III and J. H. McAndrews, "Corresponding patterns of contemporary pollen and vegetation in central North America," *Geol. Soc. Am. Mem.*, **145,** 267–299 (1976).
29. P. A. Delcourt, H. R. Delcourt, and T. Webb III, "Atlas of paired isophyte and isopoll maps for important tree taxa of eastern North America," *American Association of Stratigraphic Palynologists Contribution Series,* No. 14, 1-131 (1984).
30. T. Webb III, "The reconstruction of climatic sequences from pollen data," *J. Interdisciplinary Hist.*, **10,** 749–772 (1980).

31. H. J. B. Birks, "The Use of Pollen Analysis in the Reconstruction of Past Climates: A Review." In T. M. L. Wigley, M. J. Ingram, and G. Farmer, eds., *Climate and History,* Cambridge University Press, Cambridge, 1981, pp. 111–138.

32. L. von Post, "The prospect for pollen analysis in the study of the Earth's climatic history," *New Phytol.,* **45,** 193–217 (1946).

33. T. Webb III, P. Richard, and R. J. Mott, "Mapped summary of the Holocene vegetational history of southern Quebec," *Syllogeus,* No. 14, 273–336 (1983).

34. M. B. Davis, R. W. Spear, and L. C. K. Shane, "Holocene climate of New England, *Quat. Res.,* **14,** 240–250 (1980).

35. H. E. Wright, Jr. "Late Quaternary Vegetation History of North America." In K. K. Turekian, ed., *The Late Cenozoic Glacial Ages,* Yale University Press, New Haven, 1971, pp. 425–464.

36. W. A. Watts and T. C. Winter, "Plant macrofossils from Kirchner Marsh, Minnesota—A paleoecological study," *Geol. Soc. Am. Bull.,* **79,** 855–876 (1966).

37. M. B. Davis, "Climatic Interpretation of Pollen in Quaternary Sediments." In D. Walker and J. C. Guppy, eds., *Biology and Quaternary Environments,* Australian Academy of Science, Canberra, 1978, pp. 35–51.

38. T. Webb III, "A comparison of modern and presettlement pollen in southern Michigan (U.S.A.)," *Rev. Palaeobot. Palynol.,* **16,** 137–156 (1973).

39. M. B. Davis, "Quaternary History and the Stability of Forest Communities." In D. C. West, H. H. Shugart, and D. B. Botkin, eds., *Forest Succession,* Springer-Verlag, New York, 1981, pp. 132–153.

40. C. J. Heusser and S. S. Streeter, "A temperature and precipitation record of the past 16,000 years in southern Chile," *Science,* **210,** 1345–1347 (1980).

41. J. T. Andrews, W. M. Mode, and P. T. Davis, "Holocene climate based on pollen transfer functions, eastern Canadian Arctic," *Arctic Alp. Res.,* **12,** 41–64 (1980).

42. P. J. Bartlein and T. Webb III, "Paleoclimatic Interpretation of Holocene Pollen Data: Statistical Considerations." *American Association of Stratigraphic Palynologists Contribution Series,* in press.

43. P. J. Bartlein, T. Webb III and E. Fleri, "Holocene climatic change in the northern Midwest: Pollen-derived estimates," *Quat. Res.,* **22,** 361–374 (1984).

44. J. C. Knox, "Valley alluviation in southwestern Wisconsin," *Ann. Assoc. Am. Geogr.,* **62,** 401–410 (1972).

45. J. C. Knox, P. F. McDowell, and W. C. Johnson, "Holocene Fluvial Activity and Climatic Change in the Driftless Area, Wisconsin." In W. C. Mahaney, ed., *Quaternary Paleoclimate,* Geo Abstracts, Norwich, England, 1981, pp. 107–127.

46. P. F. McDowell, "Evidence of stream response to Holocene climatic change in a small Wisconsin watershed," *Quat. Res.,* **19,** 100–116 (1983).

47. P. A. Kay, "Multivariate statistical estimates of Holocene vegetation and climate change, forest-tundra transition zone, N.W.T., Canada," *Quat. Res.,* **11,** 125–140 (1979).

48. J. T. Andrews and H. Diaz, "Eigenvector analysis of reconstructed Holocene July temperature departures over northern Canada," *Quat. Res.,* **16,** 373–389 (1981).

49. J. T. Andrews, P. T. Davis, W. N. Mode, H. Nichols, and S. K. Short, "Relative departures in July temperatures in northern Canada for the past 6000 yrs.," *Nature,* **289,** 164–167 (1981).

50. J. T. Andrews and H. Nichols, "Modern pollen deposition and Holocene paleotemperature reconstructions, central northern Canada," *Arctic Alp. Res.,* **13,** 387–408 (1981).

51. R. W. Mathewes and L. E. Heusser, "A 12,000 year palynological record of temperature and precipitation trends in southwestern British Columbia," *Can. J. Bot.,* **59,** 707–710 (1981).

52. R. A. Bryson and A. M. Swain, "Holocene variations of monsoon rainfall in Rajasthan," *Quat. Res.,* **16,** 135–145 (1981).

53. A. M. Swain, J. E. Kutzbach, and S. Hastenrath, "Monsoon climate of Rajasthan for the Holocene: Estimates of precipitation based on pollen and lake levels," *Quat. Res.,* **19,** 1–17 (1983).

54. J. C. Ritchie, "The late-Quaternary vegetational history of the Western Interior of Canada," *Can. J. Bot.*, **54**, 1793–1818 (1976).

55. P. A. Delcourt and H. R. Delcourt, "Vegetation Maps for Eastern North America: 40,000 Yr B.P. to the Present." In R. C. Romans, ed., *Geobotany II*, Plenum, New York, 1981, pp. 123–165.

56. R. B. Davis and G. F. Jacobson, Jr., "Late-glacial and early postglacial vegetation in northern New England and adjacent areas of Canada," *Quat. Res.*, in press.

57. M. B. Davis, "Late-Glacial Climate in Northern United States: A Comparison of New England and the Great Lakes Region." In E. J. Cushing and H. E. Wright, Jr., eds., *Quaternary Paleoecology*, Yale University Press, New Haven, 1967, pp. 11–43.

58. G. M. Peterson, "Pollen spectra from surface sediments of lakes and ponds in Kentucky, Illinois, and Missouri," *Am. Midl. Nat.*, **100**, 333–340 (1978).

59. D. L. Elliot-Fisk, J. T. Andrews, S. K. Short, and W. N. Mode, "Isopoll maps and an analysis of the distribution of the modern pollen rain, eastern and central northern Canada," *Geogr. Phys. Quat.*, **36**, 91–108 (1982).

60. P. A. Delcourt, H. R. Delcourt, and J. L. Davidson, "Modern pollen-vegetation relationships in the southeastern United States," *Rev. Palaeobot. Palynol.* **39**, 1–45 (1983).

61. P. M. Anderson, "Reconstructing the past: The synthesis of archeological and palynological data, northern Alaska and northwestern Canada," Ph.D. diss. Brown University, Providence, 1982.

62. T. Ager, "Vegetational History of Western Alaska During the Wisconsin Glacial Interval and the Holocene." In D. M. Hopkins, J. V. Matthews, Jr., C. E. Schweger, and S. B. Young, eds., *Paleoecology of Beringia*, Academic, New York, 1982, pp. 75–93.

63. L. B. Brubaker, H. Garfinkel, and M. Edwards, "A late Wisconsin and Holocene vegetation history from the central Brooks Range: Implications for Alaskan paleoecology." *Quat. Res.*, **20**, 194–214 (1983).

64. J. C. Ritchie and L. C. Cwynar, "The Late Quaternary Vegetation of the North Yukon." In D. M. Hopkins, J. V. Matthews, Jr., C. E. Schweger, and S. B. Young, eds., *Paleoecology of Beringia*, Academic, New York, 1982, pp. 113–126.

65. B. Huntley and H. J. B. Birks, *An Atlas of Past and Present Pollen Maps for Europe: 0-13,000 Years Ago*, Cambridge University Press, Cambridge, 1983.

66. W. Szafer, "The significance of isopollen lines for the investigation of the geographical distribution of trees in the postglacial period," *Bull. Acad. Polon. Sci. Lett.*, Series B, **1**, 235–239 (1935).

67. R. Firbas, *Spät- und nacheiszeitliche Waldgeschichte Mitteleuropas nordlich der Alpen*, Gustav Fischer, Jena, 1949.

68. M. Ralska-Jasiewiczowa, "Isopollen maps for Poland: 0-11,000 years B.P." *New Phytol.*, **94**, 133–175 (1983).

69. M. I. Neustadt, *Isotriia Lesov i Paleogeografiia SSR v Golotseue*, Nauka, Moscow, 1957 (in Russian).

70. N. A. Khotinskii, *Golotsen Severnoi Evrazii* (*Holocene of the Northern Eurasia*), Nauka, Moscow, 1977 (in Russian).

71. G. M. Peterson, "Holocene vegetation and climate in the Western USSR," Ph.D. diss. University of Wisconsin, Madison (1983).

72. N. A. Khotinskii, "Transcontinental correlation of stages in the history of vegetation and climate of northern Eurasia in the Holocene," In *Problemy Palinologii. Trudy III Mezhdunarodnoi Palinologicheskoi Konferentsii*, Nauka, Moskva, 1973, pp. 116–123. (English translation by G. M. Peterson available from the Center for Climatic Research, 1225 W. Dayton, Madison, WI 53706.)

73. M. J. Salinger. "New Zealand climate: The last 5 million years." In J. C. Vogel, ed. *Late Cainozoic Palaeoclimates of the Southern Hemisphere*, Balkema, Rotterdam, The Netherlands, 1984.

74. C. J. Heusser, L. E. Heusser, and S. S. Streeter, "Quaternary temperatures and precipitation for the north-west coast of North America," *Nature*, **286**, 702–704 (1980).

75. G. Singh, "The Indus Valley culture seen in the context of postglacial climatic and ecological studies in northwest India," *Archaeol. Phys. Anthropol. Oceania,* **6,** 177–189 (1971).

76. G. Singh, S. K. Chapra, and A. B. Singh, "Pollen-rain from the vegetation of northwest India," *New Phytol.,* **72,** 191–206 (1973).

77. G. Singh, R. D. Joshi, S. K. Chapra, and A. B. Singh, "Late Quaternary history of vegetation and climate of the Rajasthan Desert, India." *Philos. Trans. Roy. Soc. London, B,* **267,** 467–501 (1974).

78. J. E. Kutzbach, "Estimates of past climate at Paleolake Chad, North Africa, based on a hydrological and energy-balance model," *Quat. Res.,* **14,** 210–223 (1980).

79. W. H. Hutson, "Transfer functions under no-analog conditions: Experiments with Indian Ocean planktonic foraminifera," *Quat. Res.,* **8,** 355–367 (1977).

80. W. L. Balsam and L. Heusser, "Direct correlation of sea surface paleotemperatures, deep circulation, and terrestrial paleoclimate: Foraminiferal and palynological evidence from two cores off Chesapeake Bay," *Mar. Geol.,* **21,** 121–147 (1976).

81. L. E. Heusser and W. L. Balsam, "Pollen distribution in the northeast Pacific Ocean," *Quat. Res.,* **7,** 45–62 (1977).

82. C. J. Heusser, "Modern pollen rain of Washington," *Can. J. Bot.,* **56,** 1510–1518 (1978).

83. CLIMAP Project Members, "Seasonal reconstructions of the earth's surface at the last glacial maximum," GSA Map and Chart Series, MC-36, 1–18 (1981).

84. W. L. Gates, "The numerical simulation of ice-age climate with a global circulation model," *J. Atmos. Sci.,* **33,** 1844–1873 (1976).

85. S. Manabe and D. G. Hahn, "Simulation of an ice age," *J. Geophys. Res.,* **82,** 3889–3911 (1977).

86. J. E. Kutzbach, "Monsoon climate of the early Holocene: Climate experiment with the earth's orbital parameters for 9000 years ago," *Science,* **214,** 59–61 (1981).

87. J. E. Kutzbach and B. L. Otto-Bliesner, "The sensitivity of the African-Asian monsoonal climate to orbital parameter changes for 9000 years B.P. in a low-resolution general circulation model," *J. Atmos. Sci.,* **39,** 1177–1188 (1982).

88. W. W. Kellogg and R. Schware, *Climate Change and Society,* Westview Press, Boulder, Colo. (1981).

89. W. L. Prell, "Monsoonal Climate of the Arabian Sea During the Late Quaternary: A Response to Changing Solar Radiation." In J. Hanson and T. Takahashi, eds., *Climate Processes and Climate Sensitivity,* M. Ewing, Volume 5, Geophysical Monograph 29, American Geophysical Union, Washington, D.C., 1984, pp. 48–57.

90. K. L. Cole, "Late Quaternary zonation of vegetation in the eastern Grand Canyon," *Science,* **217,** 1142–1145 (1982).

91. J. C. Bernabo, "Proxy data: Nature's records of past climates," EDS, NOAA, U.S. Department of Commerce. Washington, D.C., 1–8 (1978).

92. H. E. Wright, Jr., T. C. Winter, and H. L. Pattern, "Two pollen diagrams from southeastern Minnesota: Problems in the late- and postglacial vegetational history," *Geol. Soc. Am. Bull.,* **74,** 1371–1396 (1963).

93. P. J. Bartlein and T. Webb III, "Mean July temperature at 6000 yr B.P. in eastern North America: regression equations for estimates from fossil pollen data," *Syllogeus* (1984), in press.

94. G. M. Peterson, "Recent pollen spectra and zonal vegetation in the western USSR," *Quat. Sci. Rev.,* **2,** 281–321 (1984).

5

CLIMATE STUDIES IN OCEAN CORES

William F. Ruddiman

Lamont-Doherty Geological Observatory of Columbia University
*Palisades, New York**

The oceans are the major source of global climatic information on time scales of 10,000 years to 10 million years and at sample resolutions in the range of 1,000 to 10,000 years. The highest resolution (down to tens of years) is attained in a few oxygen-poor ocean basins scattered on the continental margins. The longest records, up to tens of millions of years, are those at some of the 600 sites sampled by the Deep-Sea Drilling Project. But for the most part, ocean cores contain climatic records averaging around 100,000 to 200,000 years in length and with details resolvable to a few thousand years.

Some areas of the ocean have proven more rewarding than others; these tend to be areas with relatively high rates of sediment deposition from overlying surface waters; a rich, diverse, and well-preserved microfossil record; and the absence of major physical or chemical alteration of the original paleoclimatic data. These areas generally lie within 10° of the equator, in mid-to-high latitude regions, and along coastal margins. The infertile subtropical gyres and deepest basins are the poorest sources of climatic data.

Deep-sea cores have accurately been described as multi-channel climatic recorders; from among their numerous climate-related signals, two are emphasized here. The oxygen isotopic composition of calcareous microfossils is widely regarded as a first-order indicator of global ice volume. The species composition of the calcareous and siliceous microplankton is similarly useful as an indicator of local sea-surface temperature (SST). Investigations of these two climatic signals (global ice volume, local SST) are leading to a revolutionary new understanding of the earth's climatic history on the ice age time scales.

*Lamont-Doherty Geological Observatory contribution number 3729.

HISTORICAL DEVELOPMENT OF CORE COLLECTIONS

Phase I: Expeditions (1880–1947)

During the first, and longest, phase of core collecting, a few large scale expeditions were launched as major national efforts to undertake exploration of the sea. Beginning with the remarkably comprehensive expedition of the H.M.S. *Challenger* (1872–1876), and continuing through other efforts such as the German South Polar Expedition in 1901–1903, the German research efforts aboard *Meteor* in 1925–1927, the Dutch *Snellius* Expedition in 1929–1930, and the Swedish Deep-Sea Expedition in 1947–1948, this exploratory phase gathered a sufficient collection of short gravity cores, dredges, and grab samples to enable scientists to characterize for the first time the nature of bottom sediments in many ocean areas. In some regions, cores a few meters in length were collected, and geologists found very different kinds of sediments just a few tens of centimeters below those at the sediment-water interface.[1-5] These underlying sediments were correctly attributed to the very different climatic regime of the last glaciation. Toward the end of this exploratory phase, the frequency of cruises and sample collecting began to increase, as a prologue to the explosion of exploration that began after World War II.

Phase II: Toward Global Coverage (1947–1972)

Spurred initially by a postwar influx of funding from the Office of Naval Research in the U.S. Department of Defense, and later by a similar increase in support from the newly created National Science Foundation, the oceanographic institutions of the United States quickly began to lead other countries toward a truly global knowledge of the sea floor. Among the dozen or so United States institutions engaged in this effort, one stands out prominently. The Lamont-Doherty Geological Observatory of Columbia University gathered within the space of 25 years some 10,000 core samples, most of which were piston cores 3" in diameter and generally 3 to 10 meters in length (figure 1). This collection, now 13,000 piston cores and 16,000 cores in all, represents about half of the U.S. total and almost that fraction of the world total. Much of this is the result of the efforts of one man, Maurice Ewing, who, as the first Director of Lamont, saw to it that the institution's two ships (*Vema* and *Conrad*) were not only kept at sea some 300 or more days per year, but were also required to take at least one core each day, regardless of the primary scientific purpose of the cruise. With two ships criss-crossing the entire world ocean in pursuit of diverse geophysical/geological studies, this collection grew to such a size that it has been designated a national repository. Other institutions sent ships to take cores at more selected scientific targets during this period, including efforts in the Atlantic and Mediterranean by the Woods Hole Oceanographic Institution, in the Pacific by the Scripps Institution of Oceanography, and in the Antarctic Ocean by Florida State University.

Phase II established both the distribution and mode of origin of surface and

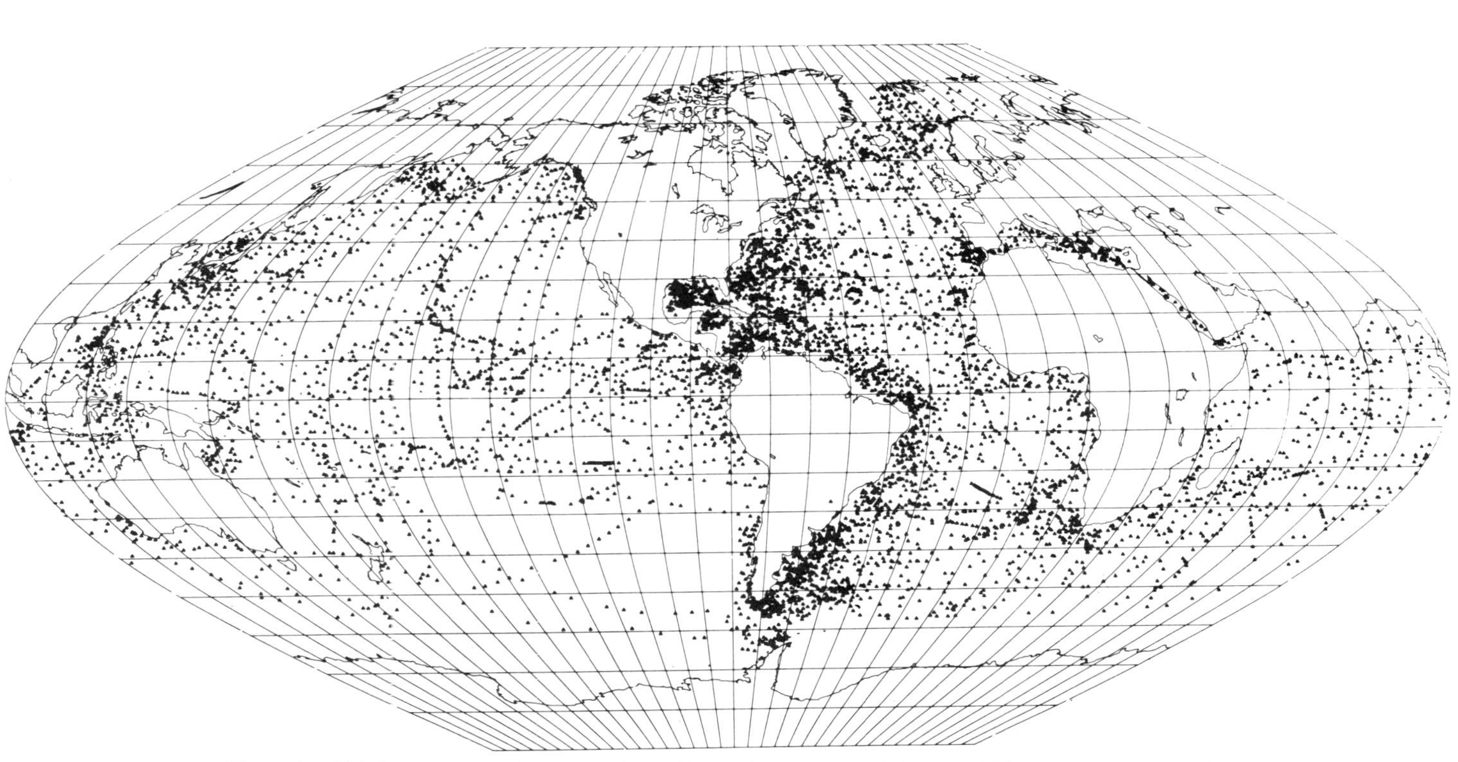

Figure 1. Global coverage of piston cores in world repositories, updated through 1978 (after ref. 6). Several hundred cores have since been added to the 15,000 cores shown.

shallow subsurface sediments across the entire world ocean. It also put in place the basic collection of cores from which the major paleoclimatic studies of the next decade would emerge.

Phase III: New Techniques (1972–present)

After the very strong U.S. Government support during the 1950s and 1960s, oceanographic funding leveled off and began to decline somewhat (in uninflated dollars) during the 1970s and early 1980s. This, and the attainment of a thorough first order global coverage of cores, brought to an end the era of broad surveys of the ocean. In its place, several new trends developed, including: smaller scale and more detailed studies of geologic processes now or once active on the sea floor, increased interest in the early (pre-Quaternary) history of the sea floor, and new techniques for retrieving sediment samples.

The process-oriented studies led to improvements in techniques for coring the uppermost layers of sediment in an undisturbed fashion, including box cores capable of sampling areas of the sea floor much larger than those retrieved in conventional coring. Interest in the early history of the sea floor led first to the rotary drill technique used aboard the Deep-Sea Drilling Project Vessel Challenger (beginning in 1966). This technique is capable of obtaining long, nearly continuous cores down to the oceanic basement, although the late Quaternary sediments were usually lost or badly disturbed during "spudding-in" of the drill string. In the late 1970s, the Hydraulic Piston Coring (HPC) device capable of collecting very long and continuous sediment sections without disturbance was developed, and initial cores were taken in 1979.

Meanwhile, during the 1970s, a small community of geologists known as "paleoceanographers" had accelerated several intensive laboratory-based efforts to unravel the late Quaternary climatic history of the oceans. Those efforts, and the factors constraining them, are the focus of this chapter.

EXISTING DATABASE FOR OCEAN-CLIMATE RESEARCH

Biological, chemical, and physical processes act upon the ocean sediments to create geologic records that differ widely in paleoclimatic usefulness, both regionally and through time.

Spatial Coverage

In general, the most informative paleoclimatic studies emerge from regions in which the sediments are dominantly pelagic in origin (falling from the surface waters); deposited at moderate to high deposition rates (1–10 cm/1,000 years); and unaffected by processes causing chemical changes, displacement, or mixing of the sediments. These constraints tend to eliminate paleoclimatic studies in the subtropical gyres (with low deposition rates due to infertile surface waters), in ocean basins deeper than 4,500–5,000 meters (where biotic components are dissolved by long exposure to corrosive bottom water), and along some continental margins (those marked by major turbidity-current flows downslope or by unusually strong contour-following bottom currents).

The major siliceous components of the shelled microplankton used for paleoclimatology (figure 2) include sand-sized zooplankton (radiolaria) and silt-sized phytoplankton (diatoms). The distributions of these components are largely controlled by surface-water productivity.[8,9] Optimal preservation occurs in regions where the rain of siliceous shells from fertile surface waters is largest, particularly the subpolar and equatorial divergences and the regions of coastal upwelling (figure 3). Foremost

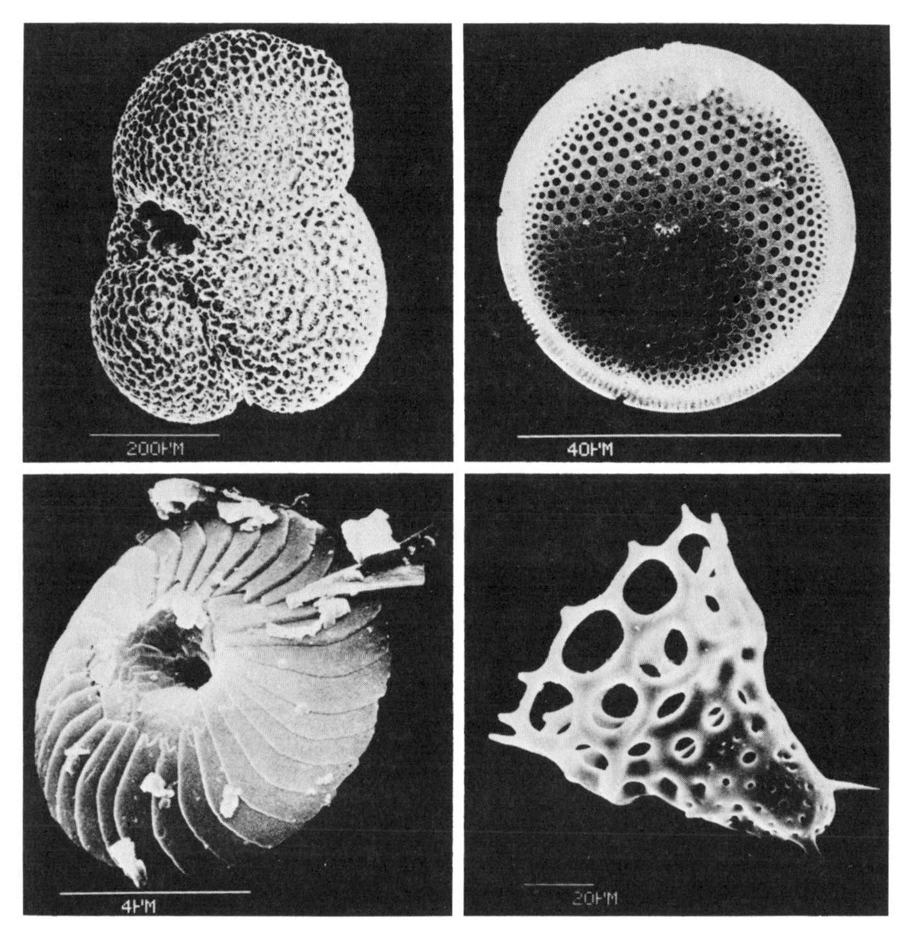

Figure 2. Examples of major biotic components used for deep-sea paleoclimatic studies. Upper left, planktonic foramisifera, sand-sized calcareous zooplankton; lower left, coccolithophorida, clay-sized calcareous phytoplankton; upper right, diatoms, silt-sized siliceous phytoplankton; lower right, radiolaria, sand-sized siliceous zooplankton. Bars indicate sizes in microns (10^{-6} meters). Benthonic foraminifera are sand-sized calcareous organisms.

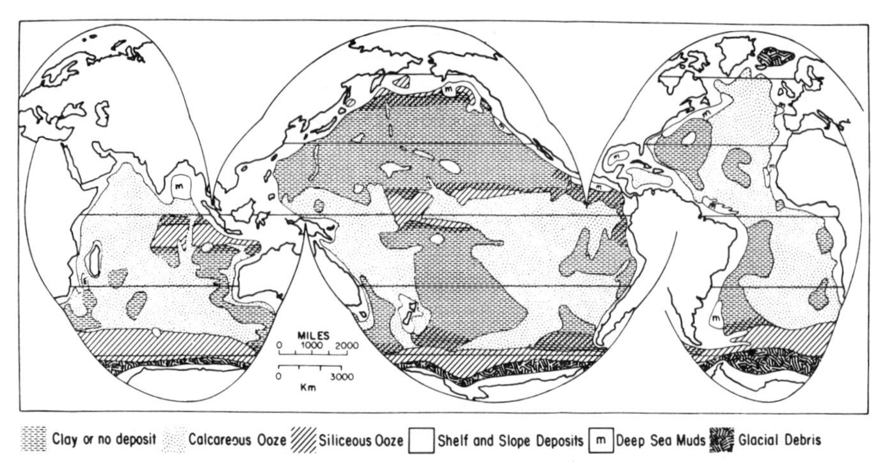

Figure 3. Global distribution of major recent sediment types on the deep-ocean floor (from ref. 10). Siliceous sediments primarily occur in high-latitude and equatorial belts, calcareous sediments on topographic highs.

among these are the Antarctic convergence, the equatorial Pacific divergence, and the boundary currents along the eastern margins of the South Pacific and South Atlantic. The Atlantic Ocean is poorest in silica, and the Pacific and Antarctic Oceans are the richest.

The calcareous components of the shelled microplankton deposited in sediments are the sand-sized zooplankton, foraminifera, and the clay-sized phytoplankton, coccolithophores (figure 2). These components are best preserved on shallow topographic highs lying above deep waters that are corrosive to the $CaCO_3$ shells. Foraminifera and coccoliths are widely used in the North and equatorial Atlantic Ocean, which are the shallowest of the ocean basins and contain the least corrosive bottom waters (figure 3). In sequential order, the South Atlantic, Indian, South Pacific, and North Pacific Oceans are progressively deeper and increasingly bathed by corrosive bottom waters, with mean $CaCO_3$ preservation declining progressively away from the North Atlantic (figure 3). Nevertheless, paleoclimatic research on calcareous plankton is still feasible on shallow topographic highs even in the Pacific Ocean.

Temporal Resolution

The second constraint on paleoclimatic research in the oceans is sediment mixing by bottom-dwelling organisms, which significantly smooth the slowly accumulating oceanic records.[11–13] In a few exceptional locations, anoxic bottom waters have suppressed the bottom fauna sufficiently to allow the accumulation of undisturbed sediments with a potential climatic resolution of hundreds or even tens of years. Generally, however, significant sediment stirring occurs to a depth of 2 to 10 cm[14,15] and records deposited at rates of 1 to 10 cm/1000 yr cannot yield resolution finer than a few thousand years. In general, resolution is highest in the areas of fastest accumulation of siliceous and calcareous sediments. In some areas near the continental margins, detritus from the continents is deposited along with microfossil-bearing pelagic sediments, thus increasing rates of deposition and improving paleoclimatic resolution. This improvement is, however, often obtained at a price: there is also an increased chance of highly nonlinear sedimentation rates, in contrast to the relatively constant rates of deposition typical of much of the open ocean.

OCEANIC PALEOCLIMATIC INDICATORS

The two kinds of climatic signals under focus here are: (1) reconstructions of surface-water environments from shelled microplankton that inhabited the upper few hundred meters of the ocean, and (2) estimates of global ice volume from the stable oxygen isotopic composition of organisms that lived either on the sea floor (benthonic foraminifera) or in surface waters in regions where environmental changes were minor (planktonic foraminifera). This treatment will necessarily omit several other signals that have shown significant climatic value: changes in sediment composition (calcium carbonate, opaline silica, ice-rafted sand, windblown quartz) and in stable carbon isotopic composition. The proxy records of sea-surface temperature and ice volume have yielded the quantitative climatic estimates most useful for spectral and time-series analysis and for time-slice modeling.

$\delta^{18}O$ As An Ice Volume Indicator

The oxygen isotopic composition of the calcitic shells of foraminifera varies primarily with two factors: (1) temperature, through the thermodynamic relationship,[16] and (2) the mean isotopic composition of ocean water. On the scale of ice age climatic changes, the isotopic composition of ocean water is most strongly affected by the storage of ^{16}O-rich water in the ice sheets. The use of oxygen-isotopic variations in the foraminiferal shells as ice volume signals thus depends critically on the degree to which temperature (and other) effects can be shown to be minimized.

The standard measure of oxygen-isotopic variations is given as the per-mil difference from a reference standard (PDB):

$$\delta^{18}O = \frac{^{18}O/^{16}O \text{ sample} - {}^{18}O/^{16}O \text{ standard}}{^{18}O/^{16}O \text{ standard}} \times 1000$$

The per-mil form is used because ^{16}O is more than 3 orders of magnitude more abundant than ^{18}O.

For over 15 years (1954–1970), the $\delta^{18}O$ analyses made by Emiliani[17,18] on planktonic foraminiferal shells comprised the only work in this field. During this phase, the $\delta^{18}O$ values were usually converted directly to a temperature scale, with a 4.2°C temperature change derived from a 1‰ oxygen-isotopic shift. This kind of use of the $\delta^{18}O$ record was only possible under the assumption that the distillation processes involved in transferring moisture from relatively low-latitude oceans to high-latitude ice sheets delivered water not greatly enriched in ^{16}O, thus leaving the $\delta^{18}O$ value of the ocean little altered. Emiliani[17] assumed a mean $\delta^{18}O$ composition of the water tied up in ice sheets of −15‰ which would demand a change in the isotopic composition of sea water of only 0.5‰ for an ice volume shift equivalent to a 125-meter change in sea level. With some isotopic curves showing glacial/interglacial amplitudes greater than 1.5‰ it appeared in the early 1960s that local sea-surface temperature changes must dominate the $\delta^{18}O$ signals.

Several developments in the late 1960s and early 1970s led to a major reassessment of the ice storage factor in $\delta^{18}O$ changes. First, measurements of the $\delta^{18}O$ composition of existing ice sheets showed values far more negative than those assumed by Emiliani: −30 to 35‰ in Greenland ice[19] and −50 to −55‰ in Antarctic Ice.[20] If representative of glacial-age ice sheets, these very negative values would require that the ice storage contribution to the observed $\delta^{18}O$ variations in foraminifera must be at least double that assumed by Emiliani (for the same assumed value of glacial ice volume).

The second factor in this reassessment was the discovery from the use of transfer functions that glacial/interglacial temperature changes at the sea surface in lower latitudes were generally too small to dominate the $\delta^{18}O$ variations.[21,22]

Finally, with the development of mass spectrometers capable of analyzing the very sparse populations of benthonic foraminifera, it was discovered that $\delta^{18}O$ signals from the deep ocean matched or even exceeded in amplitude those from surface-waters.[23] Because deep waters have very cold temperatures today (in the range of 1° to 4°C), are relatively isolated from extreme climatic fluctuations at the earth's surface, and are constrained even during full-glacial climates from cooling beyond the freezing point of sea water (−1.8°C), the temperature contribution to these large deep-ocean $\delta^{18}O$ signals was considered to be minor. The basic benthonic foraminiferal $\delta^{18}O$ signal so similar from ocean to ocean (figure 4) was attributed predominantly to ice volume.[23] This interpretation was at the same time extended to the generally similar $\delta^{18}O$ signals from planktonic foraminifera in low-latitude regions.

Because the ice volume portion of the signal is global in extent, geologists in the 1970s used $\delta^{18}O$ signals as a tool to correlate deep-sea cores.[25] By the late 1970s, climate modelers tried to simulate the $\delta^{18}O$ signals as a response to radiation changes caused by variations in the earth's orbit.[26–30]

A few words of caution on interpreting the $\delta^{18}O$ signals are worthwhile. In oceanic regions near the great ice age ice sheets, the $\delta^{18}O$ signal in surface waters may be periodically overwhelmed by ^{16}O-rich meltwater,[31,32] and thus may not primarily reflect ice volume, but its first derivative.

Also, the temperature/ice volume debate is not finished. Work on benthonic foraminifera in cores with high sedimentation rates has revealed glacial/interglacial $\delta^{18}O$ amplitudes averaging ~1.8‰, with some in excess of 2.0‰.[33,34] These large $\delta^{18}O$ changes in sea water, if paired with the usually assumed −30‰ isotopic composition of glacial ice, demand implausibly large glacial ice volumes on the

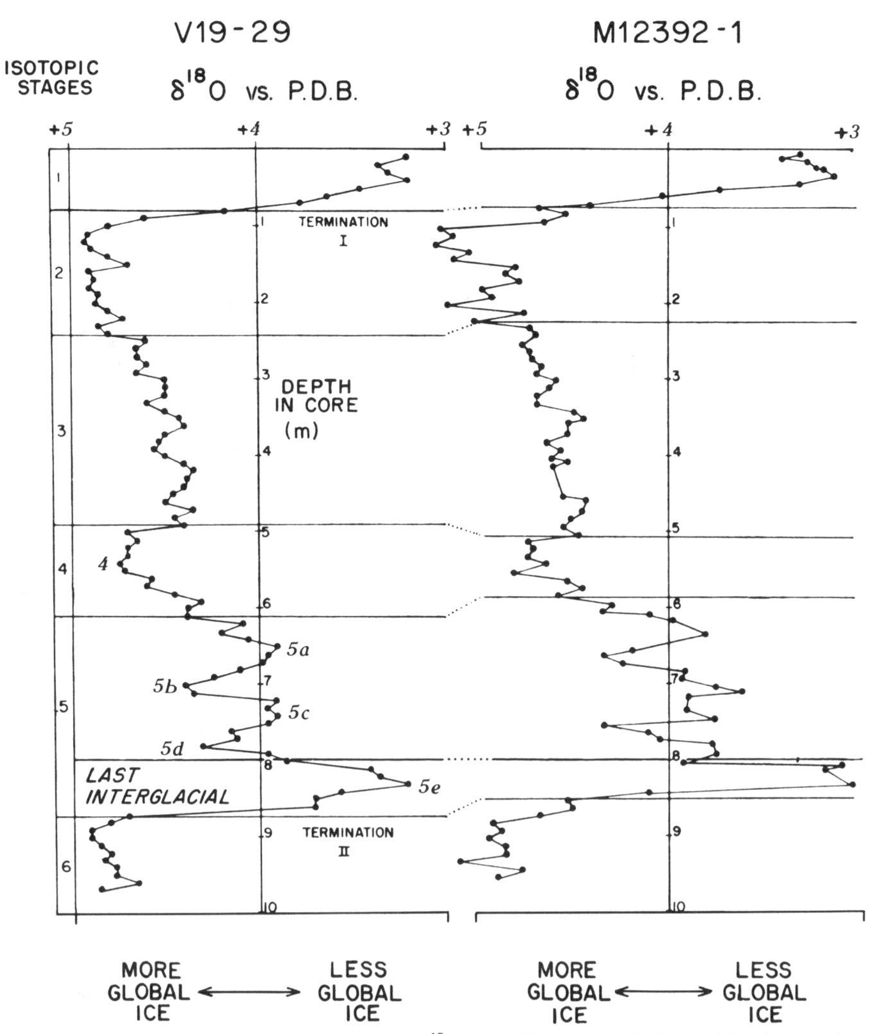

Figure 4. Per mil oxygen-isotopic composition ($\delta^{18}O$) of shells of benthonic foraminifera in two widely separated cores, one from the eastern equatorial Pacific and the other from the Canaries Current (locations in figure 10). Similarity of two curves indicates that a common factor (^{16}O storage in ice sheets) controls both curves, rather than local bottom-temperature variations (after ref. 24).

continents and unrealistically large sea-level lowerings (180 meters or more). There is now underway a reappraisal of the potentially significant (but not dominant) role of changes in bottom-water temperature through time; glacial-age bottom waters are now considered to have been colder than today.[34,35]

Also underway is an effort to improve the current simplifying assumption that glacial ice everywhere at all times has had an oxygen-isotopic composition of -30 to $-35‰$. This linear $\delta^{18}O$/ice volume assumption fails to recognize the likelihood of both geographic and temporal variations in the $\delta^{18}O$ composition of ice.[36] Mix and Ruddiman[37] conclude that, although $\delta^{18}O$ signals are a reasonable first order recorder of global ice volume, the $\delta^{18}O$ signal can lag 1000–3000 years behind true ice volume at prominent ice-growth and ice-decay transitions due to geographic and temporal inhomogeneities. This could translate into equivalent $\delta^{18}O$/ice volume lags in the long-term phase relationships and thus impact the inferred phasing of local SST with global ice volume.

PLANKTON AS INDICATORS OF SEA-SURFACE CONDITIONS

The ultimate usefulness and reliability of the four groups of microplankton in paleoclimatic work depends upon many factors, but most critically these:

1. Clusterings of individual species or assemblages within well-defined modern geographical ranges, such that clear correlations exist with oceanographic characteristics such as temperature. Ideally, this involves key species which constitute major fractions of the total biota in their optimal regions, but which may be rare or absent away from these areas.
2. The encompassing of a wide range of modern oceanographic conditions by the total array of species or assemblages, with complete coverage from equator to pole the ideal.
3. Evidence that the plankton used for paleoclimatic reconstructions of past oceans actually live today at the water depths (surface or subsurface) for which the reconstructions are sought.
4. Evidence of reasonable evolutionary stability, such that the species alive today are the same species that have existed for the relevant intervals of late Quaternary time.
5. The avoidance of no-analog situations in the past (i.e. intervals in which a fauna or flora appears for which there is no modern biotic analog anywhere in the oceans).

The planktonic foraminifera best fulfill these five criteria. The modern species distributions are both distinct and well-correlated to water-mass units and intervening faunal boundaries.[38,39] They also span most of the world oceans, although dissolution on the sea floor is a large problem for paleoclimatic studies in many areas.[40] There is a loss of resolution in waters colder than 0°C in winter and 6°C in summer, because only one species survives in such conditions.[38] Some foraminifera are known to live at depths considerably below the surface, but the most abundant species live in, or at the base of, the mixed layer that has the same oceanographic characteristics as surface waters.[41,42] Evolutionary change has affected only a few relatively scarce low-latitude species during the last 1.8 million years, with major species replacement not occurring in both the low-latitude and high-latitude records until the Plio-Pleistocene boundary.[43] No-analog faunas are rare in the late Pleistocene open ocean, although they sporadically occur in the Mediterranean during the highly stratified water-column conditions that lead to the deposition of sapropels (organic muds).

Coccoliths are more problematical in paleoceanographic reconstructions. The central problem is the high rate of evolutionary divergence during the late Quaternary. Some evolutionary appearances and disappearances occur within the Brunhes.[44] More common, however, are very large step-function shifts in abundance of the dominant species, shifts which bear no obvious relationship to the major glacial-interglacial climatic oscillations and thus are inferred to be evolutionary in nature.[45] As a result, coccolith assemblages through most of the late Quaternary have no true modern analog, even including the flora existing as recently as 10,000 years ago in the high-latitude North Atlantic. Coccoliths also lack diagnostic cold-water species, thus losing paleoclimatic utility well away from polar regions in waters at summer temperatures of ~9°C.[46,47] Coccoliths also show a much greater dominance of relatively cosmopolitan (and hence environmentally insensitive) species;[47,48] this dominance hinders precise quantification of rarer, more environmentally sensitive, species. Based on relatively meager data, most coccolithophores live within the mixed layer, and thus presumably respond directly to sea-surface changes. The primary use of coccoliths, however, is in biostratigraphic zonations.

Radiolaria are almost comparable in importance to foraminifera in Quaternary paleoceanography. Evolutionary changes occur within the Pleistocene, but are far less of a problem than for coccoliths. Radiolarian species are distributed from equatorial to polar waters, with reasonably distinctive species and assemblage groupings in different water-mass units. They provide important paleoclimatic coverage in sediments from broad regions of the Pacific (figure 3) where foraminifera are too dissolved to be useful.[49] Radiolaria appear to live deeper than any of the other four biotic groups, however, and in some areas they may respond to and record subsurface conditions not necessarily linked to those in the surface mixed layer. Two other problems with radiolaria are the very large number of rare species that cumulatively can account for well over half of the total fauna and the occurrence at no-analog abundances of one species (*Cycladophora davisiana*) in glacial-age sediments poleward of 40°–50° latitude.[50,51] Both of these situations have led to the problem species being discarded from the census counts on which paleoceanographic estimates are based.

Despite being near-surface dwellers distributed throughout the world's oceans, diatoms have been the least useful group for open-ocean paleoclimatic work. This is largely because little of the silica originally produced in surface waters is preserved on the sea floor, and diatoms tend to dissolve long before the larger, thicker shelled radiolaria.[8] This leaves only patchy regions, mostly beneath high-fertility waters of the Pacific, where diatoms are well enough preserved to delimit clearly the species and assemblage groupings in overlying surface waters.[52] As with the radiolaria, no-analog percentages of one species (*Eucampia balaustrium*) developed during glacial times in the Southern Ocean.[53]

In summary, most quantitative Quaternary paleoceanography is based on planktonic foraminifera (Atlantic and Indian Oceans) and radiolaria (Pacific and Antarctic Oceans), with more local reliance on coccoliths and diatoms in other areas.

Transfer Functions

Transfer functions are basically regression equations relating modern faunal or floral census data to oceanic properties (e.g., sea-surface temperature); they are used for the purpose of estimating past ocean properties from past census data. Among several possible techniques[54–57], that of Imbrie and Kipp[21] stands out as the most comprehensive. It consists of five basic steps.

In step 1, the raw census data from core tops are culled in preparation for subsequent processing. For some biotic groups, individually rare but cumulatively abundant species are dropped, often including a large portion of the coccoliths and radiolaria. Species which reach no-analog abundances in premodern levels are also dropped. These steps respectively streamline the counting and preclude the possibility of no-analog paleoclimatic estimates; however, it can be argued that this culling compromises the validity of the estimates by ignoring possible changes in competition through time between the species included in and those eliminated from the equations.

Step 2 is a factor analysis of the census counts. This is done to eliminate redundancies by grouping species with very similar distributions into single factors, all of which are statistically independent (mathematically orthogonal). The orthogonality of the factors insures that each transfer-function estimate from a census count is a unique solution. It also is the most economical (yet comprehensive) representation; this is an important consideration due to computer storage limitations on large data matrices.

Step 3 is the actual derivation of the transfer-function equation. At this point the critical choice is which oceanic properties to use; this choice is largely based on the correlation of the factors derived in step 2 against available oceanic properties. Surface temperature is the parameter normally estimated, although the "deep chlorophyll maximum" model[42] argues that subsurface properties may be more relevant in some areas, particularly where the permanent thermocline is relatively shallow. In such areas, the usually very high correlation of surface and subsurface temperatures may break down. Deep-dwelling preferences of individual species among the zooplankton (radiolaria and foraminifera) may complicate estimates in other areas.

In step 4, downcore counts are expressed in terms of the factors derived from the core tops in step 2. This is not a new factor analysis, but a fit of new counts into the original factor solution. The degree of fit is expressed in terms of the "communality," with low values indicating that the transfer function has encountered a faunal or floral assemblage not encompassed by modern assemblages.

Step 5 is the estimation of past oceanographic conditions. This process can be judged a success unless there are (1) no-analog estimates or (2) discordant multiple estimates. No-analog estimates are recognizable because they lie outside of the range of conditions encompassed by the modern ocean (and in extreme cases by the laws of nature). They are the consequence of extrapolating beyond the safer confines of the essentially interpolative transfer functions. Occasionally, however, estimates made on the same sample but using different biotic groups disagree by amounts larger than the respective error of estimate of the various equations. Such cases of discordant multiple estimates are a clue that one or more of the equations is yielding spurious results. Molfino et al.[58] ascribe discordant multiple estimates in the South Atlantic Ocean to increased vertical stratification in near-surface waters and to altered and intensified seasonality of upwelling.

Paleoceanographic estimates which successfully negotiate this series of internal checks stand a very good chance of being reasonably accurate reconstructions of past oceanic conditions.

TIME-SLICE CLIMATIC DATA SETS

The Last Glaciation (18,000 years B.P.)

A major international research effort called CLIMAP (Climate Long-Range Mapping and Prediction) was organized during the 1970s to reconstruct the earth's surface during the last glacial maximum 18,000 years ago.[59-60] The final oceanic reconstruction was based on data from 450 cores across the world ocean.[60]

Using foraminiferal census data from the Indian and Atlantic Oceans and radiolarian census data from the Pacific and Antarctic Oceans, Moore et al.[61] synthesized in a compact way the biotic differences between the last-glacial and modern oceans. They calculated values of the faunal similarity index between core-top and last-glacial faunas, with extreme values of 1.0 reflecting no difference between the core-top and last-glacial fauna and 0.0 indicating total faunal replacement. These values can also be visualized as the percent of overlap between histograms of species composition for the two levels.

The faunal similarity maps (figure 5) indicate that the largest changes of faunal composition in the ice age ocean generally occurred in the mid-latitude regions (35°–60°). These changes reflect the equatorward shift (and net expansion) of the polar assemblages, now far more restricted in extent (figure 6). In contrast, the faunas in the middle of the subtropical gyres and in many low-latitude areas were virtually unaltered (figure 5). As a result, the modern sub-polar and transitional-subtropical assemblages were both translated toward the equator and reduced in extent by compression against the stable faunal distributions in the centers of the subtropical gyres (figure 6).

The most prominent faunal change in the lower latitudes occurred in the eastern boundary currents along the western coasts of Africa and South America, where sub-polar and transitional species shifted equatorward on the flanks of the subtropical gyres (figure 5). In addition, the fauna characteristic of the eastern equatorial regions expanded westward along the equator. In both of these changes, cooler species typical of upwelling or mid-latitude regimes replaced the warmer tropical/subtropical assemblages.[61]

These changes in the primary faunal census data correlate directly with changes in estimated sea-surface temperature in the ice age ocean (figures 7 and 8). The largest surface-ocean coolings (>6°C) match the subpolar regions in which the

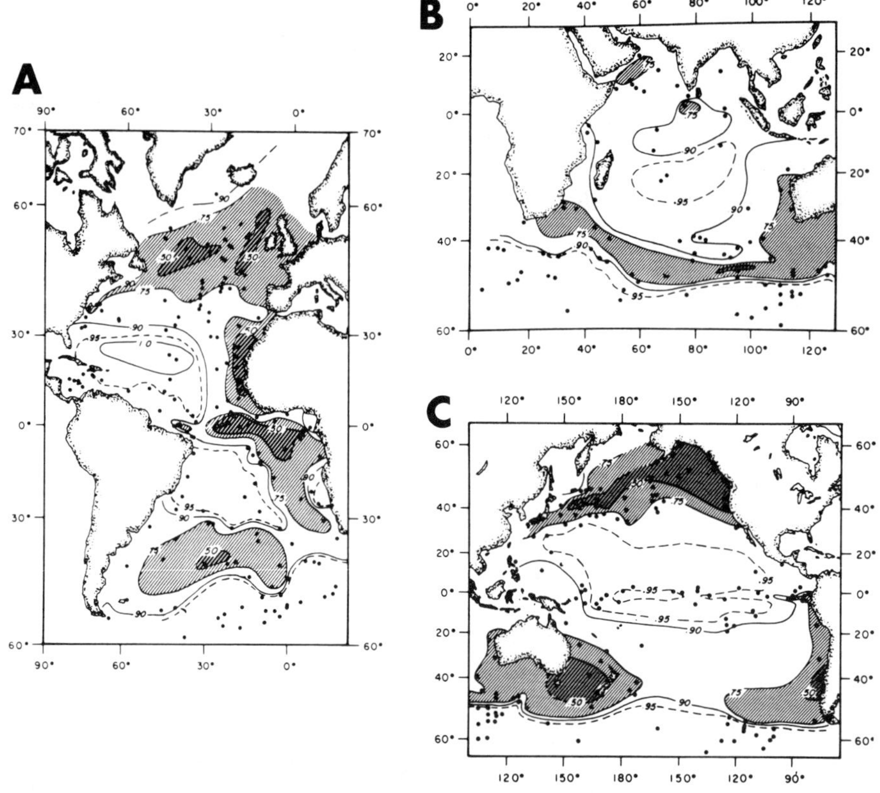

Figure 5. Faunal similarity index cos θ measured between modern and ice age samples in *(A)* the Atlantic, *(B)* the Indian, and *(C)* the Pacific Oceans. Decimal values indicate the fraction of fauna (foraminifera or radiolaria) common to both modern and ice age samples. Dots are core locations (from ref. 61; copyright 1981 by Elsevier Science Publishers).

polar faunas replaced temperate species. Surrounded on three sides by the great Northern Hemisphere complex of ice sheets, ice shelves, and sea ice, the subpolar North Atlantic cooled more than the other subpolar oceans: 10°C or more in summer, much less in winter.[60] The large summer cooling probably reflects for the most part albedo-feedback from the ice sheets;[62] the smaller winter SST anomaly may reflect the diminished albedo contrast between (ice age) glaciers and (modern) seasonal snow.

A radiolarian fauna now typical of the sea of Okhotsk spread across much of the high-latitude North Pacific, bringing a colder, harsher climate with less thermal moderation in summer and more prevalent sea ice in winter.[63] In the Antarctic, the more limited equatorward advance of the polar radiolarian fauna produced the modest SST cooling observed along 40° to 50°S (figure 8).

The equatorward expansion of the polar waters against the stable southern centers of the subtropical gyres created very strong zonal gradients in both faunal com-

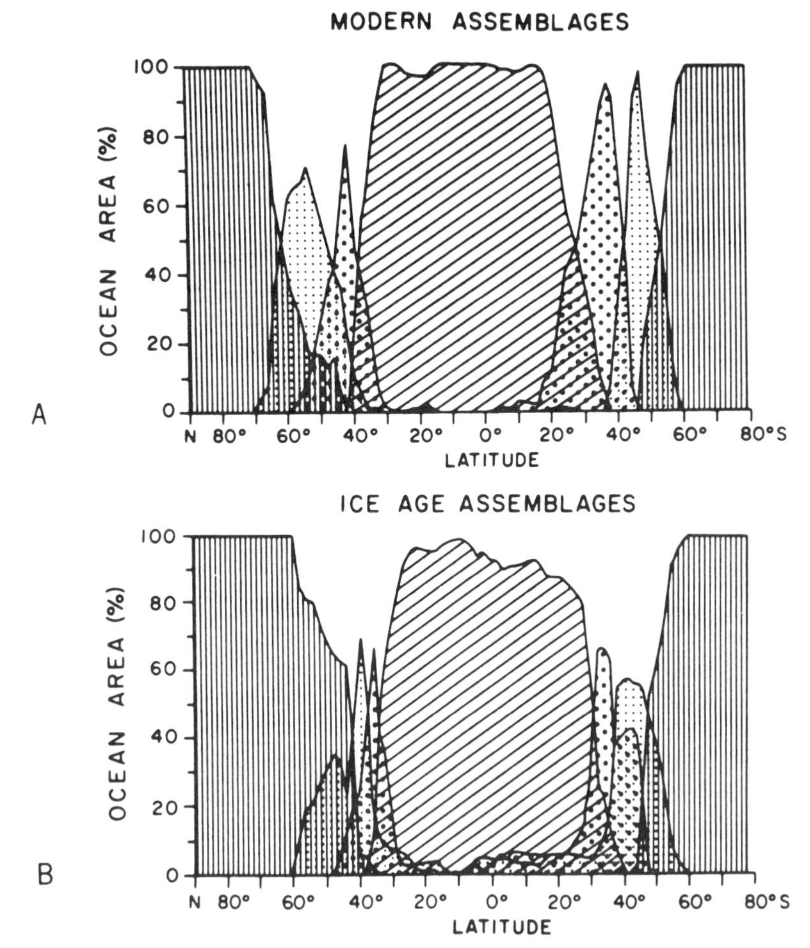

Figure 6. Percent of ocean area dominated by each of five planktonic faunal assemblages plotted as a function of latitude for (*A*) modern and (*B*) last ice age times. Polar assemblages (vertical lines) expand equatorward, displacing and compressing subpolar (small dots) and transitional (large dots) assemblages. Low-latitude tropical and southern subtropical assemblage (diagonal lines) is relatively stable (from ref. 61; copyright 1981 by Elsevier Science Publishers).

position and ocean surface temperature along latitudes 35° to 45° (figures 5 and 6). Surface and near-surface currents of considerable strength are needed to balance the strong density changes along these gradients.

Diminished surface temperature in the eastern boundary currents marked by increased percentages of cool-water species reflects both increased upwelling and equatorward advection of cool waters.[58,64,65] The cooler temperatures along the equator are more clearly indicative of enhanced divergence rather than advection.[64] Relatively minor changes in faunal composition and SST coincide with decreased upwelling in the Arabian Sea during the last glacial maximum.[65]

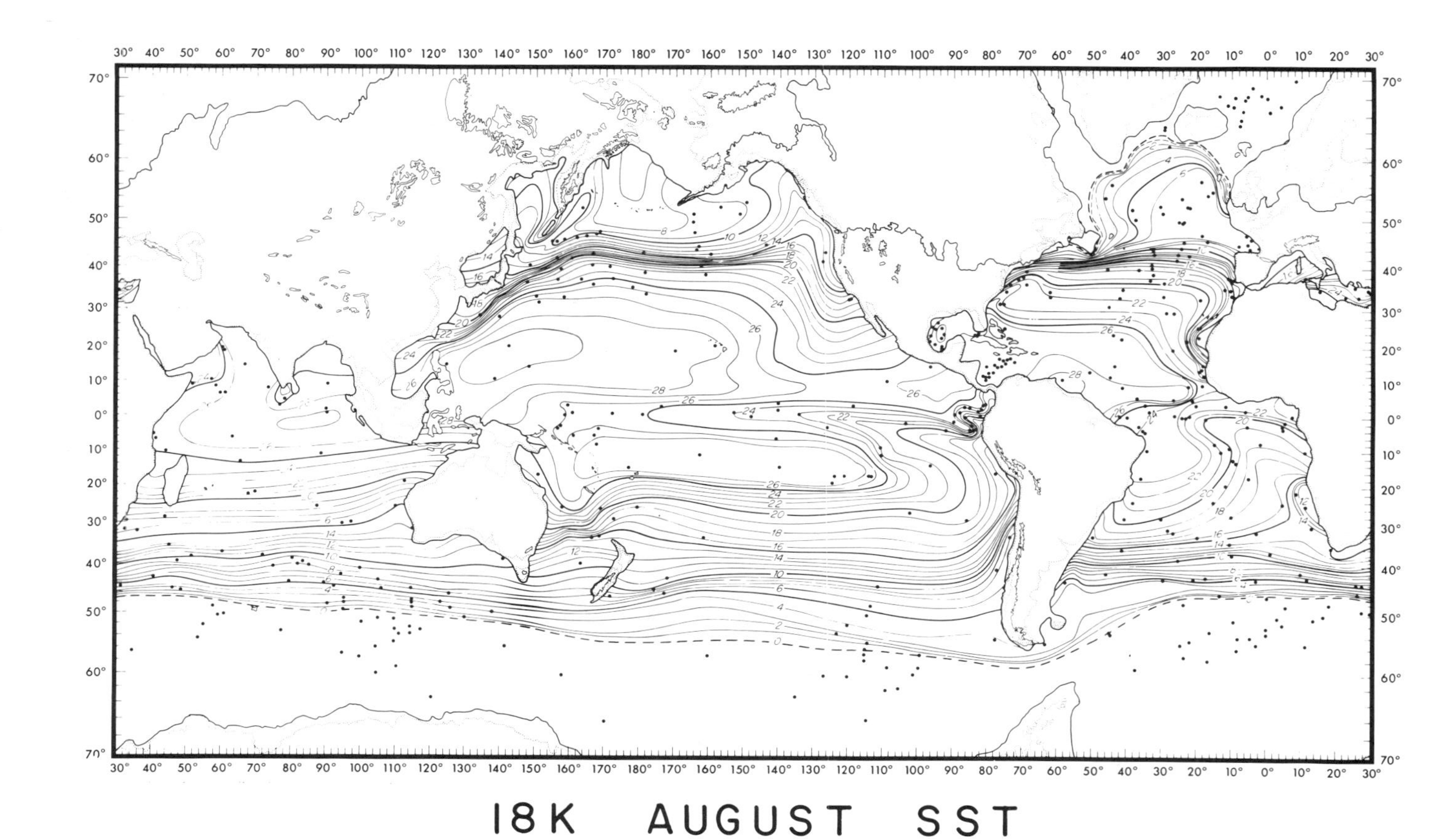

Figure 7. Transfer-function estimates of sea-surface temperature at the glacial maximum 18,000 years ago (from ref. 60).

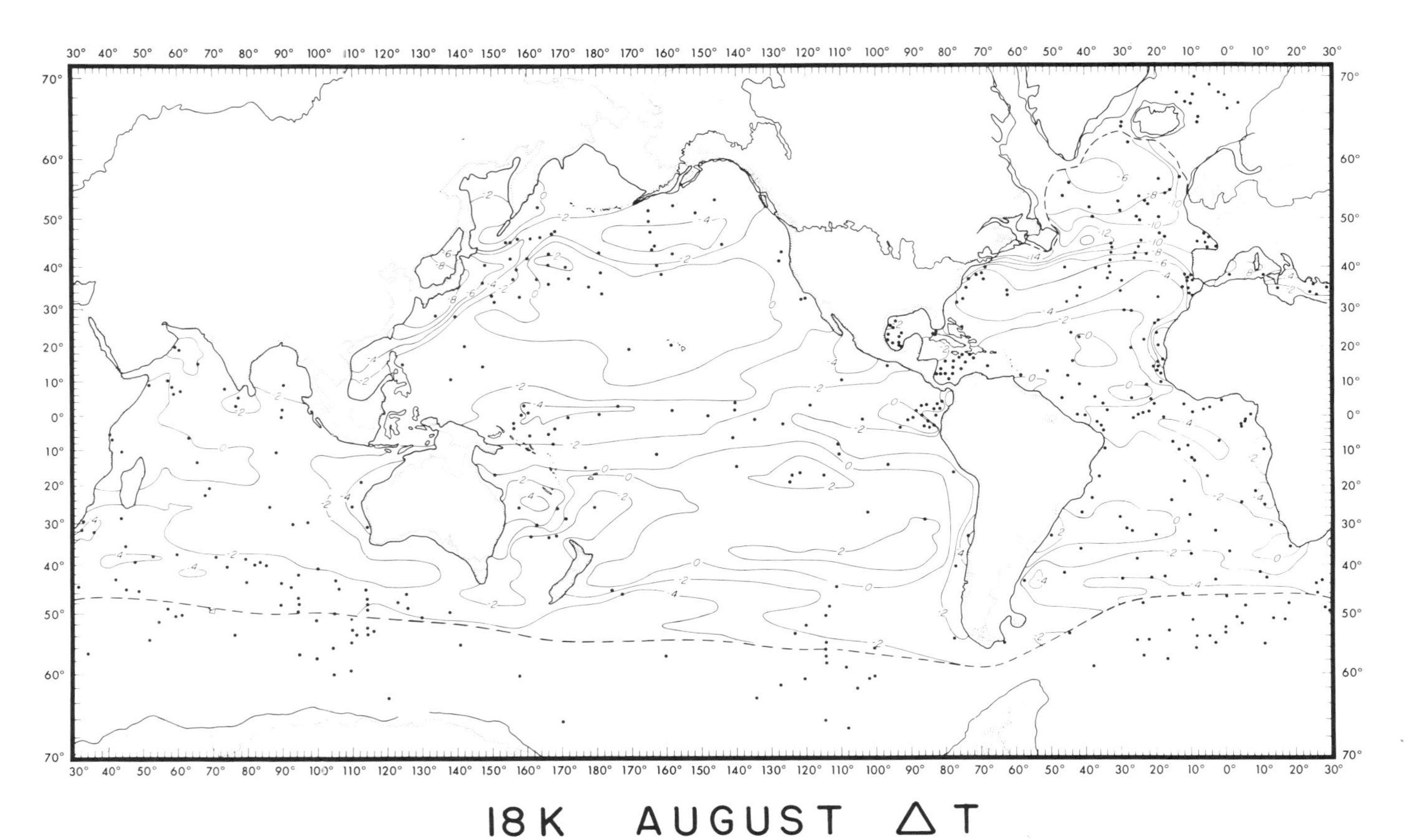

Figure 8. Difference between modern sea-surface temperature and estimated sea-surface temperature at the last glacial maximum 18,000 years ago. Negative values (in °C) mean that the ice age ocean was colder (from ref. 60).

The Last Interglaciation (122,000 Years B.P.)

The final multi-investigator effort of the CLIMAP project was a study of the Last Interglaciation, a time in which the oxygen isotopic record during substage 5e indicates that global ice volume was as small as, or smaller than, that today.[66] Based on oxygen-isotopic analyses and biotic census counts at 1,000 to 3,000-year intervals in 52 cores across the world ocean, CLIMAP compared sea-surface temperatures across the world ocean with those today (figure 9). The primary conclusion drawn was that sea-surface temperatures during the Last Interglaciation were in most regions not significantly different from those today. Roughly 60% of the estimates were offset from the modern values taken from oceanographic atlases by amounts less than the $\pm 1.5°C$ standard error of estimate derived from core-top calibrations. Many of the remaining 40% of "anomalous" estimates disappear if the comparison is made against core-top estimates of SST rather than atlas values. Still other anomalies are thrown into question by: the lack of similar anomalies in cores nearby, estimates discordant with those from other biotic groups within the same core (particularly a problem in eastern boundary currents of the South Atlantic), and severe $CaCO_3$ dissolution (especially in the Pacific Ocean).

With these complications considered, the only regions remaining with anomaly patterns coherent enough to suggest a possibly significant difference from today are the subpolar North Atlantic (slightly warmer in winter) and the Gulf of Mexico (slightly cooler in both seasons). The suggestion of warm winter anomalies in the mid-latitude North Atlantic is consistent with U-series dates of ~120,000 years B.P. on climatically indicative shallow marine and continental sequences on the European margin of the North Atlantic.[67,68] In general, however, the Last Interglacial ocean was little different from today.

Regional Time-Slice Studies

Other published time-slice reconstructions independent of the CLIMAP project cover selected portions of the Late Quaternary ocean. Some are local studies of the last glacial or interglacial maximum, but others also explore conditions at other intermediate ice volume maxima or minima, such as isotopic stages 5a and 4.[69–72]

All of the time-slice examples to this point could in a sense be considered examples of "equilibrium" climatic reconstructions. Global ice volume had in each case reached a maximum or minimum value that would be held for a few thousand years with little change. And in most areas, SST values were relatively stable. In this sense, these reconstructions will reflect the balance of circulation trends that held the world in a particular equilibrium ice volume configuration for several thousand years, but they may offer little insight into the imbalances involved in major climatic changes.

Recently some interest has begun to shift to "disequilibrium" time-slice reconstructions, that is, reconstructions of oceanic configurations during periods of rapid climatic change on land, particularly ice sheet growth or decay. Ruddiman and

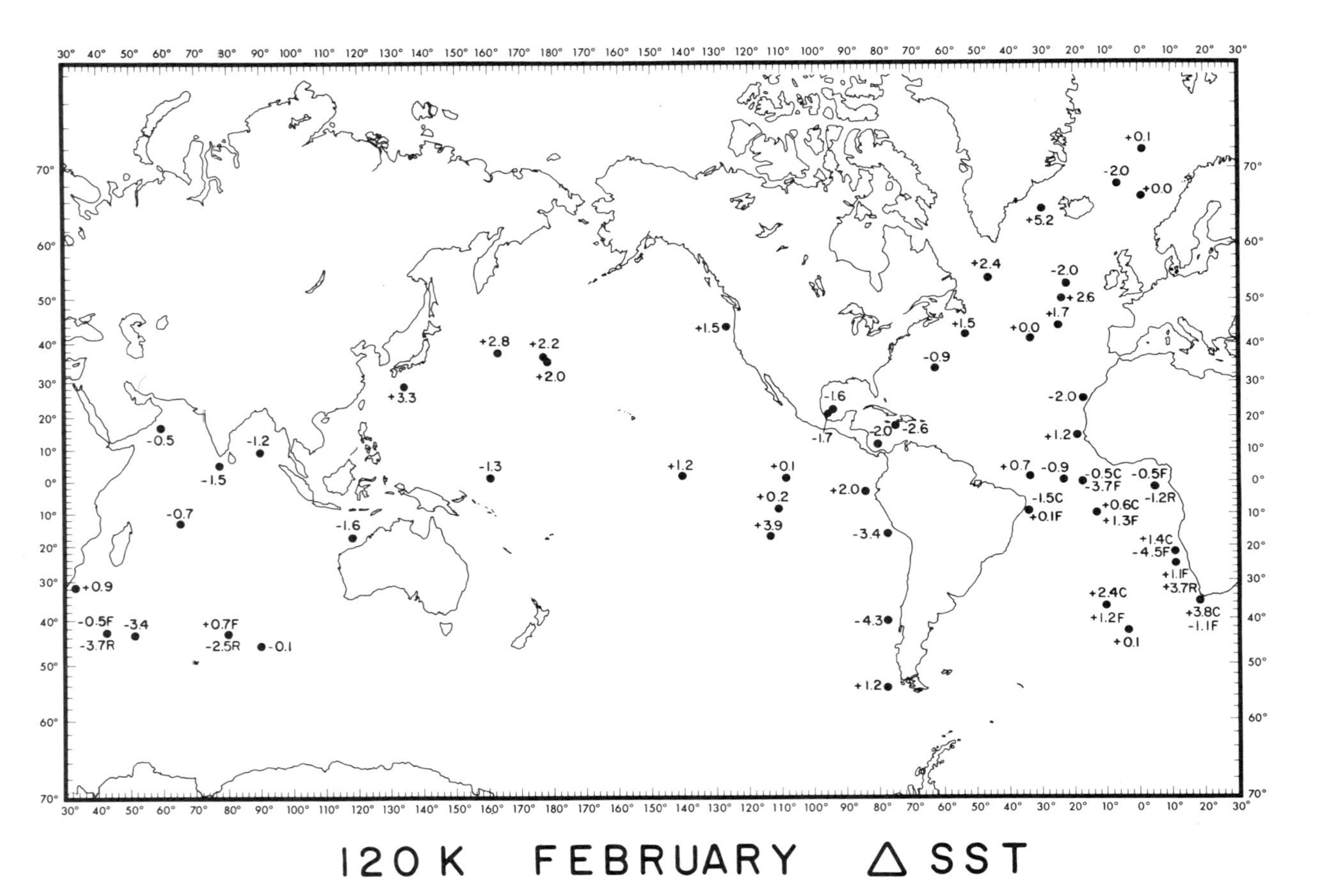

(a)

Figure 9. Difference between modern sea-surface temperature and estimated sea-surface temperature (in °C) at the last interglaciation 120,000 years ago. Negative values mean that the last-interglacial ocean was colder than today, *(a)* February and *(b)* August (from ref. 66).

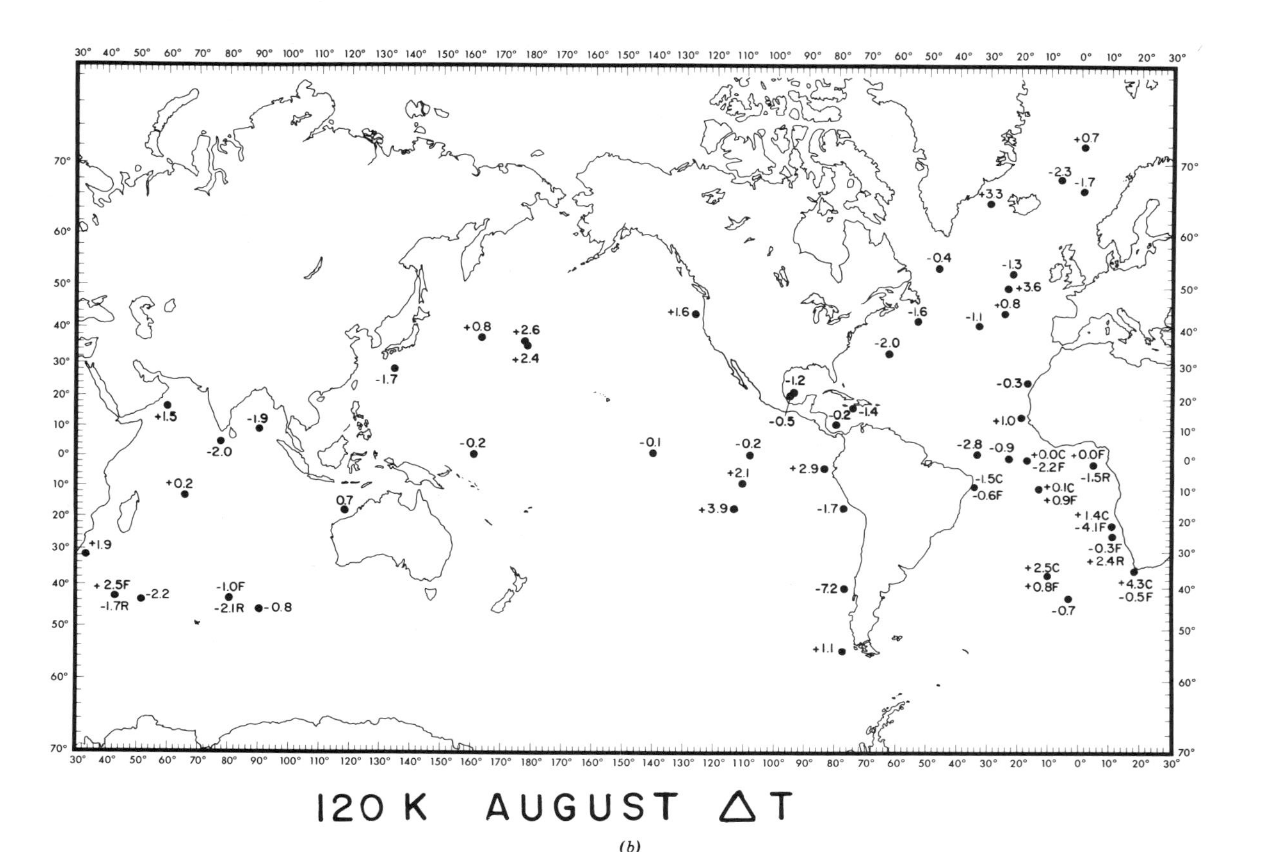

(*b*)

Figure 9. (*Continued*)

McIntyre[73] reconstructed the SST of the subpolar North Atlantic (35°–65°N) during the middle of the fastest period of ice sheet growth at the oxygen isotopic stage 5/4 boundary some 75,000 to 72,000 years ago. They found that much of the North Atlantic remained in a warm interglacial mode well into the period of fastest ice sheet growth; the longest warm-ocean lag was centered at 35° to 50°N. This lag resulted in large meridional surface temperature contrasts, with a warm interglacial ocean positioned immediately adjacent to ice-covered continents.

Ice-decay transitions have also received some attention. Ruddiman and Glover[74] used the synchronous horizon provided by a zone of volcanic ash in an attempt to reconstruct and interpret the circulation of the subpolar North Atlantic during a time slice late in the last deglaciation (~9,300 years B.P.). Later work[75] largely invalidated this earlier effort by showing that bioturbation not only smooths the deep-sea record but also causes complicated translational offsets of climatic signals. This effect, earlier noted by Peng et al.,[76] and by Hutson[77], has the largest impact when rapid climatic changes occur at the same depth in core as rapid absolute abundance changes in the climatic signal carrier.

The somewhat schematic reconstruction of deglacial North Atlantic circulation changes by Ruddiman and McIntyre[75] suggests that the most interesting ice-decay interval to examine in terms of disequilibrium climate might be 16,000 to 13,000 years B.P., when the ice sheets were rapidly disintegrating but the North Atlantic remained very cold, with widespread sea-ice cover in winter. This was interpreted as an ocean receiving abundant icebergs and meltwater from the disintegrating ice sheets.

Another oceanic region of possible future interest for time-slice reconstructions during times of climatic disequilibrium is the subpolar to subtropical portion of the Southern Hemisphere, where the oceanic response tends to lead global ice volume by several thousand years.[66,78]

TIME-SERIES CLIMATIC DATA SETS

Given the very large number of paleoclimatic studies of deep-sea sediments, this chapter cannot provide a truly comprehensive coverage of the subject. For those studies which are included, the global map in figure 10 shows the location of all cores discussed.

High-Frequency Late Quaternary Studies

In a few regions of extremely rapid deposition, it is possible to extract time series with resolution in the range of 10^1–10^2 years. Records from two such regions are discussed here: (1) the Orca Basin, a hypersaline, anoxic basin in the Gulf of Mexico about 225 km south of the Louisiana coast; and (2) the Santa Barbara Basin off California, also anoxic, with a great thickness of laminated Holocene sediments.

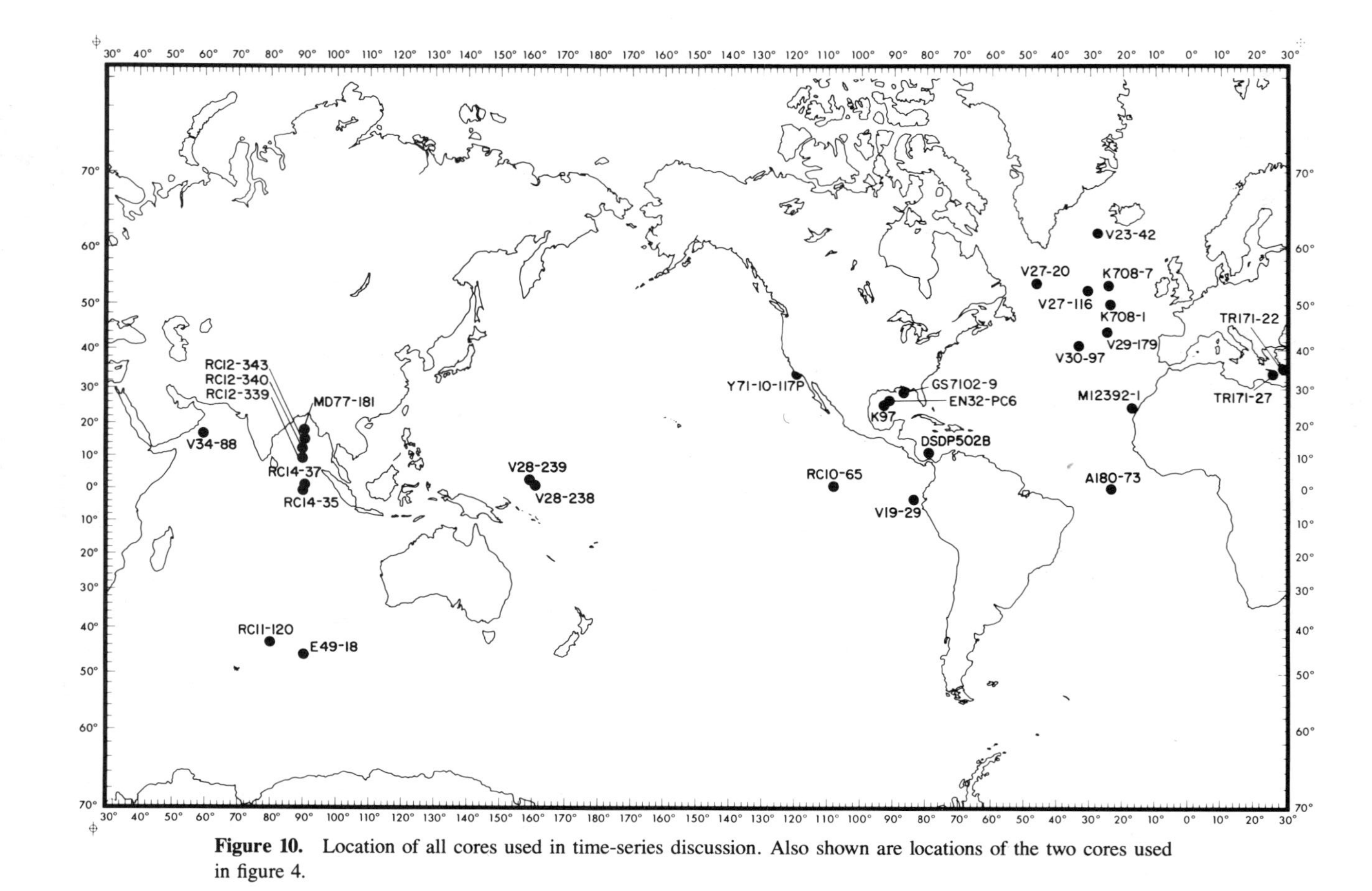

Figure 10. Location of all cores used in time-series discussion. Also shown are locations of the two cores used in figure 4.

Gulf of Mexico

The Gulf of Mexico lies in the path of one of the major meltwater drainage systems of the ice age ice sheets, the Mississippi River. Earlier publications found that oxygen-isotope signals in planktonic foraminifera recorded anomalously light values during the last deglaciation.[31,32] These anomalous values reflect the point-source delivery of ^{16}O-rich meltwater from the southern margin of the Laurentide Ice Sheet. This delivery overwhelms the background $\delta^{18}O$ signal which largely reflects the changing isotopic composition of the global ocean due to changes in ice sheet volume. The earlier studies disagreed on the timing of the meltwater "spike"; Kennett and Shackleton[31] favored an age some 2,000 years earlier than that of Emiliani et al.[32]

Leventer et al.[79] recently published an even more detailed record of the meltwater spike based on Orca Basin core EN32-PC6 (figure 11). Because Orca Basin is anoxic, this record of the last deglacial meltwater influx is unaffected by biotically induced sediment mixing. As a result, no more detailed record is likely to be found in the Gulf of Mexico. This record suggests a two-part meltwater spike, with the

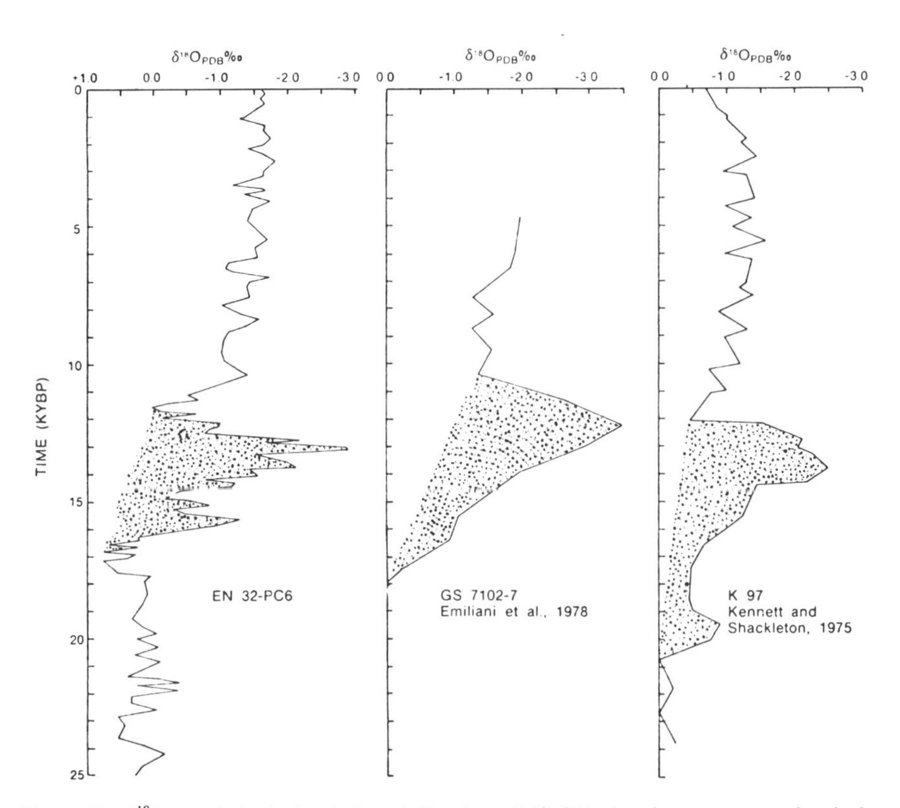

Figure 11. $\delta^{18}O$ records in planktonic foraminifera from Gulf of Mexico piston cores spanning the last 25,000 years. The stippled region in each record indicates the portion most affected by isotopically light local meltwater influx (after ref. 79; copyright 1982 by Elsevier Science Publishers).

pair of pulses occurring at 16,500 to 15,000 years B.P. and at 14,500 to 12,200 years B.P. This chronology is a compromise between the previous estimates, with the onset closer to that of Emiliani et al.,[32] but the cessation closer to that of Kennett and Shackleton.[31] This time scheme was chosen, however, after discarding numerous ^{14}C dates because of apparent unreliability due to contamination and redeposition. It is clear that questions about the timing of the earlier parts of this meltwater signal still remain.

Santa Barbara Basin

Because anoxic conditions suppress sediment mixing by bottom-dwelling organisms, the Holocene record of the Santa Barbara Basin offers a second undisturbed sequence of varved sediments for high-resolution paleoceanographic studies. Pisias[80] counted radiolarian assemblages at approximate 25-year intervals over an interval estimated as 8000 years (assuming that the varves are annual). Factor analysis of surface-sediment radiolarian counts yielded four assemblages, representing: (1) subtropical waters well south and east of the Santa Barbara Basin; (2) Baja waters just to the south; (3) central waters south and west of Santa Barbara; and (4) California Current waters to the northwest. The downcore variations of these factors are dominated by the central and subtropical assemblages (figure 12), with the subtropical factor yielding to the central factor at about 5400 years B.P.

Transfer-function estimates of paleotemperature based on these counts (figure 13) most closely resemble the subtropical factor. Pisias[80] linked several of the marked reductions in surface-water temperature in the Santa Barbara basin to the neoglacial periods of alpine ice advance.[81]

ORBITAL-FREQUENCY STUDIES

$\delta^{18}O$: Long-Term Records of Ice Volume

Over the interval of the last several hundred thousand years, many of the most detailed records of the oxygen-isotopic composition of sea water are found in Emiliani's analyses of planktonic foraminifera in the Caribbean.[17,18] The single record that clinched the argument that ice volume dominates these signals came, however, from western equatorial Pacific core V28–238 (figure 14). With this record, Shackleton and Opdyke[23] demonstrated the very close similarity of $\delta^{18}O$ signals in benthonic and (low-latitude) planktonic foraminifera due to the overriding ice volume effect on the isotopic composition of sea water.

The most detailed currently available oxygen-isotopic curves for the late Quaternary are those previously shown in figure 4. These reveal the close similarity of $\delta^{18}O$ signals measured in benthonic foraminifera from bottom waters in widely separated oceans and again confirm the importance of ice volume as a signal common to all $\delta^{18}O$ curves. With a precisely calibrated inter-ocean comparison, how-

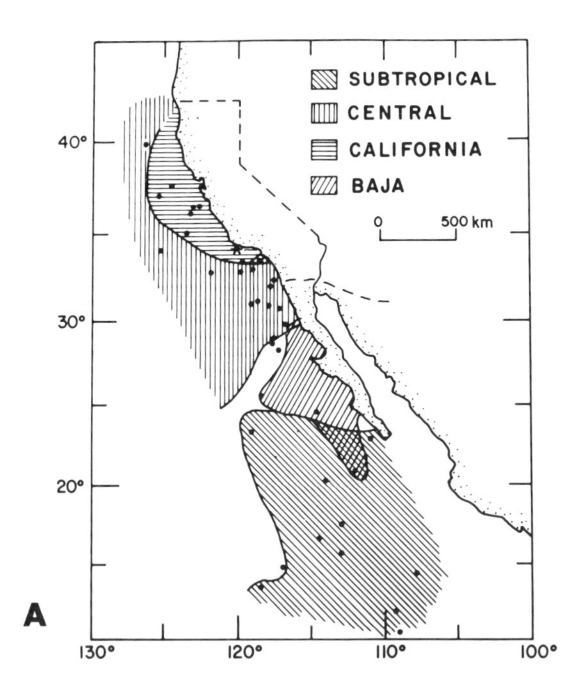

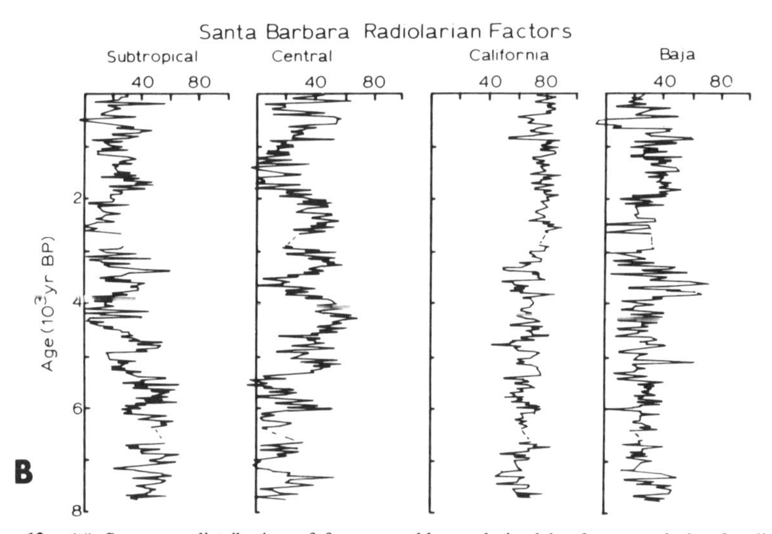

Figure 12. (*A*) Summary distribution of four assemblages derived by factor analysis of radiolarian counts in surface sediments of the eastern North Pacific (from ref. 80). (*B*) Loadings of four radiolarian factors shown in (a) through 8000-year record of downcore counts in Santa Barbara Basin core Y71-10-117P (from ref. 80).

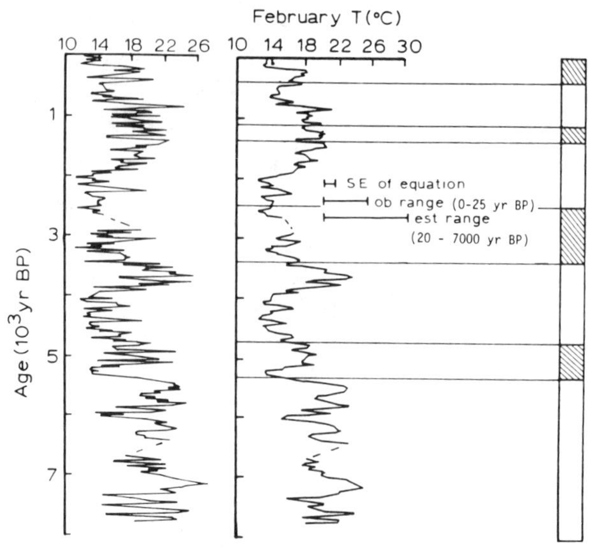

Figure 13. February paleotemperature curve for Santa Barbara Basin core Y71-10-117P (location in figure 10). SE: Standard error of the regression equation. Ob. range: observed range of February temperatures in the Southern California Borderland area. EST. range: range predicted for the entire 8000-year record. Column shading indicates neo-glacial intervals (after ref. 80).

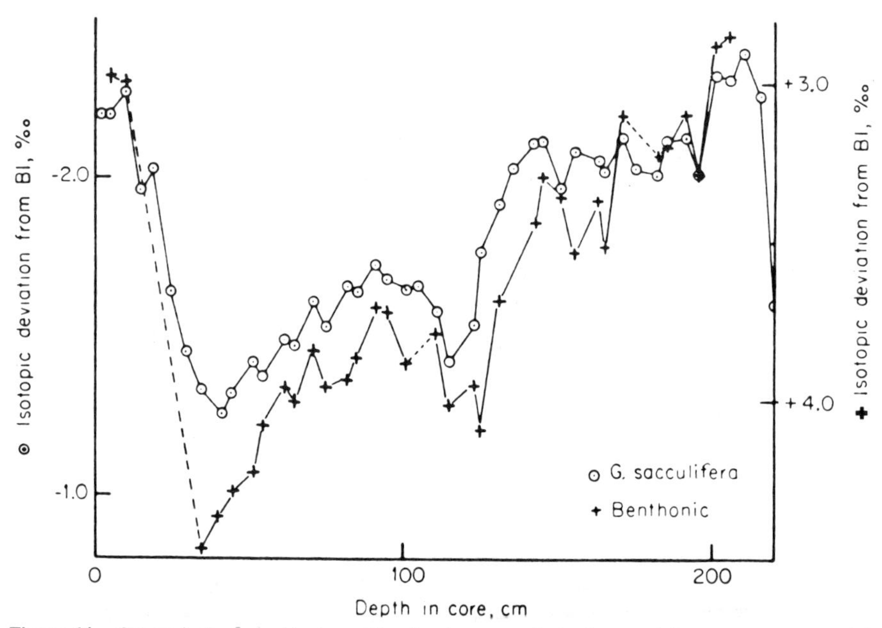

Figure 14. Comparison of planktonic and benthonic oxygen isotopic record from western equatorial Pacific core V28-238 (location in figure 10). The two sequences are plotted at the same scale of isotopic change, but with scale zero-points translated by 5.3‰, the present-day planktonic/benthonic difference. Similarity of the two records indicates common control by ice-volumn storage effect (after ref. 23).

ever, Duplessy et al.[34] found that $\delta^{18}O$ values of deep water in the Northeast Atlantic were at least 0.3‰ heavier (relative to today) than was Indian Ocean bottom water. Assuming no cooling of the already very cold Indian Ocean bottom water, this implies a deep-water cooling of at least 1.3°C in the Atlantic relative to today during glacial-isotopic stages 4 through 2.

Confirmation of the importance of orbital-frequency rhythms in the $\delta^{18}O$ record came from Hays et al.[78] who made minor but critical adjustments in the time scale of the last 400,000 years. The resulting power spectrum of the $\delta^{18}O$ signal showed a dominance of 100,000-year power, with lesser 41,000-year and still lesser 23,000-year responses. The fact that the $\delta^{18}O$ record from a high-latitude planktonic foraminifer (in this case, *Globigerina bulloides*) closely resembles other $\delta^{18}O$ signals known to be dominated by ice volume effects suggests that this species must change its season and/or depth of occurrence through time in such a way as to follow a narrow temperature regime despite large-scale local climatic changes in this area.

Within the ~20-meter length limitations imposed by conventional piston coring, western equatorial Pacific cores V28-238 and V28-239 contain the longest records with sufficient resolution to detect orbital frequency $\delta^{18}O$ changes with reasonably full amplitude.[23,82] These records (figures 14–16) suggest that the 100,000-year rhythm is appreciably weaker prior to the interval 900,000 to 700,000 years ago.[82,83] Prior to about 1.45 million years ago, all orbital rhythms appear weaker in this record, but problems of bioturbation, sample resolution, and time-scale complicate any such conclusions from these piston cores.

Detailed $\delta^{18}O$ studies of hydraulic piston cores now underway are overcoming the resolution problems inherent in older sediments from conventional piston cores.[84] Isotopic signals from planktonic foraminifera in the Caribbean confirm a change in amplitude of the variations somewhere in the interval 900,000 to 700,000 years ago (figure 17).[84]

As these HPC studies are pushed back into early Pleistocene and Pliocene intervals characterized by oxygen-isotopic variations of smaller and smaller amplitude, the question of temperature overprints on a diminishing ice storage signal becomes much more important. There is now significant uncertainty as to whether Northern Hemisphere glaciation began around 3 or 2.4 million years ago, based largely on disagreements in oxygen-isotopic signals from different regions.

Regional Paleoceanographic Studies

The Quaternary paleo-oceanographic response has been investigated in numerous regions across the globe. Three groupings that are particularly important to the global climate system and that are understood in some detail will be considered here: (1) high-latitudes, in which very large changes in temperature, salinity, and sea-ice extent occur; (2) monsoon regions, including both areas that serve as sources of moisture and those that receive runoff; and (3) equatorial regions, marked by large changes in equatorial divergence and coastal upwelling. Omitted in this somewhat selective overview are several fairly intensively studied regions with more

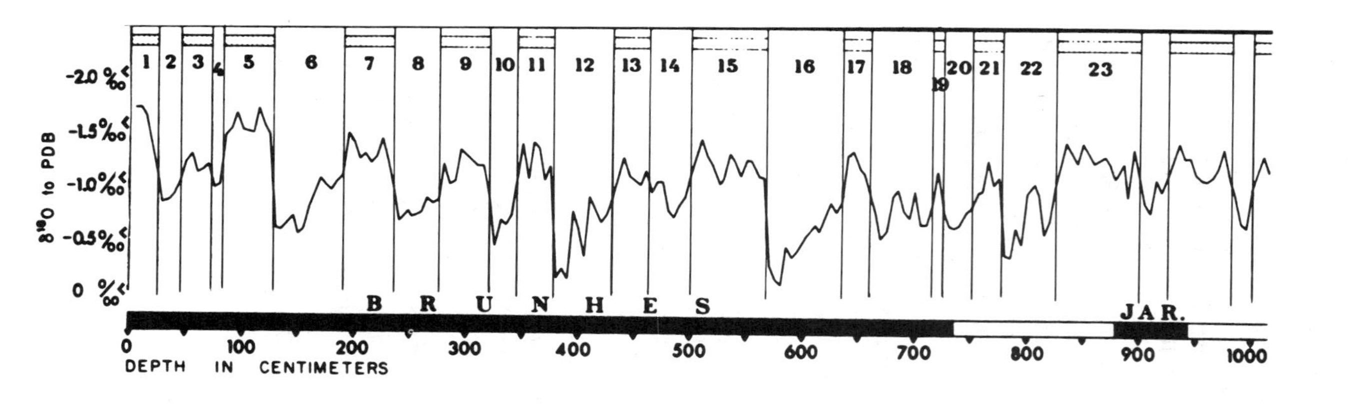

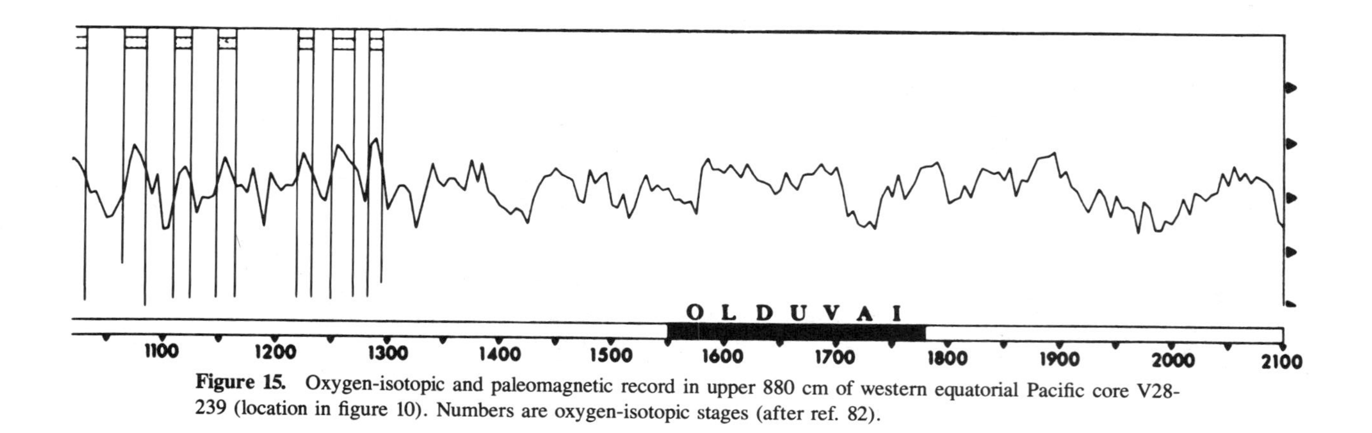

Figure 15. Oxygen-isotopic and paleomagnetic record in upper 880 cm of western equatorial Pacific core V28-239 (location in figure 10). Numbers are oxygen-isotopic stages (after ref. 82).

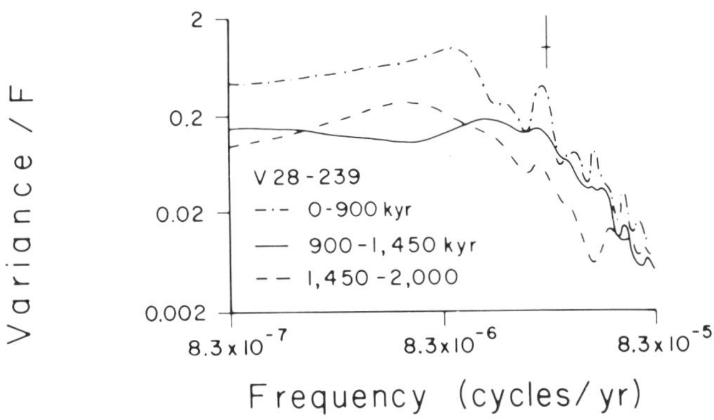

Figure 16. Log-linear variance spectra for oxygen isotopic records from western equatorial Pacific core V28-239 (location in figure 10). Confidence interval (vertical and horizontal bar) is at the 80 percent level. Spectra are calculated for three intervals: 0 to 900,000 years ago; 900,000 to 1,450,000 years ago; and 1,450,000 to 2,000,000 years ago. Power at 100,000-year and 41,000-year periods is appreciably weaker in the two earlier periods than in the last 900,000 years (after ref. 83; copyright 1981 by Elsevier Science Publishers).

muted paleoclimatic responses or with relatively local significance (subtropical gyres, western equatorial regions, eastern boundary currents, coastal upwelling regions, the Arctic Ocean, and numerous marginal seas).

High-Latitude Regions

The largest climatic changes on earth during the Quaternary were concentrated in and around the three high-latitude oceans: the North Atlantic, the Antarctic, and the North Pacific.

North Atlantic. The most intensively studied region in the world ocean is the North Atlantic between 35°N and 65°N. This interest reflects its critical location adjacent to the ice age ice sheets and astride regions of intermediate and deep-water formation. Studies through the mid-1970s laid out the basic late Quaternary patterns of faunal and floral change[45,86–87] and of estimated SST response.[88] The major result of these efforts was the discovery that a mass of frigid, ice-laden polar water filled the subpolar North Atlantic southward to a sharply defined front at 45°N during glaciations. This transition is clearly shown by the north-south transect of downcore faunal trends in figure 18, with a sharp decrease in polar faunal percentages centered at about 45°N near core V29-179.

More recent work[89,90] has dealt with the spectral character and phasing of these surface-ocean changes, especially in comparison to that of the $\delta^{18}O$ ($\sim$ ice volume) record. Transfer-function analysis of the faunal analyses shown above yielded the estimates of (summer) sea-surface temperature shown in figure 19. Spectral analysis of four of these seven records produced SST variance density plots (figure 20) which show several features: (1) dominant, or nearly dominant, 100,000-year power in all cores; (2) 23,000-year power at a maximum in core V30-97 at 41°N, and declining rapidly northward to lower values at 54°N; and (3) 41,000-year power at a minimum

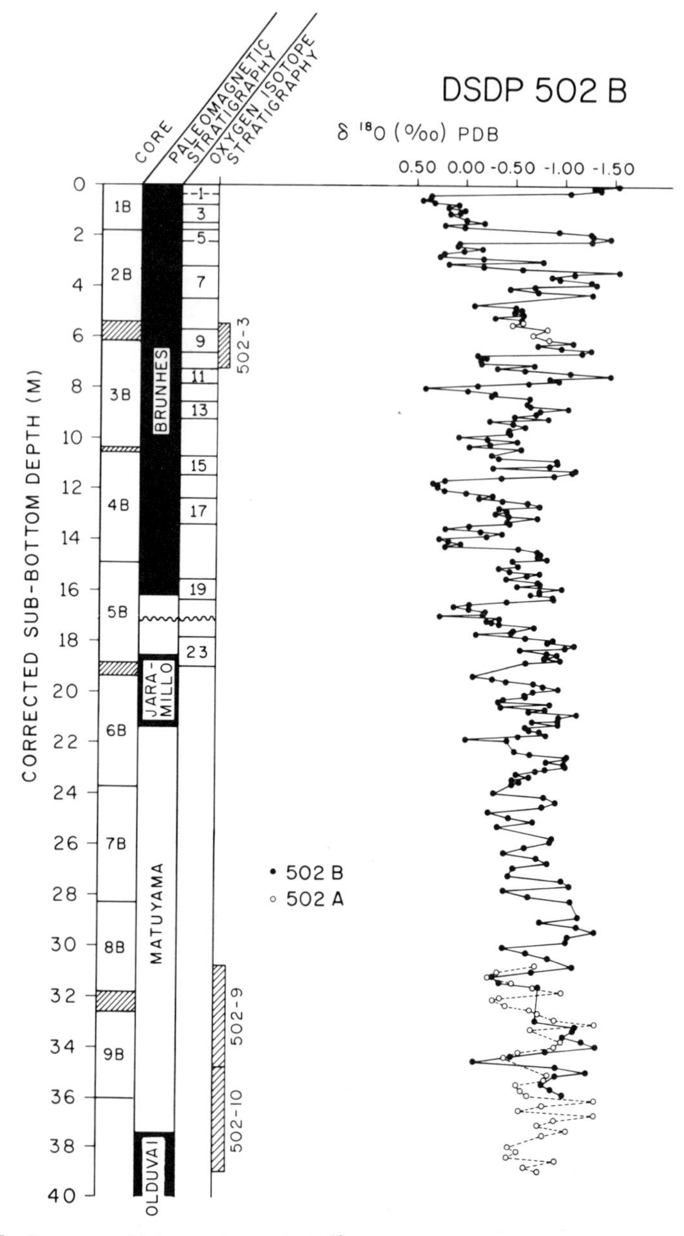

Figure 17. Long-term (Pleistocene) record of $\delta^{18}O$ in planktonic foraminifera from DSDP Hydraulic Piston Core 502B in the Caribbean (after ref. 84). Amplitude of $\delta^{18}O$ change increases, and mean value becomes more positive, after 900,000 years B.P. (top of Jaramillo normal magnetic chron).

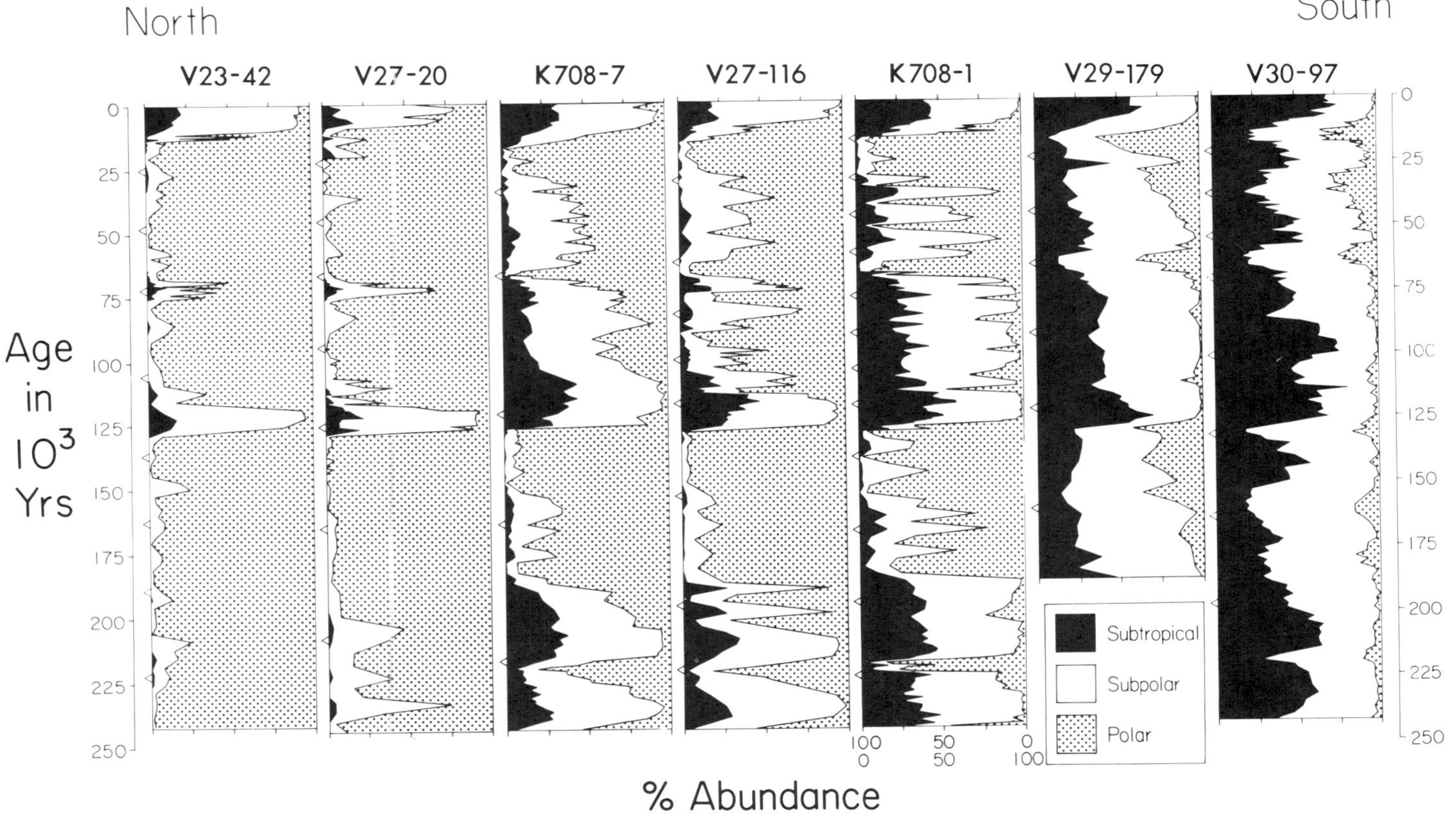

Figure 18. Time-stretched cumulative percentage plot of three major groupings of planktonic foraminifera in subpolar North Atlantic piston cores (locations in figure 10). Polar fauna dominates northern cores; subpolar and subtropical species in southern cores (from ref. 89).

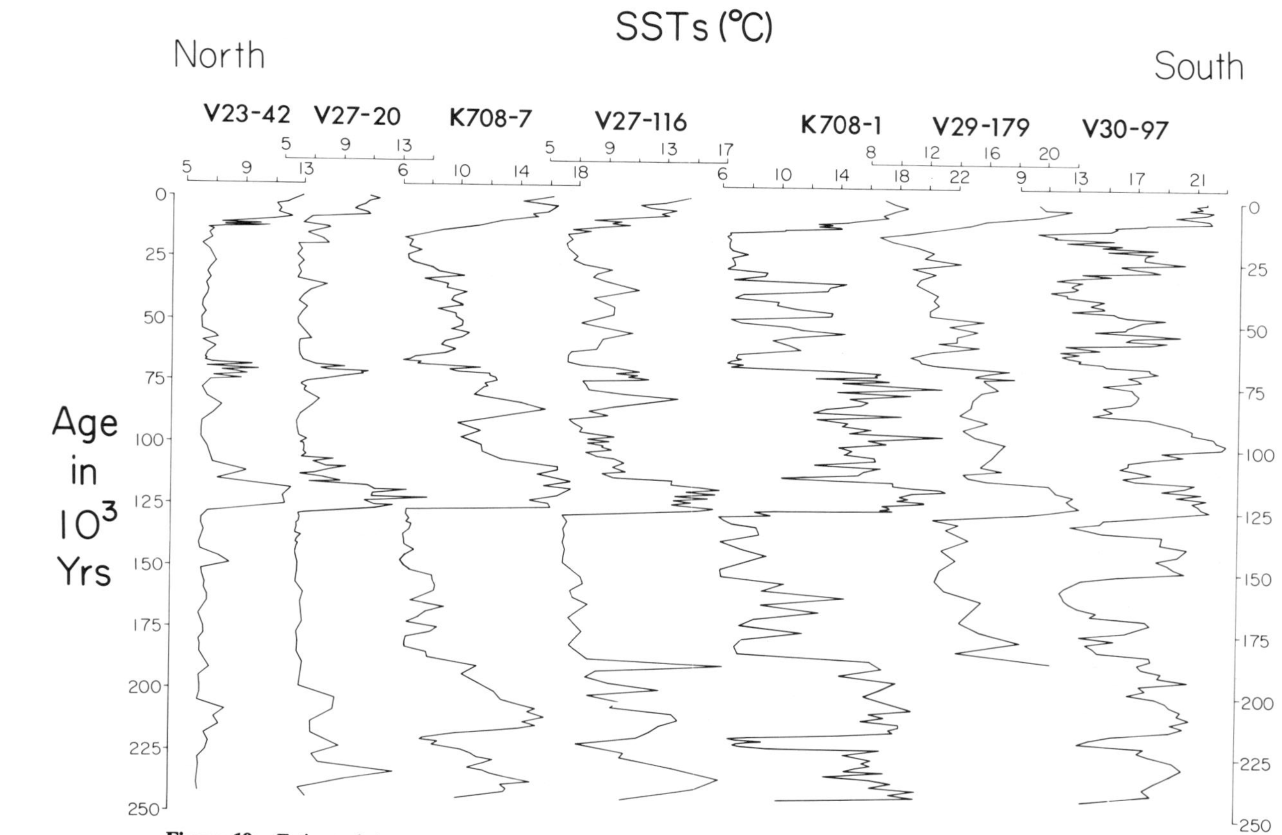

Figure 19. Estimated August sea-surface temperature based on species counts in seven subpolar North Atlantic cores shown in figure 18. Estimates are constrained to values greater than 6°C because of transfer-function limitations. Plots are stretched to time axis (from ref. 89).

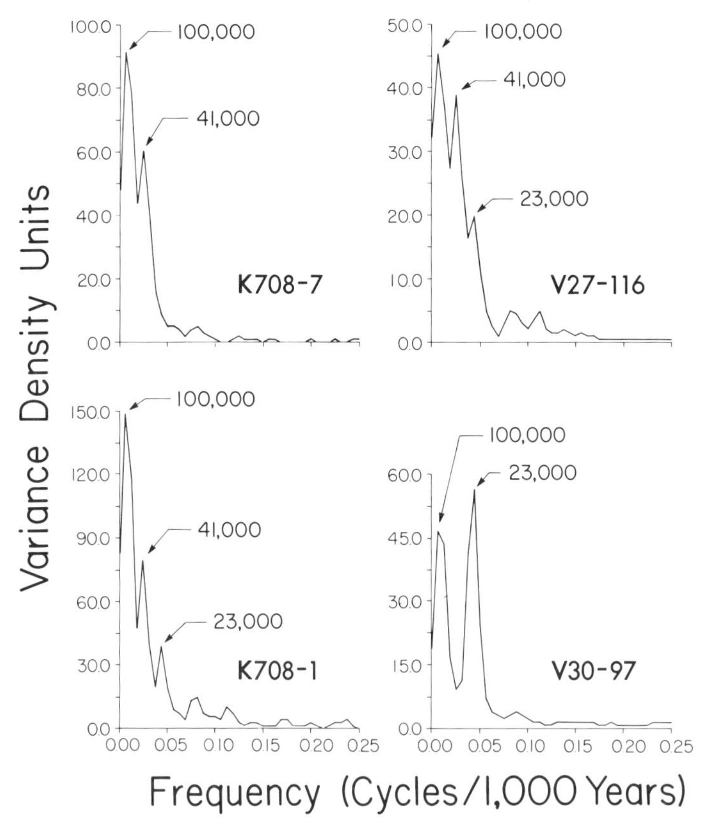

Figure 20. Variance density plots of estimated August sea-surface temperature in four of the seven cores shown in figure 19. The 100,000-year rhythm is strong in all regions; the 41,000-year power is strongest in the three northern cores but yields to 23,000-year power in mid-latitude core V30-97 (from ref. 89.

in the core at 41°N, but dramatically larger in the three cores between 50° and 54°N. Filtering of the 23,000-year and 41,000-year components from these SST records further delimits these trends (figure 21).

These results show that the cold "polar-water" gyre that periodically developed north of 45°N fluctuated with a 100,000-year and a 41,000-year rhythm, whereas a very different 23,000-year periodicity characterized the response of the northern subtropical gyre immediately to the south. This change in character of the SST spectra occurs in the middle of the region of largest ice age thermal response (figure 22). The diminished response south of 35°N is associated with the stable center of the subtropical gyre;[72,91] north of 65°N, the Norwegian Sea waters remained almost constantly frigid, with temperate water penetrating from the south only every 100,000 years or so during peak interglaciations like today.[92,93]

The 100,000-year and 41,000-year SST signals north of 50°N respond directly in phase with the same periodicities in the $\delta^{18}O$ ($\sim$ice volume) signals. This syn chronous phasing implies that the high-latitude ocean basically follows (with no

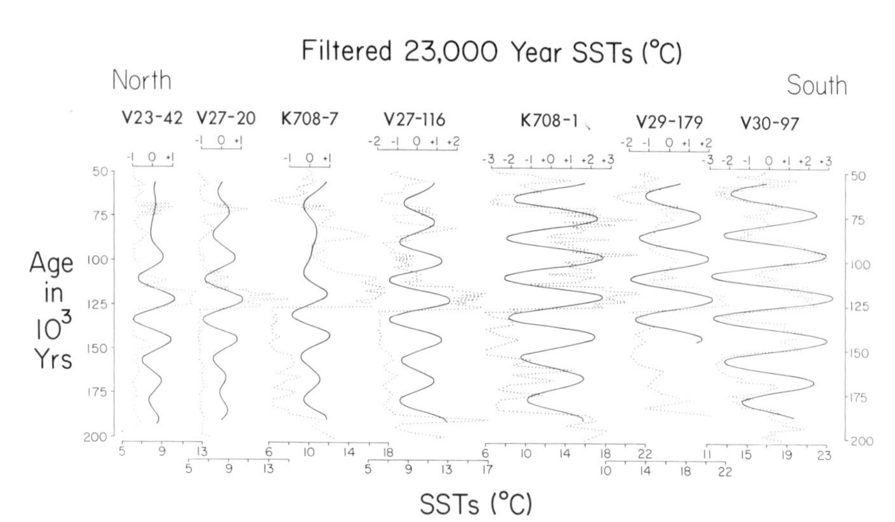

(a)

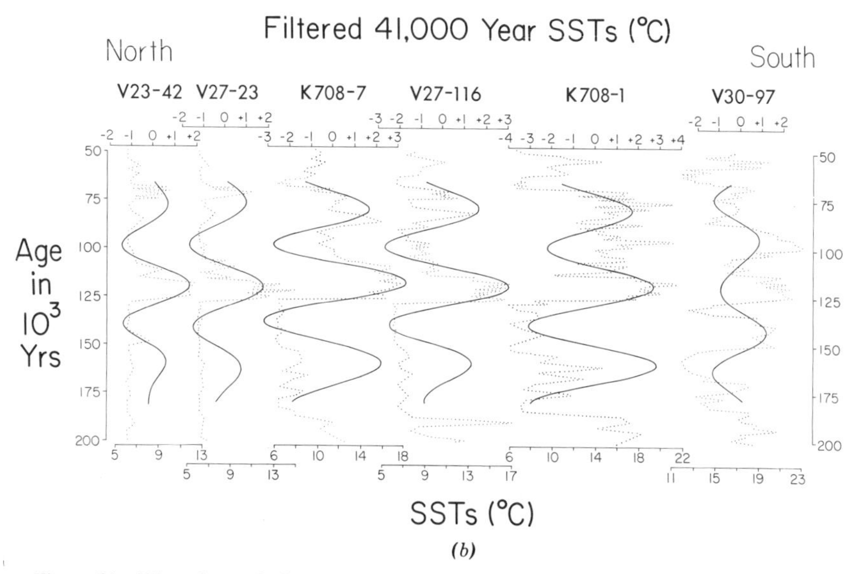

(b)

Figure 21. Filtered record of the *(a)* 23,000-year and *(b)* 41,000-year signals in estimated SST curves shown in figure 19. The 23,000-year rhythm is strongest in the southern cores; the 41,000-year signal is strongest in cores farther to the north (from ref. 89).

lag) the tempo set by the high-latitude ice sheets, which, by albedo feedback, appear to guide the circum-Arctic regions into and out of glacial conditions. The $\delta^{18}O$/SST phasing indicates that the ocean provides positive feedback to the insolation-driven oscillations in the ice sheets through changes in oceanic heat flux but negative feedback through changes in the moisture flux. Because the ocean north of 50°N carries only modest amounts of heat (figure 22, bottom), this region appears to

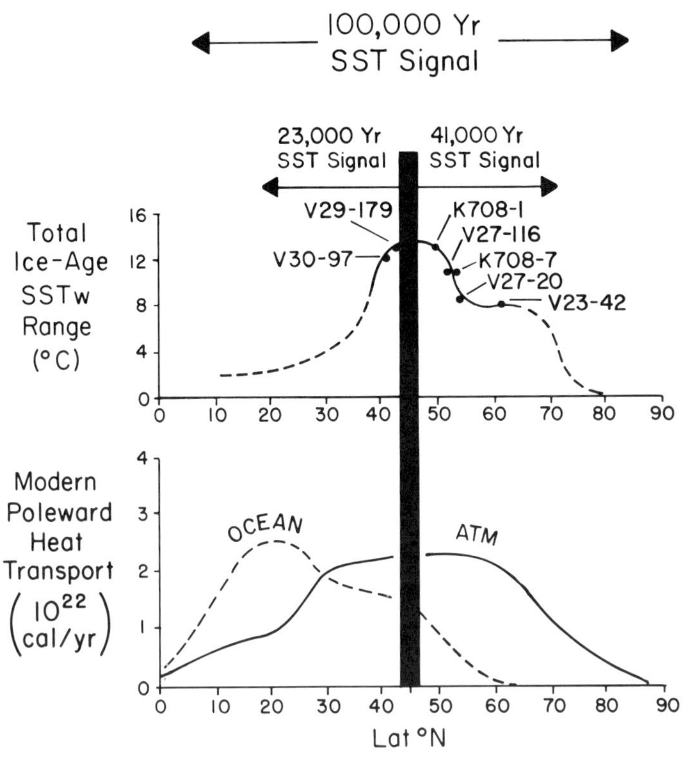

Figure 22. Summary of late Quaternary North Atlantic SST response. Top: Amplitude of glacial/interglacial SST change, with regions of spectral dominance superimposed. Cores are those in figure 19. Bar is boundary of 41,000-year and 23,000-year responses. Bottom: Estimated amount of energy carried poleward by ocean and atmosphere in Northern Hemisphere (from ref. 89).

play a minor role in the total global heat transfer. However, proximity to the ice sheets leaves open the possibility that even this small transfer could be significant to ice sheet mass balance.

The 23,000-year SST signal that is so strong south of 50°N lags roughly 6,000 years (one-quarter wavelength) behind the 23,000-year $\delta^{18}O$ signal. This phasing has been attributed to two major factors: (1) the effects of meltproducts from the mid-latitude ice sheets, which fluctuate in size with the 23,000-year periodicity; and (2) the impact of variable advection of warm water northward from the low latitudes of the North and South Atlantic.[89,90,94] Because the 23,000-year SST period dominates in the highly energetic circulation of the northern subtropical gyre (figure 22), oceanic feedback processes at this periodicity may be very significant both locally and globally. This part of the ocean provides a negative feedback of heat to ice sheet oscillations at the 23,000-year periodicity, but a positive moisture feedback that is potentially very powerful, especially in winter.

Antarctic. The Antarctic Ocean is the only body of water with uninterrupted circumpolar flow. Like the North Atlantic, it is also an area of vigorous air-sea energy exchange and of voluminous formation of waters that fill the deeper ocean basins. Unlike the North Atlantic, the glacial oscillations of the antarctic polar front were not large, reaching a maximum of 3° to 5° of latitude in the Atlantic but considerably smaller elsewhere.[60,95]

South of the narrow polar front, temperature reconstructions for the Quaternary are hindered by the no-analog abundances of *C. davisiana.* In addition, temperatures in this region are so cold today (~2°-5°C) that little additional cooling is possible before freezing of the sea surface occurs. There is also a scarcity of the calcitic organisms needed for a $\delta^{18}O$ stratigraphy. Hays and colleagues[50,96–98] interpret downcore changes south of the polar front (changes in percent *C. davisiana* and in the relative deposition rates of biogenic silica and ice-rafted debris) as indicators of variations in geographic and seasonal extent of sea ice. The timing of these changes at high latitudes is established by correlation with variations of percent *C. davisiana* in mid-latitude cores, for which oxygen-isotopic records are available.

In the sub-Antarctic latitudes (40°–50°S), the wider range of Quaternary surface-ocean changes and the resulting greater diversity in the radiolarian fauna led first to transfer function estimates of sea-surface temperature and then to the pivotal efforts in spectral analysis and phase analysis by Hays, Imbrie and Shackleton[78] (figures 23 and 24). Estimates of local sea-surface temperature were found to lead the $\delta^{18}O$ signal used as a proxy for global ice volume (figure 24). The amount of SST lead varied from 1500 to 2000 years, depending on the core interval examined and on the frequency band for which the comparison of filtered signals was made. Hays[97] concluded from this Antarctic lead that changes in size of Northern Hemisphere ice sheets cannot control changes in Southern Hemisphere (oceanic) climate, but that the Southern Hemisphere changes could influence Northern Hemisphere climate.

There is some uncertainty about the linkage of the radiolarian response to surface-water changes in the circum-Antarctic. During the course of the CLIMAP investigation of the Last Interglaciation,[66] it was discovered that foraminiferal estimates of SST disagreed significantly with radiolarian estimates in the same cores (figure 25). In various instances, the foraminiferal signals either showed no lead relative to $\delta^{18}O$ or a smaller lead than the radiolaria. A re-examination of the relationship between surface-water oceanographic properties and the foraminiferal/radiolarian assemblages derived by factor analysis suggests that the foraminiferal factors correlate with surface-water properties but the radiolarian factors may be linked to subsurface structure.[99] In this view, any Antarctic ocean lead would be a subsurface response tied to waters formed in the subtropical gyre.

This matter is not yet resolved, and investigations of other biotic components in the Antarctic are in more primitive stages. Coccoliths do not exist south of the polar front and are relatively limited in use for some distance to the North.[58] Although diatoms are very abundant, systematic analyses have not been published.

North Pacific. As in the Antarctic, Quaternary paleoceanography in the Pacific north of about 45° is hindered by the lack of calcite for $\delta^{18}O$ stratigraphy and ^{14}C dating. To some extent, this is compensated by the long-term stratigraphic control provided by paleomagnetic reversals, by widespread volcanic ash layers that

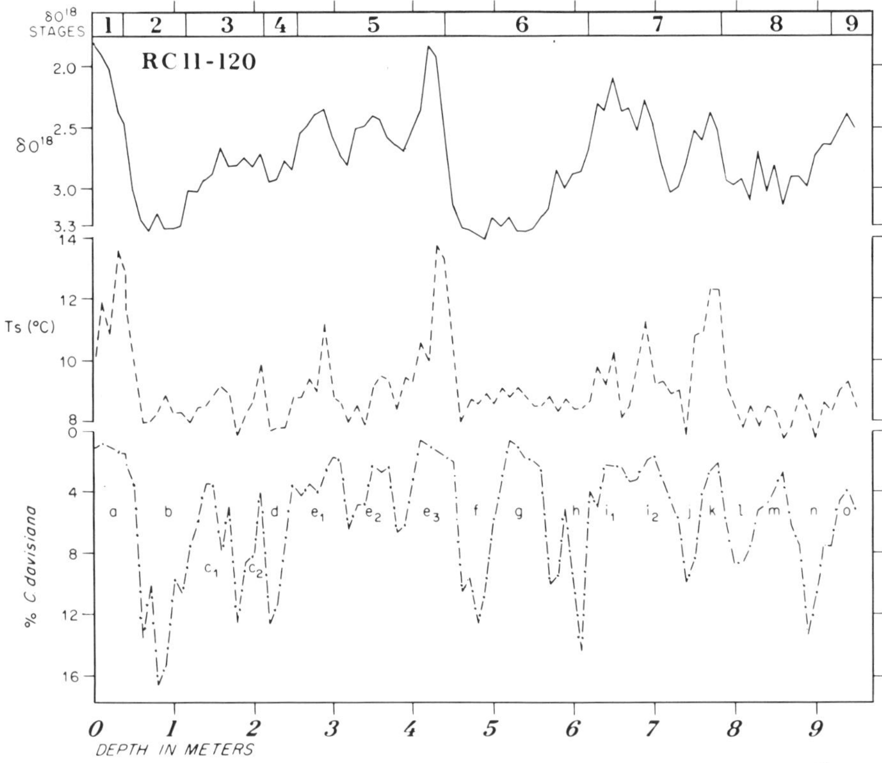

Figure 23. Plots of four parameters measured in sub-Antarctic Indian Ocean core RC11-120: $\delta^{18}O$ from planktonic foraminifer *Globigerina bulloides*, estimated summer (February) sea-surface temperature based on radiolaria, and percent of *Cycladophora davisiana* in radiolarian fauna (after ref. 97).

serve as correlative stratigraphic markers, and by extinctions of radiolaria and diatoms.[100–102] Still, however, time control as precise as in other oceans is not obtainable.

The major ice age response of the high-latitude North Pacific was an eastward translation of faunal and floral belts, with biota now characteristic of the Sea of Okhotsk and northwestern Bering Sea expanding well out into the western and central North Pacific. This pattern occurred basically in phase with ice sheet expansion and is evident both in radiolarian distributions[49,63,103] and in the diatom changes shown in figure 26.[104] It appears to indicate that the strongly seasonal, almost continental, climatic regime now characteristic of these marginal basins expanded into the Northwest Pacific. North of the strong glacial-age subarctic front, this regime probably included severe winter cooling with sea-ice formation, a delayed and probably incomplete summer mixing and warming of the upper gyres, and productivity suppression except for a brief period of open water in summer.[104] This regime has been linked to the increased impact of frigid Arctic air masses sweeping across the Pacific.

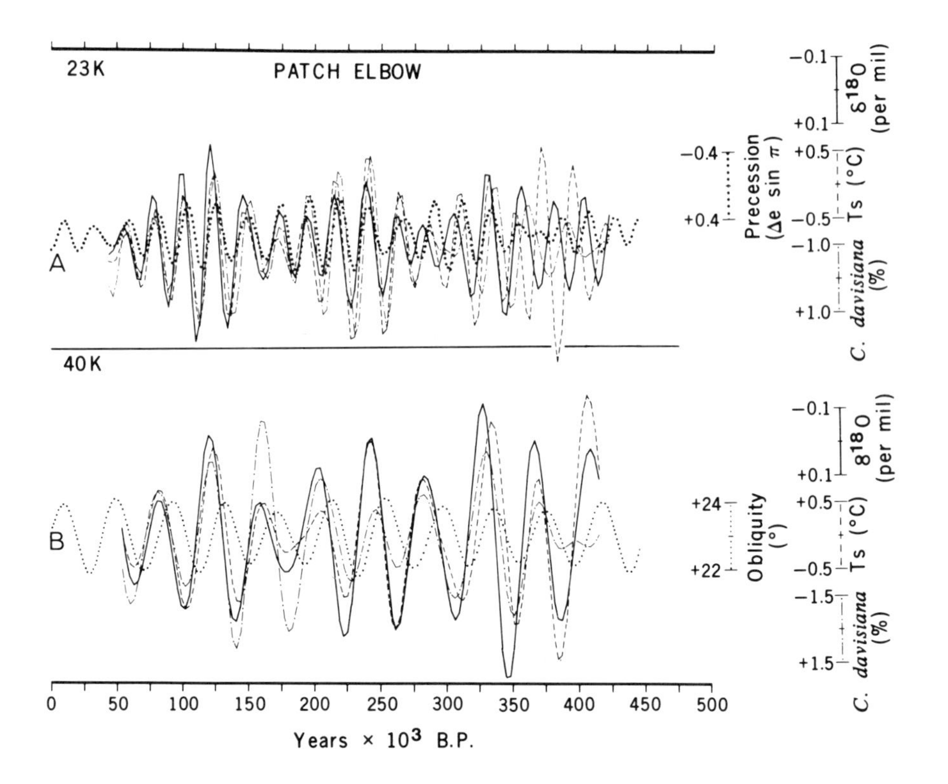

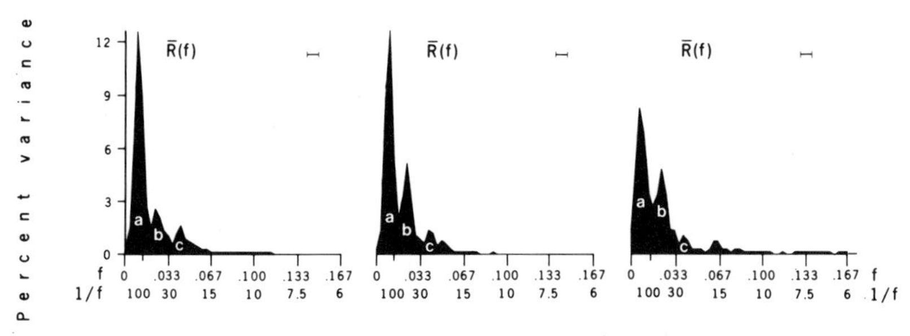

Figure 24. Top: Filtered 23,000-year and 41,000-year signals of $\delta^{18}O$, % *C. davisiana*, and radiolarian-based estimates of summer (February) SST in combined record of sub-Antarctic cores RC11-120 and E49-18, compared to obliquity and precessional signals. Cores located in figure 10. Bottom: Variance density spectra of $\delta^{18}O$ (left), % *C. davisiana* (center), and estimated summer (February) SST (right) in above cores (after ref. 78). Letter *a* on spectra marks 100,000-year peak; *b* is 41,000-year peak; *c* is 23,000-year peak. (After ref. 78. Copyright 1976 by the AAAS.).

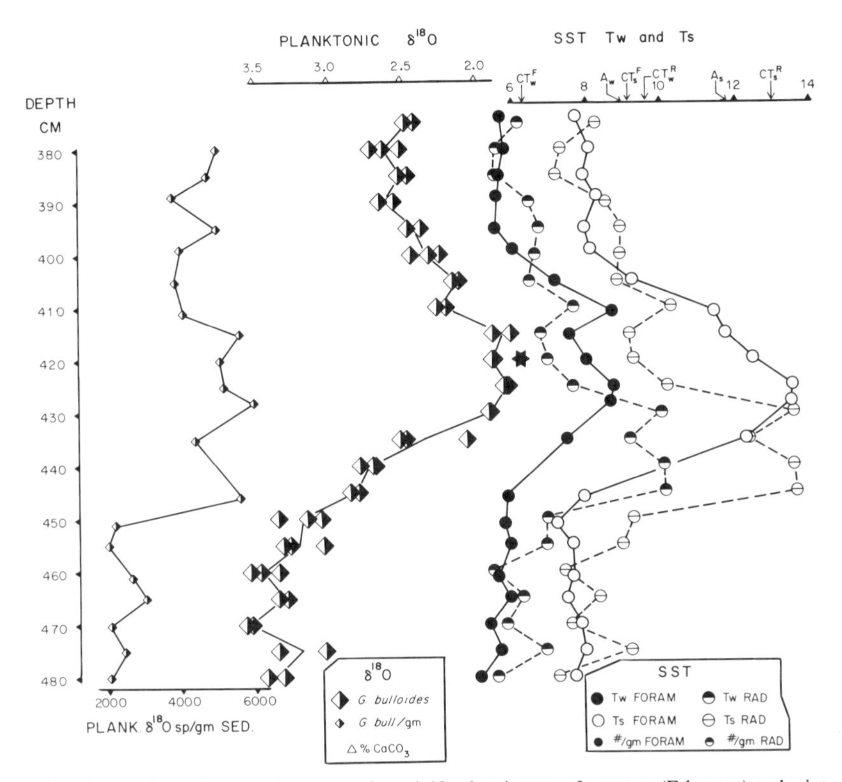

Figure 25. Comparison of radiolarian versus foraminiferal estimates of summer (February) and winter (August) SST signals in sub-Antarctic Indian Ocean core RC11-120 through last interglacial interval (from ref. 66). Radiolarian-based SST curve leads $\delta^{18}O$ signal into and out of last interglaciation; foraminifera-based SST curve disagrees and is roughly in phase with $\delta^{28}O$ signal.

Monsoon Regions

The Northern Hemisphere monsoon, centered over the Indian subcontinent, but affecting adjacent portions of Asia, Europe, and Africa, is another vital component of the climatic system. Oceanic data pertinent to the intensity of monsoon circulation are available in one source region of monsoon moisture (the Arabian Sea) and two areas impacted by monsoon runoff (the Bay of Bengal and the Mediterranean Sea).

Arabian Sea Upwelling. Along the coast of Arabia, surface sediments show high concentrations of *Globigerina bulloides,* a species normally indicative of cooler, high-latitude waters. Prell and Curry[105] found that the relative abundance of this species is correlated to August sea-surface temperature and phosphate content (figure 27). This suggests that coastal upwelling driven by strong southwesterly winds

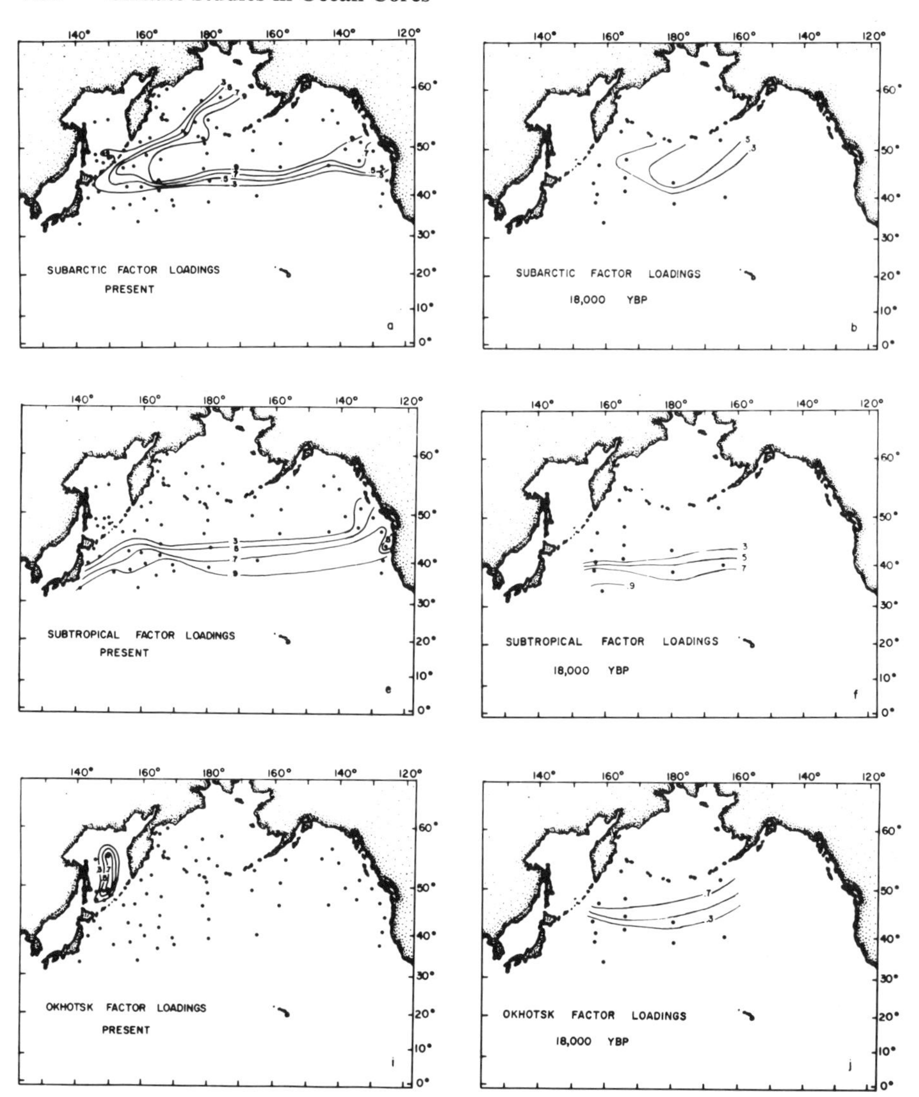

Figure 26. Diatom-based comparison (after ref. 104; copyright 1979 by Elsevier Science Publishers) of the modern world (left column) and the last glacial maximum 18,000 years ago (right column). Comparison indicates the expansion of Okhotsk fauna into the North Pacific at the expense of sub-arctic and subtropical factors. This indicates marked increase in seasonality in the western North Pacific.

during the summer monsoon season produces the cooler near-surface temperatures and high-productivity conditions in which this species flourishes. This interpretation was supported by the strong correlation between August SST and $\delta^{18}O$ values measured in several species of planktonic foraminifera from surface sediments; for any time-equivalent set of samples, local geographic variations in foraminiferal $\delta^{18}O$ will reflect gradients in oceanographic properties (in this case primarily summer

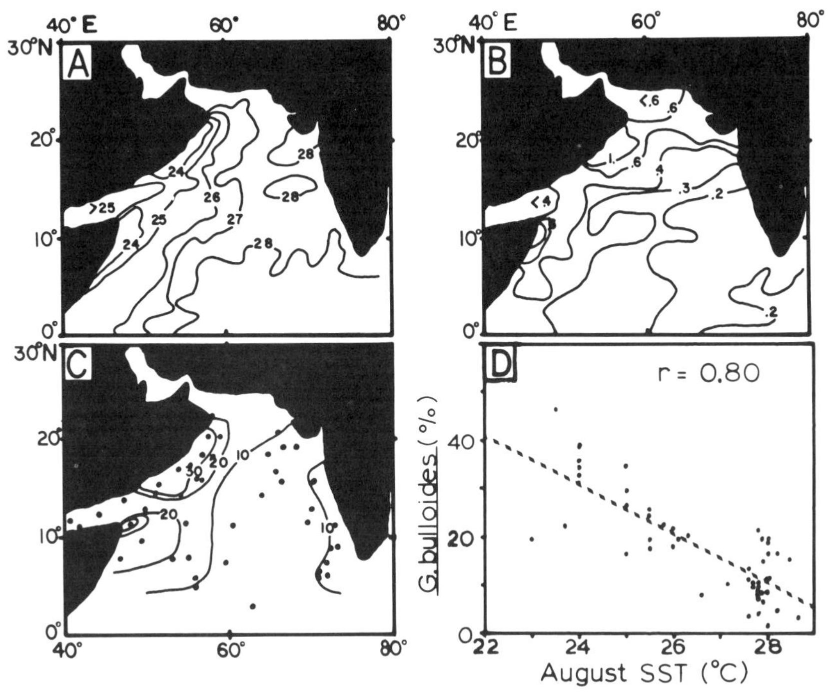

Figure 27. The geographic distribution during the southwest monsoon season of (*A*) August sea-surface temperature, (*B*) phosphate content in ug-atoms per liter during the May–October monsoon season, and (*C*) relative percent frequency of planktonic foraminifer *Globigerina bulloides* in surface sediments. (*D*) Regression of % *G. bullides* against August SST (correlation coefficient −0.80). From ref. 105; copyright 1979 by Gauthier Villars.

SST). Analog models confirm that coastal upwelling off Arabia is tightly linked to the intensity of the summer monsoon of the Northern Hemisphere.[106]

The relative abundance of *G. bulloides* during the last 130,000 years in one Arabian Sea core[107] shows a significant concentration of power at a periodicity of 23,000 years and a strong coherency of this peak with solar radiation (figures 28 and 29).

This result confirms earlier modeling efforts[108] suggesting that increased radiative heating in summer over the Tibetan Plateau affects the intensity of the summer monsoon circulation. The last maximum in Arabian Sea upwelling and monsoon intensity occurred about 9000 years ago, in the late stages of the last deglaciation. Prell[107] also found an average lag of about 5000 years between mid-summer insolation and monsoonal upwelling in this downcore record; this implies that the influence of summer insolation on the monsoonal circulation is retarded by some factor, possibly seasonal snow and ice cover over the Tibetan Plateau.

Bay of Bengal Monsoon Runoff. The modern monsoon runoff from the Indian subcontinent creates a very strong salinity gradient in surface waters of the Northern Indian Ocean. Cullen[109] found that sea-bed distributions of planktonic foraminifera

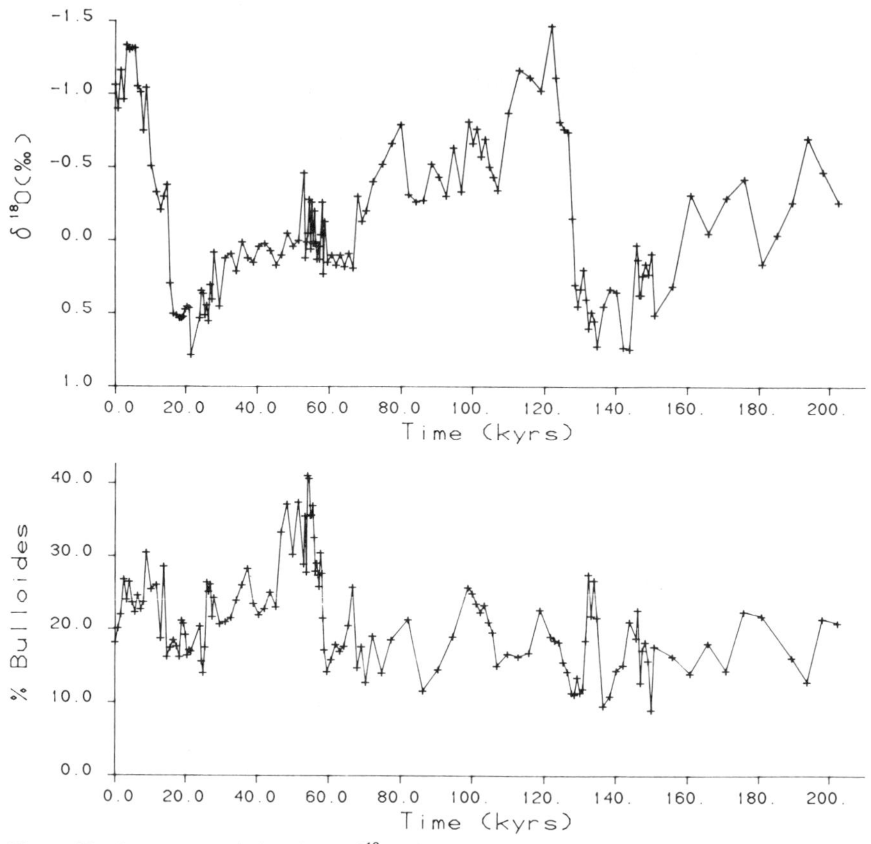

Figure 28. Downcore variations in (*a*) $\delta^{18}O$ of planktonic foraminifer *G. sacculifera* and (*b*) % *G. bulloides* in Arabian Sea core V34-88 plotted on time axis (from ref. 107). Core location in figure 10.

in this area are at least as useful for estimating sea-surface salinity as they are for the more commonly used SST parameter (figure 30). High values of one species, *Globoquadrina dutertrei*, are particularly well correlated to low-salinity water.

Using oxygen isotopes and ^{14}C for stratigraphic control, Cullen[109] examined a north-south transect of cores down the center of the Bay of Bengal along 90°E longitude. He found a maximum in percent *G. dutertrei* and a minimum in estimated sea-surface salinity occurring about 11,500 years ago (figure 31). This signal indicates that the monsoon runoff was stronger at that time than it is now or than it was during the last glacial maximum.[59,60]

Mediterranean Stagnation. There is now evidence that monsoonal effects can be felt even in the eastern Mediterranean Sea. Organic-rich muds called sapropels noted in Mediterranean cores have long been attributed to some combination of glacial meltwater runoff and increased local precipitation.[110–114] Olausson[115] hypothesized that fresh-water influxes formed a low-density surface layer that hindered normal vertical mixing by diminishing the amount of sinking of denser water formed by evaporation and chilling.

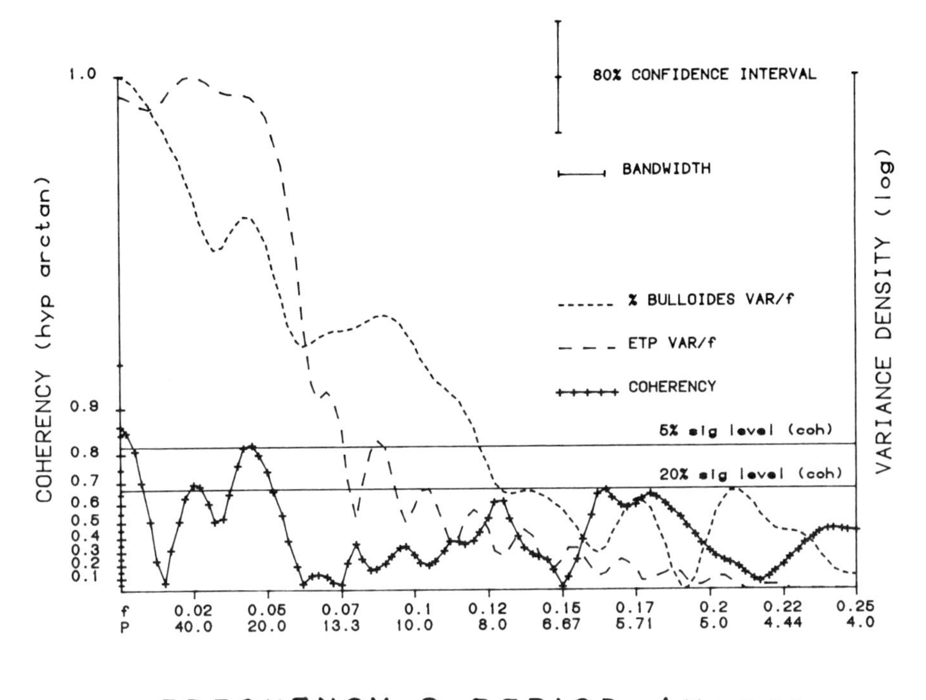

Figure 29. Variance and coherency plots vs. frequency (f) and period (p) for % *G. bulloides* curve shown in figure 28 and ETP radiation curve (from ref. 107).

Direct evidence for a fresh layer exists in the light $\delta^{18}O$ values found in surface-dwelling planktonic foraminifera in two sapropels (figure 32). Other evidence includes abnormally high (no-analog) abundances in some sapropels of planktonic foraminifera like *G. dutertrei* thought to prefer low-salinity conditions.[110,116]

Rossignol-Strick et al.[117] pointed out the correlation of the most recent Mediterranean sapropel with an interval of very heavy Nile River runoff. Because the Nile drains the Ethiopian highlands far to the south, where rainfall is controlled by the summer monsoon, it now appears that the stagnation of the Eastern Mediterranean at subtropical latitudes may be largely controlled by monsoon delivery of moisture from the Indian and Atlantic Oceans to tropical west Africa far to the south.

Equatorial Circulation

A second low-latitude region of fundamental importance to global climate includes the equatorial divergences, especially in the eastern sectors. Sensitivity tests with General Circulation Models indicate that SST changes of a few degrees centigrade in these areas can have a major impact on the atmospheric circulation over both oceanic and continental regions of the low and middle latitudes. Oceanic data

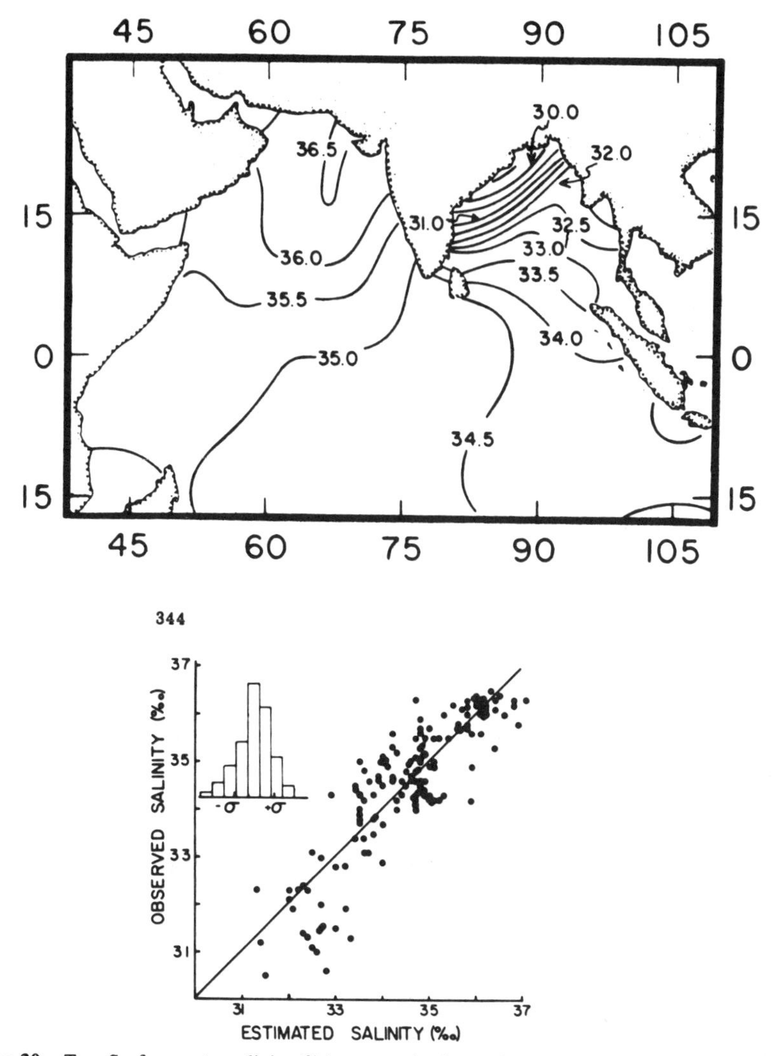

Figure 30. Top: Surface-water salinity (‰) patterns in the modern Indian Ocean during modern SW monsoon (May–October). Bottom: Scatter diagram of observed sea-surface salinity values vs. transfer-function estimates for SW monsoon period (May–October). Histogram shows frequency distribution of residuals between estimated and actual values (from ref. 109; copyright 1981 by Elsevier Science Publishers).

relevant to the long-term responses of these regions are available in two areas, the eastern equatorial Pacific and the equatorial Atlantic.

Equatorial Pacific. Particularly critical in its impact on the modern global circulation is the eastern equatorial Pacific, which is the site of the strong El Niño signal that overwhelms the more modest interannual SST changes.[118] Following earlier work in this region,[49,119–121] Romine and Moore[122] explored the response of the eastern equatorial Pacific during the last 127,000 years. Factor analysis of

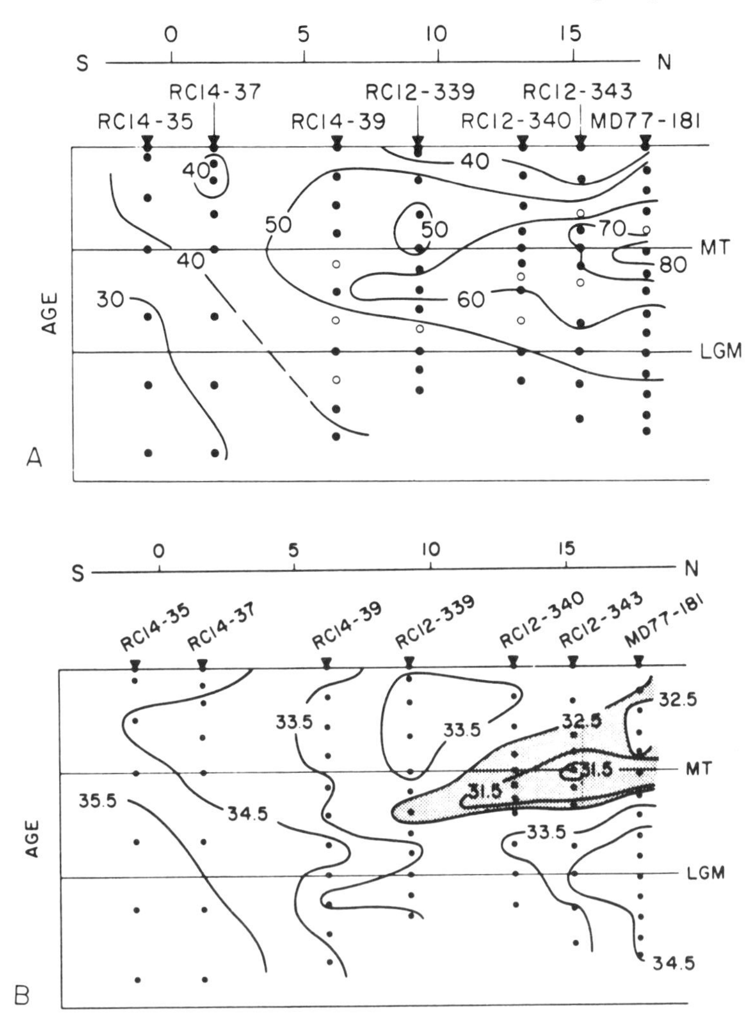

Figure 31. (*A*) Fluctuations of planktonic foraminifer *Neogloboquadrina dutertrei* as a function of relative age along north-south transect at 90°E in the Indian Ocean. Core ages aligned at core tops (~modern age), mid-termination (MT: ~11,000 years ago) and Last Glacial Maximum (LGM, ~18,000 years ago). *N. dutertrei* favors low-salinity waters in the modern ocean. (*B*) Fluctuations in estimated NE monsoon (November–April) sea-surface salinity along the same transect. The low-salinity tongue (stippled) on the deglaciation matches the period of high % *N. dutertrei* (from ref. 109; copyright 1981 by Elsevier Science Publishers).

radiolaria in surface sediments identified four major assemblages reflecting (1) western tropical mixed-layer flow; (2) coastal upwelling and the oxygen minimum layer; (3) eastern equatorial divergence; and (4) subtropical water. Modern distributions of the coastal upwelling and equatorial divergence factors are shown in figures 33 and 34, along with their factor loadings in two downcore records spanning the last 127,000 years. Both of these kinds of upwelling are related to trade winds; the coastal upwelling is stronger during increased wind strength, whereas the equa-

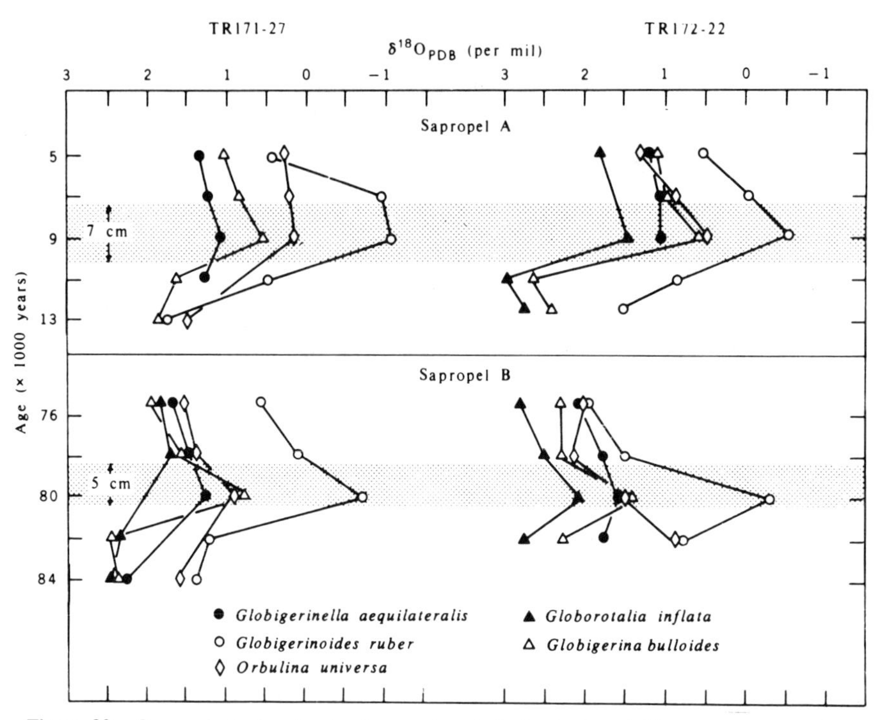

Figure 32. Oxygen-isotopic ratios for five species of planktonic foraminifera across sapropel layers A (~9000 years ago) and B (~80,000 years ago) in two Mediterranean cores (from ref. 116; copyright 1979 by Elsevier Science Publishers). Role of monsoon-related low-salinity water in creating density stratification that leads to stagnant sapropelitic muds (stippled) is evident from light $\delta^{18}O$ values in surface-dwelling species, *Globigerinoides ruber.*

torial divergence is stronger during periods of weaker wind strength when the equatorial counter-current nears or reaches the surface.

Romine and Moore[122] cite independent mineralogic data (windblown quartz abundance) from Molina-Cruz[121] as evidence that wind strength in the south equatorial Pacific was weak during interglacial isotopic stage 5 and then strong during glacial isotopic stages 4 through 2. The two radiolaria factors shown (figures 33 and 34) in part reflect these changes in trade wind strength, with weaker coastal upwelling and a stronger equatorial undercurrent in stage 5 and the converse during glacial isotopic stages 4 through 2. The trends do not, however, correlate to isotopic curves in any simple manner. Romine and Moore[122] infer that the mean latitudinal position of the trade winds, as well as their zonal/meridional orientation, are also important considerations. Cross-correlation analysis of unfiltered data suggests that changes in trade wind strength over the eastern equatorial Pacific lead $\delta^{18}O$ by 1000–2000 years.

Equatorial Atlantic. As in the equatorial Pacific Ocean, coastal upwelling and equatorial divergence are the most important components of equatorial Atlantic oceanography. In addition, there is a factor possibly unique to the Atlantic: cross-

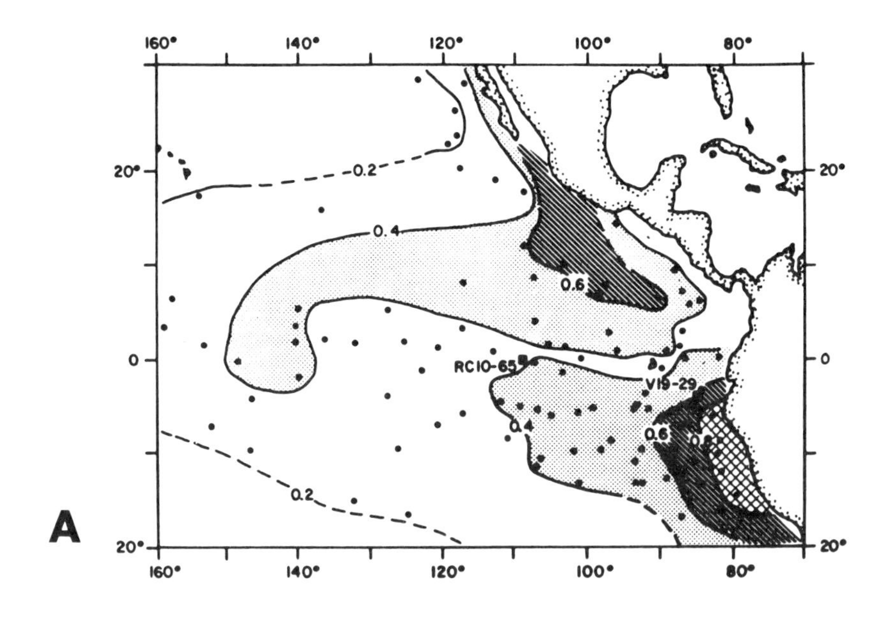

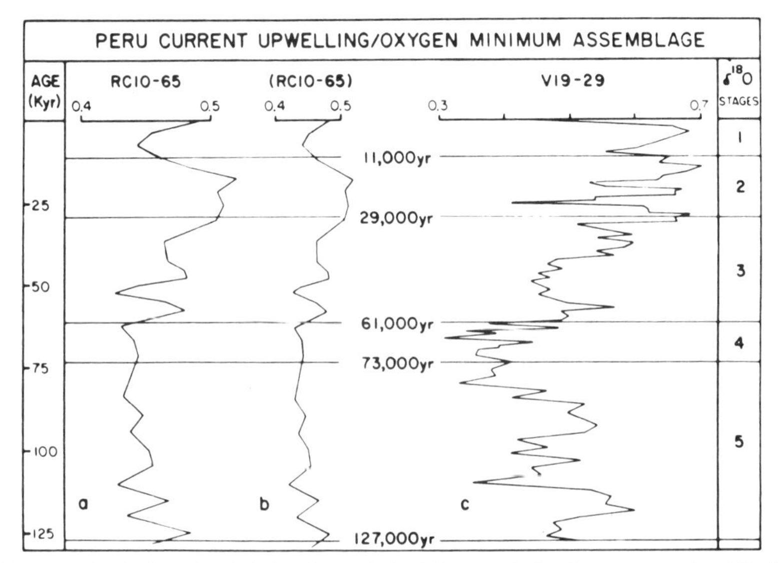

Figure 33. (*A*) Distribution of radiolarian factor derived from analysis of eastern equatorial Pacific core tops. Spatial distribution of factor loadings suggests an association with Peru Current coastal upwelling and oxygen minimum layer. (*B*) Downcore loadings of Peru Current upwelling/oxygen minimum layer assemblage in two eastern equatorial Pacific cores. Core locations shown at top and in figure 10. Largest factor loadings occur in last 60,000 years, with no obvious correlation with glacial/interglacial oxygen isotopic stages (after ref. 122; copyright 1981 by Elsevier Science Publishers).

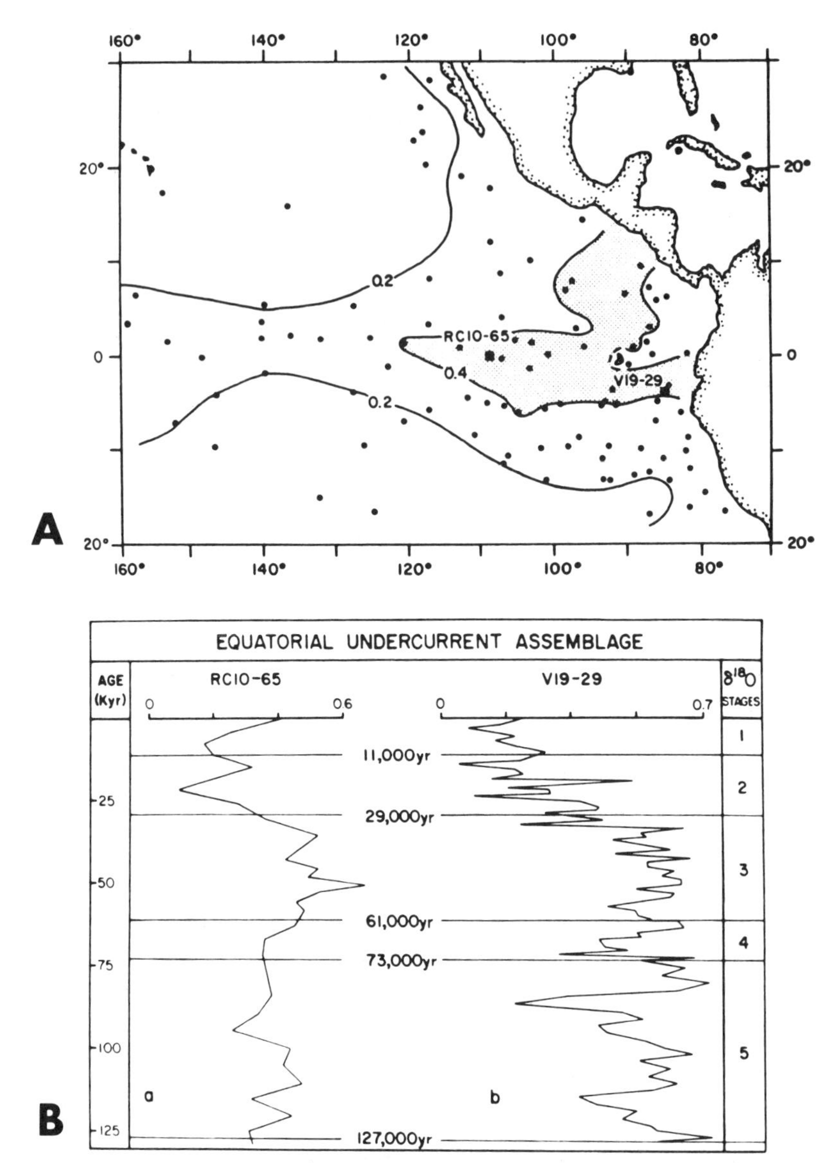

Figure 34. (*A*) Distribution of radiolarian factor derived from analysis of eastern equatorial Pacific core tops. Spatial distribution of factor loadings suggests an association with Eastern Equatorial undercurrent. (*B*) Downcore loadings of Eastern Equatorial undercurrent assemblage in two Eastern Equatorial cores. Core locations shown at top and in figure 10. Largest factor loadings occur during interval 125,000 to 30,000 years ago, with no strong correlation to glacial/interglacial isotopic stages (after ref. 122; copyright 1981 by Elsevier Science Publishers).

equatorial transport of heat from the Southern Hemisphere into the North.[123] For paleoceanographic studies, the equatorial Atlantic is blessed with moderately high deposition rates, good stratigraphic control including $\delta^{18}O$, and a diverse biota (foraminifera, coccoliths, radiolaria, and diatoms).

Because most paleoceanographic studies of the equatorial Atlantic predate the full realization that $\delta^{18}O$ signals are primarily a record of ice volume, they rely primarily on $CaCO_3$ stratigraphy or *Globorotalia menardii* biozones.[124] Gardner and Hays[125] determined the regions of major glacial/interglacial faunal and oceanographic change during the Quaternary; these include the region of the Cape Verde Islands with strong February cooling during glacials and a broader band from 0 to 5°S reaching across much of the east-central Atlantic with significant August cooling during glacials (figure 35). These changes largely reflect the hemispheric responses to increased trade winds in the respective winter seasons.

Changes in estimated SST through time in a core almost on the equator are shown in figure 36. These changes appear to be basically synchronous with $\delta^{18}O$ shifts,[125] but small leads or lags may exist. Spectral analyses and phase analysis of equatorial Atlantic records have not yet been attempted.

Changes in estimated SST during the Quaternary generally were smaller in the western equatorial Atlantic[126] and in the Caribbean.[21,127] The Caribbean response was marked by large variations in a foraminiferal factor dominated by *Globigerinoides ruber*, which presently is most abundant in the Sargasso Sea. Estimates of sea-surface salinity correlate positively with this factor and suggest that the Caribbean became markedly more salty during the glacial isotopic stages, particularly stage 4.

APPLICATION TO CLIMATE MODELING

Time-Slice Modeling. There are two general strategies for time-slice modeling on a paleoclimatic time scale. To date, studies have exclusively focused on past climatic extremes, that is, intervals when surface conditions were at a maximum difference from those today, such as the last glacial maximum. A second potential strategy which may be used in future studies, is to look at time slices centered on periods of fastest transition from one climatic regime to another. In both cases, geologic data are used to establish conditions at the surface of the earth; the models calculate atmospheric circulation.

Climatic Extremes. Three general circulation models have been used to analyze the last glacial maximum 18,000 years ago.[128–130] Despite some differences in database and in model reconstructions of the atmospheric circulation, all three studies portray a glacial world that is not just cold but also dry, with a much reduced hydrologic cycle characterized by recycling of precipitation within the oceans at the expense of the ocean-to-continent transfer. At high latitudes, the reduced hydrologic cycle is attributable to the expanded polar anticyclone over the glaciated land masses and to the increase in sea-ice cover over the colder oceans. At lower latitudes, the cooler ocean surface slows the hydrologic cycle, and the monsoon circulation was

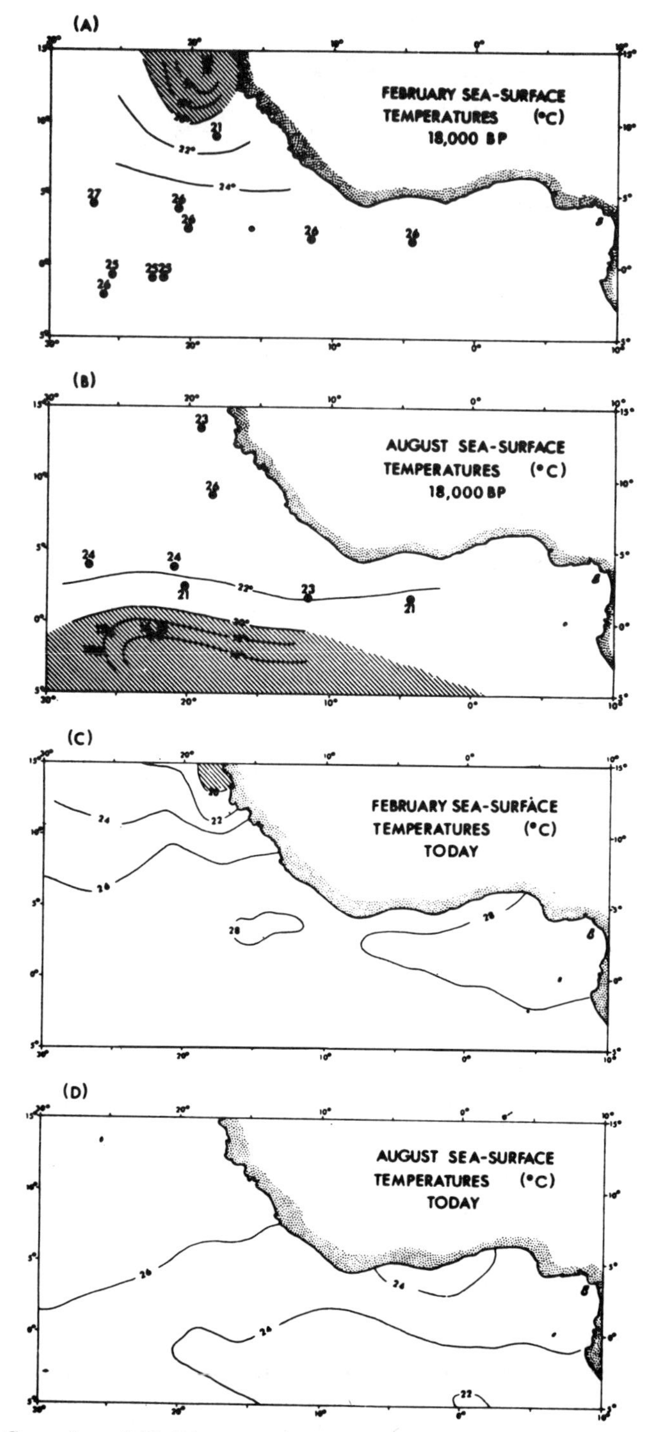

Figure 35. Comparison of (*A*) February and (*B*) August estimates of sea-surface temperature with present-day (*C*) February and (*D*) August observed values in the equatorial Atlantic Ocean (from ref. 125). Cross-hatching shows regions cooler than 20°C. Coolest glacial-age waters develop in region of coastal upwelling at 10° - 15°N due to Northern Hemisphere trade winds (*A*) and equatorial divergence at 0°–5°S due to Southern Hemisphere trade winds (*B*).

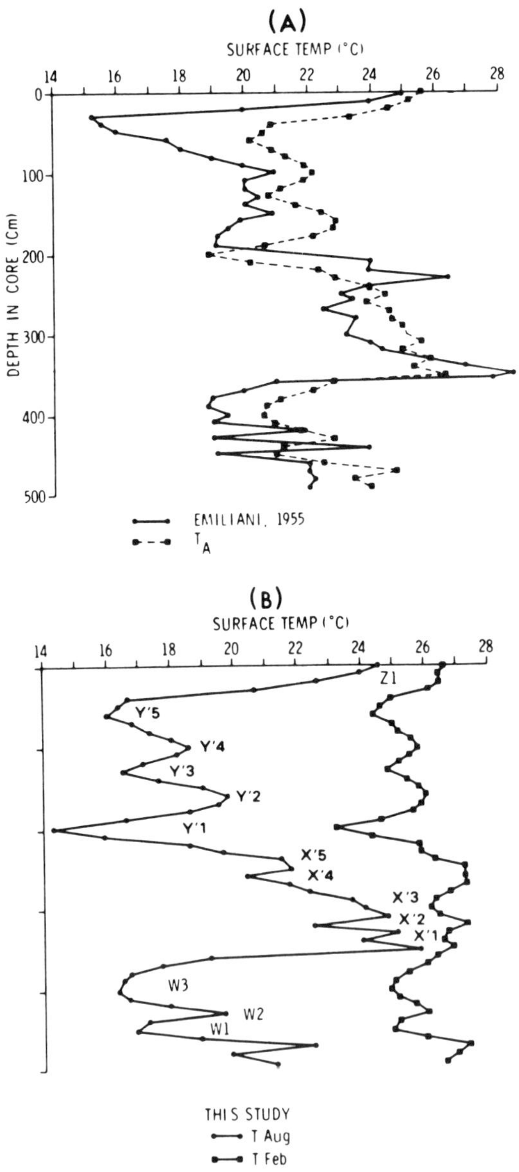

Figure 36. Plot of estimated August and February SST in equatorial Atlantic core A180-73 (from ref. 125). Time interval covered is last 150,000 years. Core location in figure 10.

also somewhat weaker than today. The extreme development of maximum low-latitude monsoonal circulation probably occurred on the deglaciation roughly 9,000 years ago rather than on the glacial maximum.

The glacial maximum axis of lower tropospheric wind strength in the general circulation models tends to have been shifted equatorward several degrees of latitude relative to today. High latitudes were characterized by a more extensive but weaker easterly wind regime, while the lower-latitude Hadley circulation was neither shifted in position nor much changed in strength by the changes at middle and high latitudes.

The modern world effectively represents the other extreme of Quaternary temperature oscillations—the warm interglacial end member. At least over the oceans, the Last Interglaciation was not sufficiently different from today for general circulation models to produce significant differences in atmospheric circulation. Much warmer conditions over the continents in summer might, however, be significant.

Climatic Transitions. Periods during which critical climatic parameters were changing at maximum rates are inherently interesting for modeling studies. To date, no such intervals have been subjected to analysis by general circulation models, which seem better suited to the short-term "equilibrium" conditions characteristic of the climatic extremes. However, some climatic transitions may also be appropriate for GCM analysis in the future. One question of perennial interest in paleoclimatic research is the trajectory taken by the mid-latitude moisture used to build ice sheets at high and middle latitudes. The interesting SST configurations of the North Atlantic ocean during ice sheet growth and decay suggest several experiments of this kind. To reconstruct the full hydrologic cycle of ice sheets during such intervals would probably require full annual simulations in order to distinguish between changes in precipitation and in ablation, the two major components of ice sheet mass balance. There might also be a need for multi-year simulations in order to achieve statistics that are significant relative to the large variability typical of high latitudes. The most recent deglaciation, with a wealth of geologic data constrained by radiocarbon dating, offers the most exciting opportunity for modeling climate transitions.

Time-Series Modeling

Time-series modeling (of ice dynamics and of local oceanic SST responses) relies on the fact that clear orbital periodicities have been detected in deep-sea records. The fact of periodic, and thus predictable, behavior, combined with knowledge of the forcing functions, has spurred the development of models which try to capture the critical long-term physics of the earth's climatic system.

Ice-Dynamics. The $\delta^{18}O$ record is used as an ice volume target in modeling ice dynamics over time spans of several hundred thousand years. Most of the models are two-dimensional (zonally averaged), and thus the flow-law physics which control ice sheet behavior are specified in a north-south transect.[26] Normally, accumulation and ablation rates are parameterized as a direct response to insolation, particularly summer insolation in the Northern Hemisphere, as emphasized by Milankovitch.[131] The underlying assumption here is that variations in summer insolation can be directly converted into lateral or vertical displacements of the snow line. The interactions of the ocean and atmosphere are reduced to an albedo/temperature feedback, in which southward advances of snow and ice (and increased vertical ice extent) lead to local cooling, with a reduction in both accumulation and ablation.

Early versions of these models simulated several significant features of the $\delta^{18}O$ curves, including the several-thousand-year lag behind insolation forcing, but consistently underrepresented the 100,000-year power associated with abrupt deglacial terminations.[26,27,36] Imbrie and Imbrie[28] reproduced some of that 100,000-year power (in a model lacking explicit ice physics) with the simplified formulation that ice decay occurs at a rate four times faster than ice growth. Subsequent efforts have largely focused on a search for a physical basis to provide the nonlinear impetus to drive deglaciations at a faster pace than ice growth. Birchfield et al.[132] achieved some 100,000-year behavior by incorporating the viscoelastic response of the earth to ice loading, but their model fell short of the 100,000-year power observed in $\delta^{18}O$ curves. Pollard[133] then included topography to represent higher ground in the northern end of the ice sheet and added a crude parameterization of calving at the equatorward ice sheet tip. Each additional extension of the model improved the simulations of 100,000-year power in the $\delta^{18}O$ record, both from time-series and spectral vantage points.

Other aspects of ice behavior remain to be incorporated. Oerlemans[134] has modeled the effects of basal sliding. Denton and Hughes[135] have modeled the linkage of marine-based and terrestrial ice sheets. The major opportunities for additional information to spur future advances in time-series modeling seem to be: (1) the acquisition of very long and detailed time series of $\delta^{18}O$ from hydraulic piston cores, and (2) the increasing availability of SST time series from oceans adjacent to the ice sheets for use in explicit formulation of ocean/atmosphere interactions with the ice. The major uncertainties in this work will remain the local temperature overprints on the ice volume portion of the $\delta^{18}O$ signal and the possibility of significant nonlinear linkage between ice volume and $\delta^{18}O$.

Regional Oceanic Responses. Modeling of surface oceanic responses to isolation forcing is still in an early stage. Most quantitative ocean modeling studies to date[136–140] have explored the possible interactions between the thermal inertia of the deep ocean and those parts of the climatic system with slower responses (the ice sheets) and faster responses (sea ice, the atmosphere, and the surface ocean). None of these studies, however, uses as a target for modeling any of the surface-ocean responses discussed previously, in part because most such SST time series have only recently become available.

Attempts to model explicitly the surface-ocean response face the general problem of the highly interactive nature of the oceanic mixed layer, which is in contact to some degree with all other parts of the climatic system (insolation, atmosphere, ice, deep ocean). Most modeling studies fall into one of two categories. The first category, involving studies of the immediate oceanic response to direct receipt of insolation, faces a specific problem similar to that addressed by Milankovitch[131] and other climatologists in explaining ice sheet responses. Summer and winter insolation changes are opposite in sign and nearly equal in magnitude; this forces the selection of a critical season and latitude whose insolation-driven climatic effects outweigh those from other latitudes and other seasons. The second modeling category assumes an indirect response to insolation forcing through a climatic intermediary: insolation-driven changes occur first on land (such as ice sheets or monsoon variations) and these changes subsequently impact the surface ocean.

Several descriptive (i.e. nonquantitative) models have been devised to explain observed SST responses. Ruddiman and McIntyre[89] ascribe the behavior of the 41,000-year and 100,000-year SST periodicities north of 50°N in the Atlantic to albedo-temperature feedback from high-latitude continental ice sheets varying with those periodicities. The equally prominent 23,000-year SST periodicity observed in the Atlantic at 40° to 50°N has been ascribed to two factors: (1) fluctuating iceberg and meltwater outflow from mid-latitude ice sheets at that period,[89,90] and (2) fluctuating oceanic heat advection northward from low latitudes and from the Southern Hemisphere.[90,94] These descriptive models are the result of using oceanic SST time series as an empirical basis first for testing and refining earlier models[141–145] and second for inferring new ocean-air-ice linkages.

Other descriptive models based on oceanic data previously discussed include: (1) attribution of Arabian Sea upwelling to the strength of the insolation-driven monsoonal winds blowing across India,[106–108] (2) interpretation of Ganges River outflow into the Eastern Indian Ocean in terms of monsoonal rain strength,[109] and (3) explanation of Mediterranean stagnation episodes by means of increased Nile runoff during insolation-driven monsoonal precipitation maxima over the Ethiopian highlands.[117]

All of these oceanic time series, and others, are ripe for quantitative modeling by the climatological community. Major opportunities for future advances in the database include the filling in of global geographic coverage, subject to the limitations noted earlier; the acquisition of long time series from HPC records; and the development of new techniques for measuring oceanic responses, especially in the deep ocean. Major uncertainties in these efforts will probably focus on the accuracy of estimates of sea-surface response.

REFERENCES

1. E. Philippi, "Die grundproben der deutschen Sudpolar-Expedition 1901–1903." In *Deutsche Sudpolar-Expedition 1901–1903*, Vol. 2, Geographie und Geologie, Berlin, 1912, pp. 415–616.
2. W. Schott, "Die Foraminiferen in dem aquatorialen Teil des Atlantischen Ozeans," *Deut. Atl. Exped. Meteorol.*, **3,** 43–131 (1935).
3. M. N. Bramlette and W. H. Bradley, "Geology and biology of North Atlantic deep-sea cores between Newfoundland and Iceland, Part I: Lithologic and geologic interpretations," U. S. Geological Survey Professional Paper 196-A, 1–34 (1941).
4. G. Arrhenius, "Sediment cores from the East Pacific," Report of Swedish Deep-Sea Expedition 1947–1948, Vol. 5, 1–228, (1952).
5. F. B. Phleger, F. L. Parker and J. F. Pierson, "North Atlantic foraminifera," Report of Swedish Deep-Sea Expedition 1947–1948, Vol. 7, 1–122, (1953).
6. Courtesy of National Geophysical Data Center, NOAA and Lamont-Doherty Geological Observatory.
7. Courtesy of Dee Breger, Joseph Morley, Constance Sancetta, and Andrew McIntyre.
8. G. R. Heath, "Dissolved silica and deep-sea sediments." In W. W. Hay, ed., *Studies in Paleo-Oceanography*, Society of Economic Paleontologists and Mineralogists Special Publication Vol. 20, 1974. pp. 77–93.
9. W. H. Berger, "Biogenous deep-sea sediments: Production, preservation and interpretation," *Chem. Oceanogr.*, **5,** 265–388 (1976).

10. W. H. Berger, "Deep-sea sedimentation." In C. A. Burke, and C. L. Drake, eds., *The Geology of Continental Margins*, Springer-Verlag, New York, 1974, pp. 213-241.
11. W. H. Berger and G. R. Heath, "Vertical mixing in pelagic sediments," *J. Mar. Res.*, **26,** 135-143 (1968).
12. W. F. Ruddiman and L. K. Glover, "Vertical mixing of ice-rafted volcanic ash in North Atlantic sediments," *Geol. Soc. Am. Bull.*, **83,** 2817-2836 (1972).
13. N. L. Guinasso and D. R. Schink, "Quantitative estimates of biological mixing rates in abyssal sediments," *J. Geophys. Res.*, **80,** 3032-3043 (1975).
14. W. H. Berger and R. F. Johnson, "On the thickness of the mixed layer in deep-sea sediments," *Earth Planet. Sci. Lett.*, **41,** 223-227 (1978).
15. T.-H. Peng, W. S. Broecker, and W. H. Berger, "Rates of benthic mixing in deep-sea sediments as determined by radioactive tracers," *Quat. Res.*, **11,** 141-149 (1979).
16. S. Epstein, R. Buchsbaum, H. A. Lowenstam, and H. C. Urey, "Revised carbonate-water isotopic temperature scale," *Geol. Soc. Am. Bull.*, **64,** 1315-1328 (1953).
17. C. Emiliani, "Pleistocene temperatures," *J. Geol.*, **63,** 538-578 (1955).
18. C. Emiliani, "Paleotemperature analysis of Caribbean cores P6304-8 and P6304-9 and a generalized temperature curve for the past 425,000 years," *J. Geol.*, **74,** 109-126 (1966).
19. W. Dansgaard, S. J. Johnson, H. B. Clausen, and C. C. Langway, "Climatic Record Revealed by the Camp Century Ice Core." In K. K. Turekian, ed., *The Late Cenozoic Glacial Ages,* Yale University Press, New Haven, 1971, pp. 37-56.
20. S. J. Johnson, W. Dansgaard, H. B. Clausen, and C. C. Langway, "Oxygen-isotopc profiles through the Antarctic and Greenland Ice Sheets," *Nature*, **235,** 429-434 (1972).
21. J. Imbrie and N. G. Kipp, "A New Micropaleontological Method for Quantitative Paleoclimatology: Application to a Late Pleistocene Caribbean Core." In K. K. Turekian, ed., *The Late Cenozoic Glacial Ages*, Yale University Press, New Haven, 1971, pp. 71-181.
22. J. Imbrie, J. van Donk, and N. G. Kipp, "Paleoclimatic investigation of a Late Pleistocene Caribbean deep-sea core: Comparison of isotopic and faunal methods," *Quat. Res.*, **3,** 10-38 (1973).
23. N. J. Shackleton and N. D. Opdyke, "Oxygen isotope and paleomagnetic stratigraphy of equatorial Pacific core V28-238: Oxygen isotope temperatures and ice volumes on a 10^5 year and 10^6 year scale," *Quat. Res.*, **3,** 39-55 (1973).
24. J. Thiede, "Aspects of the variability of the glacial and interglacial North Atlantic eastern boundary current (last 150,000 years)," *Meteor Forschungsergeb C*, 1-36 (1977).
25. R. M. Cline and J. D. Hays, eds., *Investigation of Late Quaternary Paleoceanography and Paleoclimatology*, Geological Society of America Memoir 145, 1976.
26. J. Weertman, "Milankovitch solar radiation variations and ice age ice sheet sizes," *Nature*, **261,** 17-20 (1976).
27. G. E. Birchfield and J. Weertman, "A note on the spectral response of a model continental ice sheet," *J. Geophys. Res.*, **83,** 4123-4125 (1978).
28. J. Imbrie and J. Z. Imbrie, "Modeling the climatic response to orbital variations," *Science*, **207,** 943-953 (1980).
29. J. Oerlemans, "Continental ice sheets and the planetary radiation budget," *Quat. Res.*, **14,** 349-359 (1980).
30. D. Pollard, A. P. Ingersoll, and J. G. Lockwood, "Response of a zonal climate ice sheet model to the orbital perturbations during the Quaternary ice ages," *Tellus*, **32,** 301-319 (1980).
31. J. P. Kennett and N. J. Shackleton, "Laurentide ice sheet melt water recorded in Gulf of Mexico deep-sea cores," *Science*, **188,** 147-150 (1975).
32. C. Emiliani, S. Gartner, B. Lidz, K. Eldridge, D. K. Elvey, T. C. Huang, J. J. Stipp, and M. F. Swanson, "Paleoclimatological analysis of Late Quaternary cores from the Northeastern Gulf of Mexico," *Science*, **189,** 1083-1088 (1975).

33. N. J. Shackleton, "The oxygen isotope stratigraphic record of the Late Pleistocene," *Philos. Trans. R. Soc. London, Ser. B*, **280,** 169–182 (1977).
34. J. C. Duplessy, J. Moyes, and C. Pujol, "Deep water formation in the North Atlantic Ocean during the Last Ice Age," *Nature*, **286,** 479–482 (1980).
35. W. S. Broecker, "Floating glacial ice caps in the Arctic Ocean," *Science*, **188,** 1116–1118 (1975).
36. J. F. Budd and I. N. Smith, "The growth and retreat of ice sheets in response to orbital radiation changes." In *Sea Level, Ice, and Climatic Changes, Proceedings of Camberra Symposium, December 1979*, IAHS Publication 131, 1980, pp. 369–409.
37. A. C. Mix, and W. F. Ruddiman, "Oxygen-isotope analyses and Pleistocene ice volumes," *Quat. Res.*, **21,** 1–20 (1984).
38. N. G. Kipp, "New transfer function for estimating past sea-surface conditions from sea-bed distribution of planktonic foraminiferal assemblages in the North Atlantic." In R. M. Cline and J. D. Hays, eds., *Investigations of Quaternary Paleoceanography and Paleoclimatology*, Geological Society of America Memoir 145, 1976, pp. 3–41.
39. A. W. H. Bé, "A Taxonomic and Zoogeographic Review of Recent Planktonic Foraminifera." In A. T. S. Ramsey, ed., *Oceanic Micropaleontology*, Academic, London, 1977, pp. 1–100.
40. W. H. Berger, "Planktonic foraminifera: Selective solution and paleoclimatic interpretation," *Deep-Sea Res.*, **15,** 31–43 (1968).
41. C. Emiliani, "Depth habitats of some species of pelagic Foraminifera as indicated by oxygen isotope ratios," *Am. J. Sci.*, **252,** 149–158 (1954).
42. R. K. Fairbanks, P. H. Wiebe, and A. W. H. Bé, "Vertical distribution and isotopic composition of living planktonic foraminifera in the Western North Atlantic," *Science*, **207,** 61–63 (1979).
43. W. A. Berggren and J. van Couvering, "Biostratigraphy, geochronology and paleoclimatology of the last 15 million years in marine and continental sequences," *Palaeogeogr., Palaeoclimatol., Palaeoecol.*, **16,** 1–216 (1974).
44. H. Thierstein, K. R. Geitzenauer, B. Molfino, and N. J. Shackleton, "Global synchroneity of Late Quaternary coccolith datum levels: Validation of oxygen isotopes," *Geology*, **5,** 400–404 (1977).
45. A. McIntyre, W. F. Ruddiman, and R. Jantzen, "Southward penetration of the North Atlantic Polar Front: Faunal and floral evidence of large-scale surface water mass movements over the last 225,000 years," *Deep-Sea Res.*, **19,** 61–77 (1972).
46. A. McIntyre and A. W. H. Bé, "Modern Coccolithophoridae from the Atlantic Ocean—I. Placoliths and Cyrtoliths," *Deep-Sea Res.*, **14,** 561–597 (1967).
47. H. Okada and A. McIntyre, "Modern coccolithophores of the Pacific and North Atlantic Oceans," *Micropaleontology*, **23,** 1–55 (1977).
48. S. Honjo, "Biogeography and Provincialism of Living Coccolithophorids in the Pacific Ocean." In A. T. Ramsey, ed., *Oceanic Micropalenotology*, Academic, London, 1977, pp. 951–972.
49. T. C. Moore, Jr., "The distribution of radiolarian assemblages in the modern and ice-age Pacific," *Mar. Micropaleontol.*, **3,** 229–266 (1978).
50. J. D. Hays, J. A. Lozano, N. J. Shackleton, and G. Irving, "Reconstruction of the Atlantic and western Indian Ocean sectors of the 18,000 B.P. Antarctic Ocean." In R. M. Cline and J. D. Hays, eds., *Investigation of Late Quaternary Paleoceanography and Paleoclimatology*, Geological Society of America Memoir 145, 1976, pp. 337–372.
51. J. J. Morley and J. D. Hays, "*Cycladophora davisiana*: A stratigraphic tool for Pleistocene North Atlantic and interhemispheric correlation," *Earth Planet. Sci. Lett.*, **4,** 383–389 (1979).
52. A. P. Jousé, O. G. Kozlova, and U. V. Mulhina, "Distribution of diatoms in the surface layer of sediment from the Pacific Ocean." In B. M. Funnell and W. Riedel, eds., *The Micropaleontology of Oceans*, Cambridge University Press, Cambridge, 1967, pp. 263–269.
53. L. Burckle and D. B. Clarke, "Diatom transfer functions in the southern oceans," *Nova Hedwigia Beiheft*, **54,** (1977).

54. G. W. Lynts and J. B. Judd, "Late Pleistocene paleotemperatures at tongue of the Ocean, Bahamas," *Science*, **171,** 1143–1144 (1971).

55. A. Hecht, "A model for determining Pleistocene paleotempertures from planktonic foraminiferal assemblages," *Micropaleontology*, **19,** 68–77 (1973).

56. W. H. Berger and J. V. Gardner, "On the determination of Pleistocene temperatures from planktonic foraminifera," *J. Foraminiferal Res.*, **5,** 102–113 (1975).

57. W. H. Hutson, "The Agulhas current during the late Pleistocene—Analysis of modern faunal analogues," *Science*, **207,** 64–66 (1980).

58. B. Molfino, N. G. Kipp, and J. J. Morley, "Comparison of foraminiferal, coccolithophorid, and radiolarian paleotemperature equations: Assemblage coherency and estimate concordancy," *Quat. Res.*, **82,** 279–313 (1982).

59. CLIMAP Project Members, "The surface of the ice-age Earth," *Science*, **191,** 1131–1137 (1976).

60. CLIMAP Project Members, "Seasonal reconstructions of the Earth's surface at the Last Glacial Maximum." In A. McIntyre, leader, *Geol. Soc. America Map and Chart Series*, **36,** (1982) (Text, Maps and Microfiche).

61. T. C. Moore, Jr., W. H. Hutson, N. G. Kipp, J. D. Hays, W. L. Prell, P. Thompson, and G. Boden, "The Biological record of the Ice-Age Ocean," *Palaeogeogr., Palaeoclimatol., Palaeoecol.*, **35,** 357–370 (1981).

62. A. McIntyre and N. G. Kipp with A. W. H. Bé, T. Crowley, T. Kellogg, J. V. Gardner, W. L. Prell, and W. F. Ruddiman, "Glacial North Atlantic 18,000 years ago: A CLIMAP reconstruction." In R. M. Cline and J. D. Hays, eds., *Investigation of Late Quaternary Paleoceanography and Paleoclimatology*, Geological Society of America Memoir 145, 1976, pp. 43–76.

63. J. Robertson, "Glacial to interglacial oceanographic changes in the Northwest Pacific, including a continuous record of the last 400,000 years," Ph.D. diss., Columbia University, New York, 1975.

64. T. C. Moore, Jr., L. H. Burckle, K. Geitzenauer, B. Luz, A. Molina-Cruz, J. H. Robertson, H. Sachs, C. Sancetta, J. Thiede, P. Thompson, and C. Wenkam, "The reconstruction of sea-surface temperatures in the Pacific Ocean at 18,000 B.P.," *Mar. Micropaleontol.*, **5,** 215–247 (1980).

65. W. L. Prell, W. H. Hutson, D. F. Williams, A. W. H. Bé, K. Geitzenauer, and B. Molfino, "Surface circulation of the Indian Ocean during the last glacial maximum, approximately 18,000 yr B.P.," *Quat. Res.*, **14,** 309–336 (1980).

66. CLIMAP Project Members, W. F. Ruddiman, task group leader, "The last interglacial ocean," *Quat. Res.*, **21,** 123–224 (1984).

67. M. Gascoyne, A. P. Currant, and T. C. Lord, "Ipswichian fauna of Victoria Cave and the marine paleoclimatic record," *Nature*, **294,** 652–654 (1981).

68. D. H. Keen, R. S. Harmon, and J. T. Andrews, "U series and amino acid dates from Jersey," *Nature*, **289,** 162–164 (1981).

69. J. Alvinerie, M. Caralp, C. Latouche, J. Moyes, and M. Vigneaux, "Contribution to the knowledge of the paleohydrology of the north-east Atlantic during the terminal Quaternary period," *Oceanological Acta*, **1,** 87–98 (1978).

70. T. B. Kellogg, "Paleoclimatology and paleoceanography of the Norwegian and Greenland Seas: Glacial-interglacial contrasts," *Boreas,* **9,** 115–137 (1980).

71. P. R. Thompson, "Planktonic foraminifera in the western North Pacific during the past 150,000 years: Comparison of modern and fossil assemblages," *Palaeogeogr., Palaeoclimatol., Palaeoecol.*, **35,** 241–281 (1981).

72. T. J. Crowley, "Temperature and circulation changes in the eastern North Atlantic during the last 150,000 years: Evidence from the planktonic foraminiferal record," *Mar. Micropaleontol.*, **6,** 97–129 (1981).

73. W. F. Ruddiman and A. McIntyre, "Warmth of the subpolar North Atlantic Ocean during northern hemisphere ice-sheet growth," *Science*, **204,** 173–175 (1979).

74. W. F. Ruddiman and L. K. Glover, "Subpolar North Atlantic circulation at 9300 B.P.: Faunal evidence," *Quat. Res.*, **5,** 361–389 (1975).

75. W. F. Ruddiman and A. McIntyre, "The North Atlantic during the last deglaciation," *Palaeogeogr., Palaeoecol., Palaeoclimatol.*, **35,** 145–214 (1981).

76. T. H. Peng, W. S. Broecker, G. Kipphut, and N. J. Shackleton, "Benthic mixing in deep-sea cores as determined by ^{14}C dating and its implications regarding climate stratigraphy and the fate of fossil-fuel CO_2." In N. R. Anderson and A. Malahoff, eds., *The Fate of Fossil Fuel CO_2 in the Oceans*, Plenum, New York, 1977, pp. 355–373.

77. W. H., Hutson, "Bioturbation of deep-sea sediments: Oxygen isotopes and stratigraphic uncertainty," *Geology*, **8,** 127–130 (1980).

78. J. D. Hays, J. Imbrie, and N. J. Shackleton, "Variations in the Earth's orbit: Pacemaker of the Ice Ages," *Science*, **194,** 1121–1132 (1976).

79. A. Leventer, D. F. Williams, and J. P. Kennett, "Dynamics of the Laurentide ice sheet during the last deglaciation: Evidence from the Gulf of Mexico," *Earth Planet. Sci. Lett.*, **59,** 11–17 (1982).

80. N. G. Pisias, "Paleoceanography of the Santa Barbara basin during the last 8,000 years," *Quat. Res.*, **10,** 366–384 (1978).

81. G. H. Denton and W. Karlen, "Holocene climatic variations—their pattern and possible cause," *Quat. Res.*, **3,** 155–205 (1973).

82. N. J. Shackleton and N. D. Opdyke, "Oxygen-isotope and Paleomagnetic Stratigraphy of Pacific Core V28-239 Late Pliocene to Latest Pleistocene." In R. M. Cline and J. D. Hays, eds., *Investigation of Late Quaternary Paleoceanography and Paleoclimatology*, Geological Society of America Memoir 145, 1976, pp. 449–464.

83. N. G. Pisias and T. C. Moore, Jr., "The evolution of Pleistocene climate: A time series approach," *Earth Planet. Sci. Lett.*, **52,** 450–458 (1981).

84. W. L. Prell, "Oxygen and carbon isotope stratigraphy for the Quaternary of Hole 502B: Evidence for two modes of variability." In W. L. Prell, J. V. Gardner, et al. Init. Repts. *DSP*, 68, U.S. Government Printing office, Washington, D.C., pp. 455–464, (1982).

85. W. F. Ruddiman and A. McIntyre, "Time-transgressive deglacial retreat of polar waters from the North Atlantic," Quat. Res., **3,** 117–130 (1973).

86. W. F. Ruddiman and A. McIntyre, "North Atlantic Paleoclimatic Changes Over the Past 600,000 years." In R. M. Cline and J. D. Hays, eds., *Investigation of Late Quaternary Paleoceanography and Paleoclimatology*, Geological Society of America Memoir 145, 1976, pp. 111–146.

87. W. F. Ruddiman and A. McIntyre, "Late Quaternary surface ocean kinematics and climatic change in the high-latitude North Atlantic," *J. Geophys. Res.*, **82,** 3877–3887 (1977).

88. C. Sancetta, J. Imbrie, and N. G. Kipp, "Climatic record of the past 130,000 years in North Atlantic deep-sea core V23-82: Correlation with the terrestrial record," *Quat. Res.*, **3,** 110–116 (1973).

89. W. F. Ruddiman and A. McIntyre, "Ice-age thermal response and climatic role of the surface Atlantic Ocean, 40° to 63°N," *Geol. Soc. Am. Bull.*, **95,** 381–396 (1984).

90. W. F. Ruddiman and A. McIntyre, "Oceanic mechanisms for amplification of the 23,000-year ice-volume cycle," *Science*, **212,** 617–627 (1981).

91. T. Crowley, "Fluctuations of the eastern North Atlantic gyre during the last 150,000 years," Ph.D. diss., Brown University, Providence, R.I. 1976.

92. T. B. Kellogg, "Late Quaternary climatic changes in the Norwegian and Greenland Seas." In G. Weller and S. A. Bowling, eds., *Proceedings of the 24th Alaskan Scientific Conference*, 1973.

93. T. B. Kellogg, "Late Quaternary climatic changes: Evidence from deep-sea cores of Norwegian and Greenland Seas." In R. M. Cline and J. D. Hays, eds., *Investigation of Late Quaternary Paleoceanography and Paleoclimatology*, Geological Society of America Memoir 145, 1976, pp. 77–110.

94. A. Mix, A. McIntyre, and W. F. Ruddiman, "Late quaternary paleoceanography of the tropical Atlantic: spatial variability of sea-surface temperatures," (in preparation).

95. J. J. Morley and J. D. Hays, "Comparison of glacial and interglacial oceanographic conditions in the South Atlantic from variations in calcium carbonate and radiolarian distributions," *Quat. Res.*, **12,** 396–408 (1979).

96. J. A. Lozano and J. D. Hays, "Relationship of radiolarian assemblages to sediment types and physical oceanography in the Atlantic and western Indian sectors of the Antarctic Ocean." In R. M. Cline and J. D. Hays, eds., *Investigation of Late Quaternary Paleoceanography and Paleoclimatology*, Geological Society of America Memoir 145, 1976, pp. 303–336.

97. J. D. Hays, "A Review of the Late Quaternary Climatic History of Antarctic Seas." In E. M. van Zinderen Bakker, ed., *Antarctic Glacial History and World Paleoenvironments*, A. A. Balkema, Rotterdam 1978, pp. 57–71.

98. D. W. Cooke, "Variations in the seasonal extent of sea ice in the Antarctic during the last 140,000 years determined from deep-sea sediments," Ph.D. diss., Columbia University, New York, 1978.

99. W. Howard and W. L. Prell, "The origin of the southern hemisphere SST lead: An alternate interpretation based on a comparison of radiolarian and foraminiferal paleoecology in the southern Indian Ocean," *Quat. Res.*, **21,** 244–263 (1984).

100. D. Ninkovitch and J. H. Robertson, "Volcanogenic effects on the rates of deposition of sediments in the northwest Pacific Ocean," *Earth Planet. Sci. Lett.*, **27,** 127–136 (1975).

101. D. Ninkovitch, N. Opdyke, B.C. Heezen, and J.H. Foster, "Paleomagnetic stratigraphy, rates of deposition and tephrachronology in north Pacific deep-sea sediments," *Earth Planet. Sci. Lett.*, **1,** 476–492 (1966).

102. N. D. Opdyke and J. H. Foster, "Paleomagnetism of cores from the North Pacific." In J. D. Hays, ed., *Geological Investigations of the North Pacific*, Geological Society of America Memoir 126, 1970, pp. 83–119.

103. H. M. Sachs, "Late Pleistocene history of the north Pacific: Evidence from a quantitative study of radiolaria in core V21-173," *Quat. Res.*, **3,** 89–98 (1973).

104. C. Sancetta, "Oceanography of the north Pacific during the last 18,000 years: Evidence from fossil diatoms," *Mar. Micropaleontol.*, **4,** 103–123 (1979).

105. W. L. Prell and W. B. Curry, "Faunal and isotopic indices of monsoonal upwelling: Western Arabian Sea," *Oceanol. Acta*, **4,** 91–98 (1979).

106. W. L. Prell and H. F. Streeter, "Temporal and spatial patterns of monsoonal upwelling along Arabia: A modern analogue for the interpretation of 'Quaternary SST anomalies'," *J. Mar. Res.*, **40,** 143–155 (1982).

107. W. L. Prell, "Monsoonal climate of the Arabian Sea during the Late Quaternary: A response to changing solar radiation." In A. L. Berger et al., eds., *Milankovitch and Climate*, Part I, Riedel, Chicago, pp. 349–366, 1984.

108. J. Kutzbach, "Monsoon climate of the early Holocene: Climate experiments using the Earth's orbital parameters for 9,000 years ago," *Science*, **214,** 59–61 (1981).

109. J. L. Cullen, "Microfossil evidence for changing salinity patterns in the Bay of Bengal over the last 20,000 years," *Palaeogeogr., Palaeoclimatol., Palaeoecol.*, **35,** 315–356 (1981).

110. W. H. Bradley, "Mediterranean sediments and Pleistocene sea-levels," *Science*, **88,** 376–379 (1938).

111. B. Kullenberg, "On the salinity of the water contained in marine sediments," *Medd. Oceanogr. Inst. Goteborg*, **21,** 1–38 (1952).

112. W. B. F. Ryan, "Stratigraphy of Late Quaternary sediments in the eastern Mediterranean." In D. J. Stanley, ed., *The Mediterranean Sea: A natural sedimentation laboratory*, Dowden, Hutchinson and Ross, Stroudsburg, Pa., 1972, pp. 149–169.

113. C. Vergnaud-Grazzini, "δO^{18} changes in foraminifera carbonates during the last 10^5 years in the Mediterranean Sea," *Science*, **190,** 272–274 (1975).

114. M. B. Cita, C. Vergnaud-Grazzini, C. Robert, H. Chamley, N. Claranfi, and S. D'Onofrio, "Paleoclimatic record of a long deep-sea core from the eastern Mediterranean," *Quat. Res.*, **8**, 205–235 (1977).

115. E. Olausson, "Studies of deep-sea cores," Report of Swedish Deep-Sea Expedition, 1947–1948, Vol. 8, 353–391 (1961).

116. D. F. Williams and R. C. Thunell, "Faunal and oxygen isotopic evidence for surface water salinity changes during sapropel formation in the Eastern Mediterranean," *Sediment. Geol.*, **23**, 81–93 (1979).

117. M. Rossignol-Strick, V. Nesteroff, P. Olive, and C. Vergnaud-Grazzini, "After the deluge: Mediterranean stagnation and sapropel formation," *Nature*, **295**, 105–110 (1982).

118. K. Wyrtki, "El Nino: The dynamic response of the equatorial Pacific Ocean to atmospheric forcing," *J. Phys. Oceanogr.*, **5**, 572–584 (1975).

119. A. Blackman and B. L. K. Somayajulu, "Pacific Pleistocene cores: Faunal analysis and geochronology," *Science*, **154**, 886–889 (1966).

120. B. Luz, (with appendix by N. J. Shackleton), "Stratigraphic and paleoclimatic analysis of Late Pleistocene tropical southeast Pacific cores," *Quat. Res.*, **3**, 56–72 (1973).

121. A. Molina-Cruz, "The relation of the southern trade winds to upwelling processes during the last 75,000 years," *Quat. Res.*, **8**, 324–339 (1977).

122. K. Romine and T. C. Moore, "Radiolarian assemblage distribution and paleoceanography of the eastern equatorial Pacific Ocean during the last 127,000 years," *Palaeogeogr., Palaeoclimatol., Palaeoecol.*, **35**, 281–315 (1981).

123. S. Hastenrath, "Heat budget of the Tropical Ocean and atmosphere," *J. Phys. Oceanogr.*, **10**, 159–170 (1980).

124. D. B. Ericson, and G. Wollin, "Pleistocene climates and chronology in deep-sea sediments," *Science*, **162**, 1227–1234 (1968).

125. J. V. Gardner and J. D. Hays, "Responses of sea-surface temperature and circulation to global climatic change during the past 200,000 years in the eastern equatorial Atlantic Ocean." In R. M. Cline and J. D. Hays, eds., *Investigation of Late Quaternary Paleoceanography and Paleoclimatology*, Geological Society of America Memoir 145, 1976, pp. 221–246.

126. A. W. H. Bé, J. E. Damuth, L. Lott, and R. Free, "Late Quaternary climatic record in western equatorial Atlantic sediment." In R. M. Cline and J. D. Hays, eds., *Investigation of Late Quaternary Paleoceanography and Paleoclimatology*, Geological Society of America Memoir 145, 1976, pp. 165–200.

127. W. L. Prell, J. V. Gardner, A. W. H. Bé, and J. D. Hays, "Equatorial Atlantic and Caribbean foraminiferal assemblages, temperatures and circulation: Interglacial and glacial comparisons." In R. M. Cline and J. D. Hays, eds., *Investigation of Late Quaternary Paleoceanography and Paleoclimatology*, Geological Society of America Memoir 145, 1976, pp. 247–266.

128. J. Williams, "Simulation of the atmospheric circulation using the NCAR Global Circulation Model with present day and glacial period boundary conditions," University of Colorado Boulder Institute of Arctic and Alpine Research Occasional Paper 10, (1974).

129. W. L. Gates, "Modeling the ice-age climate," *Science*, **191**, 1138–1144 (1976).

130. S. Manabe and D. G. Hahn, "Simulation of the tropical climate of an Ice Age," *J. Geophys. Res.*, **82**, 3889–3911 (1977).

131. M. M. Milankovitch, "Canon of insolation and the ice-age problems," Koniglich Serbische Akademie, Beograd. English translation by the Israel Program for Scientific Translations, published for the U.S. Department of Commerce and the National Science Foundation, Washington, D.C., (1941).

132. G. E. Birchfield, J. Weertman and A. T. Lunde, "A paleoclimate model of Northern Hemisphere ice sheets," *Quat. Res.*, **15**, 126–142 (1981).

133. D. Pollard, "A simple ice sheet model yields realistic 100 kyr glacial cycles," *Nature*, **296**, 334–338 (1982).

134. J. Oerlemans, "Glacial cycles and ice-sheet modeling," *Climatic Change*, **4,** 353–374 (1982).

135. G. H. Denton and T. J. Hughes, "Milankovitch Theory of Ice Ages: Hypothesis of Ice-Sheet Linkage between Regional Insolation and Global Climate," *Quat. Res.*, **20,** 125–144 (1983).

136. R. E. Newell, "Changes in the poleward energy flux by the atmosphere and ocean as a possible cause for ice ages," *Quat. Res.*, **4,** 117–127 (1974).

137. B. Saltzman and R. E. Moritz, "A time-dependent climatic feedback system involving sea-ice extent, ocean temperature and CO_2," *Tellus*, **32,** 93–118 (1980).

138. V. Ya. Sergin, "Numerical modeling of the glacier–ocean-atmosphere global system," *J. Geophy. Res.*, **84,** 3191–3204 (1979).

139. E. Kallen, C. Crafoord, and M. Ghil, "Free oscillations in a climate model with ice-sheet dynamics," *J. Atmos. Sci.*, **36,** 2292–2303 (1979).

140. B. Saltzman, "Global mass and energy requirements for glacial oscillations and their implications for mean ocean temperature oscillations," *Tellus*, **29,** 205–212 (1977).

141. P. K. Weyl, "The role of the oceans in climatic change: A theory of the ice ages," *Meteorol. Monogr.*, **8,** 37–62 (1968).

142. W. L. Donn and M. Ewing, "A theory of ice ages II," *Science*, **152,** 1706–1712 (1966).

143. D. P. Adam, "Ice ages and the thermal equilibrium of the Earth," *J. Res. U.S. Geol. Surv.*, **1,** 587–596 (1973).

144. D. P. Adam, "Ice ages and the thermal equilibrium of the Earth," *Quat. Res.*, **5,** 161–171 (1975).

145. R. G. Johnson and B. T. McClure, "A model for northern hemisphere continental ice sheet variation," *Quat. Res.*, **6,** 325–353 (1976).

6

SNOW AND ICE DATA

R. G. Barry

World Data Center-A for Glaciology (Snow and Ice),
Cooperative Institute for Research in Environmental Sciences and
Department of Geography, University of Colorado
Boulder, Colorado

CRYOSPHERIC PARAMETERS

Snow and ice features occur on the earth's surface in a wide variety of forms, collectively referred to as the *cryosphere.* For our purposes, the major components are snowcover, freshwater ice in lakes and rivers, sea ice, glaciers, ice sheets, and ground ice or permafrost. Solid precipitation is a meteorological phenomenon originating in the nucleation of individual ice crystals, whereas floating ice (except for icebergs) forms primarily from the freezing of water in bulk. Frozen ground need not involve any water at all in the case of bedrock at subzero temperatures. The residence time of water in each of these cryospheric phenomena varies widely. Snowcover and freshwater ice are essentially seasonal and most sea ice lasts only a few years if it is not seasonal. A given water particle in glaciers, ice sheets or ground ice, however, may remain for 10^2–10^5 years or longer and an ice sheet may have been in existence in East Antarctica for at least 10 million years.

It will be helpful to consider first the dimensions of the major cryospheric components and the characteristics that are most commonly determined in each case. Table 1 shows that the majority of the world's ice volume is held in Antarctica, principally in the East Antarctic Ice Sheet, but in terms of areal extent it is the Northern Hemisphere winter snow and ice cover that is largest—amounting to 23 percent of the hemispheric surface area in January.

Direct observations of snow and ice forms span only a few hundred years at most and until recently have been mainly limited to point measurements of snowfall, reports of the extent of sea ice or of iceberg sightings, and documentation of glacier extent. However, internal records contained in the snow and ice accumulation on ice caps and ice sheets cover time intervals up to almost 10^5 years and indirect

Table 1. Distribution of Snow and Ice (modified from Hollin and Barry,[146])

	Area (10^6 km^2)	Volume (10^6 km^3)	Sea-level Equivalent (m)
Land ice			
Antarctica[a]	12.05	29.4	73.5
Greenland	1.8	3.0	7.5
Small ice caps and mountain glaciers	0.5	0.12	0.3
Ground ice (excluding Antarctica)			
Continuous	7.6	0.2–	0.5–
Discontinuous	17.3	0.5	1.2
Sea ice			
Arctic: max.[b]	14.0		
min.	7.0		
Antarctic:[c] max.	18.4		
min.	3.6		
Total land ice, sea ice, and snow			
January: N. Hemisphere	58.0		
S. Hemisphere	18.0		
July: N. Hemisphere	14.0		
S. Hemisphere	27.0		
Global mean annual	59.0		

[a] Grounded ice sheet, excluding peripheral, floating ice shelves (which do not affect sea-level). Roughly 10 percent of the Antarctic ice volume is in West Antarctica and 90 percent in East Antarctica[1]

[b] Excluding the Sea of Okhotsk.

[c] Actual ice area excluding open water.

stratigraphic evidence of the occurrence of ice in the past is present in many periods of the geological record.

THE ROLE OF THE CRYOSPHERE IN THE CLIMATE SYSTEM

The global snow and ice cover is subject to major changes in extent and water storage on time scales of seasons (snow cover and sea ice) as well as of hundreds to thousands of years (snow and ice extent). The continental ice sheets that now cover most of Greenland and Antarctica, and formerly covered much of Europe and North America, have lifetimes of tens of thousands to tens of millions of years (in the case of Antarctica). The principal climatic roles of snow and ice relate to their high reflectivity and low thermal conductivity.[2] On long time-scales, the changes in water storage can lead to substantial changes in global sea-level which, for example, was lowered some 120 to 160 m during the last glacial maximum (about 18,000 years ago) according to different calculations.[3,4] Consequently, the cryosphere-climate interactions involve positive feedbacks such as the snow and ice albedo-temperature effect[5] and negative feedbacks such as ice sheet buildup, which eventually causes a lack of accumulation in the interior of the ice mass and therefore its contraction,[6,7] and glacio-isostasy which causes relative rise in snowline. More complex are the short-term interactions between snow and ice phenomena and the atmospheric circulation. These interactions are considered below.

Physical Processes

The ratio of reflected to incident solar radiation is termed the *albedo.* For climatological purposes we are primarily interested in the integrated value for the spectral range (approximately) 0.3–3.5 μm. Typical albedo values are: 0.80–0.90 for fresh, dry snow; 0.60 for old, melting snow and bare ice; and 0.30–0.40 for melting sea ice with puddles. Rapid shifts in surface reflectivity occur in autumn and spring[8,9] and surface albedo is large in high latitudes[10] but the climatic significance of these effects is diminished because the earth-atmosphere (or *planetary*) albedo is determined principally by cloud cover and because the total solar radiation received in high latitudes is small as a result of the large zenith angle of the incident radiation and the high average cloudiness.[11]

Snowcover has an insulating role for the ground surface and likewise sea ice for the underlying ocean. The effect is most pronounced in the latter case since a sea ice cover decouples the ocean-atmosphere interface with respect to both heat and moisture fluxes. The flux of moisture from a water surface is eliminated by even a thin skin of ice whereas the flux of heat through thin ice is substantial until it attains a thickness in excess of 30–40 cm.[12]

An extensive snowcover is usually produced by cyclonic precipitation in an outbreak of cold arctic air. Subsequently, this snow area may serve to deflect cyclonic activity southward along its margin, further expanding the snowcover through the local cooling effect of the snow on the air crossing it.[13] Despite some empirical analyses and modeling studies with general circulation models,[14–17] these relationships are not well-defined in quantitative terms.

Sea ice limits may be extended as a result of advection of ice by the wind or by the cooling of surface water below the appropriate freezing temperature (approximately −1.8°C). Both processes operate on synoptic and seasonal time scales.[18,19] The opposite mechanism of atmospheric forcing by sea ice is less well-documented. On the seasonal time scale, it appears that in summer the ice responds to forcing by the atmosphere whereas in autumn there are comparable two-way interactions.[20] On the synoptic scale, Ackley and Keliher[21] have illustrated possible links between sea ice divergence causing enhanced oceanic heat flux to the atmosphere and thereby increasing cyclonic activity. However, the synoptic interactions appear to operate in both directions, but with differences according to season and geographical location.[22]

Modeling Cryosphere-Climate Interactions

The cryosphere has been represented in climate models by a wide range of parameterizations. In simple energy budget models, for example, the ice line may be represented by a step-function albedo parameter which varies latitudinally with temperature.[23,24] Such models, however, show an oversensitivity to small changes in external conditions. Improved parameterizations of snow and ice for zonally averaged energy budget models of this type are being examined by Robock.[25] The specific response of schematic ice sheets to external (astronomical) forcing has been examined by Weertman,[26] Birchfield,[27] and Pollard et al.,[28] while Oerlemans[29] has considered stochastic forcing (associated with weather fluctuations). In more complete representations of surface type, such as that used by Ghil and LeTreut,[30] the earth's surface is segmented to include a land ice cover and an ocean ice cover. The primary interest in most of these studies lies in the fluctuations of ice extent over time, in response either to external forcing of the climate system, or to internal stochastic variability.

In higher order general circulation models (GCMs) (chapter 8), the surface hydrological cycle may be modeled to include a seasonally varying snow cover,[31,32] or a thermodynamic model of sea ice growth, such as that of Parkinson and Washington,[33] may be coupled to the GCM.[34] Two of these models have been used to examine the possible responses of the atmosphere and cryosphere to a simulated doubling or quadrupling of atmospheric carbon dioxide.[35,36] Figure 1 illustrates the change in the calculated sea ice thickness for such a CO_2 experiment. Ice sheet models have not yet been linked interactively with climate models, due to the problem of widely different time scales. However, the glaciological and climatic controls of ice sheet growth and decay have been simulated, incorporating the effects of ice dynamics. The most advanced work of this type has been performed by Budd and Smith[37] who examine the response of the North American (Laurentide) ice sheet to forcing by the (Milankovitch) astronomical radiation changes. The timing of the growth and decay of the ice sheet is found to be quite realistic if the effects of ice albedo feedback, isostatic bedrock response, and decreasing precipitation in the central area of the ice sheet are incorporated. Figure 2 shows the phase differences between changes in high-latitude radiation, climate (including albedo feedback), bedrock depression, and ice volume over the last 120,000 years for the modeled North American ice sheet.

The sensitivity of global climate to the radically different boundary conditions of a glacial regime has been examined with several GCMs,[38-40] although only rather broad climatic generalities are shown in common by the model simulations.[41] The role of snow cover in amplifying the astronomical effects of ice age radiation deficits has been illustrated by Adem[42] with a thermodynamic model.

Modeling glacial climates requires, among other things, considerable information on cryospheric phenomena both as input data to the model and for verification purposes. Budd[6] notes the limitations of existing cryospheric-related information, in terms of dating control and its calibration with reference to climatic variables, for testing the predictions of time-changes in ice extent and global ice volume during

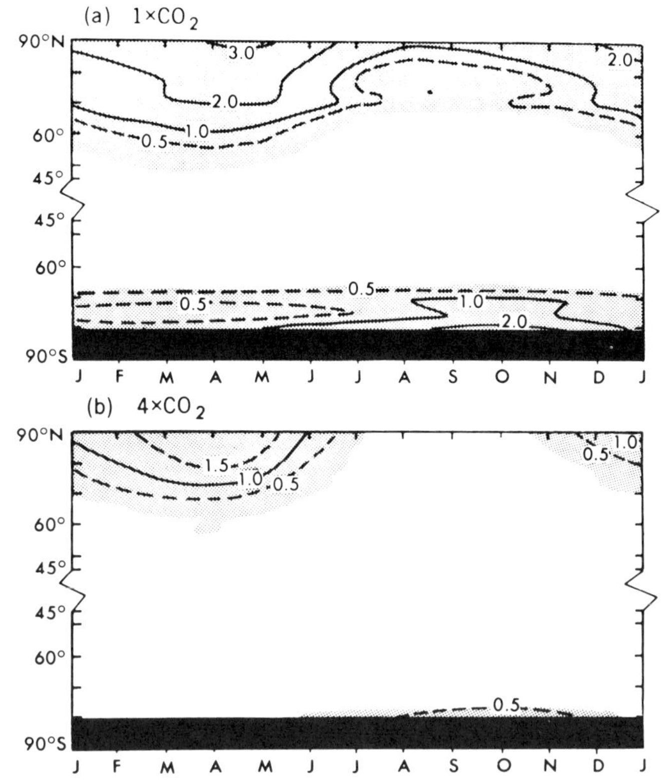

Figure 1. Changes in zonally averaged sea ice thickness (m) in response to a fourfold increase in atmospheric CO_2 (*b*) compared to a control experiment (*a*) calculated with the GFDL general circulation model incorporating a mixed-layer ocean and realistic global geography (from Manabe and Stouffer[35]).

the last glacial cycle. For GCMs used in forecasting, snow and sea ice cover information are required in the initialization of the model;[43] little work has yet been done in this direction.

HISTORICAL OBSERVATIONS

Snowcover

Snowfall is routinely measured at meteorological stations, although the solid amount is not always carefully distinguished from rainfall when the precipitation is of a mixed type. In the lowland areas, the occurrence of falling snow requires that, on average, the freezing level be not more than about 250 m above the surface, that is, the air temperature is $\leq 1\,°C$.[44] This criterion provides a check on the reliability of historical reports of snowfall occurrences in summer months; most of these cases are due to falls of soft hail. Until 1960, it was Canadian reporting practice to

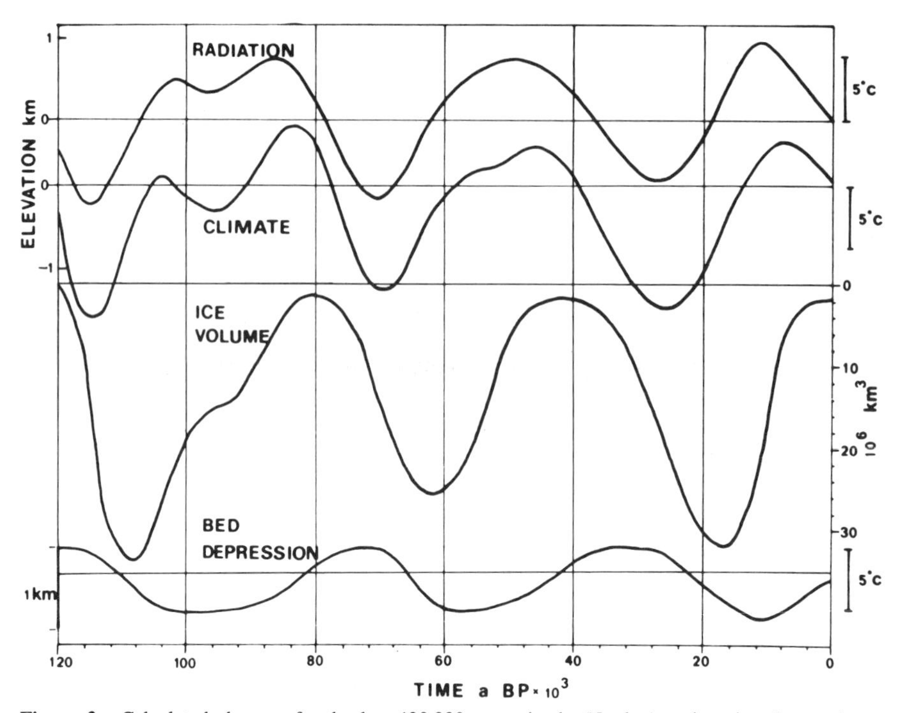

Figure 2. Calculated changes for the last 120,000 years in the North American ice sheet volume, bedrock depression and global climate (including ice-albedo feedback effects) for the specified summer radiation inputs at 70°N. The left scale shows elevation units; Radiation = 30 m/(cal cm^{-2} dy^{-1}) and climate, based on a summer temperature lapse rate, is in units of 400 m (from Budd and Smith[37]).

measure the depth of new snow on the ground and divide this figure by ten to determine the liquid precipitation amount. Subsequently, this practice was discon tinued in view of the variable density of fresh snow, and the water content is now measured by means of a weighing gauge.[45] There is a wide variety in the types of gauge, shielding devices, and installation practices adopted by national metero-logical and other agencies measuring snowfall.

The synoptic weather code has provision for measurements of depth of snow on the ground and these data are collected daily by synoptic stations and reported through the global telecommunication network. Snow depth reporting began in North America in 1941 and subsequently was extended to climatological stations. These records are not easily retrieved from the high volume synoptic meteorological data archives, but convenient extracts are available in map form. For example, weekly maps of snow depth in the United States for December and March have been published since 1935 in the weekly *Weather and Crop Bulletin*,[46] as illustrated in figure 3.

Snow depth is an indicator of storm frequency, moisture availability, and local factors such as redistribution by wind, compaction of the snowpack, and ablation. Water content is an important snow pack parameter and this is commonly measured by ground surveys in relation to streamflow and flood forecasting. Surveys at snow

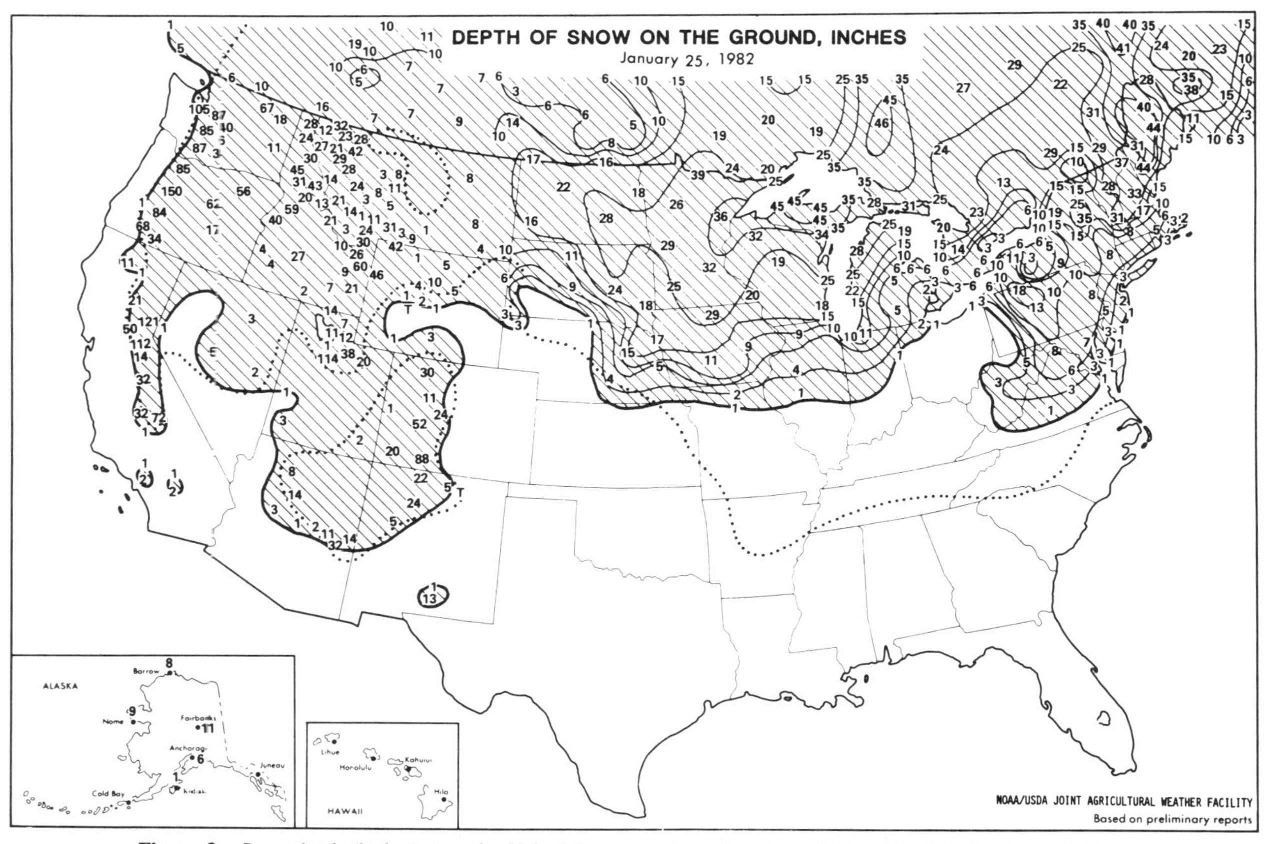

Figure 3. Snow depth (inches) over the United States on December 5, 1977 (from *Weekly Weather and Crop Bulletin*[46]).

courses were started in the United States by Mixer[47] in Maine and by Church[48] in 1909–10 in Nevada. In most western states records begin in the 1920s or 1930s (ref. 49, p. 133). Such data collection is undertaken twice monthly by the Soil Conservation Service on a state-by-state basis. More frequent data are now being retrieved by the SNOTEL (Snow Telemetry) system from about 500 remote automatic stations throughout the western United States. These data are collected centrally at Portland, Oregon, via ionospheric meteor-burst telemetry. For Canada, mean snow depth and water equivalent are published for approximately 1300 snow courses in *Snow Cover Data for Canada.*[50] Records of snow cover duration, maximum depth, and seasonal accumulation have been published for all Austrian stations for the period 1901–50[51] and subsequent decades.

Snow cover data are also collected in many mountain ranges in connection with avalanche prediction as well as for programs of winter orographic cloud seeding, although these records are not usually available in accessible archives.

There are major difficulties involved in comparing snow survey data and climatological measurements. Accumulated snowfall on forest snow courses is comparable with that measured at climatological stations, whereas snow courses in open terrain retain only about 60 percent of the accumulated station total.[52] Goodison[53] shows that compatible results are only obtained when corrections are made to the gauge measurements for wind effects and the snow survey sites are chosen to be representative of basin land use.

Long-term historical records of snowfall or snowcover are available for a few stations in Europe, east Asia, and eastern North America. Time series have been compiled by Arakawa[54] for Tokyo, by Manley[55] for central England, and by Pfister[56,57] for Switzerland (figure 4), but more data extraction and synthesis remain to be done from archives and weather diaries.

Sea Ice

There has been a longer history of interest in sea ice conditions than any other cryospheric variable. The incidence of East Greenland ice on the coasts of Iceland has been documented in the Icelandic records since A.D. 860, although with varying levels of detail and accuracy.[58] A complete annual series exists from 1780 and similar records for the "Storis" along the west coast of Greenland are available from 1821 (figure 5).[59] Records of Baltic Sea ice conditions also exist from the early eighteenth century; they are also summarized by Lamb (ref. 59, p. 586–589). By contrast, no Antarctic sea ice records exist until the twentieth century. These data sources and the problems of early records are reviewed by Barry.[60] The longest series of modern data is that prepared in chart form for the Arctic Ocean, but particularly the North Atlantic sector, by the Danske Meteorologisk Institut for 1901–56.[61] A digital data file has been extracted from these charts by Kelly.[62] Another digital data set for the Arctic Ocean (1953–1976) has been compiled by Walsh[63] and subsequently used in several climate-ice interaction studies.

The formation, persistence, and extent of sea ice is related to water temperature

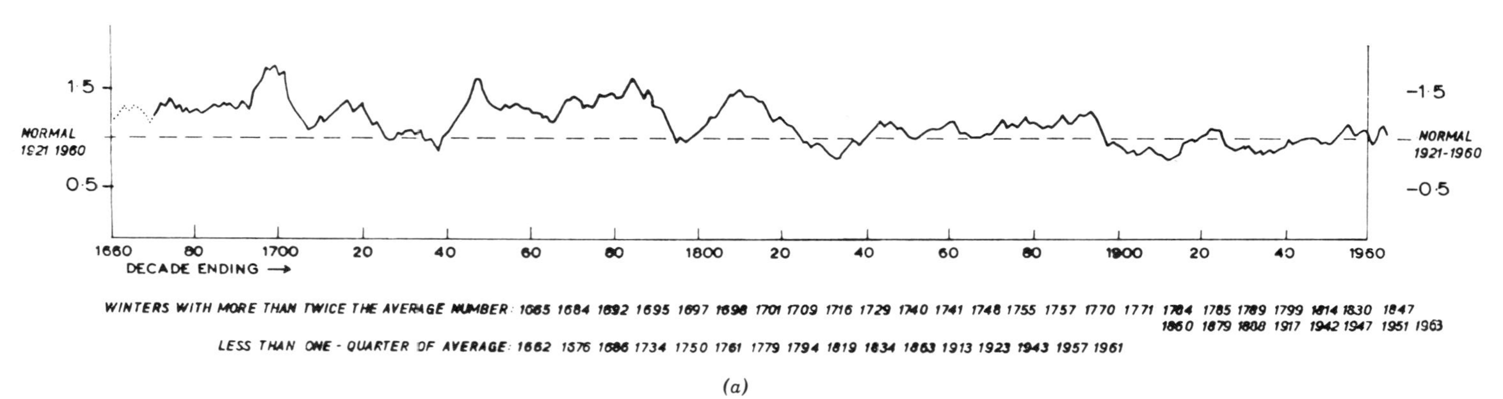

(*a*)

Figure 4. Variations in snowfall conditions since the seventeenth and eighteenth centuries. (*a*) The frequency of days with snowfall (or sleet) in London, expressed as decadal running means of winter totals as a proportion of the 1921–60 "normal" (from Manley,[55] reproduced with permission from *Weather*, Vol. 24, 1969, p. 432). (*b*) Summer snowfall and snow cover frequency on the Grand St. Bernard, Switzerland (2469 m) (from Messerli et al.,[57] reproduced with permission of the Regents of the University of Colorado from *Arctic and Alpine Research*, Vol. 10, 1978).

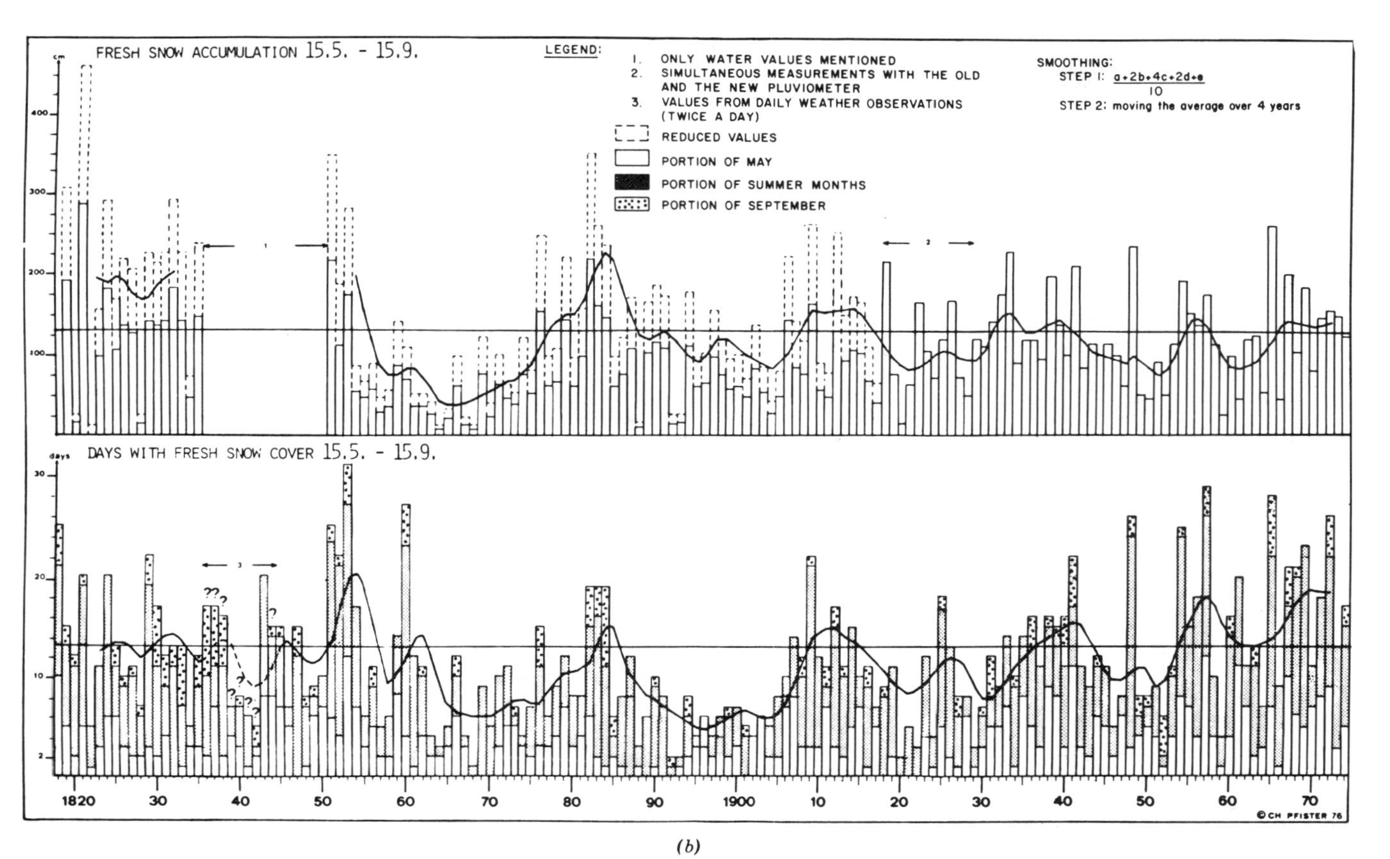

(*b*)

Figure 4. (*Continued*)

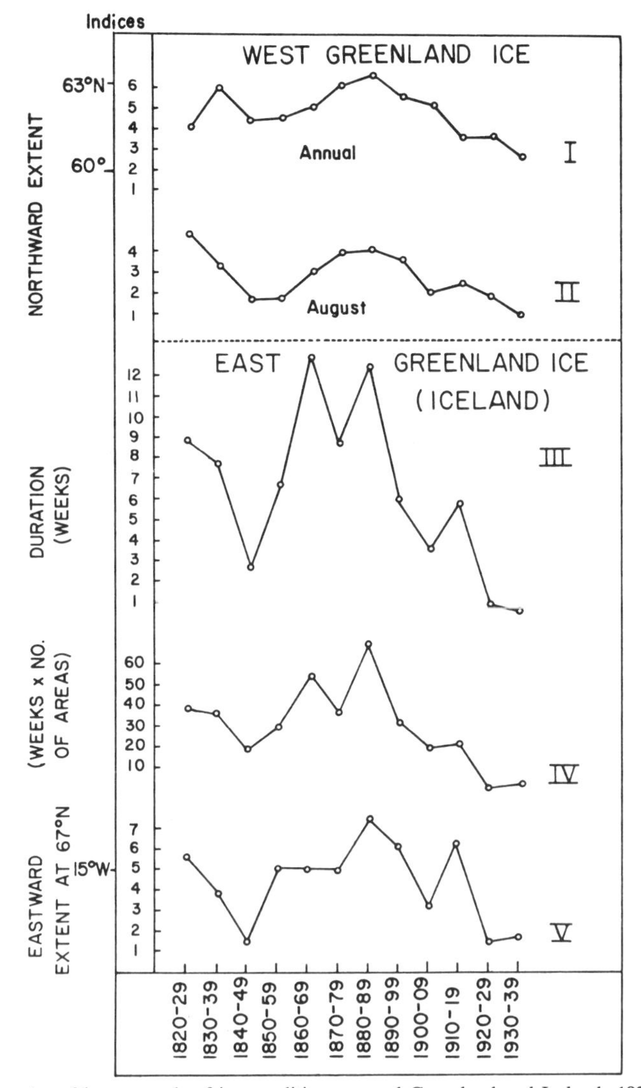

Figue 5. Examples of long records of ice conditions around Greenland and Iceland, 1820–1939. I = West Greenland annual ice (after Speerschneider), II = West Greenland August ice (after Koch), III = weeks with East Greenland ice off Iceland (after Thoroddsen), IV = weeks × number of areas with ice off Iceland (Koch), V = east-west width of the ice belt north of Iceland (Koch). Note that the vertical scale for curve II was undefined (after Koch, 1945,[60]).

and salinity as well as to surface energy fluxes and wind conditions. The ice limit is also affected by both ocean currents and surface wind drift. Moreover, the ice concentration within the limit usually mapped is seldom 100 percent. Recent sat ellite studies in Antarctica have revealed the presence of large open water areas (*polynyi*) within the ice limits even in winter.[64] The occurrence of polynyi and open leads in the ice during winter is of considerable climatic importance since they allow large turbulent heat transfers into the atmosphere at a time when the sea ice otherwise eliminates the ocean-atmosphere interface. The heat loss from young ice in refrozen leads is still many times larger than that through multiyear ice until a thickness greater than 40 cm is reached.[65]

Freshwater Ice

Only during the last 40–50 years has the date of ice formation and disappearance on rivers and lakes been regularly and systematically observed. The most complete and best documented records are available for locations in Canada,[66] although data exist in locally held files for water bodies in the United States, for example, and there are many European records, mainly from the late nineteenth-early twentieth centuries. The time of lake freeze-up and break-up is well correlated with air temperature indices, such as accumulated freezing and thawing degree-days,[67] although other factors such as runoff, wind, and winter thickness of ice and snow may cause variability in the actual dates. Shortwave radiation is the main control of ice melt, while freeze-up depends on turbulent heat losses. Ice cover on rivers is less easily related to simple climatic indices since the date of formation and disappearance of the ice is affected by channel geometry, water level, and upstream processes.

Historical records of ice formation and disappearance on estuaries around James Bay and Hudson Bay, Canada, were collected by the Hudson's Bay Company between 1714 and 1870. These have recently been analyzed by Catchpole and Ball.[68] Similar records exist in Europe. For example, van den Dool et al.[69] have utilized data on the duration of ice on the Haarlem-Leiden canal during 1657–1757 and 1814–39 to determine winter temperature conditions. Care is required in such interpretations, however, due to the possible biases introduced by non-meteorological factors (chapter 2). Lamb (ref. 59, pp. 568–570) notes that winter freezing of the River Thames in England is not a guide to winter severity after the early nineteenth century as a result of changes in the water flow due to bridges. Such freeze-up probably implies a mean temperature in at least one winter month below +0.5°C. For Japan, records of the freeze-up of Lake Suwa since A.D. 1441 have been summarized by Arakawa[70] (also ref. 59, pp. 608–10) and used to estimate mean winter temperatures in Tokyo by Gray[71] and Tanaka and Yoshino.[72]

Glaciers

The climatic interpretation of glacier fluctuations remained subjective and essentially qualitative until recently. Glaciers respond to both winter accumulation and summer ablation and it is difficult to isolate changes in one or other component. Moreover, the terminal location of a glacier is determined by its mass balance and its flow dynamics. Hence, the relationships between glacier fluctuations and climate are complex and involve variable time lags according to the size and regime of the ice body.[73]

Positions of glacier termini in the Alps have been recorded in drawings and documents since the twelfth century. A survey of such sources has recently been prepared for the Grindelwald glaciers[74] and the records for the Lower Grindelwald Glacier from the late sixteenth century have been analyzed by Messerli et al.[57] Regular measurements on Swiss and Austrian glaciers date from the 1890s. Strong retreat between about 1928–64 followed an interval (1909–28) when 75 percent of glaciers in Austria were advancing (figure 6).[75] Hoinkes[76] has related these fluctuations to atmospheric conditions, showing that the general retreat was associated with more frequent anticyclonic weather patterns and reduced cyclonic activity in summer. The terminal position of many glaciers worldwide is now determined annually and the records are assembled by the *Permanent Service on Glacier Fluctuations.*[77] Based on such data, Gamper and Suter[78] estimate that changes in mean annual temperature of 1 °C give rise to fluctuations of about 1 km in the length of Alpine glaciers. A thorough study of historical evidence on glaciers in the Ecuadorian Andes has been compiled by Hastenrath.[79] Scope for historical studies of glaciers in western North America is provided by photographs taken on geological and other expeditions in the late nineteenth and early twentieth century.[80] A collection of many such historical glacier photographs is maintained by World Data Center-A for Glaciology in Boulder, Colorado (figure 7). In recent years, repetitive air-photo coverage has also been provided by the U.S. Geological Survey.

Models of glacier response to changes in mass balance have recently been applied to the problem of climatic reconstruction.[81] Possible combinations of past climatic conditions that could have caused snowline lowering in Baffin Island during the Little Ice Age, for example, have been determined from an energy-mass balance

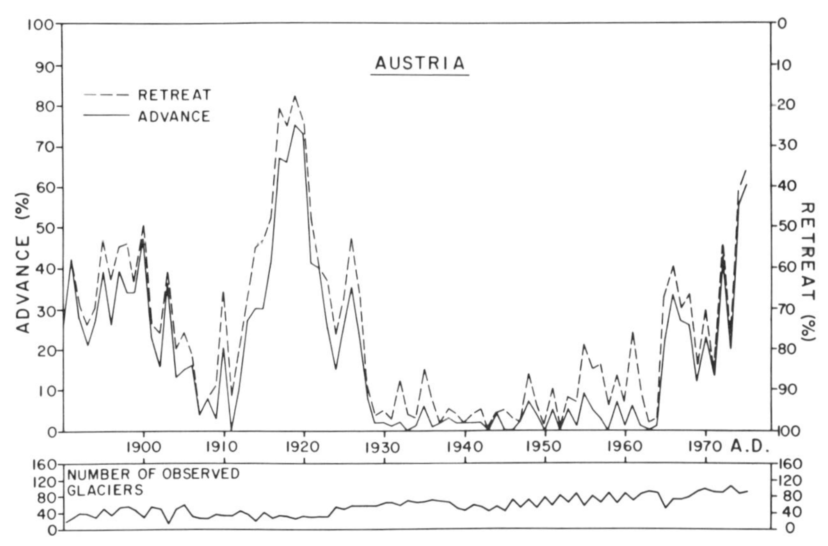

Figure 6. Percentage of Austrian glaciers showing terminal advance or retreat, 1890s–1970s (based on data provided by the Permanent Service for the Fluctuations of Glaciers analyzed by P. K. MacKinnon).

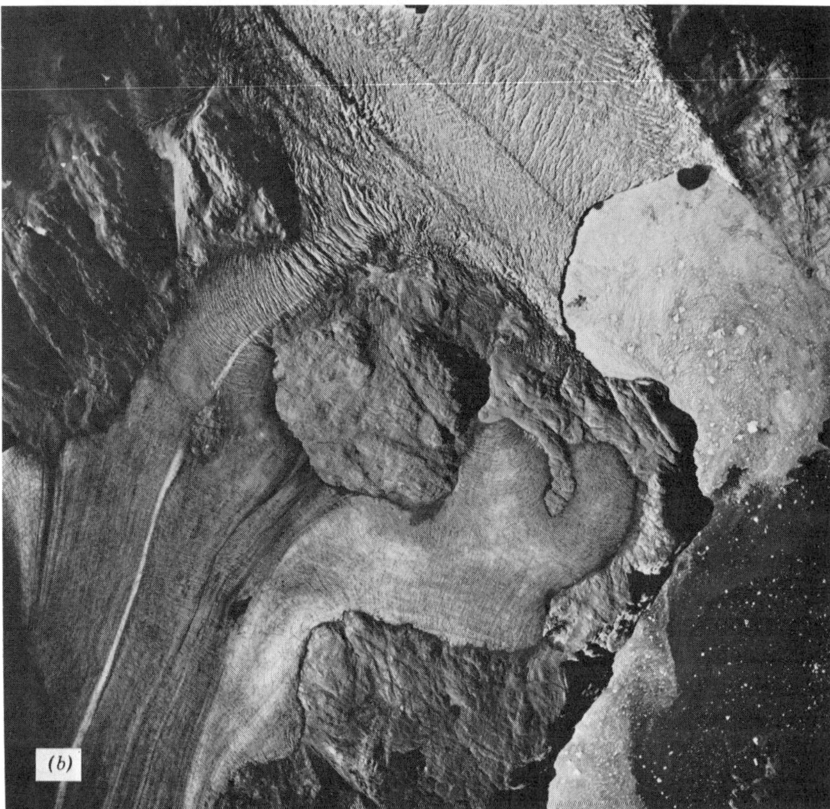

Figure 7. Photographs of the Muir Glacier, Glacier Bay, Alaska, which has been retreating since the late eighteenth century. (*a*) Photograph taken by F. LaRoche, 1893 (from the American Geographical Society collection, World Data Center-A for Glaciology, Boulder, Colorado). (*b*) Vertical air photograph taken from 4,360 m altitude by A. S. Post, 1963 (USGS/Post collection, U.S. Geological Survey, Tacoma, Washington). The glacier has retreated more than 20 km toward the head of Muir Inlet from its position in 1893.

model.[82] Glacier flow models have been applied to four European glaciers by Smith and Budd[83] and to glaciers in western New Guinea (Irian Jaya) by Allison and Kruss[84] to estimate recent climatic changes. Smith and Budd estimate an average warming since the Little Ice Age amounting to about 0.5–1.0°C in summer; quite similar results for the last century or so are also obtained by Allison and Kruss for New Guinea.

Permafrost and Ground Ice

Permanently frozen ground, or permafrost, occurs continuously where mean annual air temperatures are below about −8°C and discontinuously in areas with temperature less than −1° or −2°C.[85] Based on average latitudinal temperature gradients, a 2°C temperature oscillation might cause a 250–350 km displacement of the southern boundary of permafrost, but such shifts can seldom be determined with certainty because reliable maps of permafrost extent date only from about 1950. In Manitoba and in the upper Mackenzie Valley a general melting and retreat of permafrost has been reported,[86,87] while in western Siberia recent development of permafrost, related to a decrease in ground temperature, is inferred by Belopukhova.[88] It is known that changes in vegetation and snow depth can considerably modify the ground temperature regime, thus complicating any attempted paleoclimatic interpretations.

The vertical profile of ground temperatures in permafrost areas has been proposed as a device to estimate temperature changes during the last few centuries. In Alaska, a warming of 2–4°C over the last 75–100 years is inferred from such profiles by Gold and Lachenbruch,[89] but in coastal areas these calculations could be biased by maritime influences. Similar calculations for two boreholes in Ontario[90] indicate temperature anomalies of +1.5°C between about A.D. 900 to 1300 and/or anomalies of −1°C from about A.D. 1500 to 1800. On a longer time scale, Balobaev[91] estimates that temperatures during the last glacial maximum in Yakutia and also in western Siberia were 10–13°C lower than today.

SATELLITE DATA

The value of satellite remote sensing of snow and ice was recognized early in the development of weather satellites and such systems now provide the primary tool for routine observations of the cryosphere. Satellites afford frequent repetitive global coverage, data collection in the absence of illumination and in the presence of clouds (in certain wavelengths), and a relay and control system for data collection platforms installed in remote locations.

Snowcover mapping from satellite imagery began operationally in 1966. The National Environmental Satellite (now Data and Information) Service (NESDIS) prepares hemispheric maps of snow and ice approximately weekly, based on three reflectivity classes.[92] These classes correspond approximately to highly reflective snow/ice (category 3; albedo ≈60 percent), moderately reflective (category 2; albedo ≈45 percent), and least reflective—such as patchy mountain snowcover (category 1; albedo ≈30 percent).[93,94] The maps have recently been digitized for November 1966 through December 1980 by Dewey and Heim[95] for 7921 grid boxes covering the Northern Hemisphere, but snow and ice, and the reflectivity categories, are not separately identified. Indices of reflectivity for geographical sectors have also been compiled from the same data source by Kukla and Gavin.[96]

Some deficiencies of this data set should be noted. These include the fact that the charts from 1966 through 1974 did not consistently map Himalayan snowcover, the occasional extension of snowcover beyond the southern limits of the map, the seasonal northern limit of illumination for the satellite sensors in the visible wavelengths, the omission of scattered mountain snows due to the coarse grid resolution, and the absence (due to lost charts) of 10 individual weekly maps from the series.[95] A more detailed assessment of the NESDIS charts is provided by Kukla and Robinson.[97] Since April 1983 these charts have depicted only snow-covered and partly covered land areas.

A major problem with snowcover mapping is the difficulty of distinguishing between snow and cloud in polar and mountain areas. New sensors in the 1.6 μm channel should greatly assist in resolving this difficulty in the future when they become operational.[98,99]

Mapping of sea ice from the air was begun in the Siberian Arctic in the 1930s and in the North American Arctic in 1952. Hemispheric mapping using surface, airborne, and satellite data was initiated by the U.S. Navy on a weekly basis for the Arctic in 1972 and for the Antarctic in 1973. Until the early 1970s, sea ice analyses depended on vidicon data and thereafter these were used in conjunction with multispectral scanning radiometer data. Infrared data, in the absence of clouds, allows ice mapping during the polar night.

Since 1972 the advent of passive microwave sensors has permitted the mapping of ice extent in the presence of cloud cover as well as darkness. The brightness temperature of the ocean surface, which has a low emissivity, is about 135K, contrasting markedly with sea ice which has brightness temperatures in excess of 200K in the 1.55 cm band. First-year and multiyear ice have emissivities, respectively, of about 0.92 and 0.84, allowing their respective concentrations to be determined during the winter season. In summer, the presence of surface meltwater and reduced ice concentrations results in less definitive interpretations.[100] The resolution of microwave data at present is about 30 km. Multifrequency microwave data are now being used for ice mapping. The use of four frequencies and two polarizations permits a considerable increase in the types of information that can be obtained.[101]

Data buoys relaying information via satellite are now being used to determine ice drift in the Arctic Ocean and the Weddell Sea.[102] They were also deployed in the Fram Strait between Greenland and Svalbard in 1982–83 during the Marginal Ice Zone Experiment (MISEX). This experiment was planned to provide information on the mesoscale interactions of sea ice, ocean, and atmosphere along the ice edge.

Satellite systems are being used extensively for other glaciological applications. Radar altimetry with a vertical resolution of <1 m and a horizontal resolution of a few meters has been applied to mapping the surface topography of the southern part of the Greenland Ice Sheet[103] and parts of Antarctica. Repeated surveys of the polar ice sheets using satellite and airborne radar altimeters can be used to detect changes in the mass balance of the ice sheets. Such changes can be quantitatively related to corresponding changes in mean sea level as indicator of global climate change.[104] There is significant potential for future radar and scatterometer studies of sea ice ridging, as illustrated by the limited sea ice data obtained by the Seasat program. Specifications of sensor types, capabilities, and their actual or potential applications in cryosphere-climate studies have been fully documented in various Climate Program reports.[105]

GEOLOGICAL RECORDS

There are two major types of geological evidence bearing on paleoglaciological/paleoclimatological reconstructions. The first is the stratigraphic record indicating former presence of land, sea, and ground ice, with varying degrees of reliability. The second is the highly detailed record provided by vertical cores extracted from ice sheets.

Glacial and Periglacial Features

The identification of former glacial ages by geologists in the middle eighteenth-early nineteenth centuries resulted from the recognition that glacier advances could transport erratic blocks and deposit moraines.[106,107] The Pleistocene Epoch was subdivided into four glacial-interglacial cycles on the basis of the sequence of glacial outwash terraces on the Alpine Foreland of southern Germany by Penck and Brückner[108] and this classical scheme remained in the literature until the advent of marine sedimentological studies and paleomagnetic dating in the 1960s.

The land record left by glacial processes is still a cornerstone of Quaternary studies, but it is now widely acknowledged that it poses many problems. First, the stratigraphic record is discontinuous in both space and time. Deposits from early glacial advances are weathered and eroded in subsequent interglacial intervals, and then reworked and overlain by other deposits in later glaciations. Second, the dating of all but the most recent glaciation relies largely on potassium/argon and uranium-series dates which have wide time errors. Consequently, the land and ocean records are still not reliably intercorrelated beyond about 120,000 years B.P. Glacier fluctuations during the last 40,000–60,000 years can be dated by radiocarbon (^{14}C) given the inclusion of suitable organic materials in the deposits. Neoglacial fluctuations in the late Holocene are often dated by lichenometric methods.[109]

The extent and direction of movement of land ice can generally be determined from geologic field evidence.[110] Former ice boundaries are commonly marked by terminal moraines, and till sheets mantle the area formerly covered by ice. The past movement of ice may be determined from the orientation of glacial striae or grooves on rock surfaces as well as from the general orientation of drumlins and trains of glacial debris deposited by the ice. The former ice thickness can be estimated from theoretical ice sheet surface profiles related to the former marginal positions combined with calculations of isostatic recovery of the land surface from its depression by the ice load. This uplift forms elevated strandlines in coastal areas, for example. In mountain ranges, the former vertical extent of glaciers can be observed from vegetation trimlines and upper limits of ice abrasion.

From these types of evidence the extent, movements, and thickness of ice during the advances of at least the final stages of the last (Würm-Wisconsin) glaciation can be determined. More limited information can be obtained for earlier glaciations. Global reconstructions are detailed for 18,000 years B.P. by Denton and Hughes.[111] Figure 8 illustrates their "minimum reconstruction, representing the views of geologists who favor limited polar ice cover. An alternative maximum reconstruction is presented by Hughes et al.[112] depicting extensive marine-based ice sheets as well as the better known terrestrial domes. At present the field data are still inadequate to resolve this controversy.

Outside the former ice sheet limits, periglacial phenomena were widespread during previous glacial intervals. These include permafrost, ice wedges, patterned ground forms and solifluction features related to frost action, and wind-blown sand dunes and loess.[113,114] The past occurrence of permafrost can be inferred where frost-crack pattern and casts (pseudomorphs) of ice or sand wedges are present. Such features are widespread in Europe and in a narrow zone across North America. Péwé[115,116] considers that active ice wedges occur in regions where mean annual temperature is less than about $-7°C$, but vegetation cover and soil drainage can cause local variations in occurrence. Relic pingo forms—domed mounds caused by the intrusion and subsequent freezing of groundwater above the permafrost table—can often be identified in these same areas. The distribution of the entire suite of these landforms provides a view of the extent of periglacial conditions, but the paleoclimatic regimes can usually only be defined in general terms and it may be difficult to assign a date to many of these features.[117,118]

Sedimentary records in the oceans can yield indirect evidence of the former extent of sea ice or icebergs. Volcanic ash and detritus that have been rafted by floating ice provide an indication of cold ocean water limits and have been used to infer paleo-ocean circulation changes in the North Atlantic[119] and sea ice extent around Antarctica.[120] Ash zones are especially useful since they are readily dated. Changes in faunal assemblages in marine sediments also reflect changes in ocean temperature and therefore of ice cover. It was thought that the boundary of silty diatomaceous clay (to the south) and diatomaceous ooze (to the north), in recent sediments and in those of glacial age, marked the northern edge of summer sea ice extent, but recently Burckle et al.[121] have suggested that this sediment boundary represents the approximate limit of spring sea ice. The summer extent of Southern Ocean ice during the last glacial maximum is still uncertain, therefore.

The climatic interpretation of the evidence for past glacial and periglacial conditions is far from simple and the procedures are still largely qualitative and subjective. Snowline lowering, for example, may be caused by lower summer temperatures and/or increased winter accumulation so that a unique paleoclimatic reconstruction cannot usually be obtained from estimates of changes in the snowline or equilibrium-line altitude (ELA) of glaciers.[122] Moreover the inferred climatic

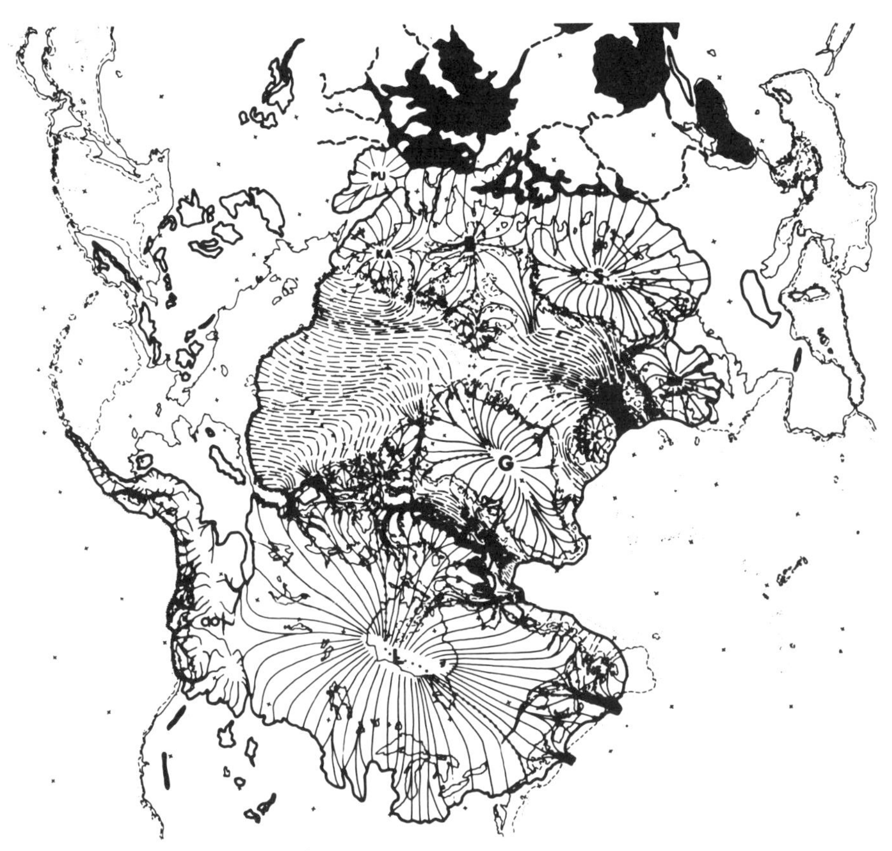

Figure 8. A minimum reconstruction of the late-Würm Arctic Ice Sheet, in which an ice shelf covering the Arctic Ocean flowed from the Canadian-European sector to the Alaskan-Siberian sector, where it grounded along the continental shelf margin. Present sea-level contour and present lacustrine shorelines ———, Present 100-m bathymetric contour and a conservative late-Würm sea level contour ....., Present 1000-m bathymetric contour and, except for island arcs and ocean ridges, the approximate continental-shelf margin, —-, Limit of late-Würm glaciation —--, Surface flowlines on the respective grounded and floating portions of the late-Würm Arctic Ice Sheet. L, Laurentide; G, Greenland; B, Barents; S, Scandinavian major domes of the Arctic Ice Sheet. IN, Innuitian; CO, Cordilleran; NE, Newfoundland; IC, Iceland; BR, British; PU, Putorana; KA, Kara minor domes of the Arctic Ice Sheet ......, portion of the Arctic Ice sheet that drained into the Atlantic Ocean. Regions flooded by ice-dammed rivers (black areas) include Mansi Lake in the Ob River basin, the Aral-Caspian Sea, and the Black Sea, which were 128 m above, 48 m above, and at least 50 m below present sea level, respectively, and were interconnected. +, Intersections of 10° longitude and latitude intervals (from Hughes et al.,[112] reprinted by permission from *Nature,* Vol. 266, p. 596, copyright © 1977 Macmillan Journals Limited).

changes could result from a variety of possible climatic processes. It is essential, therefore, that field evidence not be considered solely in a local context and that the fullest possible range of proxy data be examined. During the last few centuries, snowline lowering on many Northern Hemisphere mountain ranges was of the order of 100–200 m, representing some 10–20 percent of maximum snowline depressions during the last glacial maximum.[122] Latitudinal variations in the magnitude of glacial-age snowline depression in North Africa and the Andes have been used to examine large-scale contrasts in paleoclimate.[123,124] Specific paleoclimatic reconstructions using such paleo-glaciological information, and detailed ocean surface temperature data compiled by the CLIMAP project[125] for 18,000 years B.P., have been carried out using two different atmospheric general circulation models by Gates[39] and Manabe and Hahn (chapter 5).[40] Another approach is represented in the model studies of ice sheet growth and decay using prescribed climatic forcing functions (for solar radiation, albedo feedback, and precipitation) and isostatic bedrock response to changes in ice loading.[37] The reconstructions of the Laurentide ice sheet of North America from 120,000 years B.P. to the present can be compared with available field evidence.

Ice Core Records

Ice sheets and ice caps retain much of the snowfall and other material from the atmosphere that accumulates on their surface. Hence, vertical cores extracted from polar ice bodies contain very long proxy records of paleoclimate. The principal parameters of paleoclimatic significance that can be determined from ice cores are: variations in the annual accumulation, stable isotopes, microparticle concentration, acidity and electrical conductivity (as measures of volcanic activity), the gas content of air bubbles, and the chemistry of soluble and insoluble impurities.[126]

In areas where there is an annual accumulation of at least 25 cm/yr and no summer melt, it is feasible to identify annual snow/ice layers from seasonal differences in isotopic ratios and dustfall over the last 10,000 years or more. These annual layers become thinner with depth, however, as a result of the weight of overlying ice and lateral flow of the ice mass. The basal sections of deep cores in eastern Antarctica could yield ice several hundred thousand years old. Deep cores must be dated indirectly by a combination of radiometric techniques and calculations from ice flow models. The latter can given an accuracy of ± 3 percent for cores extending back to about 30,000 years B.P. provided that the location is not glaciologically complex.

Worldwide, almost 100 cores with lengths exceeding 100 m have so far been collected.[127] The deepest are those from Byrd Station, Antarctica (2164 m), and from two cores drilled to bedrock in Greenland, at Camp Century (1387 m) in 1966, and at Dye 3 (2035 m) in 1981. These cores have a time scale of approximately 100,000 years, according to the available estimates. A core from central Greenland or eastern Antarctica is expected to span several hundred thousand to perhaps a million years back in time (Dansgaard et al.,[128]).

Stratigraphic studies indicate long-term trends in accumulation rate. At South Pole station, for example, Mosley-Thompson and Thompson[129] find an average surface accumulation rate of 6.96 g/yr (water equivalent) for A.D. 1046–1956, with maxima of 8.03 g/yr during 1867–96 and 7.93 g/yr during 1057–86, and a minimum of 5.47 g/yr for 1657–86.

The isotopic composition of ice provides an estimate of atmospheric temperature through the per mille deviation of the concentration of ^{18}O ($\delta^{18}O$‰) in ice from that in Standard Mean Ocean Water (SMOW). The $^{18}O/^{16}O$ ratio is normally stated as a "delta" value:

$$\delta(^{18}O/^{16}O)_{sample} = \left[\frac{(^{18}O/^{16}O)_{sample}}{(^{18}O/^{16}O)_{SMOW}} - 1\right]10^3$$

The heavier isotopic component $H_2{}^{18}O$ condenses preferentially, due to a slightly lower vapor pressure than for $H_2{}^{16}O$, so that at higher latitudes and altitudes remote from the moisture source there are larger negative departures of $\delta^{18}O$ in the precipitation. The assumptions of this approach are well-known,[130,131] but the quantitative relationships between $\delta^{18}O$ and temperature are still poorly known and few detailed studies of modern variations in $\delta^{18}O$ in snow have been made. Changes in the surface elevation of an ice dome may cause spurious trends in $\delta^{18}O$ values; an analysis of this effect has been made by Jenssen[132] and it is important that the history of elevation changes of an ice body be well known in order to distinguish between such effects and real climatic trends. A further source of potential error in $\delta^{18}O$ data could be caused by variations over time in the seasonal distribution of precipitation and the related changes in the source regions that supply the moisture, but no examination of this has yet been carried out.

Several new techniques provide information on changes in atmospheric composition. The atmospheric dust load, for example, can be inferred from microparticle concentrations in the ice. The size range of microparticles is typically 0.5–16 μm. Some originate from volcanic debris that has been injected into the stratosphere and then circulated within the hemisphere or globally, according to the location of the eruption, and others are derived from local surfaces by wind entrainment. Mosley-Thompson and Thompson[129] find a large increase in particle concentration in an Antarctic ice core between A.D. 1450 and 1860, a period coinciding with the Little Ice Age, although the origin of the particles cannot be determined. A new technique enables volcanic events to be detected from their characteristic acidity (SO_2 combined with atmospheric water) by measuring the electrical "conductivity" of the ice.[133,134] The fallout of stratospheric sulfate aerosols, which is associated with enhanced current signals, has allowed well-known historical eruptions to be identified from cores in Greenland, as illustrated in figure 9. However, from central Antarctica, at least, there is no evidence of a temporal increase over the last hundred years related to global sulfate pollution and acid precipitation.[135]

Another recent technique involving infrared laser spectroscopy has permitted the measurement of atmospheric carbon dioxide content in the past. The snow matrix traps air within it and thus minute air bubbles are incorporated into the ice. Typically, there may be 100 bubbles per cm^3 with a diameter of about 0.1–0.4 mm.[126] The gas content is dependent on the temperature and air pressure (and hence, elevation) at the time of "close-off" when the firn is transformed into ice. In dry snow, this occurs at a depth of about 100 m. Studies by Neftel et al.[136] suggest that a CO_2 minimum (approximately 210 ppm, compared with 340 ppm today) was attained around 20,000 years B.P. with a subsequent 20–40 percent increase during the Holocene. This is the first specific confirmation of long-term trends in atmospheric CO_2. The causes of such a dramatic change are as yet unknown; it may, of

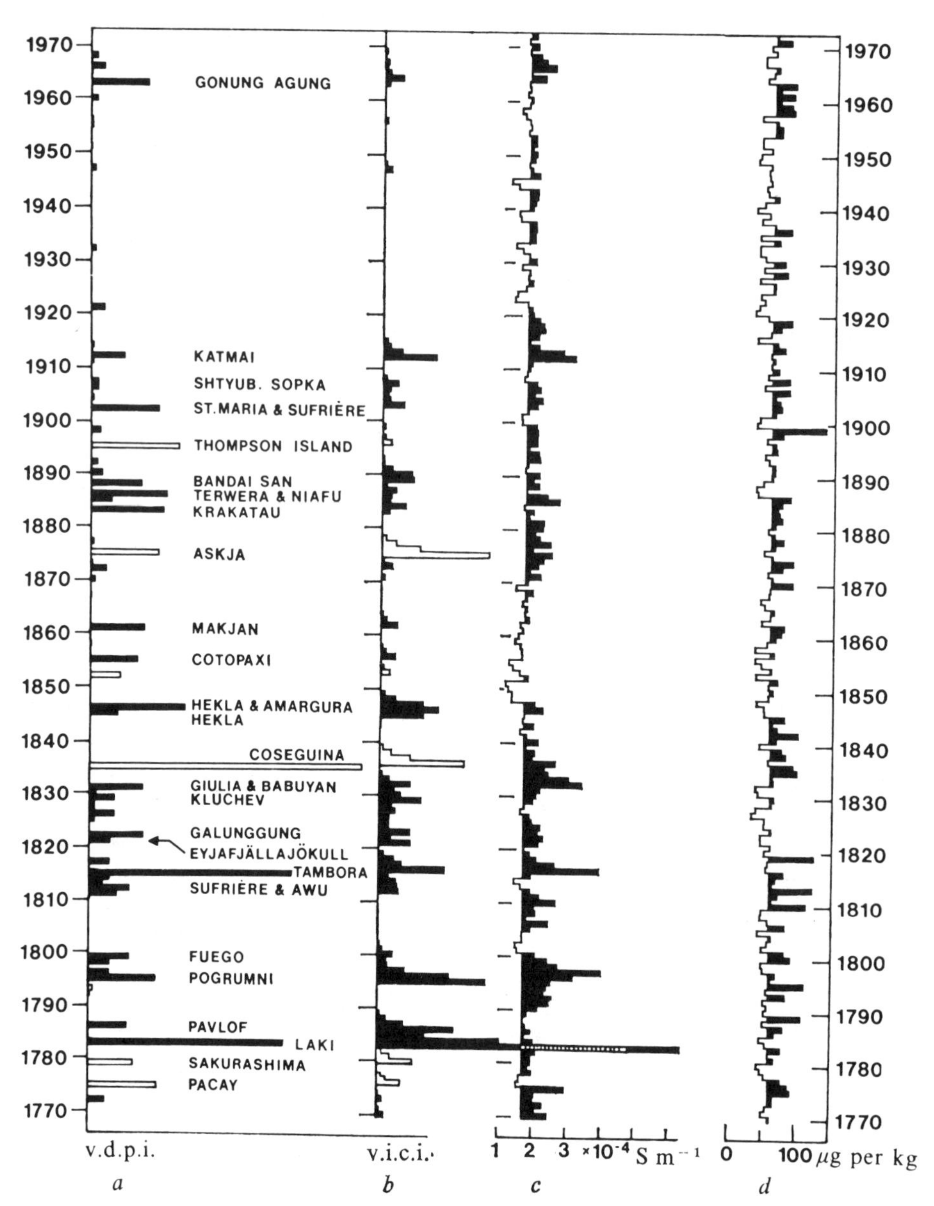

Figure 9. Volcanic activity and impurity profiles along an ice core from Crête, Greenland, spanning the period 1770–1972. (*a*) Lambs's volcanic dust production index (v.d.p.i.): open bars refer to estimates mainly based on an assumed relationship between climatic temperature anomalies and atmospheric dust veils. (*b*) Volcanic impurity concentration index (v.i.c.i.) for high northern latitude precipitation, derived from the v.d.p.i. by correction for latitudinal spreading of the eruption cloud and for residence time of aerosols in the atmosphere. The Icelandic Laki eruption, 1783, stands out. (*c*) Specific conductivity of melted samples of annual precipitation. Values exceeding the mean in the volcanically quiet period 1920–60 are set out in black. (*d*) Insoluble micro-particle mass concentration (in g per kg of ice). The bulk of insoluble impurities must be continental surface dust, because the curve has no correlation to the v.i.c.i. and no increasing trend in the industrial period (from Hammer,[133] reprinted by permission from *Nature*, Vol. 270, p. 482, copyright © 1977 Macmillan Journals Limited).

course, be a result of glacial conditions and associated changes in vegetation cover rather than a factor contributing directly to their development. Atmospheric CO_2 concentrations were in fact similar to now around 30,000–25,000 years B.P. when global climate was in a glacial mode. Nevertheless, the possible feedback mechanisms involved here need careful investigation.

Concern has been expressed over the magnitude of sampling error that may be present in individual ice cores as a result of the tendency for localized snow accumulation/ablation due to wind drifting and the formation of surface sastrugi. Comparison of two cores 27 m apart on the Devon Island ice cap indicates a highly variable correlation between annual $\delta^{18}O$ values.[137] However, the use of 50-year averages resulted in a correlation of 0.965 (excluding the lowest 10 m of core). A means of assessing the large-scale representativeness of a core site is available through the analysis of radio-echo sounding profiles. These provide two-dimensional cross-sections and the potential for three-dimensional arrays of ice thickness and bottom topography.[138] Such data also display extensive internal layering structures at least some of which are thought to represent acidic layers due to volcanic eruptions.[134] Radio-glaciology techniques can now be used to identify optimal sites for the drilling of deep cores, which in the past were usually obtained at locations chosen primarily for their logistic convenience.

The principal paleoclimatic results[128,130,139] of the research based on the three longest ice cores can be summarized as follows:

1. The temporal pattern of the response of the north and south polar regions has been established over the last 125,000 years. The major trends appear to be broadly synchronous with those in many northern and southern land areas. Figure 10 depicts the warm last interglaciation (Stage 5), a long cold glacial interval (Stages 4, 3, 2), followed by the present interglaciation (Stage 1). In contrast, planktonic foraminifera in ocean sediment cores indicate that the subtropical oceans were slightly warmer during the glacial maximum than during the interglaciations.[125]
2. They demonstrate $\delta^{18}O$ differences of -5 to -7‰ between the glacial and interglacial regimes in cores where no major effects arise due to changes in ice surface elevation. This information is needed for proper assessments of isotopic changes in benthic foraminifera in marine sediment cores which indicate variations in global ice volume.
3. They indicate that atmospheric dust loadings in the polar areas during the last glacial maximum were much higher than now, up to eight times greater than in the Holocene,[140,141] although the paleoclimatic significance of this is not yet understood.

A new result which is apparent from the Dye-3 core[128] is that snow accumulation rates in Greenland seem to have been high around 125,000–115,000 years B.P., 80,000–60,000 years B.P. and 40,000–30,000 years B.P. At least during the first two intervals, the occurrence of warm planktonic foraminifera species in the subpolar North Atlantic Ocean indicates the presence of warm surface waters that could transfer abundant moisture to the atmosphere in order to generate rapid buildup of the continental ice sheets in North America and Scandinavia.[142]

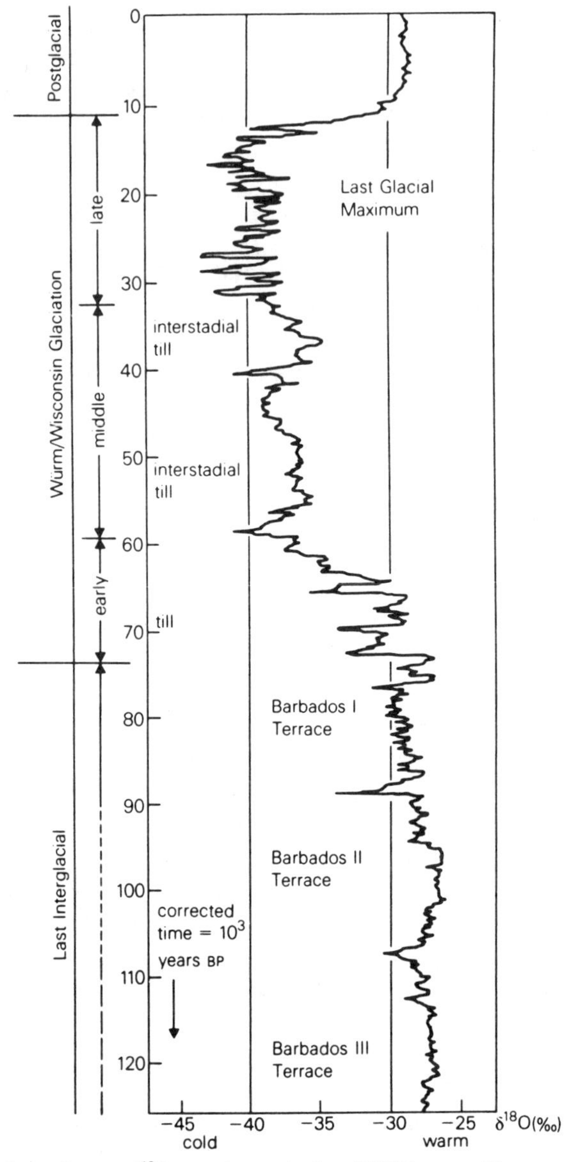

Figure 10. The Camp Century $\delta^{18}O$ record over the last 125,000 years. The record shows the rapid transitions from interglacial to glacial and back to postglacial conditions. The marine terraces in Barbados record high sea levels (after Dansgaard et al.[130]).

The deep and intermediate cores have also yielded significant paleoclimatic information on shorter time scales. The isotopic records from Greenland and Devon Island all show strong irregular fluctuations in ice formed between 30,000 and 10,000 years ago. These oscillations of 4 to 5‰ in $\delta^{18}O$ have been ascribed to large-scale climatic shifts in the North Atlantic-Greenland sector.[128] Another major result concerns the timing of the postglacial thermal maximum. While this occurred in the middle Holocene in the Northern Hemisphere, in eastern Antarctica (and in other Southern Hemisphere records) it is dated around 9,500 years B.P.[143] The detailed resolution of ice core records also allows assessment of the temporal variability of climate during glacial and interglacial intervals on time scales of about 50–100 years or more, by means of band-pass filtering analyses. A striking result from the first detailed work on the 2035 m core from Dye-3, Greenland, is the appearance of a rapid change in chemical constituents of the ice, paralleled by changes in crystal anisotropy, at the Wisconsin/Holocene boundary.[144,145] This transition apparently took place in ≤ 50 years. The reasons behind such rapid changes remain to be determined, but it is clear that the climatic signals provided by ice core data may be of the highest significance for our understanding of the mechanisms of global climate change.

FUTURE DIRECTIONS

Developments in cryospheric data for climate studies can be anticipated in several main areas. In the case of modern observations, improved coverage of basic snow and ice conditions is anticipated with sensors of higher resolution and the operational use of new techniques for snow/cloud discrimination and for mapping sea ice type and concentration. Detailed maps of ice sheet elevations will be provided through radar and laser altimetry and, by repeating such mapping every few years, it will be possible to determine mass balance changes of the Antarctic and Greenland ice sheets.

Internationally agreed formats for cryospheric data, such as the SIGRID proposed for sea ice information, will greatly facilitate preparation for data archiving. However, as with all climate data, there is a major need for the provision of digital files of past and current snow and ice data. The preparatory stages of data inventory are well advanced in some areas,[49,127] but in the case of freshwater ice and permafrost, even this stage has not yet been reached. The necessary steps in preparing data archives are described elsewhere.[60]

Over long time scales, the most detailed and continuous paleoclimatic and paleoenvironmental records will be obtained from ice cores. New techniques using linear accelerators promise reliable ^{14}C dating of small ice samples over the last 50,000–60,000 years, but for longer time scales the dating of ice cores remains an unsolved problem. The core from Dye 3 in Greenland is already providing exciting new information and in the near future a valuable equatorial mountain record should be obtained from the Quelccaya ice cap in Peru. Further ahead, there will be additional intermediate cores and, it is hoped, cores to bedrock in East Antarctica where the basal ice may have an age in excess of one million years.

ACKNOWLEDGMENTS

This contribution was finalized during a leave of absence made possible by a University of Colorado Faculty Fellowship and a Fellowship from the John Simon

Guggenheim Memorial Foundation, both of which are gratefully acknowledged. I am also indebted to Margaret Strauch for her patient preparation of the manuscript.

REFERENCES

1. D. J. Drewry, S. R. Jordon, and E. Jankowski, "Measured properties of the Antarctic ice sheet: Surface configuration, ice thickness, volume and bedrock characteristics," *Ann. Glaciol.*, **3,** 83 (1982).
2. G. Kukla, "Climatic role of snow covers." In I. Allison, ed., *Sea Level, Ice and Climatic Change,* International Association of Scientific Hydrology, Publication no. 131, 1981, pp. 79–107.
3. J. Chappell, "Relative and average sea level changes, and endo-, epi-, and exogenic processes on the earth." In I. Allison, ed., *Sea Level, Ice and Climate Change,* International Association of Scientific Hydrology, Publication no. 131, 1981, pp. 411–430.
4. T. J. Hughes, G. H. Denton, B. G. Anderson, D. H. Schilling, J. L. Fastook, and C. S. Lingle, "The Last Great Ice Sheets: A Global View." In G. H. Denton and T. J. Hughes, eds., *The Last Great Ice Sheets,* Wiley, New York, 1981, pp. 263–317.
5. W. W. Kellogg, "Climate Feedback Mechanisms Involving the Polar Regions." In G. Weller and S. A. Bowling, eds., *Climate of the Arctic,* University of Alaska, Fairbanks, 1975, pp. 111–116.
6. W. F. Budd, "The importance of ice sheets in long term changes of climate and sea level." In I. Allison, ed., *Sea Level, Ice and Climatic Changes,* International Association of Scientific Hydrology, Publication no. 131, 1981, pp. 441–471.
7. M. Ghil, "Internal Climatic Mechanisms Participating in Glaciation Cycles." In A. Berger, ed., *Climatic Variations and Variability: Facts and theories,* D. Reidel, Dordrecht, 1981, pp. 539–558.
8. D. M. Houghton, "Heat sources and sinks at the earth's surface," *Meteorol. Mag.*, **87,** 132 (1958).
9. E. C. Kung, R. A. Bryson, and D. H. Lenschow, "Study of continental surface albedo on the basis of flight measurements and structure of the earth's surface cover over North America," *Mon. Weather Rev.*, **92,** 543 (1964).
10. A. Robock, "The seasonal cycle of snow cover, sea ice and surface albedo," *Mon. Weather Rev.*, **108,** 267 (1980).
11. C. I. Jackson, "Some climatological grumbles—Part 1," *Weather,* **18,** 278 (1963).
12. G. A. Maykut, "Surface Heat and Mass Balance." In N. Untersteiner et al., eds. *The Geophysics of Sea Ice, Proceedings NATO Advanced Study Institute Maratea, Italy,* Plenum, New York, 1984.
13. H. H. Lamb, "Two-way relationships between the snow or ice limit and 1,000-500 mb thickness in the overlying atmosphere," *Q.J.R. Meteorol. Soc.*, **81,** 172 (1955).
14. J. Namias, "Influences of Abnormal Surface Heat Sources and Sinks on Atmospheric Behavior." In *Proceedings of the International Symposium on Numerical Weather Prediction, Tokyo, 1960,* Meteorological Society of Japan, 1962, pp. 615–627.
15. J. Williams, "The influence of snowcover on the atmospheric circulation and its role in climatic change: An analysis based on results from the NCAR global circulation model," *J. of Appl. Meteorol.*, **14,** 137 (1975).
16. J. E. Walsh, D. R. Tucek, and M. R. Peterson, "Seasonal snow cover and short-term climatic fluctuations over the United States," *Mon. Weather Rev.*, **110,** 1474 (1982).
17. A. M. Carleton, "Cryosphere-climate interactions based on satellite data analysis of cloud vortex systems in the northern hemisphere," Ph.D. diss. University of Colorado, Boulder, 1982.
18. S. F. Ackley, "A review of sea-ice weather relationships in the Southern Hemisphere." In I. Allison, ed., *Sea Level, Ice and Climatic Change,* International Association of Scientific Hydrology, Publication no. 131, 1981, pp. 127–159.

19. C. H. Pease, "East Bering Sea ice processes," *Mon. Weather Rev.*, **108,** (1980).

20. J. E. Walsh and C. M. Johnson, "Interannual atmospheric variability and associated fluctuations in Arctic sea ice extent," *J. Geophys. Res.*, **84,** 6915 (1979).

21. S. F. Ackley and T. E. Keliher, "Antarctic sea ice dynamics and its possible climatic effects," *AIDJEX Bull.*, **33,** 53 (1976).

22. R. G. Crane, R. G. Barry, and H. J. Zwally, "Analysis of atmosphere–sea ice interactions in the Arctic Basin using ESMR microwave data, "*Int. J. Remote Sensing,*" **3,** 259 (1982).

23. M. I. Budyko, "The effect of solar radiation variations on the climate of the earth," *Tellus,* **21,** 611 (1969).

24. W. D. Sellers, "A global climatic model based on the energy balance of the earth–atmosphere system," *J. Appl. Meteorol.*, **8,** 392 (1969).

25. A. Robock, "Ice and snow feedbacks and the latitudinal and seasonal distribution of climate sensitivity." *J. Atmos. Sci.*, **40,** 986 (1983).

26. J. Weertman, "Milankovitch solar radiation variations and Ice Age ice sheet sizes," *Nature,* **261,** 17 (1976).

27. G. E. Birchfield, "A study of the stability of a model continental ice sheet subject to periodic variations in heat input," *J. Geophys. Res.*, **82,** 4909 (1977).

28. D. Pollard, A. P. Ingersoll, and J. G. Lockwood, "Response of zonal ice sheet model to the orbital perturbations during the Quaternary ice ages," *Tellus,* **32,** 301 (1980).

29. J. Oerlemans, "A model of a stochastically driven ice sheet with planetary wave feedback," *Tellus,* **31,** 469 (1979).

30. M. Ghil and H. LeTreut, "A climate model with cryodynamics and geodynamics," *J. Geophys. Res.*, **86**C, 5262 (1981).

31. S. Manabe and J. L. Holloway, Jr., "The seasonal variation of the hydrologic cycle as simulated by a global model of the atmosphere," *J. Geophys. Res.*, **80,** 1617 (1975).

32. S. Manabe, K. Bryan, and M. J. Spelman, "A global ocean-atmosphere climate model with seasonal variation for future studies of climate sensitivity, "*Dyn. Atmos. Oceans,* **3,** 393 (1979).

33. C. L. Parkinson and W. M. Washington, "A large-scale numerical model of sea ice," *J. Geophys. Res.*, **84,** 311 (1979).

34. W. M. Washington, A. J. Semtner, Jr., G. A. Meehl, D. J. Knight, and T. A. Mayer, "A general circulation experiment with a coupled atmosphere, ocean and sea ice model," *J. Phys. Oceanogr.*, **10,** 1887 (1980).

35. S. Manabe and R. J. Stouffer, "Sensitivity of global climate model to an increase of CO_2 concentration in the atmosphere," *J. Geophys. Res.*, **85,** 5529 (1980).

36. C. L. Parkinson and W. W. Kellogg, "Arctic sea ice decay simulated for a CO_2–induced temperature rise," *Climatic Change*, **2,** 149 (1979).

37. W. F. Budd and I. N. Smith, "The growth and retreat of ice sheets in response to orbital radiation changes." In I. Allison, Ed., *Sea Level, Ice and Climatic Change*, International Association of Scientific Hydrology, Publication no. 131, 1981, pp. 369–409.

38. J. Williams, R. G. Barry, and W. M. Washington, "Simulation of the atmospheric circulation using the NCAR global circulation model with ice age boundary conditions," *J. Appl. Meteorol.*, **13,** 305 1974.

39. W. L. Gates, "The numerical simulation of ice–age climate with a global general circulation model," *J. Atmos. Sci.*, **33,** 1844 (1976).

40. S. Manabe and D. G. Hahn, "Simulation of the tropical climate of an ice age," *J. Geophys. Res.*, **82,** 3889 (1977).

41. G. R. Heath, "Simulations of a glacial paleoclimate by three different atmospheric general circulation models," *Palaeogeogr., Palaeoclimatol., Palaeoecol.*, **26,** 291 (1979).

42. J. Adem, "Numerical experiments on ice age climates," *Climatic Change*, **3,** 155 (1981).

43. D. G. Hahn, "Summary requirements of GCMs for observed snow and ice cover data." In G.

Kukla, A. Hecht, and D. Wiesnet, eds., Snow Watch 1980, *Glaciological Data, Report GD-11*, World Data Center-A for Glaciology, Boulder, 1981, pp. 45–53.

44. R. Murray, "Rain and snow in relation to the 1000–700mb and 1000–500mb thickness and the freezing level," *Meteorol. Mag.*, **81,** 5 (1952).
45. B. E. Goodison, H. L. Ferguson, and G. A. McKay, "Measurement and Data Analysis." In D. M. Gray and D. H. Male, eds., *Handbook of Snow: Principles, processes, management and use*, Pergamon, Toronto, 1981, pp. 191–274.
46. U.S. Department of Commerce and U.S. Department of Agriculture, *"Weekly Weather Crop Bull.,"* **69,** (1982).
47. C. A. Mixer "River floods and melting snow," *Mon. Weather Rev.*, **31,** 173 (1903).
48. J. E. Church, "Snow and snow surveying: Ice." In O. E. Meinzer, ed., *Hydrology*, Vol. IX, Physics of the Earth, McGraw-Hill, New York, 1942, pp. 83–148.
49. R. G. Crane, compiler, "Inventory of snow cover and sea ice date," *Glaciological Data, Report GD-7,* World Data Center–A for Glaciology, Boulder, 1979.
50. Atmospheric Environment Service, Canada, "Snow cover data for Canada" (Annual), 1955–present, Environment Canada.
51. Hydrographischer Dienst in Osterreich, "Der Schnee in Osterreich im Zeitraum 1901–1950," *Beitr. Hydrogr. Osterreich*, **34,** (1962).
51. G. A. McKay, "Relationships between snow survey and climatological measurements." In International Association of Scientific Hydrology, Publ. 63 (International Union of Geophysics and Geodetics, General Assemby, Berkeley), *Surface Water*, 1963, pp. 214–227.
53. B. E. Goodison, "Compatibility of Canadian snowfall and snow cover data," *Water Resour. Res.*, **17,** 893 (1981).
54. H. Arakawa, "Dates of first or earliest snow covering for Tokyo since 1632," *Q. J. R. Meteorol. Soc.*, **82,** 222 (1956).
55. G. Manley, "Snowfall in Britain over the past 300 years," *Weather*, **24,** 428 (1969).
56. C. Pfister, "Fluctuations in the duration of snow-cover in Switzerland since the late seventeenth century," *Klimatologiske Medd., Danske Meteorol. Inst.*, **4,** 1 (1978).
57. B. Messerli, P. Messerli, C. Pfister, and H. T. Zumbuhl, "Fluctuations of climate and glaciers in the Bernese Oberland, Switzerland, and their geoecological significance, 1600 to 1975," *Arctic and Alpine Res.*, **10,** 247 (1978).
58. L. Koch, "The east Greenland ice," *Medd. Gronl.*, **130,** (3), (1945).
59. H. H. Lamb, *Climate: Present, past and future*, Vol. 2, *Climatic History and the Future*, Methuen, London, 1977.
60. R. G. Barry, "The data base." In N. Untersteiner et al., eds., *The Geophysics of Sea Ice*, Plenum, New York, 1984, (in press).
61. Danske Meteorologisk Institut, "The state of the ice in the Arctic Seas," Annual appendix to *Naut-Meteorol., Ann.*, (1901–56).
62. P. M. Kelly, "An Arctic sea ice data set, 1901–1956." In *Workshop on Snow Cover and Sea Ice Data, Glaciological Data Report*, GD-5, World Data Center-A for Glaciology, Boulder, 1979, pp. 101–106.
63. J. E. Walsh, "A data set on northern hemisphere sea ice extent, 1953-76," *Glaciological Data Report GD-2, Arctic Sea Ice*, World Data Center-A for Glaciology, Boulder, Colorado, pp. 49–51 (1978).
64. H. J. Zwally, J. C. Comiso, C. L. Parkinson, W. J. Campbell, F. D. Carsey, and P. Gloersen, "Antarctic sea ice, 1973–1976: Satellite passive-microwave observations," NASA SP–459, National Aeronautics and Space Administration, (1983).
65. G. A. Maykut, "Energy exchange over young sea ice in the central Arctic," *J. Geophys. Res.*, **83,** 3646 (1978).
66. W. T. R. Allen and B. S. V. Cudbird, "Freeze-up and break-up of water bodies in Canada," Meteorological Branch Department of Transportation, Toronto, CLI-1-71 (1971).

67. M. A. Bilello, "Maximum thickness and subsequent decay of lake, river and fast ice in Canada and Alaska," U.S. Army Cold Regions Research and Engineering Laboratory CRREL Report 80-6, (1980).

68. A. J. W. Catchpole and T. F. Ball, "Analysis of historical evidence of climatic change in western and northern Canada." In C. R. Harington, ed., *Climatic Change in Canada*, Syllogeus No. 33, National Museums of Canada, 1981, pp. 48–96.

69. H. M. Van den Dool, H. J. Krijnen, and C. J. E. Schuurmans, "Average winter temperature at De Bilt (The Netherlands): 1634–1977," *Climatic Change*, **1,** 319 (1978).

70. H. Arakawa, "FUJIWHARA on five centuries of freezing dates of Lake Suwa in central Japan," *Arch. Meteorol., Geophys. Bioklimatol., B*, **6,** 152 (1954).

71. B. M. Gray, "Early Japanese winter temperature," *Weather*, **29,** 103 (1974).

72. M. Tanaka and M. M. Yoshino, "Re-examination of the climatic change in central Japan based on freezing dates of Lake Suwa," *Weather*, **37,** 292 (1982).

73. J. F. Nye, "The response of glaciers and ice-sheets to seasonal and climatic changes," *Pro. R. Soc. London, Ser.*, A, **256,** 559 (1960).

74. H. J. Zumbühl, *Die Schwankungen der Grindelwaldgletscher in den historischen Bild und Schriftquellen des 12, bis 19. Jahrhunderts*, Birkhauser Verlag, Basel, 1980.

75. G. Patzelt, "Die Längenmessungen and den Gletschern der österreichischen Ostalpen 1890–1969," *Z. Gletscherkd. Glazialgeol.*, **6,** 151 (1970).

76. H. C. Hoinkes, "Glacier variations and weather," *J. Glaciol.*, **7,** 3 (1968).

77. IAHS (ICSI)-UNESCO, *Fluctuations of Glaciers 1970–1975,"* International Association of Hydrological Sciences and UNESCO, Paris, 1977.

78. M. Gamper and J. Suter, "Der Einfluss von Temperaturänderungen auf die Länge von Gletscherzungen," *Geogr. Helv.*, **33,** 183 (1978).

79. S. Hastenrath, "*The Glaciation of the Ecuadorian Andes*," A. A. Balkema, Rotterdam, 1981.

80. C. W. Locke, "Pictorial records of glaciers," *Glaciological Data Report GD-10*, World Data Center-A for Glaciology, Boulder, Colorado, 1981, pp. 39–53.

81. J. F. Nye, "An numerical method of inferring the budget history of a glacier from its advance and retreat," *J. Glaciol.*, **5,** (41), 589 (1965).

82. L. D. Williams, "The variation of corrie elevation and equilibrium line altitude with respect in eastern Baffin Island, N.W.T., Canada," *Arctic Alpine Res.*, **7,** 169 (1975).

83. I. N. Smith and W. F. Budd, "The derivation of past climatic changes from observed changes of glaciers." In I. Allison, ed., *Sea Level, Ice and Climatic Change*, International Association of Scientific Hydrology, Publication no. 131, 1981, pp. 31–52.

84. I. Allison and P. Kruss, "Estimation of recent climate change in Irian Jaya by numerical modelling of its tropical glaciers," *Arctic Alpine Res.*, **9,** 49 (1977).

85. J. D. Ives, "Permafrost." In J. D. Ives and R. G. Barry, eds., *Arctic and Alpine Environments*, Methuen, London, 1974, pp. 159–194.

86. J. Thie, "Distribution and thawing of permafrost in the southern part of the discontinuous permafrost zone in Manitoba," *Arctic*, **27,** 189 (1976).

87. J. R. MacKay, "The stability of permafrost and recent climatic change in the Mackenzie Valley, N.W.T.," Geological Survey of Canada Paper 75-1, 173–176 (1975).

88. E. B. Belopukhova, "Features of the contemporary development of permafrost in western Siberia." In F. J. Sanger and P. J. Hyde, eds., *Permafrost, USSR Contribution, Second International Conference 1973*, National Academy of Sciences, Washington, D.C., 1978, pp. 112–113.

89. L. W. Gold and A. H. Lachenbruch, "Thermal conditions in permafrost—a review of North American literature." In *Permafrost, North American Contribution, Second International Conference 1973*, National Academy of Sciences, Washington, D.C., 1973, pp. 3–25.

90. V. Cermak, "Paleoclimatic significance of measuring the temperature of permafrost." In F. J. Sanger and P. J. Hyde, eds., *Permafrost, USSR Contribution, Second International Conference 1973*, National Academy of Sciences, Washington, D.C., 1978, pp. 812–815.

91. V. T. Balobaev, "Reconstruction of paleoclimate from present-day geothermal data." In *Proceedings of the Third International Conference on Permafrost 1978*, Vol. 1, National Research Council, Canada, 1978, pp. 10–14.

92. F. Smigielski, "Northern hemisphere snow and ice charts of NOAA/NESS." In G. Kukla, A. Hecht, and D. Wiesnet, eds., *Snow Watch 1980, Glaciological Data, Report GD-11*, World Data Center-A for Glaciology, Boulder, 1981, pp. 59–62.

92. E. Batten, J. Soha, and M. Merrill, "The albedo of snow covered surfaces determined from NOAA 4 and 5 satellites," *Trans., Am. Geophys. Union*, **58,** 400 (1977).

94. G. Kukla, D. Robinson, and J. Brown, "Lamont climatic snow cover charts." In G. Kukla, A. Hecht and D. Wiesnet, eds., *Snow Watch 1980, Glaciological Data, Report GD-11*, World Data Center-A for Glaciology, Boulder, 1981, pp. 87–91.

95. K. F. Dewey and R. Heim, Jr., "Satellite observations of variations in northern hemisphere seasonal snow cover," U.S. Department of Commerce NOAA Technical Report NESS 87, (1981).

96. G. Kukla and J. Gavin, "Snow and pack ice indices." In *Snow Cover, Glaciological Data, Report GD-6*,World Data Center-A for Glaciology, Boulder, 1979, pp. 9–14.

97. G. Kukla and D. Robinson, "Climatic value of operational snow and ice charts." In G. Kukla, A. Hecht, and D. Wiesnet, eds., *Snow Watch 1980, Glaciological Data, Report GD-11*, World Data Center-A for Glaciology, Boulder, 1981, pp. 103–119.

98. R. Woronicz, "The Air Force snow cover charts." In G. Kukla, A. Hecht, and D. Wiesnet, eds., *Snow Watch 1980, Glaciological Data Report GD-11*, World Data Center-A for Glaciology, Boulder, 1981, pp. 63–70.

99. G. R. Scharfen and M. R. Anderson, "Climatological applications of a satellite snow/cloud discrimination sensor," *Proceedings of the 50th Western Snow Conference* Reno, Nevada, 1982, pp. 92–101.

100. W. J. Campbell and others, "Arctic sea ice variations from time-lapse passive microwave imagery," *Boundary Layer Meteorol.*, **18,** 99 (1980).

101. D. J. Cavalieri, P. Gloersen, and W. J. Campbell, "Determination of sea ice parameters with Nimbus 7 SMMR," *J. Geophys. Res.*, **89,** 5355 (1984).

102. A. S. Thorndike and R. Colony, "Arctic Ocean Buoy Program Data Report," Polar Science Center, University of Washington, Seattle (1981).

103. R. L. Brooks, W. J. Campbell, R. O. Ramseier, H. R. Stanley, and H. J. Zwally, "Ice sheet topography by satellite altimetry," *Nature*, **274,** 539 (1978).

104. R. Etkins and E. S. Epstein, "The rise of global mean sea level as an indication of climate change," *Science*, **215,** 287 (1982).

105. NASA, "ICEX. Ice and climate experiment," Report of Science and Applications Working Group, National Aeronautics and Space Administration, Goddard Space Flight Center, Greenbelt, Maryland, (1979).

106. H. C. Hoinkes, "Glacial meteorology." In H. Odishaw, ed., *Research in Geophysics*, Vol. 2, *Solid Earth and Interface Phenomena*, M.I.T. Press, Cambridge, 1964, pp. 391–424.

107. J. Imbrie and K. P. Imbrie, *Ice Ages: Solving the mystery*, Enslow Publishers, Short Hills, N.J., 1979.

108. A. Penck, and E. Brückner, *Die Alpen im Eiszeitalter*, Tauchnitz, Leipzig, 1909.

109. P. J. Webber and J. T. Andrews, eds., "Lichenometry," *Arctic Alpine Res.*, **5**(4) (1973).

110. D. E. Sugden and B. S. John, *Glaciers and Landscape*, Arnold, London, 1976.

111. G. H. Denton, and T. J. Hughes, eds., *The Last Great Ice Sheets*, Wiley, New York, 1981.

112. T. Hughes, G. H. Denton, and M. G. Grosswald, "Was there a late-Würm Arctic ice sheet?," *Nature* **266,** 596 (1977).

113. H. M. French, *The Periglacial Environment*, Longman, London, 1976.

114. A. L. Washburn, "Permafrost features as evidence of climatic change," *Earth-Sci. Rev.* **15,** 327 (1980).

115. T. L. Péwé, "Ice wedge casts and past permafrost distribution in North America," *Geoforum* **15,** 15 (1973).

116. T. L. Péwé, "The Quaternary geology of Alaska," U.S. Geological Survey Professional Paper 835, U.S. Government Printing Office, Washington, D.C. (1975).

117. Williams, R. B. G., "The British Climate during the Last Glaciation: An interpretation based on perigacial phenomena." In A. E. Wright and F. Moseley, eds., *Ice Ages, Ancient and Modern*, Seel House Press, Liverpool, 1975, pp. 95–120.

118. Black, R. F. "Features indicative of permafrost," *Annu. Rev. Earth Planet. Sci.*, **4,** 75 (1976).

119. W. F. Ruddiman, "Late Quaternary deposition of ice-rafted sand in the subpolar North Atlantic (lat. 40° to 65°N), *Geol. Soc. Am. Bull.*, **88,** 1813 (1977).

120. D. W. Cooke and J. D. Hays, "Estimates of Antarctic Ocean Seasonal Sea-Ice Cover during Glacial Intervals." In C. Craddock, ed., *Antarctic Geoscience*, University of Wisconsin Press, Madison, 1982, pp. 1017–1025.

121. L. H. Burckle, P. Robinson, and D. Cooke, "Reappraisal of sea-ice distribution in Atlantic and Pacific sectors in the Southern Ocean at 18,000 yr BP," *Nature*, **299,** 435 (1982).

122. S. C. Porter, "Glaciological Evidence of Holocene Climatic Change." In T. M. L. Wigley, M. J. Ingram, and G. Farmer, eds., *Climate and History*, Cambridge University Press, Cambridge, 1981, pp. 82–110.

123. B. Messerli, M. Winiger, and P. Rognon, "The Saharan and East African Uplands during the Quaternary." In M. A. J. Williams and H. Faure, eds., *The Sahara and the Nile*, A. A. Balkema, Rotterdam, 1980, 87–132.

124. S. Hastenrath, "On the Pleistocene snow-line depression in the arid regions of the South American Andes," *J. Glaciol.*, **10,** 255 (1971).

125. CLIMAP Project Members, "The surface of the Ice-Age earth," *Science*, **191,** 1131 (1976).

126. A. J. Gow, "Time-priority studies of deep ice cores." In P. K. MacKinnon, compiler, *Ice Cores, Glaciological Data, Report GD-8*, World Data Center-A for Glaciology, Boulder, 1980, pp. 91–102.

127. P. K. MacKinnon, compiler, "Ice Cores," Glaciological Data, Report GD-8, World Data Center-A for Glaciology, Boulder, (1980).

128. W. Dansgaard, H. B. Clausen, N. Gunderstrup, C. U. Hammer, S. F. Johnsen, P. M. Krittinsdottir, and N. Reeh, "A new Greenland deep ice core," *Science*, **218**(4579), 1273 (1982).

129. E. Mosley-Thompson and L. G. Thompson, "Nine centuries of microparticle deposition at the South Pole," *Quat. Res.*, **17,** 1 (1982).

130. W. Dansgaard, S. J. Johnson, H. B. Clausen, and C. Langway, "Climatic Record Revealed by the Camp Century (Greenland) Ice Core." In K. K. Turekian, ed., *Late Cenozoic Glacial Ages*, Yale University Press, New Haven, 1971, pp. 37–56.

131. W. Dansgaard, S. Johnsen, H. B. Clausen, and N. Gunderstrup, "Stable isotope glaciology," *Medd. Gronl.*, **197,** 53 (1973).

132. D. Jenssen, "Climatic and Topographic Changes from Glaciological Data." In A. B. Pittock, L. A. Frakes, D. Jenssen, J. A. Peterson, and J. W. Zillman, eds., *Climatic Change and Variability: A southern perspective*, Cambridge University Press, Cambridge, 1978, pp. 77–81.

133. C. U. Hammer, "Past volcanism revealed by Greenland Ice Sheet impurities," *Nature*, **270,** (1977).

134. C. U. Hammer, "Acidity of polar ice cores in relation absolute dating, past volcanism, and radio-echoes," *J. Glaciol.*, **25,** 359 (1980).

135. R. Delmas and C. Boutron, "Are the past variations of the stratospheric sulfate burden recorded in central Antarctic snow and ice layers?," *J. Geophys. Res.*, **85**(C10), 5645, (1980).

136. A. Neftel, H. Oeschger, J. Schwander, B. Stauffer, and R. Zumbrunn, "Ice core sample measurements give atmospheric CO_2 content during the past 40,000 yr.," *Nature*, **295,** 220 (1982).

137. W. S. B. Paterson, R. M. Koerner, D. Fisher, S. J. Johnsen, H. B. Clausen, W. Dansgaard, P.

Bucher, and H. Oeschger, "Oxygen-isotope climatic record from the Devon Island ice cap, Arctic Canada," *Nature*, **266,** 508 (1977).

138. G. de Q. Robin, D. J. Drewry, and D. T. Meldrum, "Internal studies of ice sheet and bedrock," *Philos. Trans. R. Soc. London, Ser. B*, **279,** 185 (1977).

139. G. de Q. Robin, ed., *The Climate Record in Polar Ice Sheets*, Cambridge University Press, Cambridge, 1983.

140. A. J. Gow and T. Williamson, "Volcanic ash in the Antarctic ice sheet and its possible climatic implications," *Earth Planet. Sci. Lett.*, **13,** 210 (1971).

141. L. G. Thompson, "Variations in microparticle concentration, size distribution and elemental composition found in Camp Century, Greenland, and Byrd Station, Antarctica, deep ice cores," *Proceedings of the Grenoble Symposium*, International Association of Scientific Hydrology, Publication 118, 351–364. (1977).

142. W. F. Ruddiman and A. McIntyre, "Warmth of the subpolar North Atlantic Ocean during Northern Hemisphere ice sheet growth," *Science*, **204,** 173 (1979).

143. C. Lorius, L. Merlivat, J. Jouzel, and M. Pourchet, "A 30,000-yr isotope climatic record from Antarctic ice," *Nature*, **280,** 644 (1979).

144. S. L. Herron, C. C. Langway, Jr. and K. A. Brugger, "Ultrasonic velocities and crystalline anisotropy in the ice core from Dye-3, Greenland," *EOS*, **63,** 297 (1982).

145. M. M. Herron and C. C. Langway, Jr., "Chloride, nitrate and sulfate in the Dye 3 and Camp Century, Greenland, ice cores," *EOS*, **63,** 298 (1982).

146. J. Hollin and R. G. Barry, "Empirical and theoretical evidence concerning the response of the earth's ice and snow cover to a global temperature increase," *Envir. In.*, **2,** 437, (1979).

7

LAKE LEVELS AND CLIMATE RECONSTRUCTION

F. A. Street-Perrott and S. P. Harrison

School of Geography, University of Oxford
Oxford, England

> Now I conceive that as all these Lakes do receive Rivers and have no Exits or Discharge, so 'twill be necessary that their Waters rise and cover the Land, until such time as their Surfaces are sufficiently extended, so as to exhale in Vapour that Water that is poured in by the Rivers; and consequently that Lakes must be bigger or lesser according to the Quantity of the Fresh they receive.[1]

It has been recognized for more than 250 years that many lakes fluctuate in level in response to changes in climate. Most sensitive in this respect are closed basin lakes, which lack surface outlets. Their tendency to fluctuate in depth and area is often clearly apparent from staircases of abandoned shorelines or from abrupt changes of facies in sedimentary sections. Most closed lakes are brackish to saline, and alkaline. This is largely attributable to evaporative concentration. Variations in water volume resulting from climatic change cause marked shifts in geochemical equilibria and are often readily detectable using chemical, mineralogical, isotopic, or biological methods. During their more dilute phases, closed lakes may represent an important water source in otherwise arid areas. Their fluctuations can sometimes, therefore, be reconstructed from the distribution of archaeological sites, particularly where dependence on aquatic resources can be demonstrated.

Closed lakes are most widespread today in arid to subhumid regions, particularly in the tropics and subtropics. A detailed world map of areas of inland drainage was presented by de Martonne and Aufrère.[2] Many lakes outside these areas, however, ceased to overflow during the dry phases of the Quaternary. The study of their fluctuations is at last tending to offset the overemphasis on temperature changes that stemmed from the early dominance of glacial geomorphology and palynology in Quaternary studies.

The purpose of this chapter is to examine the basis of the various techniques used to reconstruct past fluctuations in the extent of closed-basin lakes and to review the results of recent attempts to derive climatic information from these data. The discussion will concentrate mainly on the paleolimnological record. Previous reviews of lake-level fluctuations have often had a narrow disciplinary or regional focus.[3–6] An attempt is made here to illustrate both the climatic and hydrological principles governing the long-term behavior of closed lakes, and the great diversity of lake types[7] and research strategies which are encountered in this field. The examples used are selected mainly from Africa, the western United States, and Australia, because of our familiarity with those areas. We conclude with a discussion of the current status and future prospects of techniques for the climatic calibration of lake-level data.

FACTORS AFFECTING THE SENSITIVITY OF LAKES TO CLIMATIC FLUCTUATIONS

The sensitivity of a lake to climatic change is governed by a number of factors, notably the presence or absence of an outlet, the water-balance type to which it belongs, the basin relief, the nature of its sedimentary record, and the extent of human disturbance. These factors are discussed in turn below.

The existence of an outlet

The mean annual water balance of a lake is given by the equation

$$\Delta V = A_L(P_L - E_L) + (R - D) + (G_I - G_O) \tag{1}$$

where
ΔV = net change in volume of the lake
P_L = precipitation on the lake
E_L = evaporation from the lake (P_L and E_L are expressed as a depth of water)
A_L = area of the lake
R and D = runoff from the catchment and the surface discharge from the lake, respectively
G_I and G_O = groundwater flows into, and out of, the lake

For a closed lake, D is zero. Under equilibrium conditions, assuming that groundwater transfers are negligible, equation (1) reduces to

$$R = A_L(E_L - P_L) \tag{2}$$

If it is further assumed that the runoff from the drainage basin can be represented by

$$R = A_B(P_B - E_B) \tag{3}$$

where A_B = area of the catchment
P_B = precipitation over the catchment
E_B = evapotranspiration over the catchment

then,

$$z = \frac{A_L}{A_B} = \frac{P_B - E_B}{E_L - P_L} \tag{4}$$

where z is the lake area/catchment area ratio.[8] This simple expression shows that the equilibrium area of a closed lake under natural conditions is strictly dependent on the precipitation and evaporation over its catchment and water surface, whereas for an open lake, the rate of discharge through the outlet must also be considered. The latter tends to vary exponentially with the height of the lake above the floor of its spillway and may be impossible to estimate in the case of paleolakes. Downcutting of the outlet may also be a problem. It is important to note that changes in lake *depth* are a consequence of adjustments in lake *area*, A_L.

Equation (4) has been verified empirically by Bowler.[9] In arid areas, $(P_B - E_B)$ is small and $(E_L - P_L)$ is large. The reverse is true under humid conditions. Mifflin and Wheat[10] have shown that the z ratio serves as a useful climatic and paleoclimatic index. In Nevada, they found that for 33 non-overflowing "pluvial" lakes, 75 percent of the variance in z was explained by latitude and altitude. On a world scale, the percentage of the land area in each 5°latitude band occupied by lakes is strongly related to the latitudinal average of $P - E$.[11] As expected, closed lakes are concentrated on either side of the subtropical deserts where $E_L >> P_L$.[3,12]

Water balance

Street,[13] following Szestay,[14] has attempted to classify the response of lakes to climatic variations in terms of their hydrological characteristics. Figure 1 displays water-balance data for 39 lakes for which information is readily available and for which G_I and G_O can safely be neglected. The y axis gives runoff as a percentage of annual inputs and the x axis, discharge from the lake as a percentage of the outputs. The position of the data points is probably only accurate to ±5–10%. Closed lakes are situated on the y axis.

Figure 1 can be divided into nine quadrants. The data points form a broad V-shaped pattern with its apex at the origin. The lakes in category $P-E$ are referred to as *atmosphere-controlled lakes.*[14] These are large inland seas for which transfers between the water surface and the atmosphere greatly exceed inputs and outputs in the form of runoff. Lake Victoria in equatorial Africa (V) is a good example.[24] The world ocean is plotted for comparison.[14]

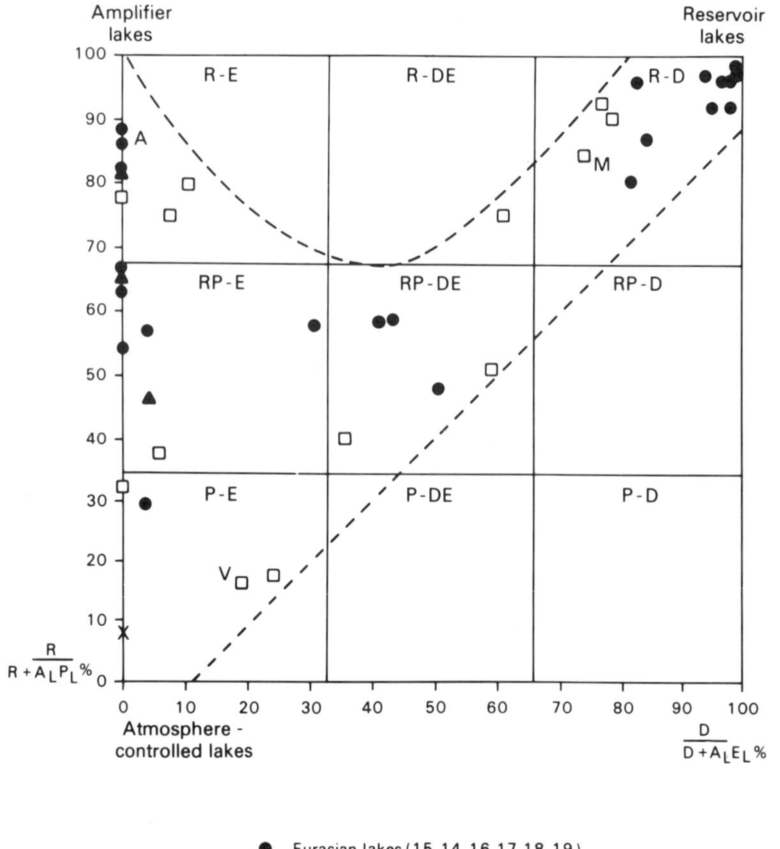

Figure 1. Water-balance classification of modern lakes for which $G_I \leq 10$ percent inputs, $G_O \leq 10$ percent outputs (from refs. 13–23).

There is a continuum from $P-E$ to the top right-hand corner of figure 1, which is occupied by *flow-dominated reservoirs* ($R-D$).[14] These are essentially "wide places in a river" with a very high rate of flushing by surface runoff. A good example is Lake Mobutu (Albert) in equatorial Africa (M). The residence time of water increases diagonally across the diagram, from <1 to 10 years for the lakes in $R-D$ to >50 years in the case of some large lakes in $P-E$ and $RP-E$.

A second continuum extends from $P-E$ towards the quadrant $R-E$. It represents increasing values of the aridity index a, where $a = E_L/P_L$.[14] This index increases from 1.0 on the diagonal $P-E/R-D$ to a theoretical value of ∞ at the top left-hand corner of figure 1. The lakes for which runoff approaches 100 percent of total inputs are known as *amplifier lakes* ($R-E$).[13] A typical example is the Aral Sea (A).

In cases where groundwater transfers are significant, such as in areas of limestone, aeolian sand, or fissured lavas, additional lake types must be recognized. *Groundwater-fed (groundwater-effluent)* lakes are partially maintained by springs or diffuse effluent seepage. Two famous cases are Lakes Asal and Afrera in the Horn of Africa.[25] A significant groundwater influx should always be suspected when the z ratio for a modern lake or paleolake exceeds about 1.0 or when the basin is underlain by permeable rocks. *Seepage* (*groundwater-influent*) lakes lose water in part by net groundwater outflow. Two well known instances in Africa are Lakes Chad[28] and Naivasha.[29]

The most sensitive group of lakes with respect to climatic variations are the closed amplifier lakes (in $R-E$ and $RP-E$), because of the numerous positive feedbacks governing runoff generation from their catchments. The lack of an outlet also means that short-term variations in precipitation are cumulated, giving rise to large fluctuations in level. In the Great Basin of the United States, for example, at least 27 lakes have varied in depth by $\geqslant$50 m.[30] Because of their large variations in level, many overflowed during the wettest periods. For these reasons, most of the best known and most detailed Late Quaternary lake-level curves are derived from closed amplifier lakes, such as Lake Abhé.[31]

Reservoir lakes are of little use as palaeoclimatic indicators, unless their surface inflows are dramatically cut off during dry phases: for example, Lake Mobutu partially dried out during the terminal Pleistocene when Lake Victoria, further upstream, ceased to overflow into the former's main affluent, the White Nile. The atmosphere-controlled lakes, which are situated in relatively humid areas, have probably overflowed quite frequently in their history, and although they react directly to small variations in P_L and E_L, are not as sensitive as amplifier lakes.

Groundwater-influent and -effluent lakes lacking surface outlets may undergo considerable fluctuations in extent, for example, Lake Naivasha.[32] There is often some doubt, however, as to whether their direct response to climatic change has not been obscured or damped by long-term variations in groundwater levels.[25,26] This problem is particularly acute where sea-level fluctuations are involved, for example in the karstic areas of Florida and Yucatán.[33]

Topography

There is an enormous variety in the origin, relief, and longevity of closed lake basins.[4] Some general principles can, nevertheless, be established regarding the sensitivity of different types to climatic change and the ease with which their fluctuations can be deciphered.[13,27] The most suitable basins according to both criteria tend to be those with moderately steep slopes, that is, those characterized by a small ratio of area to depth over a wide range of water depths.

Flat-floored basins are difficult to work with for a number of reasons, which are best illustrated by examples. The Basin of Mexico, although its total relief approaches 4000 m, has a very flat floor. Prior to the Aztec engineering works, there were several separate lakes of contrasting salinity which merged during the rainy season. As a result of the subdued relief of the basin floor, changes in lake area led to horizontal rather than vertical oscillations in the position of the shorelines. Thus the Late Quaternary sediments within the basin display very complex interdigitating relationships.[34] Moreover, the salinity at some sites actually *increased* as the brackish terminal lake Texcoco expanded to cover spring-fed freshwater marshes around the basin margins.[35]

In arid and semi-arid areas of gentle relief such as the Central Asian steppes,[36] the Texan High Plains,[26] and the Western Desert of Egypt,[37] the water table is liable to drop well below the basin floors during dry phases. The lake sediments are, therefore, subject to episodic oxidation and erosion by aeolian or fluvial processes. This is less likely to happen in basins with considerable relief.

At the opposite extreme, very steep-sided basins, especially those with a small surface area, may not be conducive to the preservation of a good sedimentary record, due to frequent sediment slumping and turbidity currents.

Large, fault-bounded troughs such as those found in the African rift valleys and the Basin and Range Province of the United States, and volcanic or meteoritic craters, often yield the best lake-level records, provided that the basin walls are not too steep and that tectonic disturbance has not been too great. Although a number of basins have been identified in which lake shorelines of Late Pleistocene or even Holocene age have been seriously affected by faulting or tilting,[38–40] the grounds for pessimism about tectonic perturbation of Late Quaternary lake-level records are not as serious as was feared a few years ago.[41]

Serious problems may, however, arise in cases where basin geometry or outlet levels have changed radically over time due to volcanic or geomorphic processes, for example through blockage by lava flows, alluvial fans, or sand dunes, or through fluvial erosion. For example, Lake Huleh in Israel, which was ponded by lava flows,[42] exhibited quite anomalous behaviour during the Late Quaternary[43] compared with other lakes in the Near East.[44] In some cases, even the reasons why an original, through-flowing drainage became blocked to form a lake may be far from clear. In several Saharan massifs, the present ephemeral wadis are lined by terraces of fine-grained lake deposits.[45] During the terminal Pleistocene–early Holocene, the streams were ponded—either by luxuriant swamp vegetation or by the growth of tufa barriers. This example does, however, illustrate the tendency of lakes to form under suitable hydrological conditions, even where no preexisting topographic barrier occurs.

The nature of the sedimentary record

Certain types of lakes leave behind an inadequate record of their fluctuations in the sedimentary column. For example, the deposits in deflation basins are often incomplete and highly variable from site to site.[46,47] Shallow, closed lakes in arid areas like the southern Mojave Desert may be so concentrated that they deposit only salts and playa muds[48] which are poor in pollen and other organic remains, and hence difficult to date by ^{14}C. Playa muds may be self-churning, which limits their stratigraphic value.[48]

Due to the heavy reliance of recent paleolimnological studies on radiocarbon dating, lakes which suffer from serious dating problems may yield only a very imprecise climatic record. For example, many lakes in the Middle East and the Great Basin have laid down massive deposits of marl and lacustrine tufa. The inaccuracy of ^{14}C dating of such carbonate materials, particularly those of Late Pleistocene age, has given rise to serious disagreements between investigators, not just over the timing of the minor fluctuations but even over the existence of the major events themselves.[49–52]

Human disturbance

In many parts of the world, human interference with lake levels has only become significant since the late eighteenth century.[53,54] Some of the great hydraulic civilizations in the Mediterranean, Near East, Central Asia, and Mesoamerica may, however, have had a limited impact on lake levels over several millennia, through the construction of dams, retaining walls, and irrigation canals.[35,55,56] The possibility of human interference should, therefore, be taken into account when historic or protohistoric fluctuations in lake level are considered.

Conclusions

The most suitable lakes to choose for a study of water-level fluctuations are closed, amplifier lakes, occupying volcanic, meteoritic, or fault-bounded basins with moderate relief. Very arid areas, areas of active tectonism, and karstic terrains are unlikely to yield a good paleoclimatic record. In the following sections, the term "closed lakes" will be used to refer to lakes which have lacked a surface outlet for most of their history, although some may have overflowed during water-level maxima.

TIME SCALES OF INVESTIGATION

The instrumental and historical periods

Closed lakes may serve as climatic indicators on time scales ranging from 10 to 10^6 years. Their versatility in this respect is probably unsurpassed. In view of the different techniques of investigation involved, the lake-level record can conveniently be divided into the instrumental, historical, and paleolimnological periods.

Lake-level measurements are very easy to make on a daily, weekly, or annual

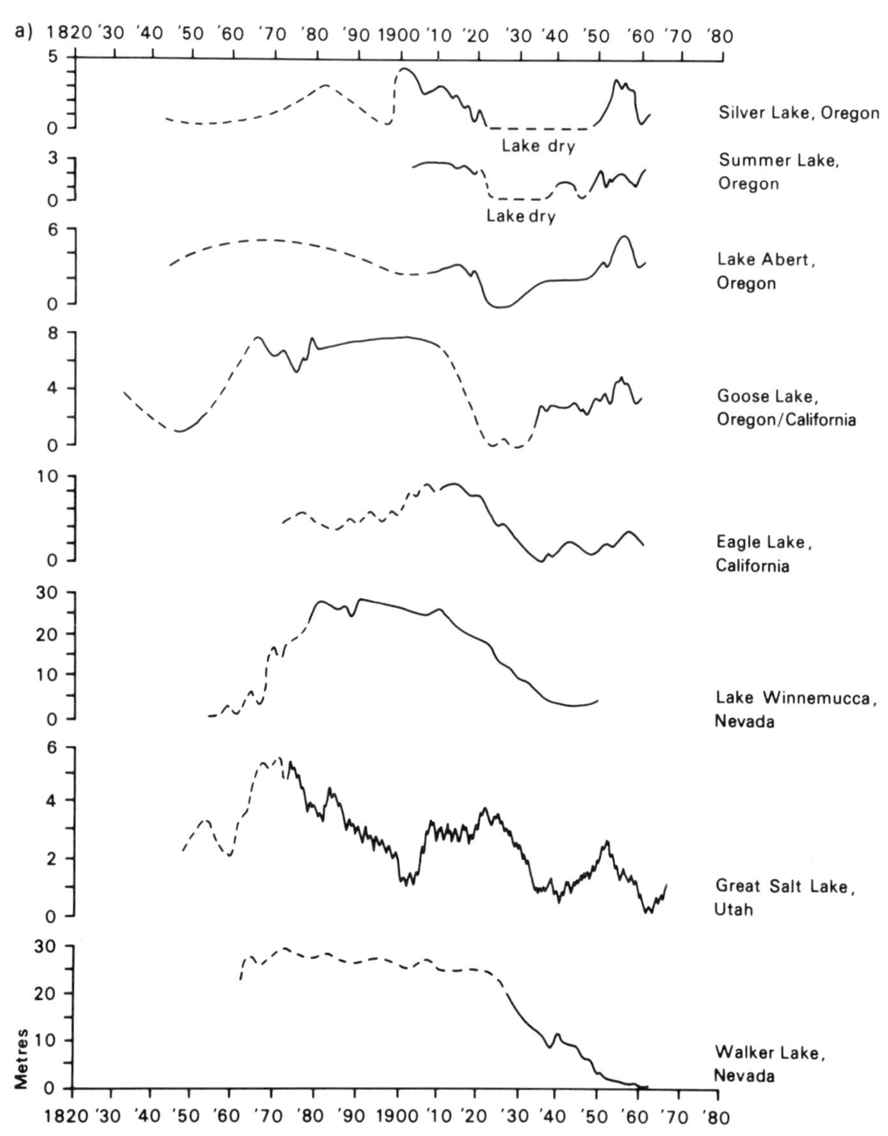

Figure 2. (*a*) Secular variations of North American lakes (solid: gauging station data; dotted: historical data) (from refs. 22, 54, 57). (*b*) Secular variations of lakes in Africa, South America and Australasia (solid: gauging station data; dotted: historical data) (from refs. 24, 58–62).

basis using a graduated staff or "gauge board." Instrumental records, therefore, often extend back to the nineteenth or early twentieth centuries, depending on the history of settlement in each area (figure 2). The dates of commencement of some important series are given in table 1a. Where detailed records exist, time-series analysis and correlations with climatic and/or hydrological data are possible.[66,58]

In many parts of the world, discontinuous records of lake-level change can be

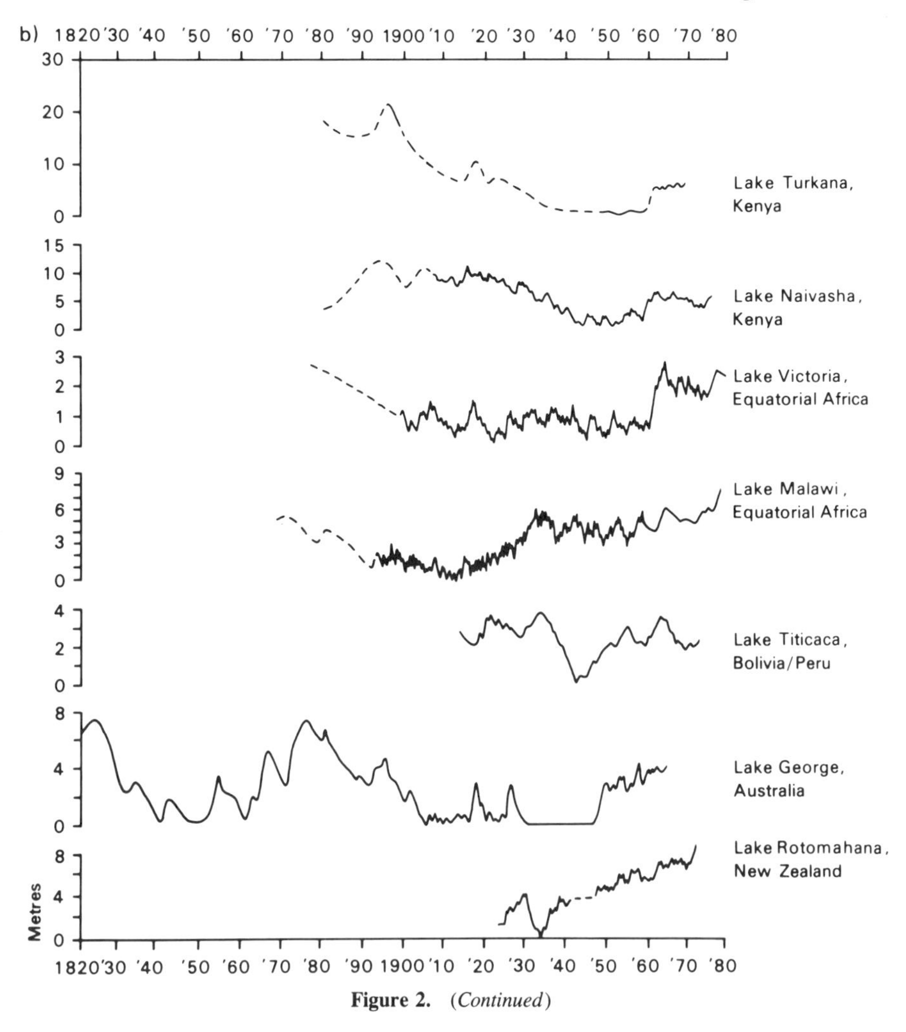

Figure 2. *(Continued)*

reconstructed from historical observations. Some long series from Europe and the USSR extend back to the early seventeenth century, for example the Neusiedlersee (1616).[17] Some other interesting records are given in table 1b.

If the historical and instrumental sequences are combined, strong resemblances between the behaviour of the lakes in individual regions become apparent[54,74,75] (figure 2*a, b*). On a world scale, it can be seen that many closed lakes that have been relatively free from human disturbance experienced high levels in the latter part of the nineteenth century, low levels in the 1920s to 1940s, and higher levels in the last few decades.[76] Certain lakes, of which the Dead Sea is the best-documented example,[77] display an almost opposite pattern of fluctuations. Nevertheless, the general similarity in behaviour to climatic records for the end of the Little Ice Age[67,78] suggests that closed lakes in many parts of the world have responded sensitively to global climatic fluctuations on a time scale of 10–10^2 years.

Table 1a. Instrumental Records of Lake-Level Fluctuations

Lake Basin	Latitude	Year	Reference
Caspian Sea, USSR	43°N	1851	63
Malheur-Harney Basin, Oregon, U.S.A.	43°N	1921	22
Eagle Lake, California, U.S.A.	41°N	1875	54
Great Salt Lake, Utah, U.S.A.	41°N	1880	57
Dead Sea, Israel	32°N	1800	51
Victoria, Equatorial Africa	1°S	1899	59
Malawi, Equatorial Africa	12°S	1895	64
Titicaca, Bolivia/Peru	16°S	1912	65
George, Australia	38°S	1819	61

Table 1b. Historical Records of Lake-Level Fluctuations

Lake Basin	Latitude	Year	Reference
Baikal, USSR.	53°N	1750	67, 68
Balkhash, USSR.	46°N	1820	67, 68
Aral Sea, USSR.	45°N	1820	67, 68
Caspian Sea, USSR.	43°N	First century A.D.	69
Malheur-Harney Basin, Oregon, U.S.A.	43°N	1826	22
Goose Lake, Oregon and California, U.S.A.	42°N	1832	22
Great Salt Lake, Utah, U.S.A.	41°N	1851	57
Pyramid Lake, Nevada, U.S.A.	40°N	1844	54
Mono Lake, California, U.S.A.	38°N	1857	54
Pang-gong Tso, Tibet	34°N	1811	70
Manasarowar, Tibet	31°N	1804	70
Basin of Mexico, Mexico	19.5°N	1325	53
Chad, Nigeria/Chad	13°N	1823	71
Valencia, Venezuela	10°N	1555	72
Tanganyika, Equatorial Africa	3°N	1846	73
Victoria, Equatorial Africa	1°S	1875	59
Malawi, Equatorial Africa	12°S	1860	64

Paleolimnological investigations

The value of fluctuations in lake levels as an indicator of climate has been recognized for more than 250 years.[1] In the mid-eighteenth century, for example, Persy and Buffon interpreted short-term fluctuations in the level of the Caspian Sea as a response to changing evaporation rates.[69] Humboldt's investigations of the Basin of Mexico[53] led him to identify climatic change as a major cause of historical variations in lake levels. Probably the earliest recognition of the former existence of a large lake in a now arid basin, however, was made by Vélez de Escalante[79] when he discovered fossil shells near Utah Lake in the course of his journey across the Bonneville Basin in 1776.

It was not until 1833, however, with the publication of Col. Julian Jackson's book *Observations on Lakes*,[80] that the study of lake-level fluctuations was placed on a firm scientific footing. Although it is difficult to estimate the impact of this book on the scientific community at the time, the now classic studies of palaeolakes in various parts of the world all postdate its publication. The existence of former lakes in Tibet and Kashmir was recognized by Strachey,[81] Godwin-Austen,[82,83] Blandford,[84] Drew,[85] Littledale,[86] Lydekker,[87] and Wynne.[88] Lartet[89] correlated high lake levels in the Dead Sea with the last glacial period, and attributed them to reduced evaporation. Early documentation of "pluvial" lakes in the western United States includes the classic studies by Whitney,[90] Simpson,[91] King,[92] Russell,[93-96] and Gilbert.[97,98]

The investigations by British geologists and surveyors in India and Tibet, which have been largely neglected in reviews of early paleolimnological work, are significant because of their recognition of various types of evidence for the past existence of more extensive lakes. Of these, "margin marks"[85] or abandoned strandlines, were probably the most important.[83,85,86] The degree of shoreline development was seen as an indication of the duration of periods of high lake levels, and of the intensity of wave action and hence the former size of the lake. The presence of lacustrine sediments,[83,85] fossils, and archaeological remains[85] was also used as evidence for variations in lake area/volume. Furthermore, it was understood that such variations were reflected in changes in water salinity and sediment chemistry[81,88] as demonstrated by the molluscan faunas associated with the palaeolake deposits.[83]

It is, however, the development of ideas about lake water balance and the relationship of lake-level fluctuations to changes in climatic parameters that makes the early studies in Kashmir and Tibet most interesting. Although Godwin-Austen[83] seems to have been the earliest writer to correlate lake-level fluctuations with the large-scale climatic changes of the glacial period in the Himalayas, Frederick Drew was the first to attempt to calculate the changes in the variation in climatic param eters necessary to produce an observed change in the z ratio. He attributed the decrease in size of the Salt Lake of Rupshu, Ladakh,[85] the waters of which "once extended (without counting the outlet valleys) over 60 or 70 square miles, having diminished their area to about eight square miles" to the fact that "the humidity of the climate must have been reduced to one-eighth of what it was formerly: that is to say, one-eighth of what it was when the lake first became a lake without effluence."[85] Moreover, he recognized that this reduction in lake area could have resulted either from decreased rainfall, or from increased temperature and hence increased evaporation, or from a combination of both factors:

> It does not follow that the reduction of precipitation between the above-defined epoch and the present time was as much as to one-eighth. One may suppose the precipitation to be one-fourth of what it was, and evaporation (area for area) to be twice what it was; this would account for the lake being one-eighth its former size. Or if the rate of evaporation has increased in the same

> proportion as precipitation has diminished, then $\sqrt{\frac{1}{8}}$, or nearly one-third, is the proportion. In general terms, however, it may be said that the climate is eight times as dry as before.[85]

The studies by Godwin-Austin and Drew, limited as they were by the lack of supporting information on the scale of global changes in the hydrological cycle, must therefore be seen as important landmarks in the development of paleoclimatology.

Methodology of investigation. Two contrasting approaches to the study of long-term fluctuations in lake levels have developed since 1890. The geomorphic approach is based on the accurate surveying and description of landforms such as wavecut cliffs and platforms, shoreline caves, beaches, spits, and deltas. Since the preservation of such features is related to their age, the range of this approach tends to be limited to the upper Late Pleistocene and Holocene. The geomorphic approach is often held to be scientifically inferior to the laboratory analysis of sediments and fossil remains collected from cores or sedimentary exposures. Thus it is important to note that all existing climatic calibration techniques require accurate measure ments of *paleolake area* (cf. equation 4). Stratigraphic techniques provide better resolution in time, but frequently only yield curves of relative variations in depth or salinity, from which variations in relative extent must be deduced. Geomorphic techniques are also more likely to reveal serious changes in basic topography due to faulting or erosion.[39]

The soundest and most rigorous method, therefore, is to combine the two approaches to achieve maximum precision in both space and time. Following the pioneering work of Russell[93] and Gilbert,[98] a number of other studies have attempted to make full use of both geomorphic and stratigraphic information.[25,99–104] The combined approach has the additional merit that numerous opportunities arise for cross-checks on the consistency of different types of evidence. But the results of these comparisons, like those of the individual techniques employed, depend heavily on the resolution and accuracy of the dating framework.

Dating techniques. Good dating control is essential to the reconstruction of detailed lake-level curves because sedimentation rates in fluctuating lakes tend to be extremely variable, and because superimposition of shorelines of different ages is the rule rather than the exception. Lakes, except in highly saline environments, offer a wide range of materials suitable for geochronometric dating compared with, for example, glaciers or dune fields.[105] Only a few of the most useful techniques are discussed here. The method most widely used for dating Late Quaternary lake-level fluctuations is still ^{14}C. At the moment, it is seldom used with success on sediments older than 30,000 years B.P. Many carbonate dates >20,000 years B.P. are highly suspect.[106,107] The advent of ^{14}C dating by mass spectrometry may eventually double the range of the technique,[108] provided that problems of contamination can be overcome. Although dates on wood, charcoal, lake mud, or peat are greatly more reliable than carbonate ages, the latter are frequently used because of the abundance of carbonate minerals in closed lake deposits (see section on Geochemistry).

Two serious problems afflict radiocarbon assays on carbonates such as shell, marl and lake tufa, and organic materials originating through aquatic photosynthesis; the incorporation of dead carbon derived from limestone or other old reservoirs, and contamination by young (particularly post-bomb) carbon carried by percolating waters.[109] Areas of calcareous sedimentary rocks are liable to be severely affected by both types of error; although contamination by young carbon can often be minimized by careful selection of samples.[110] Volcanic lithologies appear to yield the most consistent dates on carbonates,[31,111] provided that juvenile CO_2 from hot springs is not a serious problem.[112] Even under the most favorable conditions, however, the analytical and other errors associated with radiocarbon dates on sediments of mid-Holocene or greater age are such that it may not be possible to resolve the most rapid fluctuations in lake level.[111] This should be borne in mind when less precise dating techniques are considered.

Other isotopic methods are being used to an increasing extent for dating lake sequences. Uranium-series dates on shell are, in many cases, not very precise, but they have played an important role in identifying deposits which lie beyond the range of ^{14}C[52,113,114] and in assigning them to broad age categories, for example, last interglacial. The $^{230}Th/^{234}U$ technique, in particular, has recently been applied with some success to lacustrine evaporites[115] and peats.[116] Although ^{210}Pb and ^{137}Cs dating are commonly used to provide a chronology for very young ($\sim 10^2$ yr) lake sediments,[117] they have not yet been employed in the study of lake-level fluctuations. These techniques may prove valuable in the future, however, in areas with little historical data.

Volcanic areas are particularly fortunate in that a wide variety of geochronometric techniques are applicable to lava flows and pyroclastic deposits. K/Ar and $^{40}Ar/^{39}Ar$ dating of tuffs and lavas intercalated with the thick sedimentary sequences in the lake basins situated along the Eastern Rift System of Africa have provided a framework for the study of human evolution and climatic change over the last 4 m.y.[118–120] These methods, however, cannot be used with any degree of accuracy on volcanics of Late Quaternary age. Tephra horizons dated by ^{14}C, K/Ar and/or fission-track methods make excellent marker horizons.[121–123] For example, the far travelled Pearlette "O" (0.6 m.y.) and Bishop (0.7 m.y.) ashes were used as reference levels in the Burmester core from Lake Bonneville, Utah.[124] Their occurrence confirmed the presence of the Brunhes-Matuyama boundary in the core and hence, the validity of its paleomagnetic reversal chronology (see below).

Magnetostratigraphy is being increasingly applied to provide a chronology for lacustrine sequences, although it is important to note that it is not a dating technique in the strictest sense, but rather a means of making precise correlations. Reversals of the earth's magnetic field have occurred on a time scale of 10^4–10^6 years. Two important records recently dated by reference to the geomagnetic polarity sequence are the KM-3 core from Searles Lake, California,[125] which has an estimated basal age of 3.2 m.y., and a 36 m core from Lake George, Australia (3.25 m.y.).[126]

On a shorter time scale (10^2–10^4 yr), secular variations of declination and inclination *within* the major geomagnetic chron are very useful for dating continuous cores of lake sediments, particularly over the last 12,000 years.[117,127,128] It is important to note that since the earth's magnetic field at any place comprises both dipole and non-dipole components, this method can only be used to correlate sequences on a regional scale. A master curve for any region is built up by measuring several cores from each of a number of lakes. It is then calibrated by radiometric or other methods.[129] This technique holds great promise for dating cores of Late Quaternary sediments from karstic or other lakes where the radiocarbon chronology is suspect. In conjunction with mineral-magnetic measurements such as magnetic susceptibility, it can also be used to correlate sequences from one part of a lake basin to another.[117,130]

A further technique which shows considerable potential is amino-acid racemization dating of shell. In order to derive quantitative age estimates from amino-acid ratios, it is necessary first to make assumptions about past temperatures or to calibrate the technique using another geochronometric method. Because of these limitations, the technique is only being used at present as an indicator of relative age.[50]

PALEOLIMNOLOGICAL EVIDENCE USED TO RECONSTRUCT LAKE-LEVEL FLUCTUATIONS

In this section, the various paleolimnological techniques used to reconstruct lake-level fluctuations are reviewed, and their relative advantages and disadvantages are compared.

Geomorphology

The main role of geomorphic techniques is two-fold: to provide estimates of the area and volume of a paleolake for use in climatic or hydrological modeling, and to identify any changes in basin response resulting from overflow, tectonism, or other causes.[131,132] Attention, therefore, is usually focused on lake shorelines. Classic examples of Pleistocene strandlines are illustrated in Mifflin and Wheat.[10]

In lakes, as in the ocean, the character and degree of development of shoreline features are dependent on the slope, the wave-energy regime, the properties of the rocks or sediments forming the lake bed and margins, and the sediment supply.[97,111] However, since tidal effects are negligible and the fetch is often small, the transition from backshore to profundal environments usually takes place over a much narrower vertical range.

Large lakes cause the lithosphere beneath them to be isostatically depressed in a similar way to ice sheets. During low stands, isostatic rebound causes the shorelines in the center of the basin to bulge upwards. It is necessary to take this into account when interpreting the record of substantial paleolakes like Bonneville.[98,133]

Certain distinctive typcs of lake-shore features have received special attention. These include lunette dunes and lacustrine tufas. Many small lakes or "pans" in semi-arid areas are bordered on the downwind side by lunette dunes. These are crescentic ridges, consisting of either sands or clay pellets deflated from the adjacent lake basin. The sands are thought to originate from beaches during phases of high lake level and the clay pellets from exposed mud flats during low stands. Lunettes are common around lakes in western and southeastern Australia,[102,134] the Kalahari,[135] the Texan High Plains[136] and southeast Oregon.[137] A number of authors have used the orientation of lunettes to map paleowind regimes.[136,138] Bowler[138] has gone further; he advocates clay dunes as an indicator of hot, windy conditions and strong evaporation at the time of formation. This hypothesis, however, requires revaluation in view of the occurrence of clay dunes in Oregon and other temperate areas.

Another distinctive shoreline feature of many closed basins is algal tufa, which often takes the form of large, concentric growths called *stromatolites*.[49,110,139] Tufa is characteristic of sites with low rates of clastic sediment accumulation in alkaline lakes with waters that are being moderately, but not strongly, concentrated by evaporation (see section on Geochemistry). It has been extensively sampled for ^{14}C, U-series and oxygen-isotope analyses. The success of these techniques has been limited by the notorious tendency of porous tufas to re-crystallize and incorporate younger carbon.[49,110]

Other materials frequently used to date shoreline features include shells, wood, and charcoal. Care needs to be exercised to ensure that these relatively robust materials have not been reworked from older sediments.[141]

Facies relations and sedimentology

The basic rock-stratigraphic techniques developed by geologists have been widely applied to the study of closed-basin lake sediments. They are an essential tool in the reconstruction of water-level fluctuations from cores, but can also be used to investigate natural or artificial exposures. The basic principles are elegantly illus trated in the award-winning monograph on Olduvai Gorge, Tanzania, by Hay.[118]

According to Walther's Law,[142] "the facies associations which we see forming side by side today, are commonly found in the rocks both side by side and in vertical succession." Given the rapidly changing environments of closed-basin lakes, cyclic alternations of facies are to be expected in the sedimentary column (figure 3). If the sediment types represented can be attributed to particular lacustrine environments,[143-145] relative changes in water depth can be reconstructed. For example, it is common to find an upward passage from profundal muds, diatomites or marls to shallow-water sands, followed by beach or marsh sediments and ultimately, terrigenous deposits. Where an abrupt transition occurs between facies not found side by side in nature, an erosional break can be suspected.

Erosional unconformities, if traced down into the lake basin, often provide useful control on the maximum height of the lake surface during low stands.[99,111] Attempts have also been made by Morrison to use soils as stratigraphic markers in the Great Basin. The application of new dating techniques in the Bonneville Basin suggests that substantial errors may result from the miscorrelation of similar soils of different ages.[50]

Examples of the application of sedimentological and pedological criteria in the reconstruction of lake-level fluctuations from long cores are provided by Eardley et

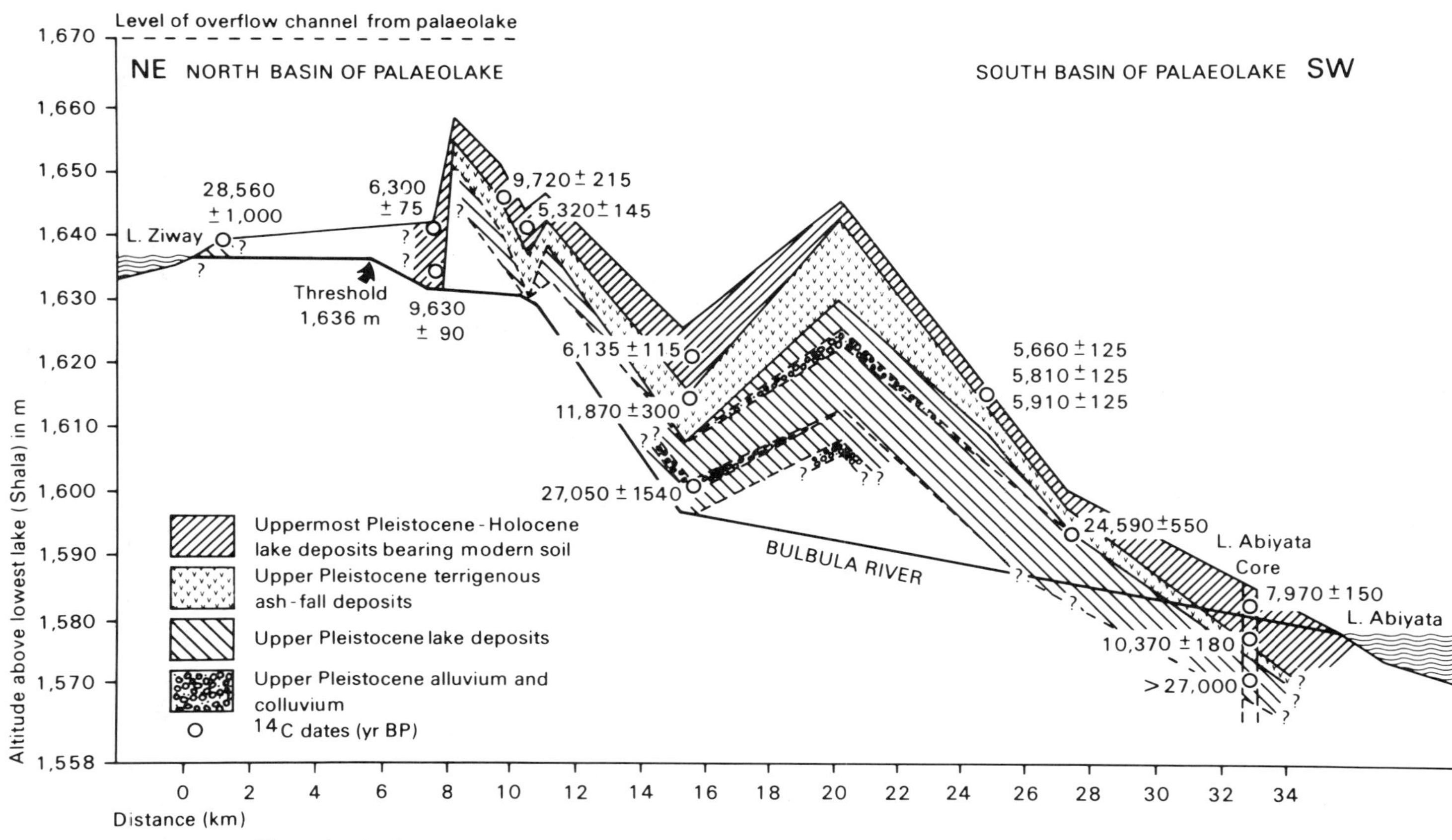

Figure 3. Cyclic alternation of facies in a closed lake basin; Ziway-Shala, Ethiopia (from ref. 111 and unpublished data).

al.[124] and Singh et al.[126] The criteria employed by Eardley et al.,[124] in their study of the Burmester core from Lake Bonneville, included grain size, color (reddish coloration was attributed to oxidation under subaerial conditions), carbonate content (sediments containing >40 percent carbonate were believed to have formed in warm, shallow water), and the presence of plant root casts. Singh et al.[126] used a combination of grain size, structural features and faunal content to distinguish lacustrine, colluvial, and pedogenic facies in a core from Lake George, Australia. Phases of past pedogenic alteration, representing dry episodes, were deduced from the following lines of evidence: obliteration of primary depositional structures; development of desiccation cracks, peds, and cutans; segregation of carbonate and iron; intrusion and preservation of plant-root structures; and soil horizonation. Because some of these features, particularly color and grain size, may be ambiguous,[146] it is important to use as wide a range of criteria as possible.

Geochemistry

Most geochemical techniques for reconstructing lake-level variations depend ultimately on the simple observation that the salinity and alkalinity of a closed lake vary with its volume[3] (figure 4a). This relationship is neither linear nor constant through time. The behavior of salinity during a lake-level cycle is frequently hysteretic (figure 4b), due to losses of salts by sedimentation and deflation[3] or to recycling from evaporite crusts and saline soils.[148] The example of Great Salt Lake[149] shows how the development of a topographic barrier across a lake basin (in this case an artificial causeway) may also influence brine evolution.

As the volume of a closed lake is reduced by evaporation, its chemical and isotopic composition evolve systematically, depending on the starting composition. The particular geochemical pathway followed is thus primarily determined by the lithology of the catchment.[148]

Figure 5 summarizes the Eugster-Hardie model. Despite its apparent complexity, the various geochemical pathways all exhibit two main evolutionary stages: first, the precipitation of alkaline earth carbonates,[150,151] and then the crystallization of other salts and silicates from the residual brine. This behavior is exemplified by the Late Quaternary record of Searles Lake, California, which lies on pathway IIIA2 (figures 5 and 6).

Searles Lake has varied in size from the present saline playa (~ 100 km^2) to a large, overflowing lake with an area of ~ 1000 km^2 and a maximum depth of 200 m.[101] Its sediments consist of a cyclic alternation of marls (predominantly Ca,Mg-carbonates) deposited under relatively dilute conditions, with beds of saline minerals such as halite, trona, and nahcolite, precipitated from brines (figure 6). The annual input of water into the lake is estimated to have increased at least 3–6 fold relative to today during episodes of marl deposition.[27]

Examples such as the one given above suggest a possible link between the sedimentation and water balance of closed lakes. Figure 7 is a preliminary attempt to test this hypothesis by plotting data on lake-sediment type on a water-balance diagram, for the few basins for which both kinds of data are readily available. Lakes

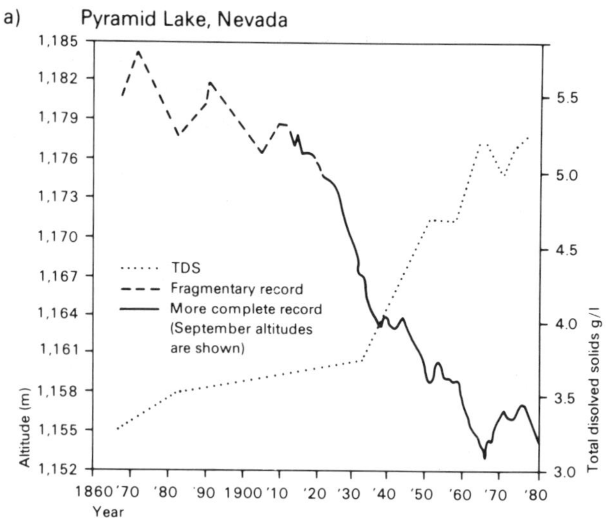

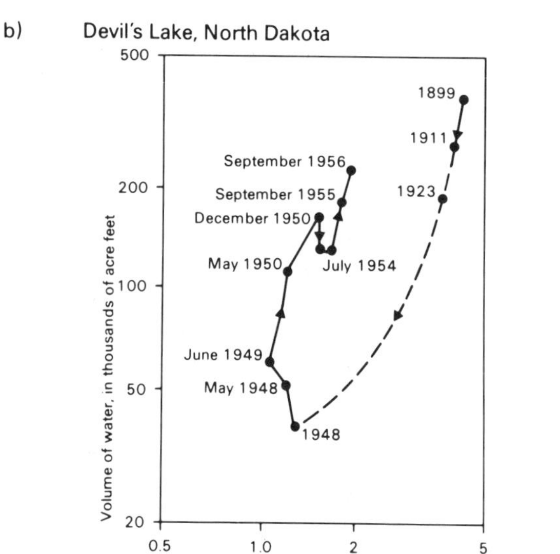

Figure 4. Secular variations in total dissolved solids as a function of changes in lake size. (*a*) Pyramid Lake, Nevada[147] (*b*) Devil's Lake, North Dakota.[3]

depositing calcareous precipitates appear to occupy a triangular field in quadrants RP-DE, P-E (upper part), RP-E and R-E. They are moderately alkaline, with intermediate flushing rates and values of the aridity index, *a*, of ~ 1-9. Dilute lakes with low-carbonate sediments fall mainly into the reservoir and atmosphere-controlled categories.[13] There is, however, considerable overlap. This may reflect different geochemical pathways, groundwater influence, or simply, inadequate data.

From the scraps of information currently available, it appears that most brine lakes which deposit saline minerals plot in quadrant R-E of figure 7, along the

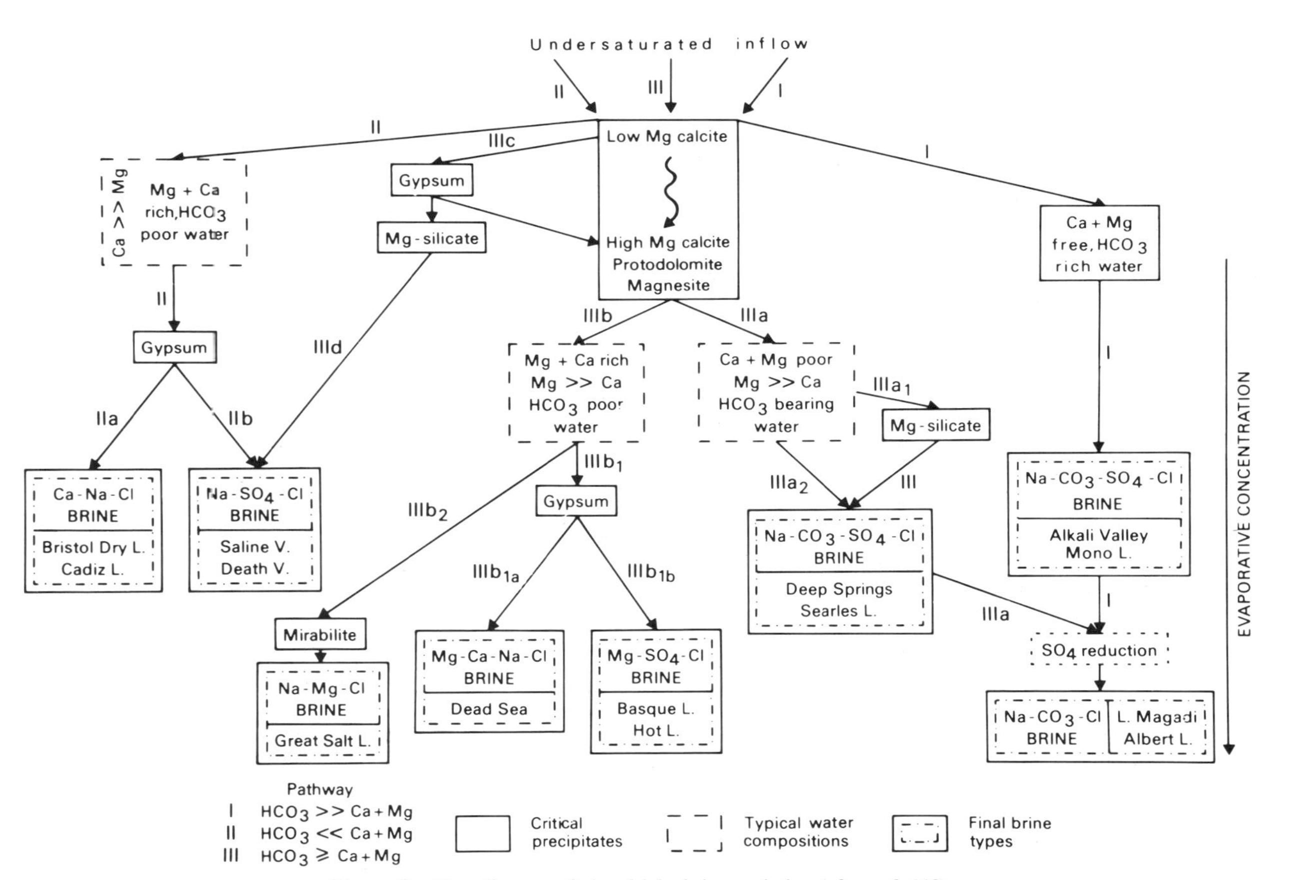

Figure 5. Flow diagram of closed lake brine evolution (after ref. 148).

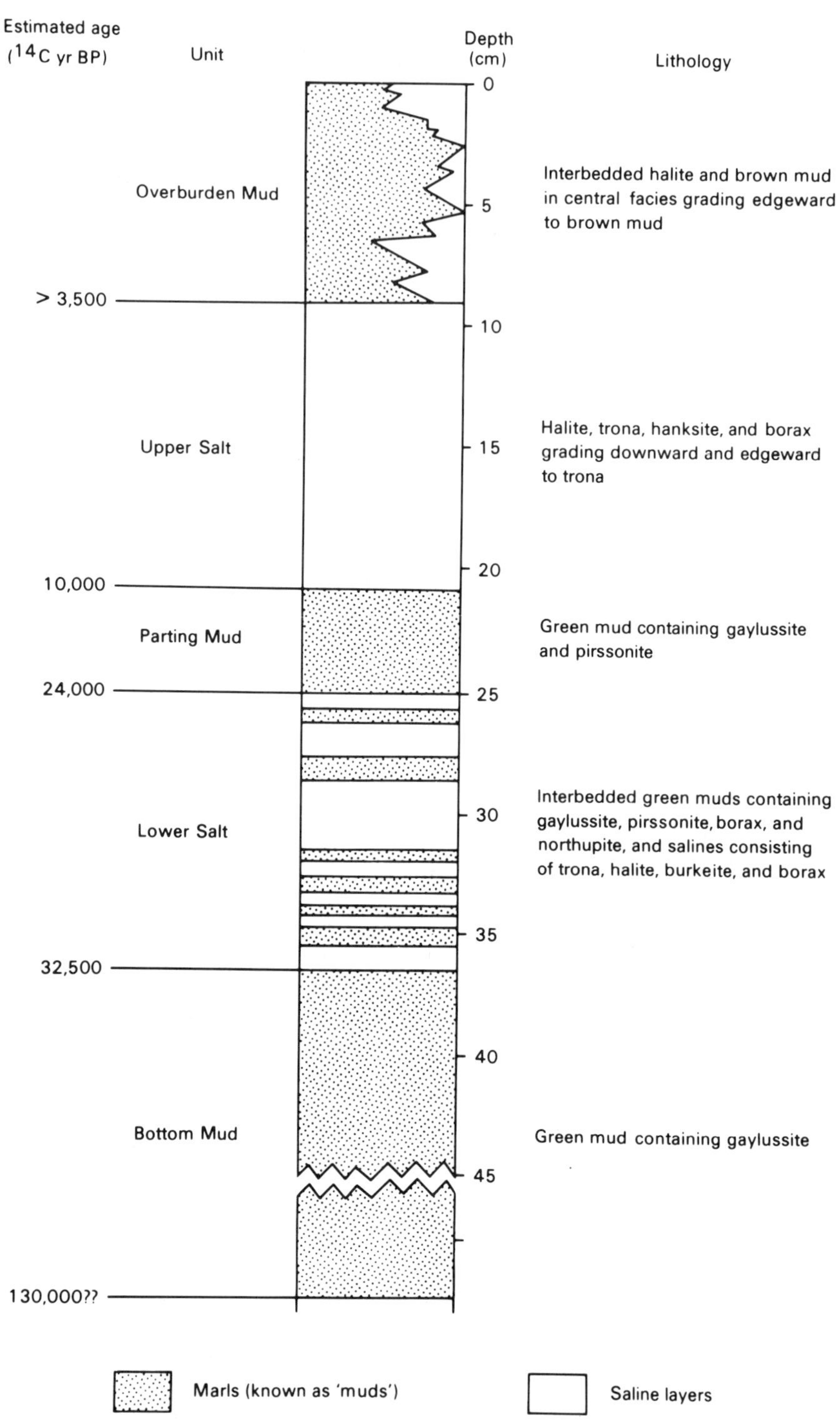

Figure 6. Late Quaternary evaporite sequence of Searles Lake, California (after ref. 101).

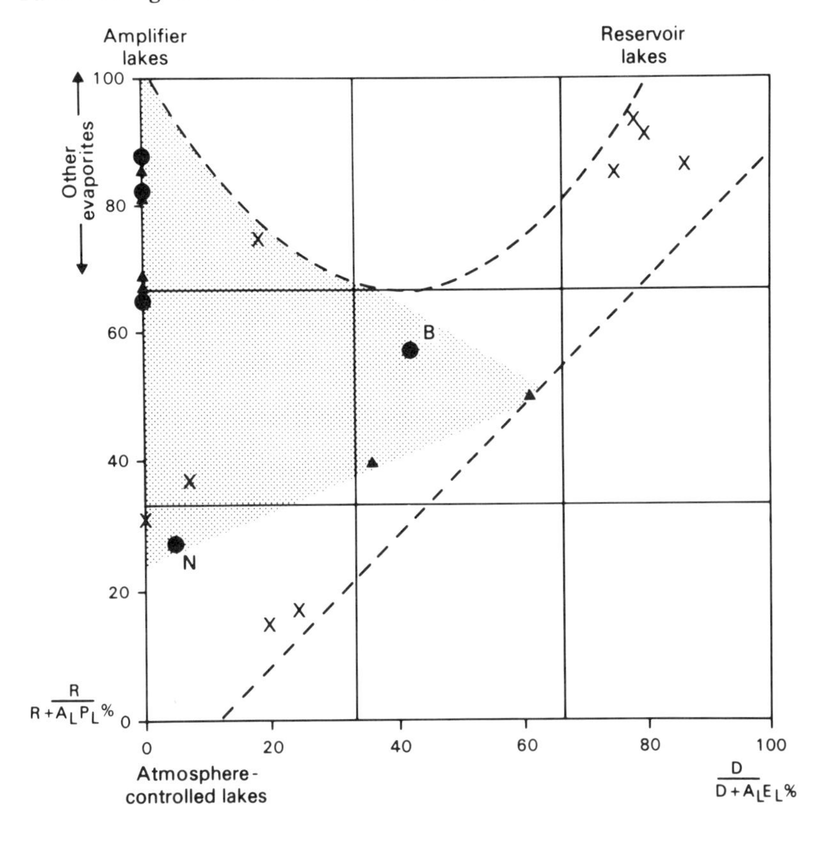

Figure 7. Water-balance field occupied by lakes which precipitate Ca, Mg-carbonate sediments (from references cited in figure 1 and numerous additional sources).

vertical axis. These are predominantly closed, amplifier lakes exhibiting a very marked degree of evaporative concentration, although closed, groundwater-effluent lakes like Tyrrell[152] and Asal[25] are also common.

In attempting to develop this model further, it will be necessary to take account not only of the different geochemical pathways followed by modern lakes, but also of the modification of sediment composition, after burial, by pore-water circulation and diagenesis.[148,153] In Searles Lake, for example, the marl layers contain the diagenetic minerals gaylussite and pirssonite (formed by the reaction of Na-carbonate brines with Ca-carbonate minerals), zeolites, K-feldspar, and borosilicates.[101] Zeolites, analcime, clay-sized mica, and chert are particularly characteristic diagenetic components of saline lake sediments.[148,154] Although the first three have been used as paleosalinity indicators in East Africa by Singer and Stoffers[155] diagenetic minerals are unlikely to reflect rapid changes in water level because of the unknown and possibly variable lag time involved in their formation. Nevertheless, Cerling[156] has attempted to use the Na:Ca ratio of exchangeable cations from the Plio-Pleistocene clays of Lake Turkana, Kenya, as a paleoalkalinity indicator. His method relies on the questionable assumption that diagenesis and even modification through exchange with groundwater have been negligible. It also assumes evolution along pathway I in figure 5.

A more widely applicable technique with a direct hydrological basis is oxygen-isotope analysis.[157] The $^{18}O/^{16}O$ ratio in closed lake waters has been observed to vary as a function of the degree of evaporative concentration.[158] In low latitudes, temperature is a less important influence then in high latitudes and the main control on the ^{18}O content of saline waters may, therefore, be the relative rates of inflow and evaporation.[33,159] At very high salinities, however, the $^{18}O/^{16}O$ ratio tends to decrease, reflecting, among other factors, the suppression of evaporation over saturated brines.[158,159]

Paleoisotope data can be obtained from $CaCO_3$ in shells or inorganic carbonates or from apatite in fish bones. For example, Covich and Stuiver[33] made $^{18}O/^{16}O$ assays on shells in a core from a closed, karstic lake in Yucatán and found that they correlated well with chemical indices of water-level fluctuation. Abell[160] has recently derived a somewhat schematic palaeoevaporation curve for Lake Turkana for the last 2 m.y. from the analysis of three different shell species.

A final, speculative technique which has been used to infer past water-level fluctuations is the geochemical reconstruction of changes in thermocline depth. Many tropical lakes deeper than 50–100 m exhibit thermal stratification.[59] Below the zone of rapid temperature decrease, or thermocline, the deep water is anoxic and strongly reducing. Mn^{2+} and Fe^{2+} accumulate there and migrate into the upper, aerated water mass, where they precipitate to form minerals such as manganosiderite [(Mn, Fe)CO_3] which is only stable over a narrow *E*h-*p*H range. Manganosiderite is accordingly only incorporated into the sediments in the vicinity of the thermocline.[161] Fluctuations in the depth of Lake Kivu in equatorial Africa, were inferred by Hecky and Degens[20] and Stoffers and Hecky[162] from the alternation of oxic and anoxic (sulphide-rich) facies in cores. High concentrations of Mn were taken to indicate the paleothermocline (figure 8). This method should not be incautiously applied to sediments where the former bathymetry is unknown, since anoxic conditions often develop in shallow, stratified brine lakes such as Bogoria, Kenya.[163] Other factors such as windiness, evaporative concentration, temperature changes, and underwater spring activity may also influence thermocline stability.

Paleoecology

A wide variety of paleoecological techniques has been applied in lake-level studies. Two approaches are common. Fossils with known depth habitats may be used to indicate past water-surface elevations directly. Alternatively, where aquatic organisms have restricted salinity ranges, fossil assemblages may provide a relative

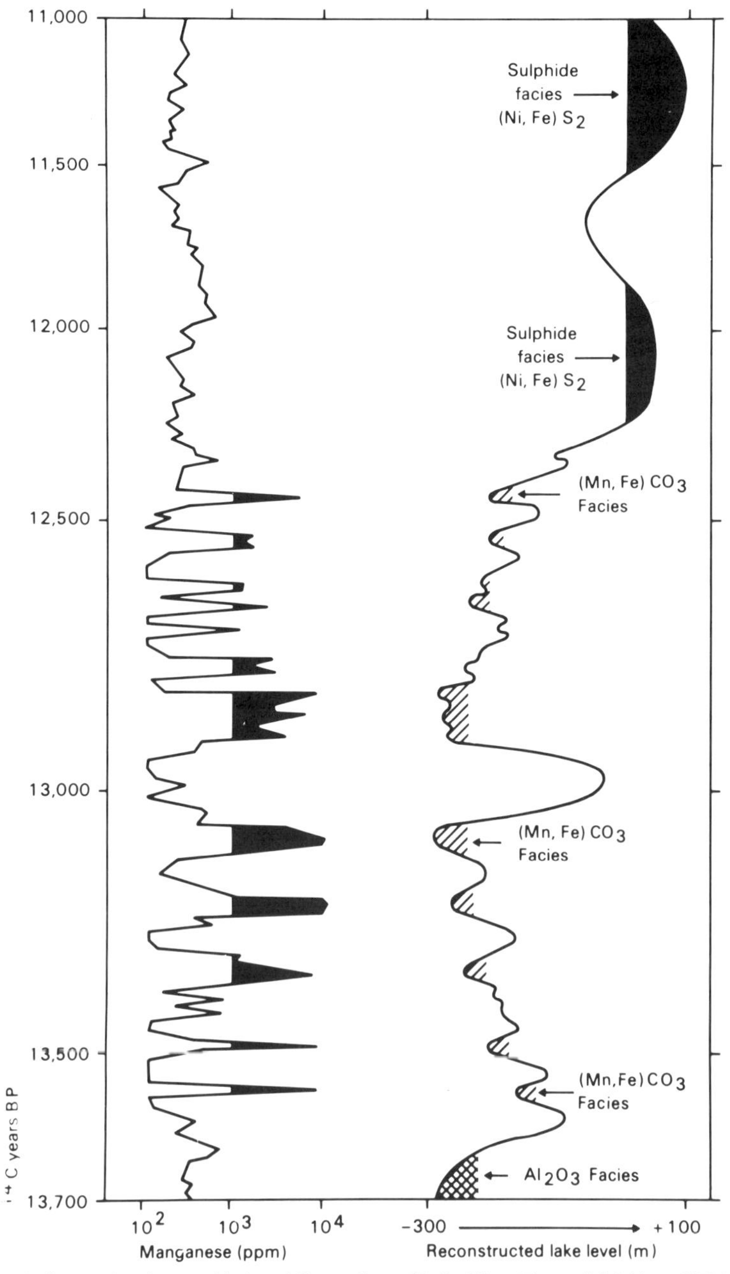

Figure 8. Sediment chemistry and inferred fluctuations of Lake Kivu, Equatorial Africa, 13,700–11,000 years B.P. (after ref. 20).

indication of lake-level change. Only a few of the more commonly used techniques will be reviewed here.

Perhaps surprisingly, pollen analysis is one of the least useful methods. Under favorable circumstances, the relative abundance of pollen grains produced by aquatic vegetation may prove an index of the proximity of the lake fringe.[164] Maxima of weedy species, particularly salt-tolerant types such as chenopods,[6] may sometimes occur during low stands when expanses of dry lake floor become available for colonization by plants. Seeds, leaves, and other plant macrofossils, where preserved, can often be identified to species level, unlike most pollen types, and therefore provide a much more powerful tool for making inferences about past water levels and salinities.[6,165]

Diatom analysis is probably the most widely used and versatile paleoecological technique. Diatoms occur in great abundance in the sediments of lakes with an adequate silica content, except for very alkaline or karstic lakes in which their frustules tend to dissolve. They have well defined habitats and salinity-alkalinity ranges, and can be identified with considerable taxonomic precision.[166] Diatom analysis, therefore, supplies mutually supporting information on water depth and chemistry. It has been widely employed in the silica-rich volcanic lakes of eastern Africa and Mexico.[31,32,35,167] In areas where the modern ecological requirements of the diatom flora are well known, quantitative or semi-quantitative calibration of fossil diatom assemblages is possible. For example, Richardson et al.[168] used the modern alkalinity ranges of diatoms in East African lakes on pathway I (figure 5) to calculate a paleoalkalinity curve for Lake Manyara, Tanzania. Calibration in terms of past water depths has even been attempted by Churchill et al.[169]

Analyses by Gasse et al.[170] suggest that modern African diatom floras are related to the water-balance categories shown on figure 1. It should eventually be feasible to use diatom assemblages in cores as a means of estimating the gross water-balance characteristics of paleolakes (cf. figure 7). The results may help to explain the surprising ecological parallelism exhibited by many tropical African lakes during Late Pleistocene and Holocene times.[167]

Charophytes, or stoneworts, are a group of submerged plants of which some species are adapted to life in saline lakes. Their calcified female reproductive organs, or oogonia, occur widely in closed-basin deposits in Australia,[138] the American Southwest[171] and the Sahara.[114] Burne et al.[172] investigated the ecology of charophytes living in NaCl-rich ephemeral lagoons in Australia. They concluded that the association of evaporites and oogonia is characteristic of brine lakes in the semi-arid, Mediterranean climatic zone that receive seasonal inputs of dilute continental groundwater. Charophytes may therefore prove to be a useful palaeohydrological indicator.

Ostracodes, molluscs, and fish remains are increasingly being employed to make quantitative estimates of paleosalinity. Of the three, only ostracodes rival diatoms in abundance and environmental sensitivity.[173,174] One effective strategy is therefore to pool information derived from different types of faunal evidence. Cohen[120] following earlier work by Cerling, has attempted to draw up a water-chemistry curve for Lake Turkana, Kenya, for the last 4.3 m.y., based on ostracode assemblages supplemented by mollusc data. This curve is in broad, but not detailed, agreement with the paleoevaporation curve published by Abell.[160]

Marine foraminifera are occasionally found in the deposits of brackish or saline lakes with a high NaCl content (pathways II and III; figure 5). Where present, forams can be used as a check on paleosalinities deduced from other indicators.[171,114]

In summary, a variety of paleoecological techniques are being used to reconstruct fluctuations in lake depth and/or salinity. By using a number of techniques in combination with geochemical data, it is possible to build up a very precise and internally consistent picture of water-level change, such as the recent multidisciplinary study of Lake Valencia, Venezuela.[175]

Archaeology

In many arid and semi-arid areas, freshwater and brackish lakes furnished important water and/or food resources for prehistoric cultures. Lake shorelines often exhibit a high density of archaeological sites compared with the surrounding hinterland.[176] The existence of a permanent water body may be confirmed by the presence of aquatic food refuse, for example fish or waterfowl bones, and artifacts such as nets, net weights, or harpoons.[34,177] A considerable amount of information on paleolake size and ecology has been derived from the distribution and character of archaeological sites, notably in the Sahara,[37] the American Southwest[176,178] and Australia.[102]

DATA NETWORK FOR THE PAST 30,000 YEARS

Problems of Coverage and Comparability

The distribution of published lake-level information is extremely uneven in its time and space coverage (figure 9), even allowing for the patchy past and present distribution of closed basins. This situation reflects not only disparities in geological awareness and scientific resources between different countries, but also the difficulties of obtaining data from regions such as Mexico, the Middle East, northern Australia, and Tibet, where problems of accessibility, stratigraphic complexity, and/or lack of good dating materials are particularly acute.

The Tropical Palaeoenvironments Research Group in Oxford maintains a large data bank of published water-level information from lakes which have been closed during all, or part, of their Late Quaternary history.[179] Only sites with a chronology based on ^{14}C, dated tephra layers, or historical sources are included. The spatial distribution of the data is summarized in tables 2 and 3.

The great predominance of Africa in this data set is immediately apparent (table 2). It has furnished by far the largest number of dated sites, a high proportion of which represent continuous records (e.g. cores, exposures) rather than discontinuous ones (e.g. dated shorelines). Many of the African basins also have long sequences

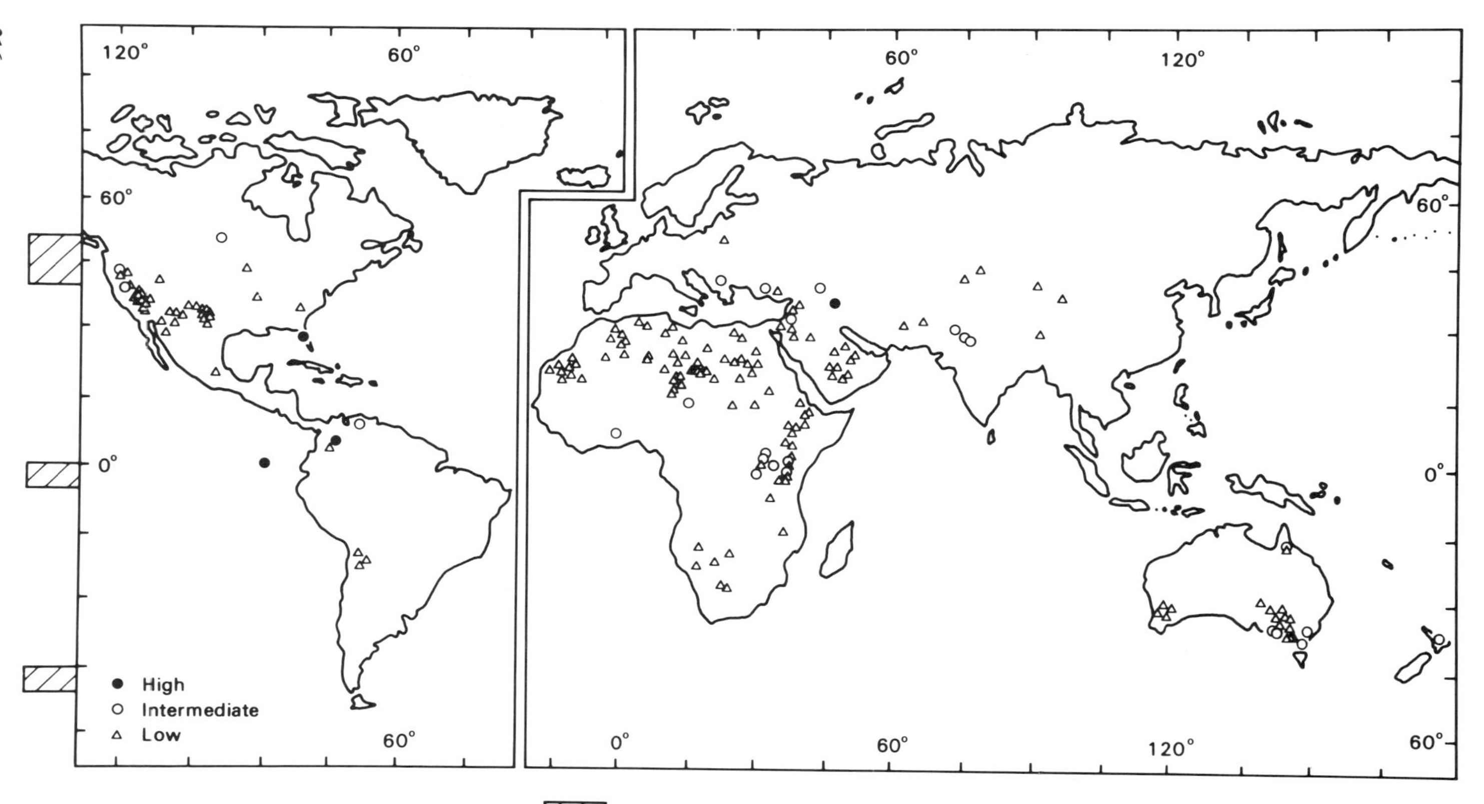

Figure 9. Global map of lake-level status of the present day. Latitudinal peaks in area extent of all lakes after Schuiling.[12]

Table 2. Composition of the Oxford Lake-Level Data Bank by Regions

Region	Total Number of Basins	Type of Record		Length of Record			Radiocarbon Control			
		Continuous	Discontinuous	< 10,000 years	10,000–20,000 years	> 20,000 years	< 5 dates	5–9 dates	10–19 dates	> 20 dates
Africa	90	23	67	46	20	24	62	13	6	9
Middle East	21	3	18	3	4	14	15	2	1	3
S. Central Asia	7	3	4	5	0	2	7	0	0	0
N. America	39	7	32	4	13	22	20	8	5	6
S. America	11	5	6	1	3	7	7	2	1	1
Europe	5	0	5	2	0	3	3	0	1	1
Australasia	25	11	14	7	7	11	14	6	2	3
Total	198	52	146	68	47	83	128	31	16	23

Table 3. Composition of the Oxford Lake-Level Data Bank by Latitude

Latitude Belt	Total Number of Basins	Type of Record		Length of Record			Radiocarbon Control			
		Continuous	Discontinuous	< 10,000	10,000–20,000	> 20,000	< 5 dates	5–9 dates	10–19 dates	> 20 dates
N of 45°N	3	0	3	2	0	1	2	1	0	0
35–45°N	28	5	23	3	6	19	14	4	4	6
25–35°N	49	10	39	16	11	22	35	6	4	4
15–25°N	49	9	40	31	7	11	36	7	2	4
5–15°N	14	7	7	3	5	6	3	4	2	5
5°N–5°S	16	8	8	3	7	6	13	2	1	0
5–15°S	4	3	1	2	0	2	3	0	0	1
15–25°S	12	3	9	4	3	5	8	2	2	0
25–35°S	10	1	9	1	5	4	8	1	0	1
35–45°S	13	6	7	3	3	7	6	4	1	2
S of 45°S	0	0	0	0	0	0	0	0	0	0
Total	198	52	146	68	47	83	128	31	16	23

with good ^{14}C control. In contrast, Central Asia, and Central and South America are very sparsely covered. There is surprisingly little data for Europe, which reflects the academic dominance of palynology and a persistent tendency to ignore lake-level fluctuations or to date them exclusively by reference to pollen sequences.[16,180]

The latitudinal distribution of published data is dominated by the zone 15–45°N (table 3), reflecting the concentration of dated sites in the Sahara, the Middle East, and the Great Basin. The Southern Hemisphere land area is still poorly covered, as is the area north of 45°N.

Attempts at Standardization

Spatial regularities in the pattern of fluctuations in the extent of closed lakes through time have been noted for more than a decade.[181–187] In order for progress to be made in the analysis of the causes of these regularities, it is necessary to standardize the data in some way. Changes in lake size are reconstructed using a wide variety of techniques (see preceding section on Paleolimnological Evidence), many of which yield data on lake depth or salinity rather than the variable most directly controlled by climate, lake area (see equation 4). Many methods provide only semi-quantitative information. Moreover, the fluctuations in each basin have to be considered in relation to its own internal range of variation.

In order to apply the numerical calibration techniques described above, it is necessary to provide, as input, quantitative estimates of paleolake area and a number of other parameters such as paleotemperature or albedo, depending on the type of model used.

In regions where sufficient data are available to compute the z ratios of lake basins for a given time slice, it is possible to map quantitatively spatial variations in the relative extent of lakes. Figure 10 is a first attempt to do this for the lakes in the Great Basin during the last glacial. It shows high z values for the basins fed by rivers from the well-watered mountain ranges of Oregon, California, and Utah, and very low values for parts of the Mojave Desert, which is the region of the United States with the lowest runoff at the present day.[188] Because of the strong dependence of lake area on P - E (see equation 4), z should also be closely correlated with the net moisture convergence[189] in the atmospheric column over the lake basin, at least in areas where groundwater transfers are negligible. The implications of this relationship for the reconstruction of paleoclimate have yet to be explored.

A great deal of information about global or regional atmospheric circulation patterns can, however, be derived solely from spatial variations in the relative extent of lakes at particular time periods. The Oxford Data Bank makes use of a very simple system of coding in which lake depth is used as a surrogate for lake area, and divided into three categories, referred to as low, intermediate, and high status, each with a broadly similar frequency of occurrence in the data set. The level of any lake at a given time is assessed from the full range of geomorphic, stratigraphic, and analytical information available. It is then allotted to a lake-status category using an arbitrary but consistent set of rules.[190] At present, lake status is coded at 1000-year intervals from 30,000 to 0 years B.P. The trend, or direction of change,

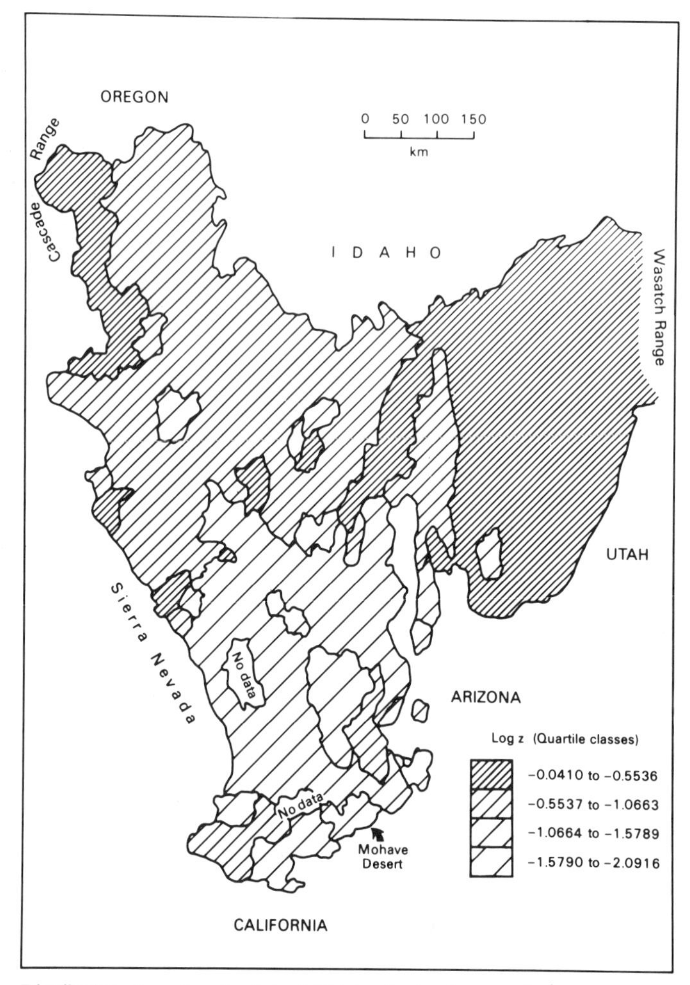

Figure 10. Distribution of z values for "pluvial" lakes in the Great Basin (SW USA) (data from refs. 10, 30).

in lake status, is also recorded whenever possible. The data on lake status and trend can either be mapped, or analyzed as a function of time. Examples of the mapped patterns are discussed below.

Reconstructed Patterns and Their Interpretation

The climatological basis for using Late Quaternary lake-level patterns to reconstruct past states of the general atmospheric circulation is the modern relationship between the global distribution of lakes and the major features of the mean annual circulation.[11] The proportion of the continental area covered by lakes shows three strong latitudinal maxima[12] (figure 9), which correspond to the Northern and Southern

Hemisphere westerlies and the equatorial trough, respectively. The subtropical anti-cyclone belts show up as pronounced minima in the extent of lakes at 20–35°N and S.

Figure 9 shows the distribution of lake status at the present day, based on the latest (1982) version of the Oxford Data Bank. It is immediately obvious that we are living in a relatively arid period compared with the rest of the Late Quaternary (figures 11–14). Three weak latitudinal maxima of lake status are, however, apparent around 40°N, 6°N, and 38°S. It is assumed here that these reflect the influence of the mid-latitude westerlies and the equatorial trough, although the distribution of data points is inadequate to define the positions of these features accurately.

Figures 11–14 illustrate the most distinctive patterns of lake status which have occurred during the last 25,000 years. At 24,000 years B.P. (figure 11), the present Northern Hemisphere subtropical arid zone was occupied by a belt of enlarged lakes, suggesting a ca. 5°N equatorward shift of the westerly storm tracks in the Northern Hemisphere. There seems to have been a narrow dry zone located on the equator, possibly indicating a displacement of the equatorial rain belt. The Southern Hemisphere westerlies are evident at ca. 36–38°S, although their position cannot be precisely located due to the lack of data from South America.

At 18,000 years B.P. (figure 12), the mean westerly storm track in the Northern Hemisphere still appears to have been situated at ca. 35°N, but an extensive dry zone had opened up in the tropics. The mean position of the equatorial rain belt cannot be specified without more data. There are slight indications of wetter conditions in Australia at 30–38°S.

It appears that 13,000 years B.P. (figure 13) was the most arid period in the Late Quaternary throughout Africa and western Eurasia. Only three data points indicate wet conditions, and these are probably spurious. Australia also appears to have been very arid at this time, with little evidence for the passage of westerly storms north of 40°S. In contrast, North America was still experiencing relatively moist conditions, indicating the existence of a mean storm track at about 36°N. Some lakes in the American Southwest were, however, just beginning to fall.

The minimum of atmospheric moisture convergence and runoff over Africa, western Eurasia, and tropical Australia at 13,000 years B.P. is one of the most intriguing features of the whole Late Quaternary record. It implies a strong suppression of summer monsoon precipitation and hence a radically different state of the Asian/Indian Ocean monsoon system from the present day. Street-Perrott and Roberts[11] discuss some of the changes in atmospheric boundary conditions which may have brought about this situation.

The pattern of lake status at 6000 years B.P. (figure 14) is almost the complete opposite of 18,000 years B.P. A very wide and strongly developed belt of enlarged lakes occupied the tropics (ca. 32°N–18°S). The mean latitude of the belt of high and intermediate lake levels in Africa and Arabia was 14 ± 3°N, the global average for 6000 years B.P. being 14 ± 4.5°N. This implies a significant northward shift in the mean position of the equatorial rain belt, as well as a greatly enhanced monsoonal transport of moisture into the tropical continents.[11]

A well developed arid zone had appeared in North America (ca. 32–51°N) by 6000 years B.P., suggesting that a large poleward displacement of the Northern

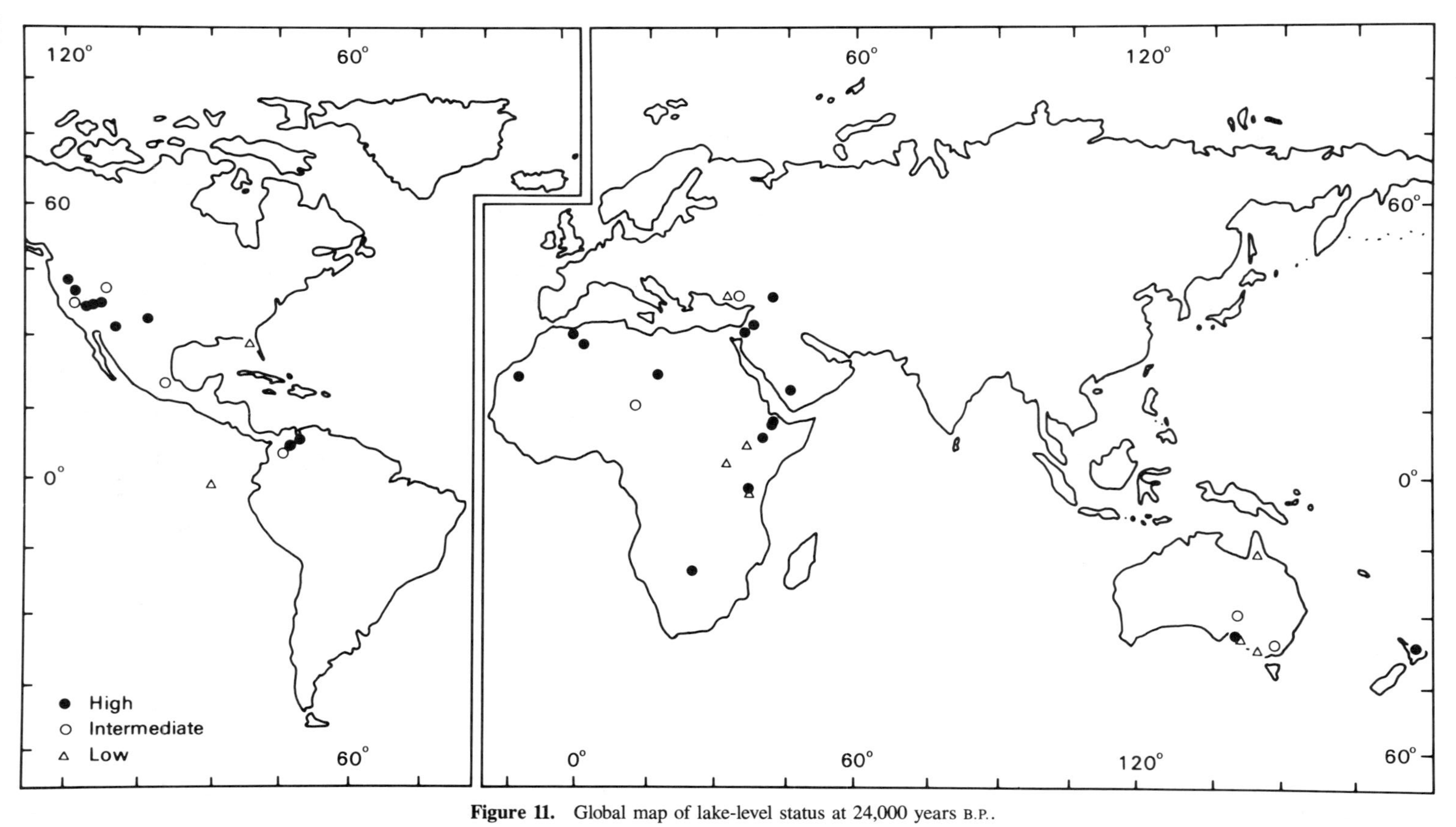

Figure 11. Global map of lake-level status at 24,000 years B.P..

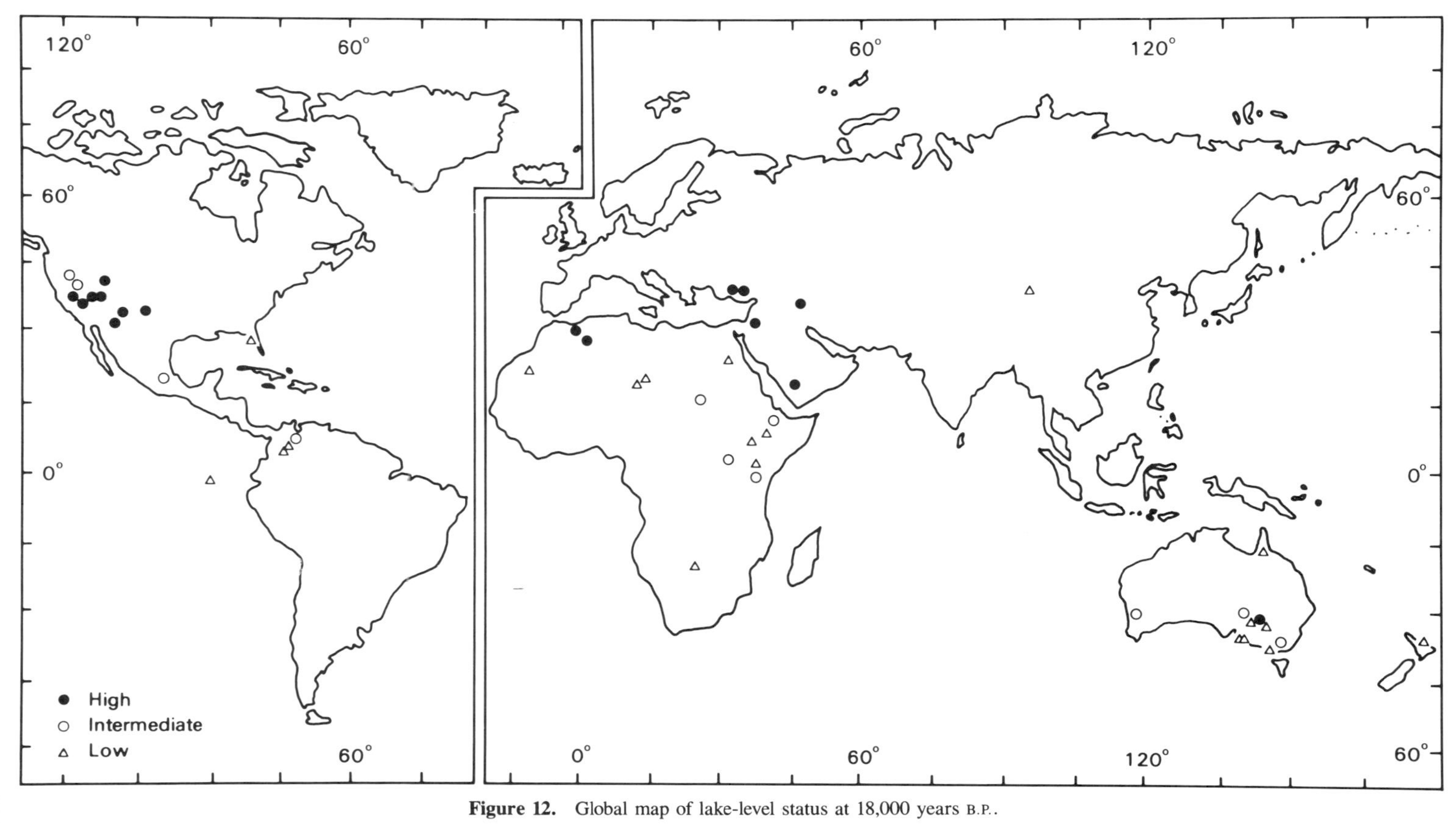

Figure 12. Global map of lake-level status at 18,000 years B.P..

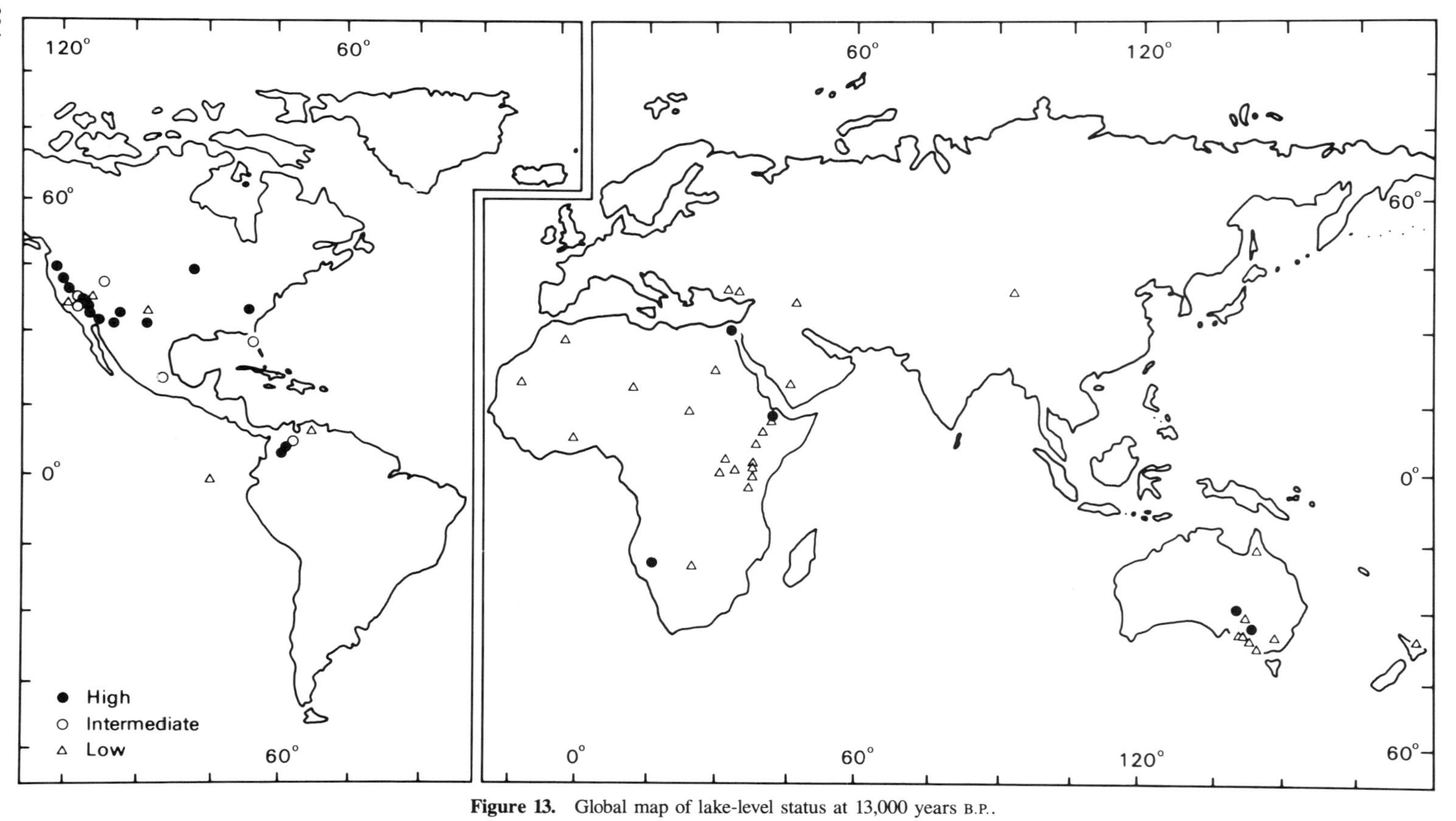

Figure 13. Global map of lake-level status at 13,000 years B.P..

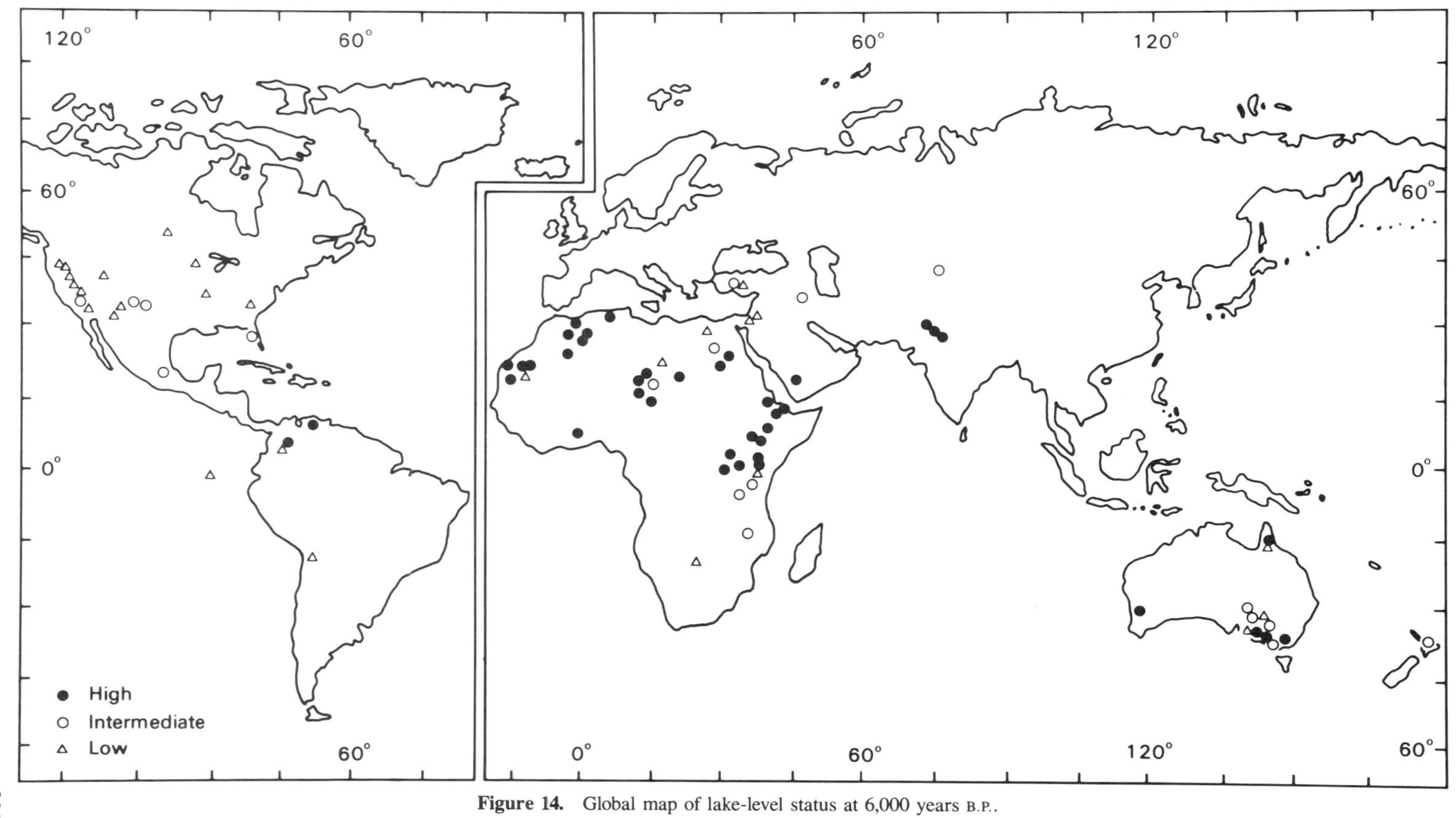

Figure 14. Global map of lake-level status at 6,000 years B.P..

Hemisphere westerly storm tracks had taken place. A less pronounced dry zone was also found in the Near East north of 26°N and east of 25°E. Kutzbach[191] has suggested that this represents a zone of enhanced northeasterly flow on the western side of an intensified Asian summer-monsoon low.

On the temperate margins of Australia, widespread signs of increased moisture at 6000 years B.P. suggest that the Southern Hemisphere westerlies were displaced about 4° equatorward of their present position (figure 9).

The above examples illustrate the kinds of inferences which may be made about past atmospheric circulation patterns, based on the spatial distribution of data on the relative extent of lakes. By analyzing the variations in lake status and trend through time in specific regions, one can also identify the times when major changes of mode in the general circulation took place.[11,190] By comparing the record thus obtained with independent data on changes in external or internal factors, such as the Milankovitch radiation curves, ocean temperatures, or the extent of snow and ice in high latitudes, it is possible to formulate inductive models of the way in which the atmospheric circulation in low and middle latitudes has responded to changes in climatic forcing.

CLIMATIC CALIBRATION OF LAKE-LEVEL DATA

In cases in which the areas of a palaeolake and its catchment can be accurately reconstructed from geomorphic data, it is possible to calculate the change in water-balance parameters necessary to maintain a closed lake in equilibrium at a given size. The variable most commonly estimated in this way is paleoprecipitation.[8,26,131,192–194]

Two principal types of models have been used to describe paleolakes: simple water-balance models[131,195] and combined water- and energy-balance models.[196–198] Combined models introduce more sophistication into their treatment of the interaction between the lake and the atmosphere, but require a wider range of assumptions about other climatic elements. No model developed so far is capable of dealing adequately with the interactions between lakes and groundwater, losses by overflow during water-level maxima, or non-equilibrium situations.

Water-Balance Models

The simple hydrological approach is based on equations (3) and (4). Paleoprecipitation can be estimated from equation 4 provided that the former evaporation rate from open water, E_L, and either paleorunoff, R, or the former evapotranspiration rate from the catchment, E_B, can be estimated. It is usual to assume in flat basins that $P_L = P_B$, and in mountainous basins that P_L is a known function of P_B, based on modern data.[8,195] E_B has commonly been estimated using the Thornthwaite method or empirical expressions for basin loss.[194] In some parts of the United States, R can be obtained from nomographs prepared by Schumm,[199] which give basin runoff as a function of mean annual temperature and precipitation. Like the Thornthwaite method,[200] which was also devised in the United States, this latter technique is likely to prove quite inaccurate if applied in regions with very different climates, for example the tropics. In East Africa, a value for R has commonly been obtained by subjectively adjusting the modern runoff coefficient (the proportion of rain falling on the catchment which ultimately reaches the lake) to allow for the effects of vegetation change as deduced by, for example, pollen analysis.[131,195]

Simple water-balance models are particularly sensitive to the value assumed for evaporation from the lake surface. It is usual to estimate E_L using information about paleotemperatures derived from treeline, snowline, or macrofossil data, in conjunction with either modern evaporation-pan data[193,195] or the data on lake evaporation in the United States compiled by Langbein.[3]

The assumed paleotemperature change is most critical in the case of colder periods, such as the last glacial, when it is difficult to distinguish the hydrological effects of increased precipitation from those of reduced evaporation. This has led to a long, fruitless, and often acrimonious debate between the adherents of the "pluvial" and "minevaporal" schools.[26,193,194] Estimates of the decrease in open-water evaporation under full-glacial conditions in the western United States, for example, range from 10 percent to 45 percent, giving rise to a large spread of estimates of the accompanying change in precipitation.[26,201] These radically different results reflect not only the mean annual temperature lowering assumed by different authors, but also the assumptions made about the seasonal cycle of temperatures under glacial conditions.

A number of other important factors are not taken into account by simple hydrological models.[111] These include changes in:

- net radiation
- wind strength
- the seasonality, intensity, and form of precipitation
- the storage of water in glacier ice
- groundwater storage and flow
- heat storage in the water mass
- salinity and
- the proportion of the basin occupied by swamp vegetation.

Despite these drawbacks, water-balance models have provided plausible and consistent estimates of changes in precipitation in certain restricted areas—for example, in eastern Africa during the Holocene. Some typical results are compared with the output of a combined water- and energy-balance model in table 4.

Combined Water- and Energy-Balance Models

Combined models attempt to calculate paleoevaporation in a more sophisticated way. They make use of the surface energy-balance equation for a closed lake, which

Table 4. Comparison of Paleoprecipitation Estimates for the Early Holocene Derived from Paleolake Data Using Simple Water-Balance and Combined Water- and Energy-Balance Models[a].

Paleolake	Reference	Present Precipitation (%)	P (mm)	Remarks
		Eastern Africa		
Ziway-Shala	195	>147	>450	W;ΔT = 0°C
	195	>128	>268	W;ΔT = −2°C
Nakuru-Elmenteita	183	>152	>505	W;ΔT = −2°C
	183	>165	>625	W;ΔT = +2–3°C
	202	>131	>280	W/E
Naivasha	183	>>125	>>225	W;ΔT = 0°C
	202	>>114	>>130	W/E
Manyara	203	133	237	W;ΔT = 0°C
		North Africa		
Megachad	196	>186	>300	W./E
		Northwest India		
Sambhar	202	>140	>190	W/E

[a] ΔP is the increase in precipitation compared with present. The inequality (>) indicates that estimates are minimum values because overflow was not taken into account. W refers to estimates made by the simple water-budget approach where ΔT is the assumed temperature difference from the present used to calculate past evaporation. W/E refers to estimates made by the combined water- and-energy-budget approach. Modified from 191.

can be stated in its simplest form as follows, assuming that advected energy and the heat flux into the lake bed can be ignored:

$$\Delta S = Q - H - lE_L \tag{5}$$

where ΔS = net heat storage in the water body
Q = net radiation flux
H = flux of sensible heat into the atmosphere
lE_L = flux of latent heat into the atmosphere
l = heat of vaporization of water

A similar expression can be written for the lake catchment (figure 15). The ingoing and outgoing radiation fluxes may be considered more explicitly than in equation (5), in order to incorporate the effects of changes in humidity, cloudiness and surface reflectivity (albedo).

The first attempt to formulate a combined model for the estimation of paleoprecipitation was by Kutzbach.[196] In order to simplify the calculations, he made use of the Bowen ratios, $\beta_L = H/lE_L$ and $\beta_B = H/lE_B$, which express the partitioning

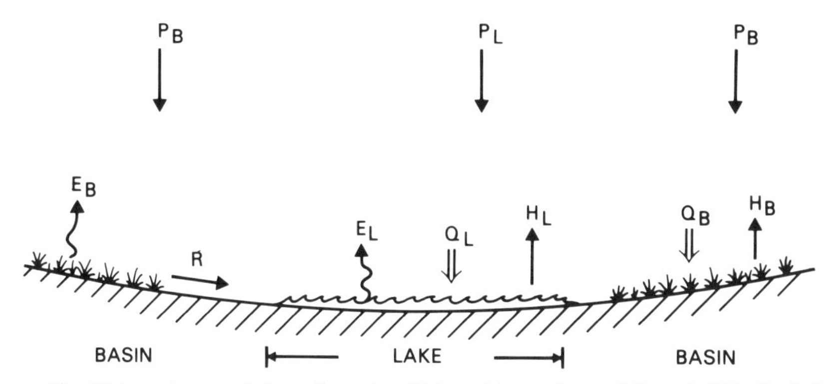

Figure 15. Water and energy balance for a closed lake and its catchment (after ref. 196). Symbols as for eqs. 1 and 5.

of energy between the sensible and evaporative heat fluxes. β is a useful climatic parameter that varies with surface moisture, temperature, and other conditions, and therefore assumes characteristic values over different types of surface. It can either be estimated from paleoecological data, or alternatively, parameterized as an empirical function of Q and P and then calculated internally. The most crucial assumption of this model is that net radiation Q, over the catchment and water surface, can be accurately specified for past periods. Although reasonable estimates of reflectivity and emissivity can be made for various surface types, it was necessary to assume that errors in estimates of cloudiness only led to relatively small errors in Q, since cloud cover modifies both ingoing and outgoing radiation fluxes. This assumption is questioned by Benson.[197] The model also ignores heat storage in the lake, ΔS, which is a valid assumption for annual-mean conditions.

The Kutzbach[196] model has now been applied to a number of paleolakes in Africa and India. In general, it yields more conservative estimates of paleoprecipitation than simpler water-balance models[191] (table 3), which may reflect a combination of two factors: the Kutzbach model tends to underestimate modern evaporation rates[198] whereas some of the hydrological models assume a substantial warming (2–3°C), and therefore high evaporation rates, during the early Holocene.

The combined model recently developed by Benson[197] considers atmospheric radiation fluxes in great detail, but treats the runoff from the lake catchment in a rather simplistic way. It once again assumes annual-mean conditions. The results are interesting in that they suggest an important role for cloud cover and cloud type, as well as paleotemperatures, in controlling fluctuations in lake area. Benson concluded that the full-glacial high stand of Lahontan, a large paleolake in the Great Basin, could have been produced by a 10 percent increase in sky cover, combined with a ≥ 10°C decrease in mean air temperature and "heap cloud" percentages in excess of 10 percent. He also demonstrated that glacial meltwater from the Sierra Nevada could not have played a significant part in maintaining the lake at high levels during the late-glacial.

Tetzlaff and Adams[198] have attempted to refine the estimation of paleoevaporation by allowing for the effect of wind strength and the seasonal cycle of heat storage in the water body. The computation of evaporation was carried out in two steps. First the water temperature T_L was calculated, based on the energy budget of the lake surface. Second, the value obtained was substituted in the Penman equation,[200] together with values of atmospheric humidity, temperature and wind speed. This method requires that a considerable number of climatic parameters be known or estimated. It is also only applicable to shallow, well-mixed lakes in arid climates. A sensitivity analysis showed that the calculated value of evaporation is most responsive to wind velocity, followed by incoming radiation and air temperature.

Tetzlaff and Adams[198] applied their model to Lake Chad during the Sahel drought (1968–73) and to the early Holocene Lake Megachad (ca. 9000 years B.P.). Although Chad experiences significant water losses by seepage to groundwater (see earlier), from the viewpoint of paleohydrology it is essentially an amplifier lake. Runoff from the catchment is therefore much more important than direct precipitation in influencing the lake area, and was probably the most significant factor during the drought. However, the results suggested that evaporation from the lake rose markedly from 1968 to 1973, due to an increase in wind speed combined with higher incoming radiation and lower humidities. For the early Holocene, these climatic data are not available. The modern meteorological data set was, therefore, modified in accordance with a 5° northward shift in the mean position of the ITCZ as suggested by other paleoclimatic data.[204] Clearly, there is some danger of circular reasoning at this stage. The results indicated that the total annual input of water into the lake was about 1200 percent greater than today during the early Holocene. The calculated evaporation rate was also 15 percent lower, in response to an assumed reduction of 5 percent in incoming radiation and to lower air temperatures.

FUTURE OUTLOOK

Lake-level studies are at present in an exciting stage, involving both the diversification and refinement of the techniques used to reconstruct changes in lake extent, and the development of theoretical models of lake behavior. Rapid progress is likely to be made during the next few years on the climatic calibration of lake-level data. Even where precise estimates of paleolake area are unobtainable, for example in the case of pre-Quaternary lakes, improvements in our understanding of the way in which lakes, as geological and biological systems, respond to climate will make it possible to derive useful information on the spatial and temporal pattern of past climatic changes. Some promising topics for future research are suggested below.

A high priority at present is the development of a wider range of models for making quantitative estimates of paleoclimatic variables. The models currently available (see Climatic Calibration section earlier) differ in the extent to which they try to simulate the complex physical processes occurring over a closed lake and its catchment. The treatment of drainage-basin hydrology is almost invariably simplistic. There is a need for models that handle the annual cycle, including such seasonal processes as heat storage in the lake and snowmelt in mountain areas, both of which may have been important in temperate areas of high relief such as the Great Basin.

Another priority is to develop better means of handling non-equilibrium situations in order to understand how lakes respond to sudden shifts in climate.

Where the stringent requirements of numerical models cannot be met, for example in regions with poor or nonexistent climatic data, a considerable amount of information on the spatial pattern of atmospheric moisture convergence and surface water balance can still be derived by mapping the distribution of z ratios at particular times in the past.

Figures 9–14 indicate the potential of lake-level data for use in reconstructing atmospheric circulation patterns during Late Quaternary time. This approach has not yet been fully explored. In particular, it would be of considerable interest to apply similar methods to instrumental and historical lake-level data.

In this paper, we have suggested ways in which mineralogical and paleoecological analyses of lacustrine sediments may be used in the future to define the gross water-balance characteristics of individual paleolakes. Obviously, the wider the range of indicators employed, the more precisely the latter can be defined. Such semiquantitative estimates serve as a check on the results of paleohydrological models. For example, the computations for the early Holocene lakes Ziway-Shala and Megachad, summarized in table 3, suggest that at the point of overflow, both lakes were situated on the left-hand edge of quadrant RP-E (figures 1 and 7) in the domain of precipitation of Ca,Mg-carbonates. This inference can be checked using sedimentological data.[103,111]

In the absence of lake-area estimates, it is still possible to map the distribution of different *lake types* at any given time.[205] For example, the penetration of the tropical summer-monsoon rains into the Sahara and Arabia during the early and middle Holocene is recorded by a northward gradient from large lakes like Megachad, to small interdune lakes, and finally to marshes. Gradients of lake characteristics also existed during the last glacial in the American Southwest.[27] Careful study of the available sedimentological and paleoecological information may make it possible to define the underlying spatial trends in climate and water balance more precisely.

At the moment we know very little about the response time of large Quaternary paleolakes to climatic variations. In some cases, the lakes may have varied in size so rapidly that their fluctuations cannot be completely resolved by ^{14}C dating.[111] In other cases, the basin response time may be quite long.[162] One of the most pressing tasks facing the paleolimnologist is the development of numerical models of the way in which closed lakes, situated on particular geochemical pathways (figure 5), vary in size, chemistry, and isotopic characteristics as a function of specific changes in water balance through time.

ACKNOWLEDGMENTS

This work was supported by the U.S. Department of Energy Carbon Dioxide and Climate Research Program (contract no. DE-ACO$_2$-79 EV10097 to T. Webb III at Brown University). We thank F. Gasse, M. J. Salinger, G. I. Smith, and M. J. Talbot for helpful discussion.

REFERENCES

1. E. Halley, "On the cause of the saltness of the ocean, and of the several lakes that emit no rivers; with a proposal, by help thereof, to discover the age of the world," *Philos. Trans. R. Soc. London,* **29,** 296–300 (1715). See page 298.
2. E. de Martonne and L. Aufrère, "L'extension des régions privées d'écoulement vers l'océan," *Ann. Géog.*, **38,** 1–24 (1928).
3. W. B. Langbein, "Salinity and hydrology of enclosed lakes," *U.S. Geological Survey Professional Paper* **412** (1961).
4. C. C. Reeves, Jr. "*Introduction to Paleolimnology,*" Elsevier, Amsterdam, 1968.
5. J. L. Richardson, "Former lake-level fluctuations—their recognition and interpretation," *Mitt. Int. Vereiningung Limnol.*, **17,** 78–93 (1969).
6. T. C. Winter and H. E. Wright, "Paleohydrologic phenomena recorded by lake sediments," *EOS,* **58,** 188–196 (1977).
7. W. D. Williams, "Inland salt lakes: An introduction," *Hydrobiologia,* **82,** 1–14 (1981).
8. C. T. Snyder and W. B. Langbein, "The Pleistocene lake in Spring Valley, Nevada and its climatic implications," *J. Geol. Res.*, **67,** 2385–2394 (1962).
9. J. M. Bowler, "Australian salt lakes: A palaeohydrologic approach," *Hydrobiologia,* **82,** 431–444 (1981).
10. M. D. Mifflin and M. M. Wheat, "Pluvial lakes and estimated pluvial climates of Nevada," *Nev. Bur. Mines Geol. Bull.*, **94,** 1–57 (1979).
11. F. A. Street-Perrott and N. Roberts, "Fluctuations in Closed Lakes as an Indicator of Past Atmospheric Circulation Patterns." In F. A. Street-Perrott, M. A. Beran, and R. A. S. Ratcliffe, eds., *Variations in the Global Water Budget,* D. Reidel, Dordrecht, 1983, pp. 331–345.
12. R. D. Schuiling, "Source and Composition of Lake Sediments." In H. L. Golterman, ed., *Interactions Between Sediments and Freshwater,* PUDOC, Wageningen 1977, pp. 12–18.
13. F. A. Street, "The relative importance of climate and hydrogeological factors in influencing lake-level fluctuations," *Palaeoecol. Afr.*, **12,** 137–158 (1980).
14. K. Szestay, "Water balance and water level fluctuations of lakes," *Hydrol. Sci. Bull.*, **19,** 73–84 (1974).
15. K. W. Butzer, "Russian climate and the hydrological budget of the Caspian Sea," *Rev. Can. Géogr.*, **12,** 129–139 (1958).
16. G. Müller and F. Wagner, "Holocene carbonate evolution in Lake Balaton (Hungary): A response to climate and impact of man." In A. Matter and M. E. Tucker, eds., *Modern and Ancient Lake Sediments*, Blackwell Scientific Publications, Oxford, 1978, pp. 55–81.
17. H. Löffler, *Neusiedlersee: The Limnology of a Shallow Lake in Central Europe*, Dr. W. Junk, The Hague, 1979.
18. G. G. Varbumyan, "Modern conditions and future of Sevan Lake," *Meteor. Hydro.*, **10,** 116–132, (1973).
19. S. Kempe, F. Khoo, and Gürleyik, "Hydrography of Lake Van and Its Drainage Area." In E. T. Degens and F. Kurtman, eds., *The Geology of Lake Van*, M.T.A. Publication no. 169, Ankara, 1978, pp. 30–48.
20. R. E. Hecky and E. T. Degens, "Lake Pleistocene-Holocene chemical stratigraphy and paleolimnology of the rift valley lakes in Central Africa," *Woods Hole Oceanographic Institute Technical Report,* **73–28,** (1973).
21. M. Kalk, A. J. McLachlan, and C. Howard-Williams, eds. *Lake Chilwa: Studies of Change in a Tropical Ecosystem*, Dr. W. Junk, The Hague, 1979.
22. K. N. Phillips, and A. S. Van Denburgh, "Hydrology and geochemistry of Abert, Summer and Goose Lakes, and other closed-basin lakes in south-central Oregon," *U.S. Geological Survey Professional Paper,* **502B,** B1–B86 (1971).

23. J-P. Carmouze, C. Arze, and J. Quintanilla, "Circulación de materia (agua-sales disueltas) através del sistema fluvio-lacustre del Altiplano: La regulación hidrica é hidroquímica de los lagos Titicaca y Poopó," *Cah. O.R.S.T.O.M. Sér. Geol.*, **10,** 49–68 (1978).
24. G. W. Kite, "Recent changes in level of Lake Victoria," *Hydrol. Sci. Bull.*, **26,** 233–243 (1981).
25. F. Gasse and F. A. Street, "Late Quaternary lake-level fluctuations and environments of the northern Rift Valley and Afar region (Ethiopia and Djibouti)," *Palaeogeogr., Palaeoclimatol., Palaeoecol.*, **24,** 279–325 (1978).
26. G. R. Brakenridge, "Evidence for a cold, dry full-glacial climate in the American Southwest," *Quat. Res.*, **9,** 22–40 (1978).
27. G. I. Smith and F. A. Street-Perrott, "Pluvial lakes." In S. C. Porter, ed., *Late Quaternary of the United States: Volume 1. The Late Pleistocene*, University of Minnesota Press, Minneapolis, 1983, pp. 190–212.
28. M. A. Roche, "Lake Chad: A subdesertic terminal basin with fresh waters." In D. C. Greer, ed., *Desertic Terminal Lakes*, Conference on Desertic Terminal Lakes, Utah Water Research Laboratory, Logan, Utah, 1977, pp. 213–223.
29. J. J. Gaudet and J. M. Melack, "Major ion chemistry in a tropical African lake basin," *Freshwater Bio.*, **11,** 309–333 (1981).
30. C. T. Snyder, G. Hardman, and F. F. Zdenek, "Pleistocene lakes in the Great Basin," *U.S. Geological Survey Miscellaneous Geological Investigations Map* **1416,** (1964).
31. F. Gasse, "Evolution of Lake Abhé (Ethiopia and T.F.A.I.) from 70,000 yr B.P.," *Nature (London)*, **265,** 42–45 (1977).
32. J. L. Richardson and A. E. Richardson, "History of an African Rift Lake and its climatic implications," *Ecol. Monogr.*, **42,** 499–534 (1972).
33. A. Covich and M. Stuiver, "Changes in oxygen-18 as a measure of long-term fluctuations in tropical lake levels and molluscan populations," *Limnol. Oceanogr.*, **19,** 682–691 (1974).
34. C. Niederberger, "Early sedentary economy in the Basin of Mexico," *Science*, **203,** 131–142 (1979).
35. J. P. Bradbury, "Paleolimnology of Lake Texcoco, Mexico," *Limnol. Oceanogr.*, **16,** 180–200 (1971).
36. L. A. Zemljanitzyna, "Inflow to Lakes of the Semi-arid Zone of the USSR from Groundwater," IAHS Publication No. 109, *Hydrology of Lakes,* 185–190 (1974).
37. F. Wendorf and R. Schild, *Prehistory of the Eastern Sahara*, Academic Press, New York, 1980.
38. R. B. Hooke, "Geomorphic evidence for late-Wisconsin and Holocene tectonic deformation, Death Valley, California," *Geol. Soc. Am. Bull.*, **83,** 2073–2098 (1972).
39. M. M. Clark, A. Grantz, and M. Rubin, "Holocene activity of the Coyote Creek fault as recorded in the sediments of Lake Cahuilla," *U.S. Geological Survey Professional Paper* **787,** 112–129 (1972).
40. M. E. Wilson and S. H. Wood, "Tectonic tilt rates derived from lake-level measurements, Salton Sea, California," *Science*, **207,** 183–186 (1980).
41. W. W. Bishop, "The Late Cenozoic History of East Africa in Relation to Hominoid Evolution." In K. K. Turekian, ed., *Late Cenozoic Glacial Ages*, Yale University Press, New Haven, 1971, pp. 493–527.
42. G. E. Hutchinson and U. M. Cowgill, "The waters of Merom: A study of Lake Huleh 3: The major chemical constituents of a 54 m core." *Arch. Hydrobiol.*, **72,** 145–185 (1973).
43. J. W. Sherman and R. Patrick, "The waters of Merom: A study of Lake Huleh 7: Diatom stratigraphy of a 54 m core," *Arch. Hydrobiol.*, **92,** 199–221, (1981).
44. N. Roberts, "Lake-levels as an Indicator of Near Eastern Palaeoclimates: A Preliminary Appraisal." In J. L. Bintliff and W. Van Zeist, eds., *Palaeoclimates, Palaeoenvironments and Human Communities in the Eastern Mediterranean Region in Later Prehistory*, British Archaeological Reports International Series 133, 1982, pp. 235–267.

45. D. Jäkel, "Run-off and fluvial formation processes in the Tibesti mountains as indicators of climatic history in the central Sahara during the late Pleistocene and Holocene," *Palaeoecol. Afr.*, **11,** 13–44 (1979).
46. F. Wendorf, ed. *Paleoecology of the Llano Estacado*, Fort Burgwin Research Centre Publication 1, Museum of New Mexico Press, Santa Fe, 1961.
47. J. Harbour, "General Stratigraphy." In F. Wendorf and J. J. Hester, eds., *Late Pleistocene Environments of the Southern High Plains*, Fort Burgwin Research Centre Publication 9, 1975, pp. 33–55.
48. C. R. Handford, "Sedimentology and evaporite genesis in a Holocene continental-sabkha playa basin—Bristol Dry Lake, California," *Sedimentology*, **29,** 239–253 (1982).
49. W. S. Broecker and A. Kaufman, "Radiocarbon chronology of Lake Lahontan and Lake Bonneville II, Great Basin," *Geol. Soc. Am. Bull.*, **76,** 537-566 (1965).
50. W. E. Scott, W. D. McCoy, R. R. Shroba, and M. Rubin, "Reinterpretation of the exposed record of the last two cycles of Lake Bonneville, Western United States," *Quat. Res.* **20,** 261–285 (1983).
51. D. Neev and K. O. Emery, "The Dead Sea," *Isr. Geol. Sur. Bull.*, **41,** 147 pp. (1967).
52. A. Kaufman, "U-series dating of Dead Sea Basin carbonate," *Geochim. Cosmochim. Acta*, **35,** 1269–1281 (1971).
53. A. von Humboldt, *Voyage aux Régions Equinoxiales du Nouveau Continent, 1799-1804 par Al de Humboldt et A. Bonpland*, 4 vols., Paris, 1814.
54. S. T. Harding, "Recent variations in the water supply of the western Great Basin," *Water Resources Center Archives, University of California Report* **16,** (1965).
55. A. S. Kes', "The causes of waterlevel changes of the Aral Sea in the Holocene," *Sov. Geogr.*, **20,** 104–113 (1979).
56. P. J. Mehringer, K. L. Petersen and F. A. Hassan, "A pollen record from Birket Qarun and the recent history of the Fayum, Egypt," *Quat. Res.*, **11,** 238–256 (1979).
57. E. L. Peck and E. A. Richardson, "Hydrology and Climatology of Great Salt Lake." In W. L. Stokes, ed., *Guidebook to the Geology of Utah (20): The Great Salt Lake*, Salt Lake City, Utah, 1966, pp. 121–134.
58. C. E. Vincent, T. D. Davies and A. K. C. Beresford, "Recent changes in the level of Lake Naivasha, Kenya, as an indicator of Equatorial westerlies over East Africa," *Climatic Change*, **2,** 175–189 (1979).
59. L. C. Beadle, *The Inland Waters of Tropical Africa*, Longman, London, 1974.
60. K. Cehak and A. Kessler, "Varianzspektrumanalyse der Seespiegelschwankungen des Titicaca-Sees (Südamerika)," *Arch. Met. Geoph. Biokl. Ser. B*, **24,** 201–208 (1976).
61. J. N. Jennings, L. C. Noakes, and G. M. Burton, "Notes on the Lake George and Lake Bathurst excursion," In Commonwealth of Australia Department of National Development, Bureau of Mineral Resources, Geology and Geophysics, ed., *Geological Excursions: Canberra District*, Australian and New Zealand Association for the Advancement of Science, Canberra Meeting, January 1964, pp. 25–34.
62. J. Healy, "The gross effect of rainfall on lake levels in the Rotorua district," *J. R. Soc. N. Z.*, **5,** 77–100 (1975).
63. G. E. Hutchinson, *A Treatise on Limnology*, Vol. 1, 2nd ed.: *Geography, Physics and Chemistry. Part 1: Geography and Physics of Lakes*, Wiley-Interscience, New York, 1975.
64. J. G. Pike and G. T. Rimmington, *Malawi: A Geographical Study*, Oxford University Press, Oxford, 1965.
65. A. Kessler, "Atmospheric circulation anomalies and level fluctuations of Lake Titicaca, South America," Proceedings International Tropical Meteorology Meeting 1974, Nairobi Kenya (1974).
66. H. T. Mörth, "Investigations into the meteorological aspects of the variations in the level of Lake Victoria," *East Afr. Meteorol. Depart. Mem.*, **4,** 1–23 (1967).

67. H. H. Lamb, *Climate: Present, Past and Future*, Vol. 2 Methuen, London 1977.

68. A. V. Šnitnikov, "Some material on intrasecular fluctuations of the climate of Northwestern Europe and the North Atlantic in the 18th–20th centuries," *Geogr. Obs. S.S.S.R.* pp. 5–29 (1969) (In Russian).

69. L. Mofakham-Pâyân, "Etude Géographique de la Mer Caspienne," *Soc. Geogr. Khorâssán, P.* **6,** 1–234 (1969).

70. H. De Terra and G. E. Hutchinson, "Evidence of recent climatic changes shown by Tibetan highland lakes," *Geogr. J.*, **84,** 311–320 (1934).

71. A. T. Grove and R. H. Pullan "Some aspects of the Pleistocene palaeogeography of the Chad Basin," In F. C. Howell and F. Bourlière, eds., *African Ecology and Human Evolution*, Aldine Publishing Co., Chicago, 1963. pp. 230–245.

72. C. Schubert, "Paleolimnología del lago de Valencia: Recopilación y proyecto,"*Bol. Soc. Venez. Ciencias Nat. No. 136*, **36,** 123–155 (1979).

73. C. Camus, "Fluctuations du niveau du lac Tanganyika," *Acad. R. Sci. Outre-Mer, Bull. Séances, Nouvelle Sér.*, **11,** 1242–1256 (1965).

74. H. H. Lamb, "Climate in the 1960's: Changes in the world's wind circulation reflected in prevailing temperatures, rainfall patterns and the levels of the African lakes," *Geogr. J.*, **132,** 183–212 (1966).

75. S. E. Nicholson, "Saharan climates in historic times." In M. A. J. Williams, and H. Faure, eds., *The Sahara and the Nile*, Balkema, Rotterdam, 1980 pp. 173–200.

76. D. B. Lawrence and E. G. Lawrence, "Response of enclosed lakes to current glaciopluvial climatic conditions in middle latitude western North America," *Ann. N. Y. Acad. Sci.*, **95,** 341–50 (1961).

77. C. Klein, "On the fluctuations of the level of the Dead Sea since the beginning of the 19th century," *Hydrological Service Israel, Hydrological Paper* **7,** (1961).

78. T. M. L. Wigley, M. J. Ingram, and G. Farmer, *Climate and History*, Cambridge University Press, Cambridge, 1981.

79. S. Vélez de Escalante, "Journal and itinerary ... of the route from Presidio de Santa Fe del Nuevo-México to Monterey, in Northern California," English translation in *Utah His. Q.*, **11,** 27–113 (1776).

80. Col. J. Jackson, *Observations on Lakes, Being an Attempt to Explain the Laws of Nature Regarding Them; The Cause of their Formation and Gradual Diminution; The Different Phenomena They Exhibit etc. With a View to the Advancement of Useful Science*, Bossange, Barthés and Lovell, London, 1833.

81. H. Strachey, "Physical Geography of Western Tibet," *J. R. Geogr. Soc. London*, **23,** 1–69, (1853).

82. H. H. Godwin-Austen, "On the lacustrine or Karéwah deposits of Kashmere," *Q. J. Geol. Soc. London*, **15,** 221–229 (1859).

83. H. H. Godwin-Austen, "Notes on the Pangong Lake District of Ladakh from a journal made during a survey in 1863," *J. R. Geogr. Soc. London*, **37,** 343–363 (1867).

84. W. T. Blandford, "Notes on route from Poona to Nagpur, via Ahmednuggur, Jalna, Loonar, Yeotmahal, Mangal and Hingunghat," *Rec. Geol. Surv. India*, **1,** 60–65 (1868).

85. F. Drew, *The Jummoo and Kashmir Territories*, London, 1875. Reprinted Oriental Publishers, Delhi, 1971.

86. G. R. Littledale, "A journey across Tibet, from north to south, and west to Ladakh," *Geogr. J.*, **7,** 453–483 (1876).

87. R. Lydekker, "The geology of the Kashmir and Chamba Territories and the British district of Khágan," *Mem. Geol. Surv. India*, **22,** 1–344 (1883).

88. Mr. Wynne, "Memoir on the geology of the Salt Range," *Mem. Geol. Surv. India*, **14,** 1–46 (1878).

89. L. Lartet, "Note sur la formation du bassin de la Mer Morte ou lac Asphaltite et sur les changements survenus dans le niveau de ce lac, *Bull. Soc. Géol. Fr.*, **22,** 420–466 (1865).
90. J. D. Whitney, "Report of progress and synopsis of the fieldwork from 1860–1864," *Geol. Surv. Calif.*, **1,** 1–452 (1865).
91. J. H. Simpson, *Explorations Across the Great Basin of the Territory of Utah*, U.S. Army Engineering Department, Washington, D.C., 1876.
92. C. King, "Systematic geology," In *U.S. Geological Exploration of the Fortieth Parallel*, Washington, D.C., 1878.
93. I. C. Russell, "Sketch of the geological history of Lake Lahontan," *Ann. Rep. U.S. Geol. Surv.*, **3,** 189–235 (1883).
94. I. C. Russell, "A geological reconnaissance in southern Oregon," *Ann. Rep. U.S. Geol. Surv.*, **4,** 431–464 (1884).
95. I. C. Russell, "Geological history of Lake Lahontan, a Quaternary lake of Northwestern Nevada," *U.S. Geol. Surv. Monogr.*, **11,** 1–288 (1885).
96. I. C. Russell, "Quaternary history of Mono Valley, California," *Ann. Rep. U.S. Geol. Surv.*, **8.** 261–394 (1889).
97. G. K. Gilbert, "The topographic features of lake shores," *5th Report U.S. Geological Survey,* 69–123 (1885).
98. G. K. Gilbert, "Lake Bonneville," *U.S. Geol. Surv. Monogr.*, **1,** 1–438 (1890).
99. R. B. Morrison, "Predecessors of Great Salt Lake." In W. L. Stokes, ed, *The Great Salt Lake*, Utah Geological Society, Salt Lake City, 1966, pp 75–104.
100. G. I. Smith, "Late Quaternary geologic and climatic history of Searles Lake, southeastern California," In R. B. Morrison and H. E. Wright Jr. eds., *Means of Correlation of Quaternary Successions,* Proceedings VII INQUA Congress, Vol. 8, University of Utah Press, Salt Lake City, 1968, pp. 293–310.
101. G. E. Smith, "Subsurface stratigraphy and geochemistry of Late Quaternary evaporites, Searles Lake, California," *U.S. Geol. Surv. Prof. Pap.*, **1043,** 1–130 (1979).
102. J. M. Bowler, "Late Quaternary environments: A study of lakes and associated sediments in south-eastern Australia," Ph.D. diss. Australian National University, Canberra, 1970.
103. M. Servant and S. Servant, "Les formations lacustres et les diatomées du Quaternaire récent du fond de la cuvette tchadienne," *Rev. Géogr. Phys. Géol. Dyn.*, **12,** 63–76 (1970).
104. M. R. Talbot and G. Delibrias,"Holocene variations in the level of Lake Bosumtwi, Ghana," *Nature (London)*, **268,** 722–724 (1977).
105. D. Q. Bowen, *Quaternary Geology: A Stratigraphic Framework for Multi-Disciplinary Work*, Pergamon Press, Oxford, 1978. See chapter 5.
106. I. U. Olsson, "Modern aspects of radiocarbon dating," *Earth-Sci. Rev.*, **4,** 203–218 (1968).
107. G. E. Williams, and H. A. Polach, "Radiocarbon dating of arid-zone calcareous paleosols," *Bull. Geol. Soc. Am.*, **82,** 3069–3086 (1971).
108. R. A. Müller, "Radioisotope dating with a cyclotron," *Science*, **196,** 489–494 (1977).
109. D. L. Thurber, "Problems of dating non-woody material from continental environments." In W. W. Bishop and J. A. Miller, eds, *Calibration of Hominoid Evolution*, Scottish Academic Press, Edinburgh, 1972, pp. 1–17.
110. L. V. Benson, "Fluctuation in the level of pluvial Lake Lahontan during the last 40,000 years," *Quat. Res.*, **9,** 300–318 (1978).
111. F. A. Street, "Late Quaternary lakes in the Ziway-Shala Basin, Southern Ethiopia," Ph.D. diss. University of Cambridge 1979.
112. R. Johnson, T. Kaufman, and R. Berger, "Radiocarbon dating in volcanic environments," In W. G. Mook and H. G. Waterbolk, eds., Proc. 1st International Symposium on C^{14} and Archaeology, V 8, Groningen, PACT, 1983, 329–336.
113. A. Kaufman and W. Broecker, "Comparison of Th^{230} and C^{14} ages for carbonate materials from Lakes Lahontan and Bonneville," *J. Geophys. Res.*, **70,** 4039–4054 (1965).

114. C. Gaven, C. Hillaire-Marcel, and N. Petit-Maire, "A Pleistocene lacustrine episode in southeastern Libya," *Nature (London)*, **290,** 131–133 (1981).

115. T-H. Peng, J. G. Goddard, and W. S. Broecker, "A direct comparison of ^{14}C and ^{230}Th ages at Searles Lake, California," *Quat. Res.*, **9,** 319–329 (1978).

116. J. C. Vogel and J. Kronfeld, "A new method for dating peat," *S. Afr. J. Sci.*, **76,** 557–558 (1980).

117. F. Oldfield, "Peats and Lake Sediments: Formation, Stratigraphy, Description and Nomenclature." In A. S. Goudie, ed., *Geomorphological Techniques*, George Allen and Unwin Ltd., London, 1981, pp. 306–326.

118. R. L. Hay, *Geology of the Olduvai Gorge—A Study of Sedimentation in a Semiarid Basin*, University of California Press, Berkeley, 1976.

119. D. C. Johanson, M. Taieb, B. T. Gray, and Y. Coppens, "Geological Framework of the Pliocene Hadar Formation (Afar, Ethiopia) with Notes on Palaeontology, Including Hominids." In W. W. Bishop ed., *Geological Background to Fossil Man*, Scottish Academic Press, Edinburgh, 1978, pp. 549–564.

120. A. Cohen, "Paleolimnological research at Lake Turkana, Ethiopia," *Palaeoecol. Afr.*, **13,** 61–82 (1981).

121. N. M. Kennedy, W. A. Pullar, and C. F. Pain, "Late Quaternary land surfaces and geomorphic changes in the Rotorua Basin, North Island, New Zealand," *N. Z. J. Sci.*, **21,** 249–264 (1978).

122. P. T. Davis, C. W. Barnosky, and M. Stuiver," A 20,000 yr. record of volcanic ashfalls, Davis Lake, Southwestern Washington," *Abstr., Seventh Bi. Conf. Am. Quat. Assoc.*, **87** (1982).

123. F. H. Brown and T. E. Cerling, "Stratigraphical significance of the Tula Bar Tuff of the Koobi Fora Formation," *Nature (London)*, **299,** 212–215 (1982).

124. A. J. Eardley, V. Gvodetsky, W. P. Nash, M. D. Picard, D. C. Grey, and G. J. Kukla, "Lake cycles in the Bonneville Basin, Utah," *Bull. Geol. Soc. Am.*, **84,** 211–216 (1973).

125. G. I. Smith, "Late Tertiary-to-present sequence of pluvial events in Searles Valley, California, and their relation to global ice volumes," *Abstr. Prog. Sixth Bi. Meet. Am. Quat. Assoc.*, August (1980).

126. G. Singh, N. D. Opdyke, and J. M. Bowler, "Late Cainozoic stratigraphy, palaeomagnetic chronology and vegetational history from Lake George, N.S.W." *J. Geol. Soc. Aust.*, **28,** 435–452 (1981).

127. C. E. Barton and H. A. Polach, "^{14}C ages and magnetic stratigraphy in three Australian maars," *Radiocarbon*, **22,** 728–739 (1980).

128. C. E. Barton and M. W. McElhinny, "A 10,000 yr. geomagnetic secular variation record from three Australian maars," *Geophys. J. R. Astron. Soc.*, **67,** 465–485 (1981).

129. K. M. Creer, "Lake sediments as recorders of geomagnetic field variations—applications to dating post-Glacial sediments," *Hydrobiol.* **92,** 587–596 (1983).

130. R. Thompson, R. W. Battarbee, P. E. O'Sullivan, and F. Oldfield, "Magnetic susceptibility of lake sediments," *Limnol. Oceanogr.* **20,** 687–698 (1975).

131. C. K. Washbourn, "Late Quaternary lakes in the Nakuru-Elmenteita Basin, Kenya," Ph.D. diss. University of Cambridge, 1967.

132. C. K. Washbourn-Kamau, "Late Quaternary lakes in the Nakuru-Elmenteita Basin, Kenya," *Geogr. J.*, **137,** 522–535 (1971).

133. M. D. Crittenden Jr., "New data on the isostatic deformation of Lake Bonneville," *U.S. Geol. Surv. Prof. Pap.*, **454E,** E1–E31 (1963).

134. J. M. Bowler, Clay dunes: Their occurrence, formation and environmental significance, *Earth-Sci. Rev.*, **9,** 315–338 (1973).

135. I. N. Lancaster, "Evidence for a widespread Late Pleistocene humid period in the Kalahari," *Nature (London)*, **279,** 145–146 (1979).

136. C. C. Reeves, Jr., "Pluvial lake basins of West Texas," *J. Geol.*, **74,** 269–291 (1966).

137. K. D. Gehr and T. M. Newman, "Preliminary note on the Late Pleistocene geomorphology and archaeology of the Harney Basin, Oregon," *Ore Bin* **40,** 165–170 (1978).

138. J. M. Bowler, "Aridity in Australia: Age, origins and expression in aeolian landforms and sediments," *Earth-Sci. Rev.*, **12,** 279–310 (1976).

139. G. D. Johnson, "Cainozoic lacustrine stromatolites from hominid-bearing sediments east of Lake Rudolf, Kenya." *Nature (London)*, **47,** 520–523 (1974).

140. K. W. Butzer, "The Holocene lake plain of North Rudolph, East Africa," *Phys. Geogr.*, **1,** 42–58 (1980).

141. R. B. Owen, J. W. Barthelme, R. W. Renaut, and A. Vincens, "Palaeolimnology and archaeology of Holocene deposits north-east of Lake Turkana, Kenya," *Nature (London)*, **298,** 523–529 (1982).

142. A. O. Woodford, "Johannes Walther's law of the correlation of facies: Discussion," *Geol. Soc. Am. Bull.*, **84,** 3737–3740 (1973). See page 3737.

143. M. D. Picard and L. R. High Jr., "Criteria for recognizing lacustrine rocks." In J. K. Rigby and W. K. Hamblin, eds., *Recognition of Ancient Sedimentary Environments.*, Soc. Econ. Paleontologists Mineralogists Special Publication, 16, 1972, pp. 108–45.

144. L. A. Hardie, J. P. Smoot and H. P. Eugster, "Saline lakes and their deposits: A sedimentological approach." In A. Matter and M. E. Tucker, eds., *Modern and Ancient Lake Sediments*, Blackwell Scientific Publications, Oxford, 1978, pp. 7–41.

145. P. G. Sly, "Sedimentary processes in lakes," In A. Lerman, ed., *Lakes: Chemistry, Geology, Physics*, Springer–Verlag, New York, 1978, pp. 65–90.

146. J. L. Richardson, "Changes in level of Lake Naivasha, Kenya, during post-glacial times," *Nature (London)*, **209,** 290–291 (1966).

147. D. L. Galat, E. L. Lider, S. Vigg, and S. R. Robertson, "Limnology of a large, deep, North American terminal lake, Pyramid Lake, Nevada, U.S.A." *Hydrobiologia*, **82,** 281–317 (1981).

148. H. P. Eugster and L. A. Hardie, "Saline Lakes." In A. Lerman ed., *Lakes: Chemistry, Geology and Physics*, Springer–Verlag, New York, 1978, pp. 237–293.

149. R. J. Madison, "Hydrology and chemistry of Great Salt Lake." In M. Le R. Jensen, ed., "Guidebook of North Utah," *Utah Geol. Mineral. Surv. Bull.*, **82,** 141–157 (1969).

150. G. Müller, G. Irion, and U. Förstner, "Formation and diagenesis of inorganic Ca–Mg carbonates in the lacustrine environment," *Naturwiss*, **59,** 158–164 (1972).

151. K. Kelts and K. J. Hsü, "Freshwater carbonate sedimentation." In A. Lerman, ed., *Lakes: Chemistry, Geology, and Physics*, Springer-Verlag, New York, 1978, pp. 295–323.

152. J. T. Teller, J. M. Bowler, and P. G. Macumber, "Modern sedimentation and hydrology in Lake Tyrrell, Victoria," *J. Geol. Soc. Aust.*, **29,** 159–175 (1982).

153. R. S. Rosich and P. Cullen, "Lake sediments: Chemical composition and some aspects of their formation and diagenesis." In Trudinger *et al.*, eds., *Biogeochemistry of Ancient and Modern Environments*, Australian Academy of Science, Canberra, 1980, pp. 105–115.

154. R. C. Surdam and R. A. Sheppard, "Zeolites in saline, alkaline-lake deposits." In L. B. Sand and F. A. Mumpton, eds., *Natural Zeolites: Occurrence, Properties, Use*, Pergamon, Oxford, 1978, pp. 145–174.

155. A. Singer and P. Stoffers, "Clay mineral diagenesis in two East African lake sediments," *Clay Minerals* **15,** 291–307 (1980).

156. T. E. Cerling, "Paleochemistry of Plio-Pleistocene Lake Turkana, Kenya," *Palaeogeogr., Palaeoclimatol., Palaeoecol.*, **27,** 247–285 (1979).

157. A. Zuber, "On the environmental isotope method for determining the water balance components of some lakes, *J. Hydrol.*, **61,** 409–427 (1983).

158. J-C. Fontes and R. Gonfiantini, "Comportement isotopique au cours de l'évaporation de deux bassins sahariens," *Earth Planet. Sci. Lett.*, **3,** 258–266 (1967).

159. J. R. Gat, "Isotope hydrology of very saline lakes." In A. Nissenbaum, ed., Hypersaline Brines and Evaporitic Environments, *Dev. Sedimentol.*, **28,** 1–7 (1980).

160. P. I. Abell, "Palaeoclimates at Lake Turkana, Kenya, from oxygen isotope ratios of gastropod shells," *Nature (London)*, **297,** 321–323 (1982).

161. E. T. Degens and P. Stoffers, "Stratified waters as a key to the past," *Nature (London)*, **263,** 22–27 (1976).

162. P. Stoffers and R. E. Hecky, "Late Pleistocene–Holocene evolution of the Kivu-Tanganyika Basin." In A. Matter and M. E. Tucker, eds., *Modern and Ancient Lake Sediments,* Blackwell Scientific Publications, Oxford, 1978, pp. 43–44.

163. J. J. Tiercelin, R. W. Renart, G. Delibrias, J. Le Fournier, and S. Bieda, "Late Pleistocene and Holocene lake level fluctuations in the Lake Bogoria basin, northern Kenya rift valley," *Palaeoecol. Afr.*, **13,** (1981).

164. A. P. Kershaw, "Local pollen deposition in aquatic sediments on the Atherton Tableland, North-Eastern Australia," *Aust. J. Ecol.*, **4,** 253–263 (1979).

165. M. A. Brock, "The ecology of halophytes in the south-east of South Australia," *Hydrobiologia*, **81/82,** 23–32 (1981).

166. A. M. Mannion, "Diatoms: Their use in physical geography," *Prog. Phys. Geogr.*, **6,** 233–259 (1982).

167. F. Gasse, "Late Quaternary changes in lake levels and diatom assemblages on the southeastern margin of the Sahara," *Palaeoecol. Afr.*, **12,** 333–350 (1980).

168. J. L. Richardson, T. J. Harvey, and S. A. Holdship, "Diatoms in the history of shallow East African lakes," *Pol. Arch. Hydrobiol.*, **25,** 341–353 (1978).

169. D. M. Churchill, R. W. Galloway, and G. Singh, "Closed Lakes and the Palaeoclimatic Record." In A. B. Pittock, L. A. Frakes, D. Jenssen, J. A. Peterson, and J. W. Zillman, eds., *Climatic Change and Variability: A Southern Perspective*, Cambridge University Press, Cambridge, 1978, pp. 97–108.

170. F. Gasse, J. F. Talling and P. Kilham, "Diatom assemblages in East Africa: Classification, distribution and ecology," *Rev. Hydrobiol. Trop.*, **1,** 3–34 (1983).

171. F. W. Bachhuber and W. A. McClellan, "Paleoecology of marine foraminifera in the pluvial Estancia Valley, central New Mexico," *Quat. Res.*, **7,** 254–267 (1977).

172. R. V. Burne, J. Bauld and P. De Deckker, "Saline lake charophytes and their geological significance," *J. Sediment. Petrol.*, **50,** 281–293 (1980).

173. L. D. Delorme, "Paleoecological determinations using Pleistocene freshwater ostracodes." In H. Oertli, ed., *Colloquium on the Paleoecology of Ostracodes, Pau, July 1970,* Société Nationale des Pétroles d'Aquitaine, Pau, (1974) pp. 341–347.

174. L. D. Delorme, S. C. Zoltai, and L. L. Kalas, "Freshwater shelled invertebrate indicators of paleoclimate in northwestern Canada during late Glacial times." *Can. J. Earth Sci.*, **14,** 2029–2046 (1977).

175. J. P. Bradbury, B. Leyden, M. Salgado-Labouriau, W. M. Lewis, C. Schubert, M. W. Binford, D. G. Frey, D. R. Whitehead, and F. H. Weibezahn, "Late Quaternary environmental history of Lake Valencia, Venezuela," *Science*, **214,** 1299–1305 (1981).

176. S. F. Bedwell, *Fort Rock Basin: Prehistory and Environment*, University of Oregon Press, Eugene, 1973.

177. J. E. G. Sutton, "The Aquatic Civilisation of Middle Africa," *J. Afr. His.*, **15,** 527–546 (1974).

178. D. L. Weide and M. L. Weide, "Time, space and intensity in Great Basin paleoecological models." In D. D. Fowler, ed., *Models and Great Basin Prehistory: A Symposium*, Desert Research Institute, Publications in the Social Sciences, 12, 1977, pp. 79–111.

179. F. A. Street and A. T. Grove, "Global maps of lake-level fluctuations since 30,000 B.P.," *Quat. Res.* **12,** 83–118 (1979).

180. G. Digerfeldt, "A Pre-Boreal water-level lowering in Lake Lyngsjö, Central Halland," *Geol. Föreningens Förhandlingar*, **98,** 329–336 (1976).

181. H. Faure, "Lacs quaternaires du Sahara," *Mitt. In. Vereiningung Limnol.*, **17,** 131–146 (1969).

182. A. T. Grove and A. S. Goudie, "Late Quaternary lake-levels in the Rift valley of Southern Ethiopia and elsewhere in Tropical Africa," *Nature (London)*, **234,** 403–405 (1971).

183. K. W. Butzer, G. L. Isaac, J. L. Richardson, and C. Washbourn-Kamau, "Radiocarbon dating of East African lake-levels," *Science*, **175,** 1069–76 (1972).

184. M. A. J. Williams, "Late Pleistocene tropical aridity synchronous in both hemispheres," *Nature (London)*, **253,** 617–618 (1975).

185. F. A. Street and A. T. Grove, "Environmental and climatic implications of Late Quaternary lake-level fluctuations in Africa," *Nature (London)*, **261,** 385–390 (1976).

186. J. M. Bowler, G. S. Hope, J. N. Jennings, G. Singh, and D. Walker, "Late Quaternary climates of Australia and New Guinea," *Quat. Res.*, **6,** 359–394 (1976).

187. P. Rognon and M. A. J. Williams, "Late Quaternary climatic changes in Australia and North Africa: A preliminary investigation," *Palaeogeogr., Palaeoclimatol., Palaeoecol.*, **21,** 285–327 (1977).

188. A. Baumgartner and E., Reichel, *The World Water Balance*, Elsevier, Amsterdam, 1975.

189. J. P. Peixoto and A. H. Oort, "The atmospheric branch of the hydrological cycle." In F. A. Street-Perrott, M. Beran and R. A. S. Ratcliffe, eds., *Variations in the Global Water Budget*, D. Reidel, Dordrecht, 1983, pp. 5–65.

190. F. A. Street-Perrott, N. Roberts, and S. Metcalfe, "Geomorphic implications of Late Quaternary hydrological and climatic changes in the Northern hemisphere tropics," In I. Douglas, ed., *Geomorphology and Environmental Change in Tropical Latitudes*, Allen and Unwin, London, (in press).

191. J. E. Kutzbach, "Monsoon rains of the Late Pleistocene and Early Holocene: Patterns, intensity and possible causes of changes." In F. A. Street-Perrott, M. Beran, and R. A. S. Ratcliffe, eds., *Variations in the Global Water Budget*, D. Reidel, Dordrecht, 1983, pp. 371–389.

192. L. B. Leopold, "Pleistocene climate in New Mexico," *Am. J. Sci.*, **249,** 152–168 (1951).

193. R. W. Galloway, "The full-glacial climate in the southwestern United States," *Ann. Assoc. Am. Geogr.*, **60,** 245–256 (1970).

194. G. H. Dury, "Paleohydrologic implications of some pluvial lakes in northwestern New South Wales, Australia," *Bull. Geol. Soc. Am.*, **84,** 3663–3767 (1973).

195. F. A. Street, "Late Quaternary precipitation estimates for the Ziway-Shala Basin, Southern Ethiopia," *Palaeoecol. Afr.*, **11,** 135–143 (1979).

196. J. E. Kutzbach, "Estimates of past climate at Paleolake Chad, North Africa, based on a hydrological and energy-balance model," *Quat. Res.*, **14,** 210–223 (1980).

197. L. V. Benson, "Paleoclimatic significance of lake-level fluctuations in the Lahontan Basin," *Quat. Res.*, **16,** 390–403 (1981).

198. G. Tetzlaff and L. J. Adams, "Evaporation of Lake Chad in the present and in the past," In F. A. Street-Perrott, M. Beran, and R. A. S. Ratcliffe, eds., *Variations in the Global Water Budget*, D. Reidel, Dordrecht, 1983, pp. 347–360.

199. S. A. Schumm, "Quaternary paleohydrology." In H. E. Wright Jr. and D. G. Frey, eds., *The Quaternary of the United States*, Princeton University Press, Princeton, 1965, pp. 783–794.

200. R. G. Barry, "Evaporation and transpiration." In R. J. Chorley ed., *Water Earth and Man*, Methuen, London, 1969, pp. 169–184.

201. D. L. Weide, "Great Basin Holocene climates: A gradient of equability," *Abstr. Geol. Soc. Am., 93rd Ann. Meet.*, November (1980).

202. J. E. Kutzbach, Unpublished data.

203. S. Holdship, "The Paleolimnology of Lake Manyara, Tanzania: A diatom analysis of a 56 meter sediment core," Ph.D. diss. Duke University, Durham, N.C. 1976.

204. S. E. Nicholson and H. Flohn, "African environmental and climatic changes and the general atmospheric circulation in Late Pleistocene and Holocene," *Climatic Change*, **2,** 313–348 (1980).

205. Fan Yun-qi, "Chemical characteristics of Xizang lakes." In Liu Dong-Sheng, ed., "Geological and Ecological Studies of Qinghai-Xizang Plateau, Vol. 2: Environment and Ecology of Qinghai-Xizang Plateau," *Proceedings of the Symposium on Qinghai-Xizang (Tibet) Plateau, Beijing, China*, Science Press, Beijing; Gordon and Breach, New York, 1981, pp. 1705–1711.

8

PALEOCLIMATIC MODELING

Barry Saltzman

Department of Geology and Geophysics, Yale University,
New Haven, Connecticut

The various lines of evidence from which we can infer the past climatic changes on the earth have been discussed in detail in the previous chapters. A brief review of some of the salient features of climate variability over intervals of 10^6, 10^5, 10^4, and 10^{2-3} years, respectively, is given pictorially in figure 1 *a–d*. A qualitative, schematic representation of the variance spectrum for climatic change is shown in figure 2. Given the resolution with which the values shown in figure 1 are determined, these curves can only represent "running average conditions" over finite sub-intervals of no less than 10 years and, in the case of the 10^6 year record, probably close to 1000 years.

An ideal goal of climate modeling is to demonstrate that variations such as those in figure 1 follow deductively from a simple set of mathematical statements (i.e., a "model") which is assumed to be rooted in the conservation laws for mass, momentum, and energy. The success of such theoretical modeling endeavors will rest on the degree to which the models yield solutions that are in accord with all known simultaneous lines of evidence, and have the additional quality of correctly predicting other simultaneous variability as yet unperceived. To correctly predict *future* variability would, of course, be a great triumph, but, unfortunately, such glories will be possible only for the very shortest time scales of climatic variability.

Since we hold as an article of faith that our models must be expressions of the known conservation laws, our particular task is to isolate from the myriad of potential feedbacks those that are essential to account for the variability observed on a given time-scale. We are in fact dealing with what may truly be termed a 'complex system' involving the many physical domains schematically portrayed in figure 3 (atmosphere, hydrosphere, cryosphere, lithosphere, and biosphere), each possessing different, heterogeneous, properties describable by many variables, and varying with vastly different time-constants over vastly different spacial scales. While the mathematical form of the governing thermo-mechanical laws for each domain or sub-domain may be known, these same equations may be practically useless when

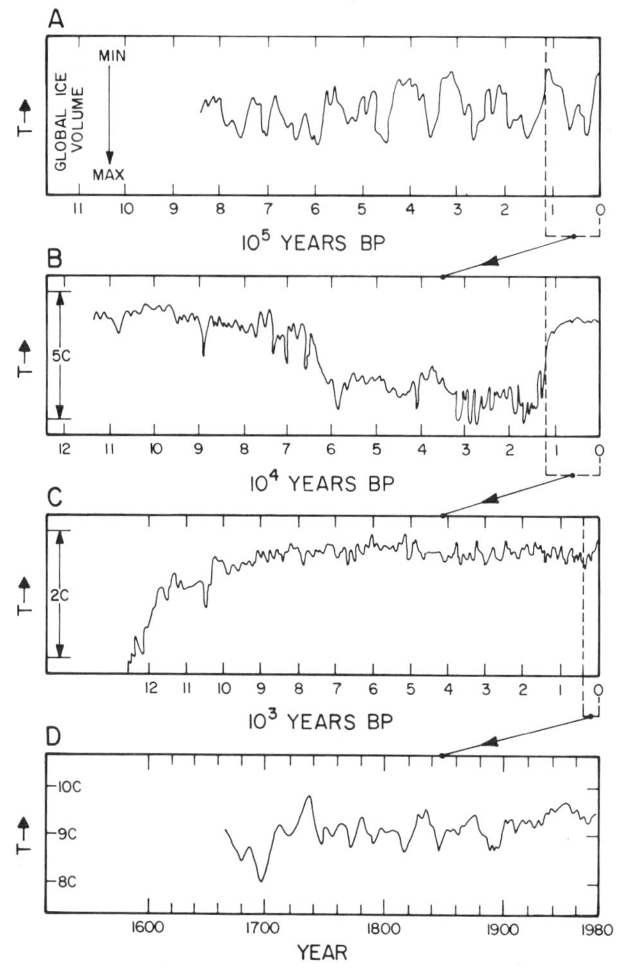

Figure 1. Selected climatic time series for the past one million years. (*A*) Global ice volume deduced from oxygen isotope variations of plantonic foraminifera in a deep sea core (Shackleton and Opdyke[122]; (*B*) and over the last 10,000 years (*C*), based on oxygen-isotope variations in a Greenland ice core (Dansgaard et al.[143]; (*D*) thermometric measurements in England over the past 300 years (Mason[144]). From Saltzman.[2]

applied to the whole 'complex' system on paleoclimatic time scales. The development of completely new mathematical forms that still satisfy the conservation principles may be required.

To begin any discussion of climate modeling we must adopt at least a "working" definition of what we seek to account for. Briefly stated, we consider *climate* to be the running-average statistical state of the atmosphere and ocean together with those portions of the bio-lithosphere that can vary freely with these fluid environments. As a minimum, we should take the running-average interval (which we shall call δ) to be greater than the time span of a succession of about 10 synoptic weather

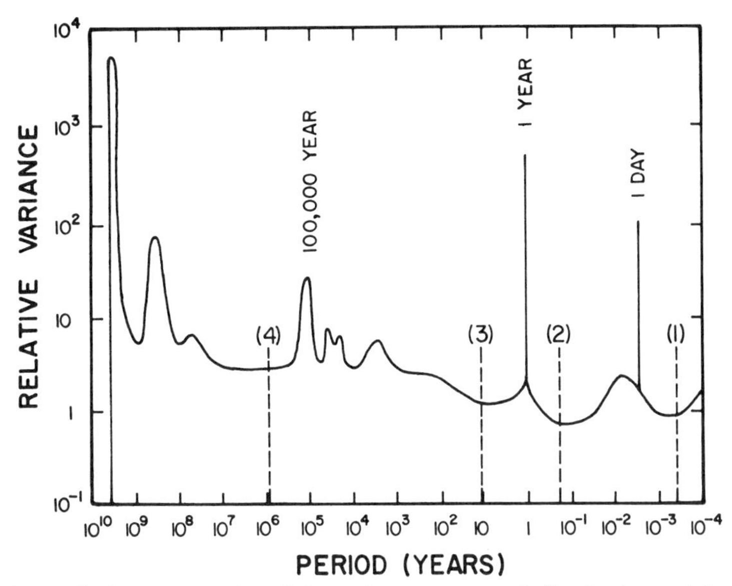

Figure 2. Qualitative representation of the variance spectrum of climatic change (adapted from Mitchell[145]). Dashed lines (1)–(4) are located at relative variance minima (i.e., gaps), representing optimal time periods over which to define average states (see table 1).

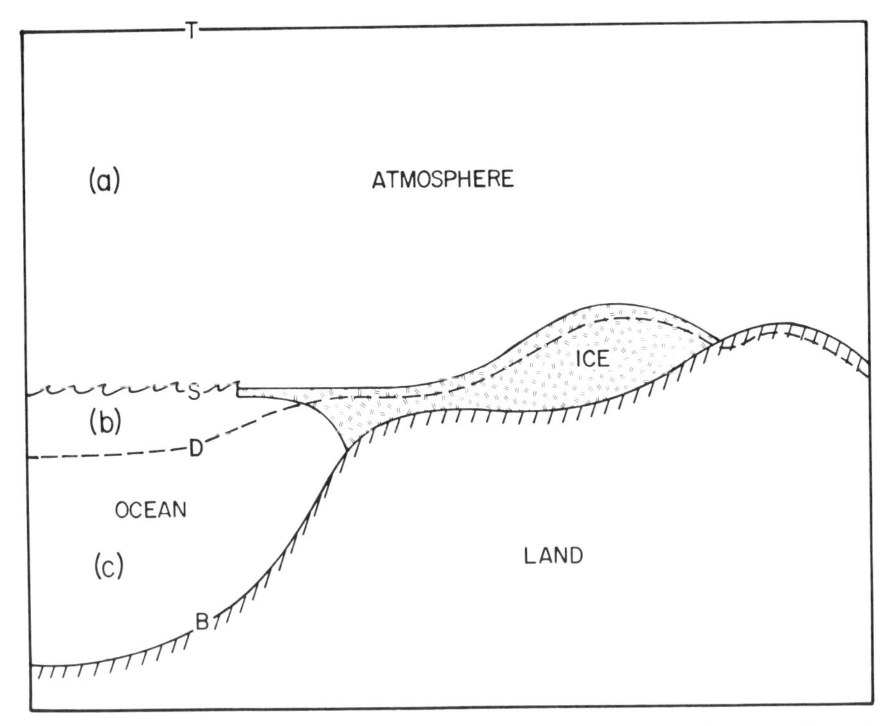

Figure 3. Schematic cross-section of the climatic system showing (*a*) the domain of atmosphere, (*b*) subsurface boundary layer including both mixed ocean layer and "active" land layer, and (*c*) deep ocean and cryosphere. Their boundary surfaces are denoted by T, S, D, and B.

Table 1. A Resolution of Climatic Variability[2]

Averaging Period (δ)	Spectral Band	Physical Phenomena	Known Forcing
	(0–1 hour)	micro and meso-meteorological eddies: turbulence and convection	
1 hour		synoptic average	
	(1 hour-3 months)	Diurnal cycle, cyclone waves, blocking, and index variations	Diurnal solar radiation cycle
3 months		Seasonal average	
	(3 months-10 years)	Annual cycle, year-to-year (interannual variability)	Annual solar radiation cycle
10 years		Climatological average	
	10 year-1 million years	Historical and Paleoclimatological variations, ice ages	Milankovitch earth-orbital radiation cycles
1 million years		Tectonic average	
	1 million years-4.5 billion years	Ultra-long-term climate variations influenced by global tectonics and planetary evolution	Continental drift
4.5 billion years		Age of the earth	

systems (e.g., a few months) which would leave the seasonal cycle as a component of climatic *change*. However, we shall here adopt a decade (i.e., $\delta = 10$ years) as our averaging interval (in conformity with the resolution of most geologic data on longer term climatic change), thereby relegating the seasonal cycle to the role of a forced climatic "eddy" that is part of the variance structure. In table 1 we list other possible choices of δ, all of which are located near what we believe to be relative minima, that is, "gaps," in the spectral variance of climatic change (cf., figure 2). By "statistical state" we mean not only the *average* values of all the variables defining the condition of the fluid envelope (e.g., temperature, motion, water phase, and trace substance mass distributions), but also the *variances* and higher moments of the probability distributions of these variables.

Fundamental Equations

Individually, the atmosphere, oceans, and ice masses can be treated as continua describable by field variables representing an infinite number of degrees freedom.

The behavior of such continua are generally governed by *partial* differential equations which are the classical fluid dynamical equations expressing conservation of mass (continuity equations), momentum (Navier-Stokes equations of motion), and energy (first law of thermodynamics), along with diagnostic relationships for internal energy and the thermodynamic state.

Let us assume that such a set of equations governing the nearly "instantaneous" values of all the variable describing the climatic system can be written in an "exact" form. In our discussion that follows, we shall start by taking these fundamental partial differential equations, governing the "instantaneous," continuum, variations of an arbitrary mass of material within any domain of the climatic system, as the most fundamental building block of climate theory—but it will become apparent that when we apply the necessary averaging to these equations to make them govern the measureable, running-average quantities we are really interested in, these equations can change into new forms that bear little structural similarity to their generic forebears. Complementary discussions of these questions and expanded formal developments of some of these ideas are found in the references,[1,2] from which we shall occasionally draw material.

The following general set of equations is commonly adopted for the behavior of any unit volume in the atmosphere and ocean:

(mass)

$$\frac{\partial \rho}{\partial t} = -\nabla \cdot \rho \mathbf{V} \tag{1a}$$

$$\frac{\partial \rho \xi}{\partial t} = -\nabla \cdot \rho \xi \mathbf{V} + \mathcal{S}_\xi \tag{1b}$$

(momentum)

$$\frac{\partial \rho \mathbf{V}}{\partial t} = -\nabla \cdot \rho \mathbf{V}\mathbf{V} - \nabla p + \rho \nabla \Phi - 2\rho \Omega \times \mathbf{V} + \rho \mathbf{F} \tag{2}$$

(energy)

$$\frac{\partial \rho e}{\partial t} = -\nabla \cdot \rho e \mathbf{V} - p \nabla \cdot \mathbf{V} + \rho q \tag{3}$$

plus the diagnostic relations

$$e = cT \tag{4}$$

$$\rho = \begin{cases} p/RT & \text{(atmosphere)} \quad (5a) \\ \rho^*[1 - \mu_T(T - T^*)] & \text{(ocean)} \quad (5b) \end{cases}$$

where ρ = density
ξ = mixing ratio of an arbitrary trace substance
p = pressure
T = temperature
$\mathbf{V} = \mathbf{v} + w\mathbf{k}$
$\mathbf{v} = u\mathbf{i} + v\mathbf{i}$
$u = a \cos \varphi \, d\lambda/dt$ (eastward speed)
$v = ad\varphi/dt$ (northward speed)
$w = dz/dt$ (vertically upward speed)
$\mathbf{i,j,k}$ = unit vectors eastward, northward, and upward, respectively
λ = longitude
φ = latitude
z = vertical distance
t = time
a = radius of the earth
$\nabla = \mathbf{i}\partial/a \cos \varphi\partial\lambda + \mathbf{j}\partial/a\partial\varphi + \mathbf{k}\partial/\partial z$
$\Phi = gz$
g = acceleration of gravity
Ω = angular velocity of the earth
$\mathbf{F} = \mathfrak{F}_\lambda \mathbf{i} + \mathfrak{F}_\phi \mathbf{j}$ = frictional force per unit mass
$\mathcal{S}_\xi$ = source function for trace substance
q = rate of heat addition per unit mass due to radiation, conduction, phase changes, and viscosity
e = internal energy
c = specific heat at constant volume
R = gas constant for air
x^* = standard value of x
μ_T = coefficient of thermal expansion

$$\frac{d}{dt} = \frac{\partial}{\partial t} + \mathbf{V} \cdot \nabla$$

These equations are not adequate to treat the complete climatic system. For example, other equations for all the modes of heating included in q, for the rate of change of ice mass at any point, and for the rates of change of all the important trace substances in the climatic system (e.g., atmospheric water vapor and CO_2 content, oceanic salinity) are also required. The modeling of water vapor and CO_2 content of the atmosphere leads inevitably to a consideration of the full hydrologic and carbon cycles involving all parts of the climatic system.

In order to obtain dynamical equations governing the evolution of the running time-average variables in which we are interested, these "exact" instantaneous equations (e.g., 1–5) must be averaged over the relevant time interval δ. As is well known from turbulence theory[3,4] there are at least two major consequences of such an averaging process: (1) eddy stress terms are introduced that must be parameterized to effect closure in terms of the averaged variables, and (2) stochastic forcing terms are introduced due to the effects of non-systematic random departures of the instantaneous values from the mean values and the impossibility of achieving exact parameterizations. A more formal discussion of the questions involved in achieving a set of partial differential equations governing the time-average variables is given in ref. 2 (pp. 177–181).

Suffice it to say that in order to solve these partial differential equations they must be reduced either to a finite system of *ordinary* differential equations governing the amplitudes of a truncated orthogonal spatial expansion of the variables, or to a system of difference equations governing values at a discrete space-time grid. Thus, in practice we are forced to deal with a more approximate, reduced, system that can usually be written in the form

$$\frac{dx_j}{dt} = f_j(x_j, F_j) + \mathcal{R}_j \tag{6}$$

where x_j denotes any time-averaged climatic variable, $f_j\ (x_j,\ F_j)$ is the deterministic part that, in addition to linear and nonlinear terms in x_j, can contain time-dependent non-autonomous forcing components F_j, and $\mathcal{R}_j$ is the stochastic part. Depending on whether this system is written for grid point values or orthogonal (e.g., Fourier) components, the variables x_j will be functions of (λ, ϕ z) or wave numbers (k, l, m) in the case of a Fourier expansion.

Such a set of time dependent ordinary differential equations governing averaged quantities and containing stochastic forcing terms constitutes a stochastic-dynamical system characterized by a very large but discrete number of degrees of freedom. If the system governs the synoptic average variables ($\delta \approx$ 1 hr.) with the synoptic spatial resolution (thereby requiring parameterization only of sub-synoptic frequencies and spatial scales) we speak of the system as a "general circulation model" or GCM. In this case the solution can be iterated forward in time to generate a full set of statistics (e.g., means, variances, frequencies of rarer events) that constitute the climate. This, of course, is a demanding procedure in time and resources.

For studies of long term climatic change, that is, evolution over hundreds of years and more, it is natural to consider a "statistical-dynamical model" (SDM) comprised of equations governing a longer term average such as 10 years. In this case we must parameterize not only sub-synoptic phenomena but all phenomena up to and including inter-annual variations, a most difficult and challenging task. While the stochastic amplitude may be relatively small in the GCM, it will tend to be large in the SDM, though hopefully not quite of the same order as the deterministic terms. Moreover, additional equations, including the parameterization formulas, will be necessary to deduce the higher order statistics (such as the spatial and temporal variances and amplitude of the seasonal cycle) that are needed for a full description of climate. This constitutes a form of "inverse problem of climatology"[5] wherein one seeks to infer the statistics of the higher frequency, higher wave-number behavior from a knowledge of the low frequency, low wave number distribution.

It seems likely that to describe the macro-behavior of the climatic system with the accuracy that observations can reveal will require a much smaller number of variables (and their governing equations) than are represented by the full set (6) applied to all the variables, in all the domains, at a synoptic spatial grid. Thus, a major challenge in developing the theory of long-term climatic fluctuations will be to learn how to truncate (or space-average) the system in ways that capture the main variabilities at the same time permitting closure of the system by physically valid parameterizations. Related to this is the need to identify groupings of variables that are "coherent" or diagnostically related so that they can be represented by much reduced sets of variables or even a single variable. A formal procedure for accomplishing such reductions by successive spatial integrations[1] starts with the very simplest one-variable model obtained by integrating over the entire climatic system[6–8] and proceeds by systematically expanding the model with added variables and equations. A partial test of the success of this process is the degree to which the unexplained variability is of a random white-noise variety on which no further determinism can be brought to bear.

In any event, systems of the form of equation 6 relevant for the study of climate can be developed with degrees-of-freedom (i.e., levels of complexity) ranging from one to the huge number involved in GCMs. At this point in the development of a theory of climate there is ample justification for studies involving all levels of complexity. It would appear that we are now at the rather primitive stage of simply "getting a feel" for the role of all the competitive feedbacks and forcings that are possible in the system depending on the choice of time scale of variability.

The main developments of a theory of climate variation as described here are summarized in the flow diagram, figure 4, which shows two routes to a theory of climatic variations. Emphasized so far is the *deductive* method, in which a model is derived from what we believe to be the fundamental statements of conservation of mass, momentum, and energy by successive averaging and parameterization of flux processes. The *inductive* method derives from a recognition that, because of the difficulty or impossibility of achieving adequate parameterization and the accuracy required, it may be necessary (or equally satisfactory) to formulate the governing statistical-dynamical model system on more heuristic grounds (i.e., to design the model, *ab initio,* with a view to generating an output that agrees with all the known observational evidence). This latter method can encompass purely statistical and stochastic models, "feedback models" suggested by qualitative physical reasoning, and the general "mean-field" methods.

Time Scales: Equilibration in the Different Climatic Domains

The climatic system is extremely heterogeneous, containing domains (or subsystems) that are all interactive to some extent but have vastly different properties and modes of behavior if considered alone. One important property distinguishing each of the various domains is its "time-scale," that is, the time it would take for dissipative processes acting alone in the absence of continued forcing to remove departures from equilibrium. A short time-scale indicates that the system (or subsystem) has the inherent capability of responding quickly to any perturbations from its equilibria. Thus, the time-scale is directly related to the so-called "equilibration

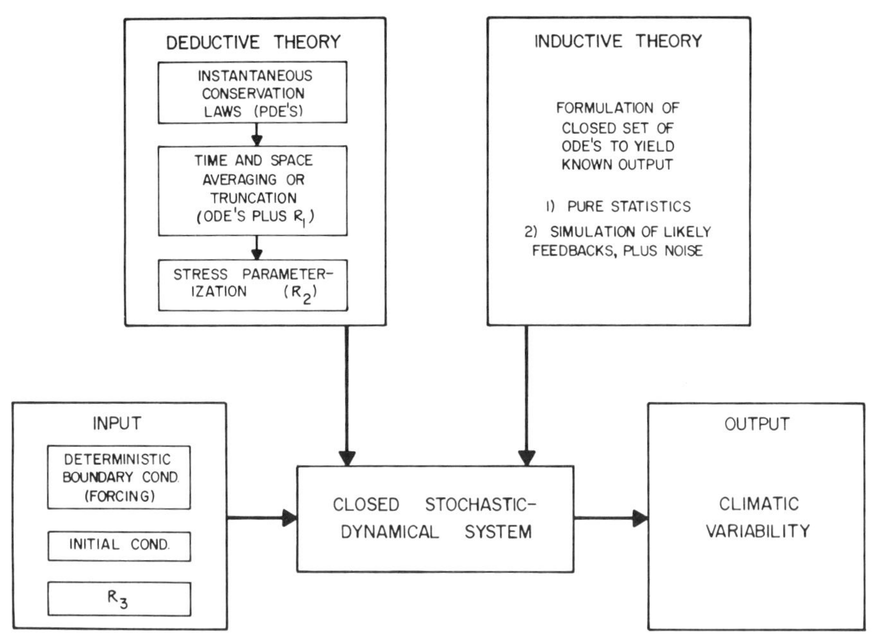

Figure 4. Schematic flow chart showing the components of a theoretical model of climatic variability. Note the alternative "deductive" and "inductive" routes to a statistical-dynamical system governing the climatic variables. R_1, R_2, and R_3 denote stochastic components introduced due to sub-climatic average aperiodic fluctuations, random errors in parameterization, and random external inputs, respectively.

time" (ϵ) (also called the "response," "relaxation," or "adjustment" time), which is a measure of the time it takes for the system (or subsystem) to reequilibrate after a small change in its boundary conditions or forcing.

Some useful relationships between the nature of the governing equation for a given climatic variable x and its equilibration time ϵ_x, averaging period δ_x, and dominant periods of variability P_x are discussed in ref. 2, pp. 184–188, 215–218. For example, if δ_x lies within a "spectral gap" and is greater than ϵ_x the equation governing the variations of x is almost certain to be "diagnostic," that is, we can assume that a quasi steady-state, $dx/dt \approx 0$, prevails.

In table 2 we list the main climatic subsystems and their characteristic present dimensions and masses, thermal constants, estimates of the thermal equilibration time ϵ_T (δ), and rough estimates of the ice mass equilibration times ϵ_M. In the atmosphere, for example, the atmospheric thermal response is so fast that the climatic mean temperature can be considered as an equilibrium state. A schematic portrait showing the linkages and equilibration times for δ_c of all the domains comprising the complete climatic system is given in figure 5. In general, as we proceed from the top of this figure (i.e., the atmosphere) to the bottom (deep ocean, and ice sheets with associated bedrock) we encounter increasingly longer equilibration times (i.e., slower response times).

In forming table 2 we have considered each climatic domain separately. Actually, all adjacent domains are linked by complex physical processes involving cross-

Table 2. Properties of the Climatic Domains

Characteristic values of the following quantities are given (some of which are derived from a table given by Hoffert et al.[146]): present climatic mean horizontal area A, present climatic mean depth D, density ρ, present climatic mean mass $M = \rho DA$, specific heat c_h, heat capacity (Mc_h), molecular thermal diffusivity κ, vertical eddy thermal diffusivity $K_v(\delta)$, horizontal eddy thermal diffusivity $K_h(\delta)$, horizontal eddy length scale $L(\delta)$, vertical eddy diffusive time scale (D^2/K_v), horizontal eddy diffusive time scale (L^2/K_h), long wave radiation time scale for the atmosphere and surface layers of the land and ocean b^{-1}, thermal equilibration time $\epsilon_T(\delta) = [b + (\kappa/D^2) + (K_v/D^2)_\delta + (K_h/L^2)_\delta]^{-1}$, mass equilibration time ϵ_M for the ice domains, and observed or inferred major periods of variation P. The averaging periods considered are $\delta = \delta_s \approx 1$ hour $\approx 10^3$s (the synoptic average), and $\delta = \delta_c \approx 10$ years $\approx 10^8$s (the climatic average). Sources other than Hoffert et al.[146] used in constructing this table include Nace[147] for water inventory, Flint[97] for ice inventory, Priestley[148] for κ and K, and Birchfield[77] and Sergin[103] for estimates of ϵ_M.[2]

Variable	A	D	ρ	M	c	(Mc)	κ	K_v		K_h		L		$\left(\frac{D^2}{\kappa}\right)$	$\left(\frac{D^2}{K_v}\right)$		$\left(\frac{L^2}{K_h}\right)$		b^{-1}	ϵ_T		ϵ_M	P_{TM}
							Exponent of 10																
Units	10^{12}m^2	m	$\frac{kg}{m^3}$	10^{18}kg	$\frac{10^3 J}{kg\ K}$	$\frac{10^{21} J}{K}$	$\frac{m^2}{s}$	$\frac{m^2}{s}$		$\frac{m^2}{s}$		m		s	s		s		s	s		s	s
Averaging Period	δ_c							δ_s	δ_c	δ_s	δ_c	δ_s	δ_c		δ_s	δ_c	δ_s	δ_c		δ_c	δ_c	δ_c	
Climatic Domain																							
Atmosphere																							
1) Free	510	10^4	0–1	5	1	5	−5	1	2	1	6	6	7	13	7	6	11	8	6	7	6	—	5,7,9–12
2) Boundary layer	510	10^3	1	0.5	1	0.5	−5	1	1	1	6	6	6	11	5	5	11	6	6	5	5	—	5,7,9–12
Ocean																							
3) Mixed Layer	334	10^2	10^3	34	4	10^2	−7	−3	−2	−1	3	5	6	11	7	6	9	9	8	7	6	—	6,7,9–12
4) Deep	362	$4{\cdot}10^3$	10^3	1400	4	$5{\cdot}10^3$	−7	−4	−3	−2	2	(6–7)	6	13	11	10	14	10	—	11	10	—	7–12
5) Sea-Ice (pack-shelf)	30	$(1–10^2)$	10^3	0.5	2	$(10^{-1}–1)$	−6	—	—	—	—	—	—	(6–10)	—	—	—	—	—	(6–10)		6–10	7,9–12
Continents																							
6) Lakes and Rivers	2	10^2	10^3	0.2	4	1	−7	−3	−3	—	−2	—	2	11	—	6	—	6	6	6	6	—	7–12
7) Litho-Biosphere	131	2	$3{\cdot}10^3$	1	0.8	1	−6	—	—	—	—	—	—	6	—	—	—	—	6	6	6	—	5,7,9–12
8) Snow and Surface Ice Layer	80	1	$5{\cdot}10^2$	10^{-1}	2	10^{-1}	−7	—	—	—	—	—	—	5	—	—	—	—	5	5	5	5	5,7,9–12
9) Mountain Galciers	1	10^2	10^3	0.1	2	10^{-1}	−6	—	—	—	—	—	—	10	—	—	—	—	—	10	10	9	9–12
10) Ice Sheets	14	10^3	10^3	10	2	10	−6	—	—	—	—	—	—	12	—	—	—	—	—	12	12	11	10–12

boundary fluxes of mass, momentum, and energy that constitute "forcing" and "feedback" in the system. These linkages are indicated by the connecting lines in figure 5. Domains with large ϵ (slow response time) will tend to 'carry along' the domains with smaller ϵ which because of their fast response times tend to adjust quasi-statically to the changing boundary conditions imposed by the slow-response domain. In this sense the deep ocean and ice sheets can be considered as the "carriers" of long-term climatic variability, the prognostic equations for these domains governing the non-equilibrium evolution of the system. The nature and path of this evolution, however, can be markedly influenced by all the higher frequency effects of the fast response domain, the deterministic parts of which must be parameterized and the random parts of which must be included as stochastic forcing. By the same token, the low frequency, long response time phenomena can be considered as fixed conditions in deducing the separate behavior of the domain with higher frequencies and shorter response times. As a general rule, if two domains have similar response times the behavior in both domains must be solved for simultaneously.

One consequence of the above remarks that we have already mentioned, is that the decadal climatic mean ($\delta = \delta_c$) state of the *atmosphere* should be deducible as a "steady-state" problem subject to fixed boundary conditions imposed by a neighboring domain. From the values of ϵ given in table 2 we see that the response of the surface of the earth (i.e. land temperature, surface snow, and ice masses, and, to a slightly lesser extent, the mixed layer ocean temperature) is fast enough to be of an equilibrium nature comparable to the atmosphere. Therefore these aspects of the surface state should be deduced simultaneously with the atmospheric state,

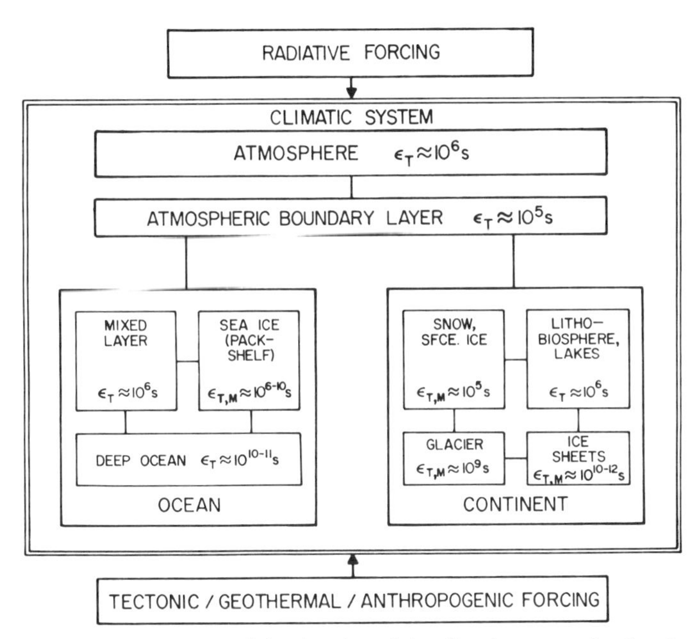

Figure 5. Schematic representation of the domains of the climatic system showing the estimated response times.

especially insofar as there are significant feedbacks between the atmosphere and these domains (e.g., the ice-albedo feedback). If we arbitrarily prescribe the surface snow-ice distribution, for example, one can indeed compute the atmospheric state that is required for equilibrium with it, but this equilibrium may not be consistent with the prescribed snow-ice field (e.g., the equilibrium may require surface temperatures that are well below freezing in regions of ample moisture wherre no ice is prescribed). Moreover it is important to realize that even though the atmosphere and its lower boundary surface layer may be in equilibrium, judging from the geologic record of long-term climatic change this equilibrium most likely entails a small but significant *dis*equilibrium of the total ice mass field and the deep ocean temperature. In a later section we point out that this disequilibrium is so small for the major ice age variations that it is impossible to measure it or compute it from physical parameterizations.

A complete model of climatic change should require the specification only of the purely *external* factors that affect the climate system (i.e., those factors that influence the system but are not themselves influenced by any climatic variable—e.g., solar radiation, earth-orbital variations, tectonic processes) with all internal variables being free or self-determining. However, the total climatic system is so complex that a practical course to follow is to first develop *in*complete models that can illustrate and give physical insight into the processes and interactions that must ultimately be encompassed by a complete model. The complexity of the system is immediately apparent from the spectrum shown in figure 2, in which one finds considerable climatic variability at frequencies far removed from any known forcing. The implication is that the internal system contains much instability and nonlinearity. It would appear that the observed "output" of the climatic system is a rich mixture of forced and free responses.

This time-scale discussion provides one of several possible bases for subdividing the climate problem into meaningful pieces. A first suggested division is between: (1) the *fast-response* parts of the system (e.g., the atmosphere and surface layers of the oceans and continents) for which we can require an essentially steady-state or equilibrium theory, and (2) the *slow-response* parts of the system (e.g., the deep ocean and ice masses) that are the time-dependent carriers of long-term climatic change. We shall discuss modeling efforts in each of these time-scale ranges in the next two sections, respectively.

We note, also, that each domain of similar time constant can be further resolved by spatial averaging or spectral truncation. A hierarchal resolution of this sort has been outlined in reference 1. The most elemental distinctions are between the zonal-average climatic state and the zonally-asymmetric state (which must be due to the presence of lithospheric topography) and between a vertical-average state and the vertically-structured state.

FAST RESPONSE CLIMATIC EQUILIBRIUM OF THE ATMOSPHERE AND SURFACE LAYERS OF THE EARTH

We have noted that the atmosphere and upper layer of the earth's surface [i.e., domains (*a*) and (*b*) of figure 3] can respond so quickly to perturbations that we can consider the climatic-average (i.e., δ = 10 y) properties of these domains to be in a quasi-equilibrium state with the much more slowly varying deep oceans and ice sheets. Numerous models have been formulated and solved to deduce such equilibrium states, with relatively high degrees of fidelity to the presently observed climate. Some of these solutions, ranging from the most detailed calculations based on GCMs to the more heavily parameterized SDMs that include, as the simplest special case, the so-called Budyko-Sellers "energy balance" model (EBM) are reviewed in this section.

Since in any theory of climatic variability we are usually most interested in the changing state of the atmospheric and surface ocean and land boundary layers where human activities are concentrated, all models of climate must ultimately include such equilibrium solutions either explicitly or implicitly. Moreover, as previously discussed, the equilibrium solution for the atmosphere and upper layers of the ocean and land will generally entail a small but significant imbalance of the deep ocean temperature and ice mass that can be viewed as a forcing function for longer-term changes in these domains. For this reason the ultimate model of long-term paleoclimatic variations will inevitably require an interactive calculation involving time-dependent changes in ocean and ice states in quasi-static equilibrium with the atmosphere.

In deducing the climatic response of the atmosphere and surface boundary layer we must recognize the necessity for deducing not only the 10 year mean "climatic average" state, but also the forced *seasonal* "eddy." This can be obtained directly from the GCMs outlined below, but must be obtained by a special calculation based on the seasonal (e.g., three-month) average equations (which are prognostic for this domain) in the SDMs. Aside from its intrinsic importance as a part of a climatic description, the seasonal cycle is of great significance in determining the 10 year mean state and all longer-term climatic changes. These effects are particularly relevant in determining the full effects of earth-orbital fluctuations in long-term climatic change.[9]

The General Circulation Models (GCMs)

General circulation models (sometimes also called *explicit-dynamical models*) are based on equations of the form (1) to (5) governing the synoptic-average variables with a spatial resolution comparable to the global synoptic network. As seen from table 2, these equations must be *prognostic* equations. The time-dependent solutions of these equations represent a sequence of weather maps that are allowed to evolve numerically until the statistically steady state is achieved. The output is then averaged to yield the model "climate" in much the same manner as real weather records are processed to determine climatic norms. It is assumed that while the individual details of the evolving numerically generated maps cannot be accurate by the standards of 'weather prediction,' their *ensemble statistics* can be fairly accurate. Because these models treat a good deal of phenomena included in the high-energy frequency band between 1 hour and 1 week (see figure 2) by *explicit* deterministic physics, this approach represents the most rigorous and complete theory of the equilibrium of the atmosphere and surface layers to the quasi-steady boundary conditions imposed by the deep ocean and ice sheets. There are now many examples of such models and solutions, good reviews of which are given in references 10–12.

However, it must also be recognized that in spite of all the detail with which synoptic variations are treated explicitly, there is still a great deal of variability on smaller scales that must still be parameterized. Errors in these parameterizations as well as in the representation of the synoptic processes can be amplified in the iterative numerical process to such a degree that the final output may be no more accurate than that of much simpler models in which the synoptic scale phenomena is also parameterized. Moreover, because of the great amount of phenomena included simultaneously in a highly nonlinear system, a "cause and effect" understanding of the processes involved is difficult to diagnose.

Since we are most concerned here with paleoclimatic modeling, we shall discuss briefly a few examples of GCM solutions for past boundary conditions. At least four GCM studies have been made thus far to determine the response to the sea surface temperatures and ice conditions that prevailed at 18,000 years B.P. (see chapter 5) in comparison with corresponding solutions for the present boundary conditions which serve as a "control."[13–17] The most general results obtained in all these studies are the expected ice age reductions in atmospheric temperature and in water vapor content and precipitation. In figure 6, we have reproduced a sample solution obtained by Gates for one of many climatic variables, that is, for the global sea-level pressure distribution. Similar distributions are obtained by Manabe and Hahn[17] based on the same CLIMAP July boundary conditions. The agreement of the results using the two different GCMs is generally good, but there are many differences in the details. Comparative discussions of all the models are given in the above papers, as well as by Williams.[18,19]

Another example of the application of a GCM to paleoclimatic modeling is the study of Kutzbach[20]. Using the low-resolution model of Otto-Bliesner et al.[21], he demonstrates that an enhanced monsoonal circulation forced by the unusually large seasonal radiation variations that prevailed 9000 years B.P., due to earth-orbital changes, can explain the observed paleoclimatic evidence for stronger rainfall in Africa, Arabia, and India at that time.

As discussed by Smagorinsky,[12] however, given the enormous amount of computer time required for even a single experimental run to an asympotic equilibrium for a high-resolution GCM, it is prohibitive to do very much experimentation and long-term integration with these models. The above examples give only the "snapshot" equilibrium solutions for a single time in the past. To simulate long-term climatic change, for which the carrier time-dependent equations are those governing the deep ocean and ice masses, it would be necessary to calculate such equilibria at some regular interval, say every few hundred years. Even this would seem to be prohibitive for an analysis of the Quaternary glaciations.

The most significant result of such an atmospheric equilibrium calculation with reference to the problem of long-term climatic change is the determination of (1) the net flux of heat into or out of the oceans (that must be balanced by release or consumption of the latent heat of fusion and/or by a net radiative imbalance at the top of the atmosphere), and (2) the net rate of accumulation or melting and ablation

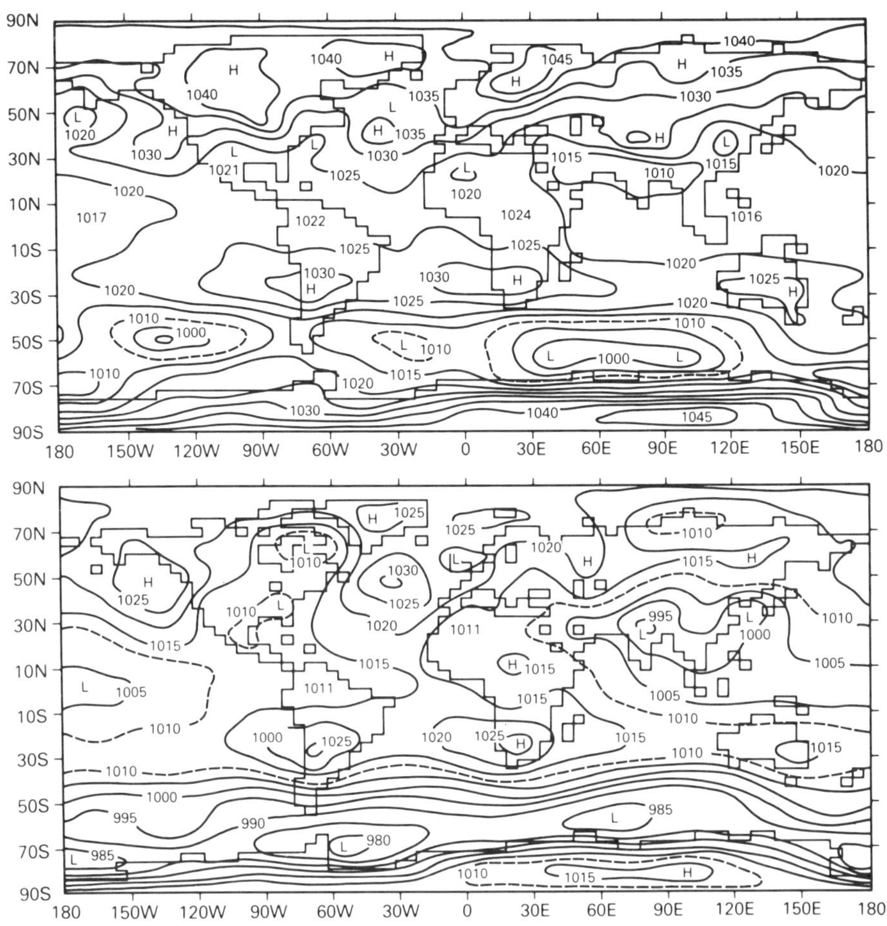

Figure 6. July sea-level pressure (mb) deduced for 18,000 years B.P. (above) and for present surface boundary conditions (below). After Gates.[16]

of ice mass. At present GCMs are generally constrained by the requirements that the net globally-averaged fluxes of heat across the top of the atmosphere and the lower boundary are identically zero. Although this would seem to be a reasonable constraint since the departures to be expected are small and perhaps not within the resolving power of the methods of computation now used, it is probably fair to say that unless some effort is made to relax these constraints and make such estimates (albeit crudely and subject to large errors or noise) the GCM will never be capable of being used to simulate long-term climate change in which the bulk of the ocean participates thermodynamically. For the very long-term changes on the scale depicted in figure 1*A* it is already clear that, because of the high levels of accuracy needed, it will be impossible to use GCMs to compute the observed variations in ice mass and the inferred variations in mean ocean temperature.[2,149]

Finally, we must recognize that the GCM typically contains a huge number of degrees of freedom as represented for example by the thousands of different equations for each dependent variable at each grid point in the three-dimensional spatial lattice. It seems reasonable to expect, however, that the macro-turbulent behaviour represented by these many equations is not altogether "free," but possesses some coherent organization to effect the the bulk transports of mass, momentum, and energy required by the conservation principle applied to the climatic-average state.

All these considerations point to the desirability of advancing the development of statistical-dynamical models in which the task of generating the synoptic eddy statistics by continuous hour-by-hour integration of a huge dynamical system is supplanted by the development of physically-based, deterministic representations of the parameterized effects of these eddies on the climatic-mean state. These SDMs will be the main focus for the remainder of this chapter.

Statistical-Dynamical Models (SDMs)

Statistical-dynamical models are based on equations governing the climatic-mean variables, which from table 2 are essentially diagnostic when applied only to the atmosphere and underlying surface layer. As noted above, SDMs are characterized by the need to parameterize not only the effects of fluctuations of frequencies less than about an hour as in the GCMs but also the effects of all frequencies up to the climatic averaging period chosen (e.g., 10 y). This can include the diurnal and seasonal cycles, meso-scale phenomena, synoptic weather waves, and even inter-annual variations. In view of the complexity and high energy level of all this sub-climatic variability, it is clear that the development of physically sound parameterizations must constitute a major challenge of dynamical meteorology. It is also likely that the stochastic terms in the governing SDM equations will be significant so that one could not expect to observe that the climatic system resides at any deterministic equilibrium even for fairly good parameterizations. Thorough reviews of the general foundations of SDMs are given in cited references.[1,22] Despite the difficulties in parameterization, SDMs based on rather crude approximations can account for a good deal of the large-scale spatial variability of the climatic mean state of the atmosphere and its underlying boundary. A most recent example of such a model is presented by Vallis[23] who also gives a brief review of earlier models.

General SDM equations derived from (1) to (5), governing the three-dimensional fields of the climatic (i.e., 10 year) average variables, for all domains, are given in reference 1. From these equations other sets can be derived governing the vertically-averaged horizontal fields, the zonal-average fields, and the "standing-wave" departures from the zonal-average fields.[1] We next review briefly some special models for these latter two fields and discuss some results relevant for paleoclimatic modeling.

Zonal-average SDM's. Using the simplest vertical resolution capable of representing mean poloidal motions and the hydrologic cycle, both of which are fundamental to a description of the atmospheric climatic state, Saltzman and Vernekar[24,25] constructed a model from which most of the presently observed zonal-mean, winter and summer, climatic statistics can be deduced to roughly the same accuracy as can be deduced from a GCM. This model is based on a two-layer representation of the atmospheric wind field, superimposed on the vertically averaged fields governed by equations given in reference 1.

In brief review, the deduced dependent variables of the model include the climatic means and variances of the temperature and wind, the humidity and all components of the surface and atmospheric heat, momentum, and water balances (e.g., precipitation and evaporation), horizontal and vertical fluxes of sensible and latent heat, and momentum due to eddies and mean poloidal motions. The vertical fluxes are determined within the atmosphere and at the interface between the atmosphere and subsurface media (ocean, land, ice), by means of a full set of parameterizations of short and longwave radiation, convection of sensible heat, and latent heat processes. The temperature at the base of the seasonal thermocline, T_D, is prescribed as a lower boundary condition and the average winter and summer insolation is prescribed as an upper boundary condition. To effect closure of the system, baroclinic and barotropic wave theory is used in a crude manner to parameterize the horizontal transports of heat, momentum, and water vapor in the atmosphere. More detailed and comprehensive reviews of the model specifications and results have been given.[1]

In figure 7 we show some sample comparative results of applying this model to present lower boundary conditions, and to the conditions that prevailed at 18,000 years B.P., the ice age maximum.[26]

One further aspect of the above two equilibrium solutions, for present and 18,000 years B.P. conditions, is of potential significance for modeling long-term climatic change. When present lower boundary ice coverage and sea surface temperatures are specified, it is found that the deduced equilibrium atmospheric climatic state is in reasonably good agreement with present observations, but this solution entails a net imbalance of the climatic system as a whole, that is, a net upward radiative flux at the top of the atmosphere, that goes along with a net loss of heat from the oceans of the same magnitude (as required by the condition of equilibrium). This result is not completely without observational support (cf. satellite radiation measurements reported[27,28]), but in view of the many possible errors in such net radiation measurements and the many inadequacies of the S-V model, it is probable that this agreement is fortuitous. Moreover, the magnitude of this imbalance seems inordinately large since the implied change in mean ocean temperature is at the rate of 0.01 Ky^{-1}. However, it is perhaps of greater significance that this same model requires a *reverse* set of fluxes for atmospheric equilibrium to be achieved when the ice age conditions that prevailed at 18,000 years B.P. are specified. This latter result is largely due to the fact that large regions of the high-latitude ocean, which under present ice-free conditions are losing heat rapidly, are "insulated" by sea ice in an ice age.[29] Thus, even if only the *direction* of the theoretically computed changes in the flux is valid it would appear that the "sea ice-insulator effect" may be of importance. The possibility exists that in a relatively ice-free state, the ocean as a whole tends to lose heat, while in a heavily glaciated state, the ocean as a whole tends to gain heat, suggesting that there can be free oscillations between mean ocean temperature and sea ice extent existing if mean ocean temperature can affect the ice extent. These ideas will be of relevance in discussion of the time-dependent theory of longer-term climatic variations.

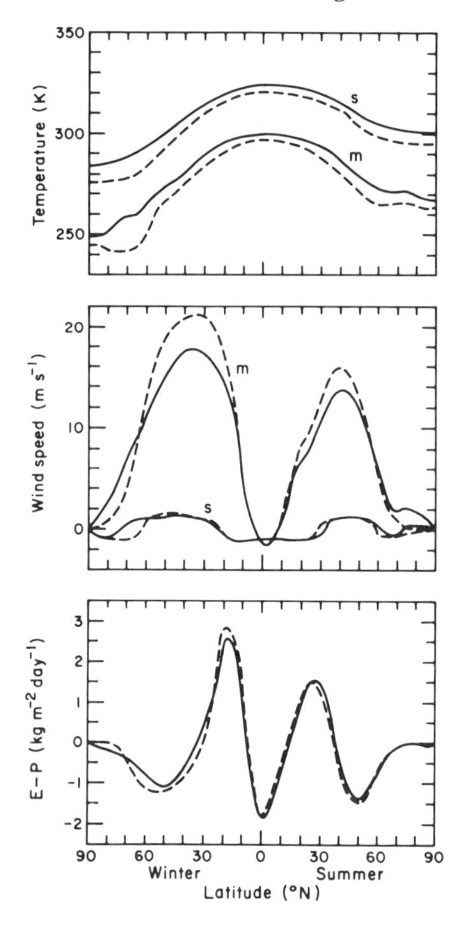

Figure 7. Zonally-averaged potential temperature (top), zonal wind (middle), and difference between evaporation and precipitation (bottom), deduced for present (full line) and 18,000 B.P. glacial (dashed line) boundary conditions. *S* and *M* denote surface and mean atmospheric values, respectively. (Adapted from Saltzman and Verneker[26] by Gates.[11]

Zonally-Asymmetric SDMs: The Complete Time-Average State. It is clear from any map of surface climatic fields (e.g., temperature) that the departures from zonal symmetry associated with the continent-ocean distribution are of the same order as the meridional variations of the zonally-averaged climatic fields. Moreover, when one examines the present pattern of glaciation, and the patterns of glaciation that probably prevailed during the last great ice age at 18,000 years B.P.,[30] a marked asymmetry around latitude circles is apparent. For these reasons it follows that a theory for the axially-asymmetric climatic state is at least as important as the theory of the axially-symmetric climatic state. Although reasonably good deductions of asymmetric surface temperatures are possible with purely thermodynamic models,[31] a fuller theory by means of which winds, the hydrologic cycle, and the upper air climate can also be deduced, requires a consideration of the equations of motion also (i.e., momentum models). This is because all these features are intimately related to *circulations* in vertical latitudinal as well as meridional planes (e.g., the "monsoons") and to the associated standing horizontal wave motions.

Reviews of linear statistical-dynamical models of the mean asymmetric flow, in which the symmetric mean flow is prescribed, are referenced.[32–35] Most of the studies on this topic have been based (1) on *prescribed* diabatic heat sources (or prescribed surface temperatures from which heat sources were obtained from a "Newtonian"-type parameterization), and (2) on prescribed asymmetric transient eddy sources of heat and momentum. As such, these models are not full-fledged, internally "self-determining," climate models. Moreover, in contrast to most of the zonally-averaged climate studies discussed thus far, these asymmetric studies have been concerned almost exclusively with deducing the wind field rather than the temperature field (i.e., the emphasis has been on the dynamics rather than the thermodynamics).

In spite of their shortcomings, the above studies have been of great value as diagnostic guides to a more valid equilibrium climate theory for the atmosphere. For example, these studies show that the effects of mountains and heating in generating the asymmetric climate patterns are of comparable magnitude. One implication is that the topographic blocking effect of ice sheets (such as the Laurentide of 18,000 years B.P.) on the prevailing atmospheric global climate may have been as large as the thermal effects associated with the increased ice coverage. Also, it is shown that quasi-resonant horizontal modes are possible that can become excited or damped depending on the distribution of continent and ocean. This will be of interest in considering the terrestrial climates that prevailed millions of years ago when land-sea distributions were much different.

First attempts, within the framework of a statistical-dynamical model, to deduce the asymmetric mean *surface* temperature field along with the free atmospheric wave structure, forced by an internally determined (not prescribed) heating field, have been made recently by Ashe,[36] and Vernekar and Chang[37] who use a vertical resolution very similar to that employed in the zonal-averaged model of Saltzman and Vernekar.[24] In treating the heating as a parameterized function of the other climatic variables, these studies open up for consideration a wide variety of possible feedbacks that have hitherto been ignored. Of special interest will be the feedbacks between atmospheric temperature or snow cover mentioned above, though there are many other interesting possibilities.[38]

In the last analysis, we are interested not in the separate axially-symmetric (i.e., zonally-averaged) climatic component or the axially-asymmetric component, but in the sum of these two representing the complete climate (i.e., the temporal means and variances) prevailing at all geographic locations as obtained in GCM solutions. A theory for this complete time-average state can be achieved by coupling models for the two separate components discussed above in the manner suggested in[35] and illustrated schematically in figure 8. In this figure, A is the set of free-atmospheric climatic statistics, B is the set of surface and boundary layer climatic statistics near the atmosphere-ocean-land interface, O is the set of free-oceanic climatic statistics, and τ denotes the appropriate sets of governing equations and boundary conditions. The bold-line arrows indicate solutions, and the required feedbacks between the components are indicated by the crossed light-line arrows. As yet no attempts have been made to combine the two problems into a comprehensive statistical-dynamical theory of the global climate in the manner indicated. It may in fact turn out to be more convenient to bypass this scheme entirely in favor of solving the complete

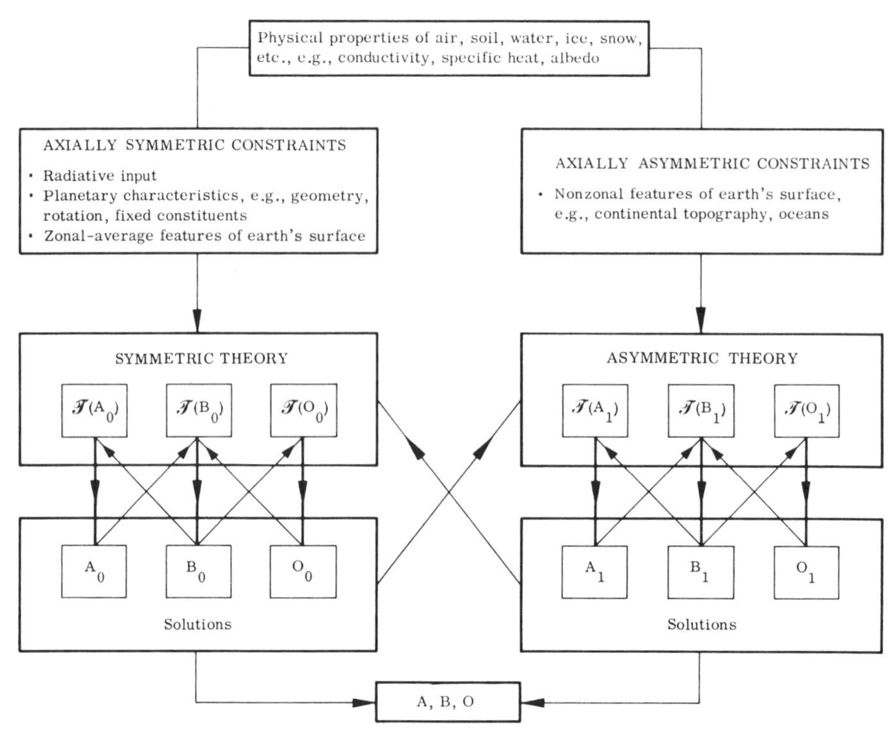

Figure 8. Schematic structure for an overall climate theory based on coupled symmetric and asymmetric models. Bold arrows indicate solutions and crossed light arrows indicate feedbacks between the components. See text for definitions of *A, B, O,* and *J.* From Saltzman.[35]

time-average equations directly, by numerical process, for a global grid network. Steps in this direction have already been taken by Sellers[39] using a thermodynamic model, and Webster and Lau[40] using a more complete momentum model. However, it should be recognized that the information gained from the separate problems will be invaluable in formulating and interpreting such a "primitive" climatic-mean model for the quasi-equilibrium state of the atmosphere and its lower boundary surface layer over the globe.

One aspect of the atmospheric climatic equilibrium state that cannot be deduced from models based on the vertically-averaged equations, or two-level extensions of these, is the continuous vertical profile of temperature constituting the "standard atmosphere." A series of models, based mainly on the thermodynamical energy equation, known variously as "vertical column," "radiative-convective," and "one-dimensional, vertical-coordinate" models, have been developed with this special aim in mind and are reviewed and developed further in references 41 and 42. Aside from accounting for the lapse rate structure of the atmosphere these models have proved to be of great value in perfecting the atmospheric radiation and convection algorithms used in GCMs and for preliminary testing of the sensitivity of the climatic state to changes in atmospheric composition and cloud structure.

In the next subsection we discuss a more significant group of *purely thermody-*

namic equilibrium models than the above "vertical column" models; these are the vertically integrated "energy-balance" models in which the degrees of freedom for a description of the global equilibrium state of the atmosphere and its underlying surface is reduced to minimal numbers.

Vertically integrated, purely thermodynamic models. By the direct application of either the vertically integrated energy equation for the atmosphere or the surface boundary layer, it is possible to form models for the equilibrium surface temperature of the earth, providing that information from the other domain is specified.[31] A more satisfactory procedure is to treat these two domains simultaneously, coupling them by the surface heat flux $H_S^{\uparrow}$ at their interface. One can thereby solve for both the surface and atmospheric temperature. Early examples of such models are referenced.[43–45]

A noteworthy application of this type of model to paleoclimatology is found in the study of Donn and Shaw.[46] Using Adem's[43] model they obtained equilibrium solutions for a sequence of five epochs ranging from the early Triassic (200 million years B.P.) to the present. According to their results, it was not until roughly 10–15 million years B.P., when land masses had drifted poleward in the Northern Hemisphere to roughly their present positions, that polar surface temperatures could drop below freezing and permit the formation of ice. Examples of their solutions are shown in figure 9. More recently, Adem[47,48] has applied his model, with modifications, to obtain an equilibrium solution for the ice age boundary conditions at 18,000 years B.P..

In another completely different application, Kutzbach[49] used the surface energy budget equation, together with a water balance equation, to estimate the paleorainfall rates in the area of a North African lake, given the geologically determined lake levels that prevailed in the past.

The above models require the parameterization of all modes of heat flux at the surface comprising $H_S^{\uparrow}$. By adding the separate energy equations for the atmosphere and subsurface "active layer" [domains (a) and (b) in figure 3] to form a single equation for the rate of change of the total energy in the atmosphere and its underlying surface boundary layer, and neglecting fluxes at the lower boundary, we can eliminate $H_S^{\uparrow}$ to obtain[1]

$$\frac{\partial E}{\partial t} = -\nabla \cdot \mathbf{F} + N^{\downarrow} \tag{7}$$

where E is the total energy per unit area (including latent energy), $\mathbf{F}$ is the horizontal flux vector for total energy, and $N^{\downarrow} = (H_T^{(1)\downarrow} - H_T^{(2)\uparrow})$ is the *net radiation* at the top of the atmosphere (i.e., $H_T^{(1)\downarrow}$ is the net incoming shortwave radiation and $H_T^{(2)\uparrow}$ is the net outgoing longwave radiation).

Given the fast thermal response time of this combined domain, it is appropriate in considering the 10-year running-average climatic state to set $\partial E/\partial t \approx 0$, from which we obtain the so-called energy balance equation

$$H_T^{(1)\downarrow} - H_T^{(2)\uparrow} - \nabla \cdot \mathbf{F} = 0 \tag{8}$$

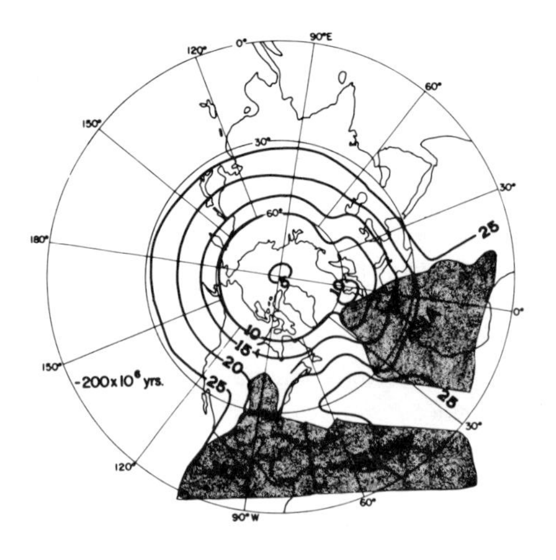

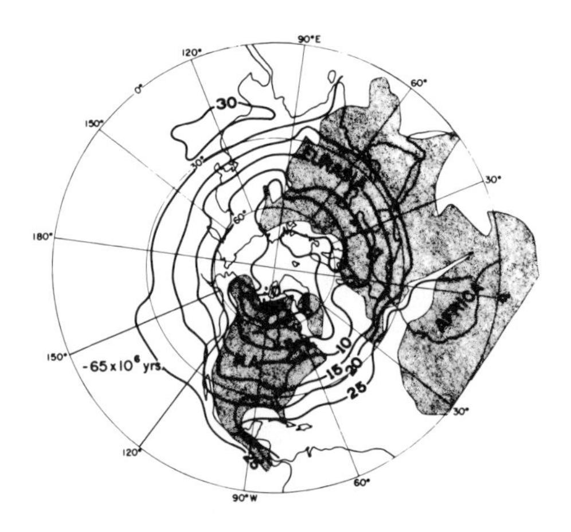

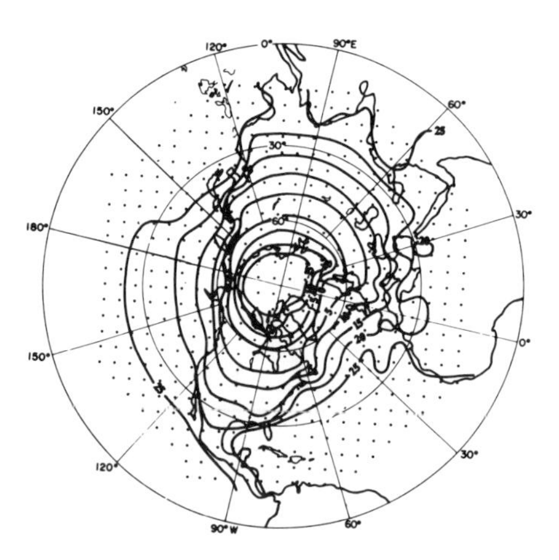

Figure 9. Isotherms computed for the early Triassic Period, 200 m y B.P. (top), early Cenzoic Era, 65 m y B.P. (middle), and present (bottom), showing by shaded regions the reconstructed position of continents superimposed on the present positions. Reproduced from *Geological Society of America Bulletin*, vol. 88, p. 390 (1977) and reproduced with permission of publisher and author.

which has been the basis of the simplest and most thoroughly studied of all climate models.

This equation has been applied two-dimensionally over the earth[39] but the more common application has been made to zonal-average conditions governed by

$$\langle H_T^{(1)\downarrow} \rangle - \langle H_T^{(2)\uparrow} \rangle - \frac{\partial[F_\phi \cos \phi]}{a \cos \phi \; \partial\phi} = 0 \tag{9}$$

where F_ϕ is the poleward component of the combined vertically integrated energy flux in the atmosphere and oceanic mixed layer, and the angular brackets denote a zonal average.

Following Budyko[6] and Sellers[8] the further approximation is often made that

$$\langle H_T^{(1)\downarrow} \rangle = Q(\phi)\{1 - \alpha[T_s(\phi)]\} \tag{10}$$

where $Q(\phi)$ is the incoming annual mean solar radiative flux at the top of the atmosphere and α is the planetary albedo which is a function of the surface temperature T_S in a manner that can represent the relatively sharp decrease in surface albedo as one proceeds equatorward across the ice edge latitude (ϕ_i), that is, from regions where $T_S < T_S(\phi_i)$ to regions where $T_S > T_S(\phi_i)$. In the usual approximation $T_S(\phi_i) = 263$ K.

For the long wave radiation it is assumed that

$$\langle H_T^{(2)\uparrow} \rangle = a + bT_S(\phi) \tag{11}$$

where a and b are empirically determined constants derived from present observations, and for the poleward flux of energy the diffusive approximation, or its equivalent, is commonly applied to the surface temperature in the form

$$F_\phi = -K \frac{\partial T_s}{a\partial\phi} \tag{12}$$

These are clearly extremely crude parameterizations, but nonetheless, with their use one can deduce the zonal mean surface temperature profile with good fidelity to the observations regarding both surface temperature and surficial ice *extent* (see figure 10). The parameterization constants used in obtaining figure 10 cannot be universally valid, however, because it is believed that for roughly the same (or even a lower) solar inputs, ice-free, warmer conditions prevailed on the earth during long periods of the earth's history. This seems to be illustrated in a recent application of a model based on equation (9) to the paleogeographic surface state believed to have prevailed 100 million years ago in the mid-Cretaceous[50] (see next chapter). Although geological evidence suggests ice-free conditions at this time, the equilibrium solution obtained is characterized by sub-freezing polar temperatures.

If equation (9) is further averaged over the entire globe, we obtain the statement

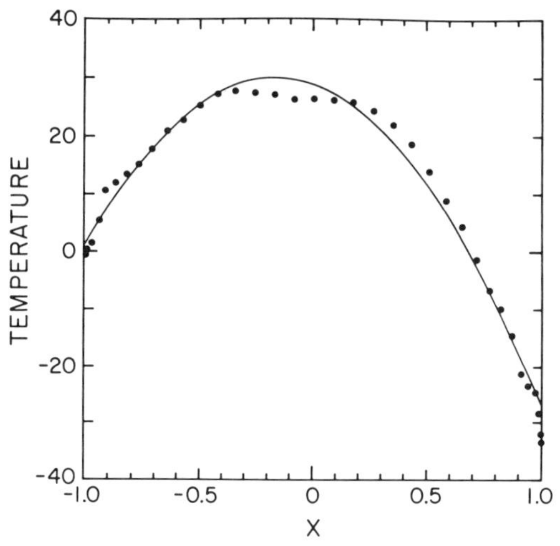

Figure 10. Zonal-average mean surface temperature for December, January, and February, deduced from an energy balance model (solid curve) compared with observed values (dots), plotted against the sine of the latitude, *X*. From North and Coakley.[57]

of radiative equilibrium for the planet

$$\tilde{H}_T^{(1)\downarrow} - \tilde{H}_T^{(2)\uparrow} = 0 \tag{13}$$

where the wavy bar denotes the horizontal global average. Together with the additional constraint that the total mass of water in each of its forms is a constant, equation (13) represents a "master equilibrium" condition for the whole climatic system. As such it provides the basis for obtaining a first estimate of the temperature of the planet, given the incoming radiation as a boundary condition. With the parameterizations,

$$\tilde{H}_T^{(1)\downarrow} = \frac{(1 - \alpha_p)S_0}{4} \tag{14}$$

$$\tilde{H}_T^{(2)\uparrow} = \sigma T_e^4 \tag{15}$$

where α_p is the "planetary albedo," and σ is the Stefan-Boltzmann constant, we obtain the following formula for the "effective" planetary radiative equilibrium temperature,

$$T_e = \left[\frac{(1 - \alpha p)S_o}{4\sigma}\right]^{1/4} \tag{16}$$

For the earth, where $S_o = 1340\ \mathrm{Wm}^{-2}$ and $\alpha_p \approx 0.34$, (16) gives $T_e \approx 246$ K (a value representative of the mid-troposphere), indicating that a large part of the long wave radiation originates from the atmosphere and not the earth's surface which has a mean temperature of 288 K. This illustrates the critical role of the atmospheric-"greenhouse" effect in maintaining a relatively warmer, more hospitable, surface temperature, which in its absence would be close to T_e (assuming the same value of α_p).

If, instead of equations (14) and (15), we use the Budyko-Sellers type parameterizations, equations (10) and (11), which implicitly include the greenhouse effect, we obtain the simplest "zero-dimensional" model of the planetary climate governing the mean surface temperature of the earth and its ice-coverage as free dependent variables.[51]

As we have noted, in spite of the extreme crudeness of the parameterizations, the Budyko-Sellers models can yield acceptable first order estimates of the annual zonal mean surface temperature. The main results, however, concern the *sensitivity* of the equilibrium response solutions for *ice-coverage* associated with the surface temperature, to altered incoming radiation (e.g., an altered solar constant, S_o). It is shown that because of the destabilizing effect of the positive ice-albedo feedback, relatively small changes in the solar constant can lead to relatively large changes in equilibrium ice coverage. For a critical decrease of S_o of roughly 2 percent an equilibrium solution corresponding to total glaciation of the planet is obtained (i.e., a "runaway glacier"). Such possibilities are generally absent in models not containing this feedback. Although this effect is undoubtedly exaggerated in these simple models, with their limited degrees of freedom and limited number of negative feedbacks that can act as buffers, the models do suggest the importance of including this particular feedback in assessing climatic thermal responses. For example, it seems clear that the response to earth-orbital radiation variations should be underestimated to some degree if this feedback is not incorporated.[9,52,53] On the other hand, we must recognize that the growth of ice not only affects the surface albedo leading to a positive feedback, but also affects other surface state parameters, that is, the subsurface conductive capacity, and the water available for evaporation, both of which can lead to negative feedbacks. Sea ice, for example, acts as an effective insulator preventing the vertical flux of sensible and latent heat from the oceans possibly leading to a net warming of the oceans. From another viewpoint, the Ewing and Donn[54] scenario regarding ice age oscillations rests largely on a negative feedback between ice coverage and water availability for supplying the hydrologic cycle.

Comprehensive reviews of energy balance models and their extensions have recently been made by Ghil[55] and North et al.[56] emphasizing the sensitivity of the equilibria obtained from these models to changes in external forcing, for example, the solar constant and the earth's orbit.

The seasonal cycle. As a final point in this section we note again that a complete model of the atmospheric and surface boundary layer climate should include a treatment of the forced *seasonal cycle* which is a major "eddy" to be superimposed on the 10-year climatic average state. These cyclic fluctuations can be deduced easily from GCMs, and also from SDMs designed to govern average conditions over several months rather than several years. In this latter case one deals with a prognostic, non-equilibrium model in which lags between the seasonal forcing and response are introduced. There are many examples of such SDM calculations, which have been reviewed.[1,56] In a recent study by North and Coakley[57] this cycle is elegantly included within the framework of the simplest energy balance model.

PALEO-EVOLUTION OF THE COMPLETE CLIMATIC SYSTEM: SLOW-RESPONSE, NON-EQUILIBRIUM DOMAINS

While the above studies of the quasi-equilibrium response of the combined atmosphere and surface boundary layer to both external forcing and the impressed conditions at *D* in figure 3 seem to be able to give an adequate account of the climatic state of these domains, they cannot, of course, account for the state and evolution of the deep ocean and deep cryosphere together with its deformable bedrock. To a large degree it is these new domains that provide the main signals of the longer-term climatic change of our planet. Because of the relatively slow response times of deep ocean and ice (cf. table 2), it is likely that we must deal with a time-dependent, or "prognostic," system to determine their variations. In turn, this prognostic system must be coupled with a diagnostic system of the type described in the previous section to account for the atmosphere and surface boundary layers.

We have already discussed some examples of the equilibration of the atmosphere and surface boundary layers to altered states of ice coverage,[18,19] to altered earth-orbital conditions,[20] and to altered states of the continent/ocean distribution.[46,52] In addition there have been numerous other "sensitivity studies" in which estimates are deduced of the new equilibrium state of the atmosphere and surface boundary layers that would follow from changes in other internal parameters and external parameters (primarily the solar constant, atmospheric CO_2, and volcanic dust levels). These studies are extremely instructive in revealing the properties of these fast response parts of the system, and may even provide good first-guess predictions (of the "second kind", according to the terminology of Lorenz[58]) of the changes that would occur due to *fast* changes in any of the above parameters. In all these studies, however, the deep ocean and deep cryosphere are fixed so that the solution cannot represent the ultimate long-term equilibration of the complete system. This long-term equilibration of the atmosphere and boundary layer can be much different than one would get from the short-term, fast response solution in which this domain is artificially held constant.

Moreover, even if one could deduce an accurate estimate of the true long-term equilibrium of the complete system, it is unlikely that at any particular time the system will actually be a realization of that equilibrium state, even if the equilibrium is stable. This will be true if only due to the ubiquitous presence of noise generated by higher frequency phenomena that cannot be accounted for in the model. We are led, therefore, in considering paleoclimatic fluctuations on all scales, to consider the full stochastic-dynamical problem posed by the complete set of equations including those governing the deep ocean and deep ice masses.

Before proceeding to a more general discussion of models treating the dynamics of these slow-response domains, we next review the nature of the external *forcing* of the system and review a few earlier studies in which climate change is viewed as a continuous fast response only to this forcing.

Climatic Forcing

We can refer to studies aimed at exploring differences in equilibrium climates obtained from different models, and with different values of the parameters for a particular model (i.e., sensitivity studies), as exercises in *comparative equilibria.* As we have indicated, in such studies the goal is to deduce the climate without concern for the *path* leading to the equilibria or from one equilibrium to another, or concern for the temporal evolution of the quasi-equilibrium over longer time periods. However, the terrestrial climate varies significantly and continuously on many time scales, ranging from several decades to the age of the earth, and our interest in this book lies more with this time-dependent path of climatic evolution than with the "snapshot" equilibria we have mainly been discussing up to now.

There are two basic modes of climatic variability:

1. *Forced* variability due to changing *external* factors that affect the climatic state but are themselves unaffected by the climatic state (i.e., astronomical factors such as the solar output and earth-orbital parameters; solid earth tectonic factors such as continental drift, mountain building and volcanic activity; human factors such as some forms of "pollution" and altered land use).
2. *Free* variations due to internal instabilities and feedbacks, usually involving nonlinear interaction among different components of the climatic system, that can occur even if there are no forcing changes.

Although it is natural to look first for possible changes in external forcing as a cause of climatic change, it is entirely possible that the swings in terrestrial climate are more of the *free* type, reflecting the behavior of an inherently nonlinear and unstable system that never achieves a true equilibrium between all its components. As we already noted, an inspection of the spectrum shown in figure 2 suggests that the observed climatic variability probably results from a complex interaction of forced and free effects (i.e., we are dealing with a forced, nonlinear, system with dissipative damping and many possible sources of instability). *Thus, it will not be possible to accurately determine the climatic response to a known forcing without knowing the "free" properties of the system.* Before we discuss this free behavior, we now discuss the nature of the external forcing and some of the quasi-equilibrium calculations made to obtain the response of the system to this forcing, neglecting the long-term dynamics of the deep hydrosphere and cryosphere.

There are at least two major *terrestrial* sources of forcing of the climatic system: (1) variations in atmospheric composition (e.g., CO_2, aerosols) due to volcanic and human activity, and (2) very long-term variations in land-sea and topographic distributions due to continental drift, polar wandering, and mountain building processes. Only a few attempts appear to have been made up to now to deduce the *continuous secular change* in mean surface temperature due to the changing levels of CO_2 and particulates originating from volcanic and anthropogenic causes. For example, Schneider and Mass[59] find from a highly simplified thermodynamic model that reasonable good agreement with observed temperature trends over the past three centuries can be obtained. Similarly, Bryson and Dittberner[60] find such agreement for the past 80 years. This indicates that these external forcing mechanisms are at least competitive with other possible mechanisms in accounting for secular climatic changes, though questions remain regarding the approximations used in the model.[61]

The most regular, and best known climatic forcing variations result from *astronomical* sources (i.e., the earth's changing orbital relation to the sun); these are primarily (1) the familiar annual—or seasonal—variations which we have considered part of the fast-response equilibration, and (2) the much longer-term so-called Milankovitch variations assocciated with changes in the earth's eccentricity e, obliquity ϵ, and precession index π ($= e \sin \omega$, where ω is the longitude of perihelion).

It was suggested qualitatively by Adhemar[62] and Croll,[63] and later determined quantitatively by Milankovitch[64–66] that systematic variations of the earth orbital elements (e, ϵ, π) having periods of the order of 10–100 thousand years would lead to varying meridional distributions of insolation that could force climatic changes on the earth and possibly account for the late Cenozoic ice ages. More comprehensive, computer based, calculations of the past and future changes of these elements, and of the implied insolation variations, have been made recently by Vernekar[67] and in an improved form by Berger.[68–72] In particular, Berger[69,71] has tabulated the *mid-monthly* insolation as a function of latitude for the past one million years, noting the importance of treating these, rather than the more highly smoothed "caloric half year" values used previously. This importance is due to the fact that seasonally-tuned feedbacks and correlations with other variables such as ice cover may produce amplified responses.[22,73,74]

Of the orbital periods, only the near-100,000 and 400,000 year eccentricity variations (which has an amplitude of less than 0.03) can lead to a change in the net annual irradiance of the earth, a low value of e (nearly circular orbit) corresponding to a less than 0.2 percent reduction of the solar constant from its value at a maximum value of e. Assuming the solar constant is 1340 Wm^{-2}, this represents a swing of about 2 Wm^{-2} over the 100,000 year period. The near-40,000 and 20,000 year periods due to variations in ϵ and π, respectively, lead to comparatively high amplitude swings in the latitudinal distribution of incoming radiation as a function of the season (e.g., at 9000 years B.P., when the earth was closest to the sun in Northern Hemisphere summer, there was 7 percent more summer radiation and 7 percent less winter radiation in the Northern Hemisphere, with important consequences for the strength of the monsoon[20]) but the *overall* radiation over the globe is a constant. The signature of the effects of these relatively high amplitude 40 and 20 thousand year cycles seems to be imprinted on the climatic record of the past one million years obtained from the deep-sea cores, indicating their probable importance for climatic change (see also chapter 5).[75]

A review of models treating the equilibrium response of the surface ice coverage to earth-orbital forcing is given in reference 1, ranging from Milankovitch's early discussion[66] to that of Suarez and Held.[9] In all these equilibrium studies, the ice is considered to have only a horizontal geographic dimension. However, the past ice ages were in fact characterized by huge accumulations of ice, in the form of ice sheets possessing a complex dynamics and inertia of their own that is essentially neglected in all the above models. Two of the earliest studies to examine the response of such ice sheets to the Milankovitch insolation changes, assuming these insolation changes are parametrically related to the latitudinal distribution of snow accumulation, have been made by Weertman[76] and Birchfield.[77] It was concluded that the insolation changes are large enough to have bearing on the growth and decay of the ice sheets, and perhaps of more significance, point to the importance of inertial and other nonlinear effects that can introduce phase shifts and strong periodic behavior at frequencies not containing the maximum forcing (e.g., the 100,000 year eccentricity-induced cycle).[75]

There are many other possible terrestrial and astronomical forcing mechanisms, the past or future variations of which we know too little to study in a time-dependent framework. These include such possibilities as changing solar output,[78] earth collision with interplanetary matter leading to "meteor showers,"[79] and changing geothermal and volcanic heat fluxes. Also, a progressive decrease in the rate of the earth's rotation has been inferred from tidal theory,[80] and paleontological studies;[81] an interesting assessment of the consequences of such a rotation change for the atmosphere only, based on a GCM, has been made by Hunt,[82] and even cruder first-order estimates can be obtained from similarity arguments relating planetary dynamics to external forcing.[83]

It is also clear that climatic forcing can take the form of a non-continuous process, that is, events of finite duration, perhaps of a "catastrophic" nature such as volcanic eruptions may occur that can trigger either longer-term responses,[84] or simply a marked climatic aberration over a short time span (e.g., the Krakatoa volcanic explosion).

Similarly, the free internal behavior of the climatic system may include events in one domain that are of a catastrophic nature (e.g., ice sheet surges[85]) and it can be of value to view these events as a quasi-external forcing of all the other components of the climatic system. More generally, it is always possible to relax our definition of "forced climatic change" to include the changes imposed by one climatic domain on those of another. Thus, for example, we may consider the "atmospheric climate" to be forced by prescribed changes in sea-surface temperature or snow coverage, or consider changes in glaciers to be forced by atmospheric conditions and thereby represent a sort of "thermometer" for the global atmospheric mean temperature. In the discussion that follows we shall generally adhere to the stricter view that forcing of climate can only come from purely external sources.

Free Long-Term Behavior: Problems of Modeling

As we have noted, the deep ocean and ice masses (together with their deformable bedrock) constitute the main "carriers" of long-term climatic variability. Prognostic equations for these domains, governing the non-equilibrium evolution of the mean temperature of the entire ocean, θ, and total ice mass of the planet M_i are developed[1,2] in the following approximate forms, respectively,

$$\frac{d\theta}{dt} \approx \frac{1}{c_w M_w}\left(H_{sw}^{\downarrow} + G_w\uparrow\right) \tag{17a}$$

$$\approx \frac{1}{c_w M_w}\left(N\downarrow + G\uparrow + L_f \frac{dM_i}{dt}\right) \tag{17b}$$

and

$$\frac{dM_i}{dt} = P_i + F \tag{18a}$$

$$\approx -\frac{1}{L_f}\left(N\downarrow + H_s^{\uparrow}\right) \tag{18b}$$

where c_w is the specific heat of ocean water, L_f is the latent heat of fusion, M_w is the total mass of ocean water, $H_S^{\downarrow}$ and $H_{SW}^{\downarrow}$ are the net downward fluxes of heat at the earth's surface for the whole globe and ocean, respectively, including those due to all water phase transformations (evaporation, melting, and freezing), $G\uparrow$ and $G_W^{\uparrow}$ are the net upward geothermal fluxes of heat for the whole globe and ocean, respectively, P_i is net snowfall rate over the globe, and F is the net rate of freezing of surface water (or, if negative, melting of ice) over the globe.

Equations (18a) and (18b) are the basis for all models that take the deep cryosphere into account, representing the most general relation governing the ice volume variations deduced from the oxygen-isotope analyses of the deep-sea sediment cores. With additional approximations, based on continental glacial mechanics, an equation relating the ice mass to the shape and hence the 'extent' of ice sheets can be derived.[86] Weertman's equation, which can be expressed in the form:[2]

$$M_i = \frac{4}{3}\pi a^2 \alpha \rho_i \left(\frac{\tau_0 a}{\rho_i g}\right)^{1/2} (1 + \epsilon)(\gamma - \delta)^{3/2}$$

(where α is the fraction of latitude occupied by the ice sheet, ρ_i is the density of ice, τ_o is the shear stress at the bed of the ice sheet, $\epsilon/(1 + \epsilon)$ is the fraction of the total ice depth that is submerged below ground level due to the ice-load effect, γ and ζ are the sines of the latitudes of poleward and equatorward extent of the ice sheet, respectively) has been applied with (18a) in many models.[77,87,93]. We should note, however, that a good deal of the total ice volume could be in the form of marine ice shelves, deep grounded ice, as well as continental ice sheet.[94,95] Equation (18b) could form the basis for a model of long term ice mass variations in which the difficulties of parameterizing $(P_i + F)$ are replaced by the probably comparable difficulties of parameterizing $N\downarrow$ and $H_s^{\downarrow}$.[2,96]

Given the estimate that ice mass M_i decreased by about 50×10^{18} kg from 18,000 to about 8000 years B.P.,[97] we can calculate that the *mean* rate of energy consumption due to melting over this period was about 10^{-1} Wm^{-2}. From (17b) we can show that if $N_s^{\downarrow} = G\uparrow = 0$ this value of 10^{-1} Wm^{-2} would lead to a decrease in mean ocean temperature of 3K over the 10,000 years. In fact, the oceanic geothermal flux is always a positive value, currently estimated[98] to be about 10^{-2} to 10^{-1} W/m^2 and it is also likely that $N_{sw}^{\downarrow}$ is not zero. A counter-intuitive possibility even exists that at a glacial maximum, when sea-ice "insulation" of the ocean in high latitudes is a maximum, $N^{\downarrow}$ ($\approx H_{sw}^{\downarrow}$) is positive, leading to a warming of the deep ocean.[1,29] This can be accomplished physically by the shutting off of deep water production in high latitudes allowing slow downward thermal diffusion and upward geothermal fluxes to dominate. This possibility is speculative, but nonetheless poses critical questions which must be resolved for a complete understanding of long-term climatic change. What is fairly certain is that *the departures from equilibrium of the rate of energy flux and the rate of the hydrologic ice forming processes (expressed in terms of latent heat release) accompanying climatic variations of the scale of the major Pleistocene ice ages, is of the order of* 10^{-1} Wm^{-2} *or at most 1* Wm^{-2}. This is well below the resolution of our observations and the possible accuracy of our parameterizations, representing the behavior of a system exhibiting an extremely "slow physics;"[99] a fundamental problem is thus exposed regarding our ability to model long-term climatic change by straightforward, purely physical, deductive process starting with the fundamental equations. Briefly stated, *we may never be capable of representing the deterministic physics inherent in climatic imbalances, except by some rough physical insights coupled with our knowledge of the observations to be deduced*—that is, except by an "inductive" rather that purely "deductive" process. The difficulties are surely compounded by the fact that higher frequency aperiodic fluctuations of both an external nature (e.g., volcanicc and geothermal activity) and internal nature (e.g., interannual sea-ice activity) are likely to require relatively high stochastic amplitudes for their representation.

Stated differently, for non-equilibrium (i.e., prognostic) equations of the form (6) governing the variations of climatic mean variables x_i, that is,

$$\frac{dx_i}{dt} = f_i(x_j,F_i) + \mathcal{E}$$

(where $\mathcal{E}$ includes both systematic and random errors in the parameterizations embodied in f_j) it is unreasonable to expect that the expressions f_j can provide anything more than a "suggestive" representation of the relevant feedback processes if $\mathcal{E}$ is greater than the observed estimates of $|dx_j/dt|$. For the slow variations of climate this is bound to put our ability to parameterize to a severe test that probably cannot be passed.

It has sometimes been argued that the problems engendered by the smallness of the observed rates of change of ice mass involved in the main (near-100 kyr) Quaternary glacial oscillations can be avoided, and even put to advantage, by treating these oscillations as quasi-equilibrium states responding to external forcing. In this view the total ice mass bears an essentially *diagnostic* relation to a purely external agent, the most reasonable candidate for which appears to be the radiative variation induced by the earth-orbital (Milankovitch) cycles. This conjecture implies that the continental ice mass, ζ, is a function only of the earth-orbital parameters:

$$\zeta = f(e, \mathcal{E}, \pi). \tag{19}$$

Although the main variations of ζ over the past 700 kyr are of a period of about 100 kyr, near which only a seemingly weak forcing due to the e-variations prevails ("weak" in comparison with the forcing due to $\mathcal{E}$ and π variations of near-40 kyr and 20 kyr periods), it can be argued further that f is in all likelihood a highly nonlinear function that can lead to some "rectification" of the input with a consequent amplified near-100 kyr output.[77]

At least one simple test is available to judge this possibility: if (19) is valid, then at any two times when the set of values $(e, \mathcal{E}, \pi)$ is nearly identical there should be nearly identical amounts of ice on the planet unless (19) admits multiple solutions. In the latter case, the realization of one particular equilibrium will depend on either stochastic events or prior evolution, both of which would negate the applicability of (19) alone.

As it turns out, one can easily find nearly identical values of $(e, \mathcal{E}, \pi)$ occurring at times when vastly different planetary ice masses prevail, e.g., at $t = 0$, 20, and 59 kyr BP when $(e, \mathcal{E}, \pi) \approx (0.017, 23°40, 0.016)$[70] (see figure 1*a*). The implication of this demonstration that ζ is not a function of orbital parameters alone as represented by (19) is, of course, that

$$\frac{d\zeta}{dt} \neq \frac{df(e, \mathcal{E}, \pi;t)}{dt} \equiv g(t)$$

meaning that the evolution of ζ can be determined only by consideration of a nonequilibrium problem with all the attendant difficulties that lead to the desirability of an inductive approach.[149] A general conclusion is that although $d\zeta/dt$ may be very small in relation to our ability to measure or to parameterize the fluxes that determine it, $d\zeta/dt$ is not so small as to be negligible compared to the dissipative process for ζ measured by its system response time.

We can, in fact, pose a somewhat deeper question of whether in practice we can determine a longer term average state, e.g., over the last million years, that *might* constitute the equilibrium about which the time-dependent ice ages evolve subject to the free properties of the climatic system, stochastic perturbations, and the departures of $(e, \mathcal{E}, \pi)$ from their mean values over this period, $(e_0, \mathcal{E}_0, \pi_0)$. It would be to this "equilibrium" state that the climate would relax if the equilibrium were *stable* or about which the climate would fluctuate if the equilibrium were *unstable,* providing we removed variable forcing due to stochastic perturbations (e.g., irregular volcanic activity) and earth-orbital cycles. As one test, if such an equilibrium is to be calculable, it should be possible to show, say with a "deluxe" GCM applied to both present-day near-minimjum ice mass and 18 kyr BP near-maximum ice mass, that for the same values of $(e, \mathcal{E}, \pi)$ a *nonequilibrium* state of net ice melting or freezing must exist in at least one or possibly both of these cases. However, since a net rate of ice mass change corresponding to even a 10 cm yr^{-1} sea-level change (corresponding to a net latent heat flux of 1 W m^{-2}) would be huge in comparison with the inferred Quaternary glacial changes, it seems highly unlikely that, given the known accuracy of the GCM flux parameterizations, we will be able to make such a determination. As we noted, there have, in fact, been several such dual solutions for present and 18 kyr BP ice states using the same orbital parameters that prevail at *present* rather than (e_0, $\mathcal{E}_0$, π_0).[16,17] In these cases the results were reported to represent acceptable asymptotic equilibrium states to the accuracy that such equilibria can be established, and, justifiably, no attempts were made to discuss any possible disequilibrium of the prescribed glacial ice mass. Thus if (19) is valid for some very long term averaging period, it is far from clear how one would ever be able to determine whether an ice-maximum or ice-minimum solution, or something in between, was the "correct" equilibrium solution. This state of affairs calls into question our ability to determine a "master equilibrium" state about which time-dependent (nonequilibrium) variability might be taking place, especially if it is an unstable equilibrium.

In spite of the difficulties it behooves us to press on with attempts to construct "deductive" deterministic models, if only to expose all the physical possibilities and feedbacks necessary to account for the climatic record. Many interesting studies have been made to account for the ice volume record, some of which we shall attempt to summarize in the next section.

As portrayed in figure 4, we shall divide the formulation of the deterministic component of paleoclimate models into two groups: a quasi-*deductive* group in which we proceed from the fundamental laws by attempting explicitly to parameterize specific flux processes; and a more openly *inductive* group in which we attempt to construct a dynamic system of equations from which the known paleoclimatic output can be deduced, guided by some understanding of the physical feedbacks we believe to be important.

Quasi-Deductive Deterministic Models

Within the last decade a growing number of physically-based models have been presented that are based on prognostic equations for (1) continental ice, (2) its deformable bedrock, (3) thick marine ice shelves and grounded ice, and (4) deep ocean temperature. Some of these models are reviewed below.

Continental Ice Sheet-Bedrock Models. The prototype model of this type was developed in two-dimensional form by Weertman.[86] (Other three-dimensional models have been developed more recently by Mahaffy[100] and Budd and Smith.[101]) These equations are generally combined with approximate relations for snow accumulation and ablation (i.e., for P_i and F), often expressed as a function of surface temperature variations governed by a diagnostic EBM (of the Budyko-Sellers or Suarez and Held types), that can equilibrate as a fast response to earth-orbital radiative forcing. The results of these types of models have been reported in a series of papers (e.g., Weertman,[76] Birchfield,[77] Birchfield and Weertman,[87] Birchfield et al.[88,102] Sergin,[103] Pollard,[104,105] Pollard et al.,[93] Oerlemans, [91,92,106,107,108], Källén et al.,[90] Ghil and LeTreut,[89] LeTreut and Ghil [109]). These studies can be summarized as follows:

1. Of the above papers, Weertman,[76] Birchfield,[77] and Birchfield and Weertman[87] simply prescribe the snowfall and ablation as a direct forcing (i.e. there is no intermediate EBM). The important result obtained is that the nonlinerity of the time-dependent ice sheet equation will permit a subharmonic response of 100,000 years given an impressed forcing of the near-20,000 and 40,000 year precessional and obliquity periods, which are of much higher amplitude than the near-100,000 year eccentricity period, a result similar to that obtained by Wigley.[110] However, this 100,000 year response seems to be weaker than is required by the observations.
2. Most of the remaining references above, starting with Pollard,[105] use an EBM to compute surface temperature, which in turn can be assumed to determine the snow accumulation-ablation profile on the ice sheet. A general consensus seems to emerge, that the near-10,000 year time constants that seem appropriate for ice sheet dynamics, together with the seasonal obliquity/precessional forcing, is adequate to account for the observed minor peaks of near-20 and 40 kyears in the ice volume spectrum. However, the ice sheet dynamics acting alone cannot account for the major 100,000 year peak.
3. As we have noted, the response time for the surface domain governed by the EBM is relatively fast so that the EBM provides only a "diagnostic" equation to the system. However, Källén et al.[90] use their EBM in a somewhat more general way than the others in that the thermal relaxation time is artificially permitted to be much longer than would be appropriate for the atmosphere and subsurface boundary layers alone. This leads to another prognostic equation, governing what might be identified with the deep ocean rather than surface temperature. The resulting coupled prognostic system can sustain *free* as well as forced harmonic behavior leading in the limit to stable limit cycle behavior. A similar free solution involving deep ocean variations as a second prognostic variable was attained earlier by Sergin.[103] The periods derived for such auto-oscillations seems to be of the order of 10,000 years.
4. Another way to introduce a second relevant prognostic variable within the context of an ice sheet model is to consider the slow response of the *bedrock* underlying the ice sheets. This idea has been developed in the papers by Oerlemans[92], Birchfield et al.[88], Ghil and LeTreut[89], LeTreut and Ghil[109], Peltier[111], and Pollard[104]. According to this latter study; the inclusion of the bedrock dynamics seems to lead to an improved capability for accounting for the observed 100,000 year cycle, particularly if additional features such as topography and calving are taken into account. In figure 11 we reproduce some of the results obtained by Pollard.[104]
5. All of the above discussions seem to rule out the possibility that the direct effects of the 100,000 year eccentricity forcing can play a role in producing the observed 100,000 year ice volume cycle during the Pleistocene.[56] This is because the thermal response of an *equilibrium* EBM to the less than 0.2 percent change in solar constant induced by the eccentricity cycle can only be negligibly small (estimated to about 0.4°C for global mean temperature), with little effect therefore on

the "snowline." However, recognizing that 0.2 percent of the solar constant is more than 2 Wm^{-2} (larger than the mean energy fluxes involved in the last major ice age) we must at least leave open the possibility that in a non-equilibrium model involving the deep ocean reservoir and ice sheets there can be some significant effect of this neglected mode of forcing, if only to determine the phase of the variations.[152]

In any event, for this latter reason, as well as for the more pervasive reason that the deep ocean temperature and the thick ice forms that can be created in the ocean can have response times no less than that of continental ice, it will be necessary to consider the forced and free behavior of these domains in considering long-term climatic change.

Marine Ice-Deep Ocean Temperature Models. Some attempt to explore the possibilities for interactions between slowly varying deep ocean temperature and marine ice, treated as a closed system independent of land ice formation, has been

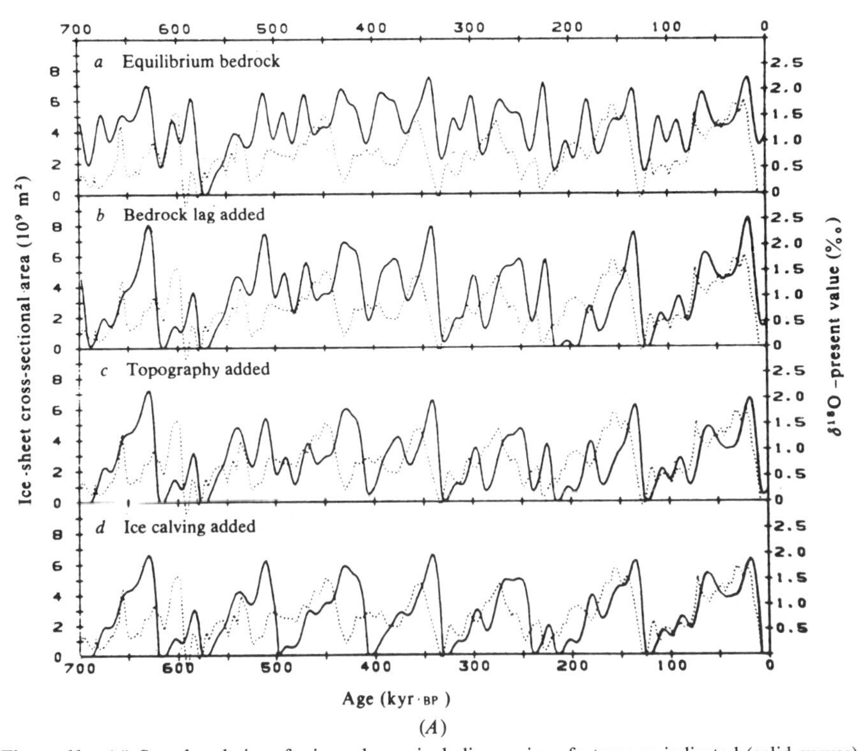

(*A*)

Figure 11. (*A*) Sample solutions for ice volume, including various features as indicated (solid curves), compared with observed ice volume records deduced from $\delta^{18}O$ records (dashed curve). (*B*) This sequence of height-latitude cross sections of the ice sheet (heavy dashed curve) and bedrock surface (solid line) corresponding to run (*d*) above. Also shown are ablation-accumulation equilibrium line (light dashed), the undisturbed topography (dotted), and regions of water incursion (shaded). From Pollard,[104] reprinted by permission from *Nature*, Vol. 296, pp. 334, 336, copyright © 1982 Macmillan Journals Limited.

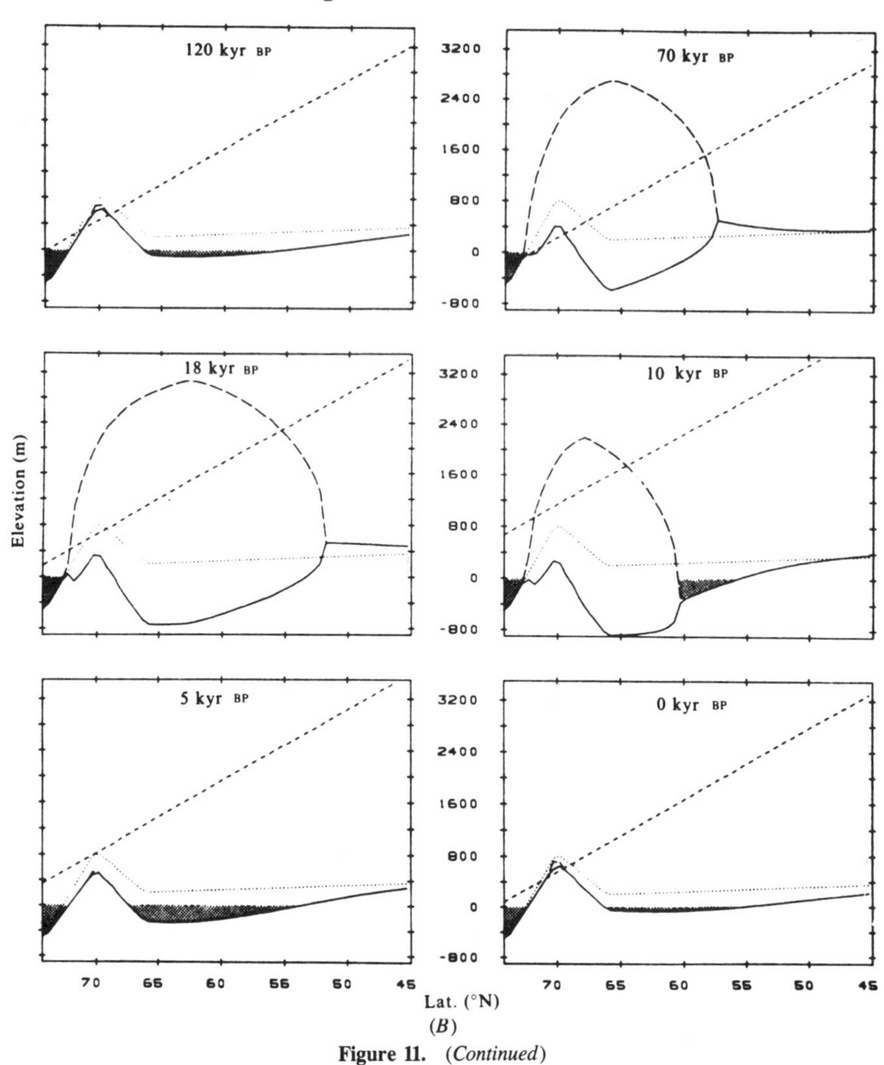

(B)

Figure 11. (*Continued*)

made in a series of papers (Saltzman,[1,112] Moritz,[113] Saltzman and Moritz[114]). These studies stemmed largely from the suggestion of Newell that sea ice formation may "insulate" the ocean thereby providing a negative feedback with deep ocean temperature and also from the desire to explore the possible modes of climatic behavior engendered by coupled, two-prognostic variable, systems (e.g., systems capable of yielding stable limit cycles). Although in forming these models a good deal of effort was made to parameterize $H_{sw}^{\downarrow}$ (i.e., the processes whereby heat is exchanged across the ocean-atmosphere boundary and at the ice edge), relatively cruder parameterizations were employed to relate the resulting changes in sea ice-edge location to the consequent changes in the mean ocean temperature. More detailed modeling treatments of pack and shelf ice growth are in references.[94,115,116] Included in the simple models is an hypothesized diagnostic relation between atmospheric CO_2 and

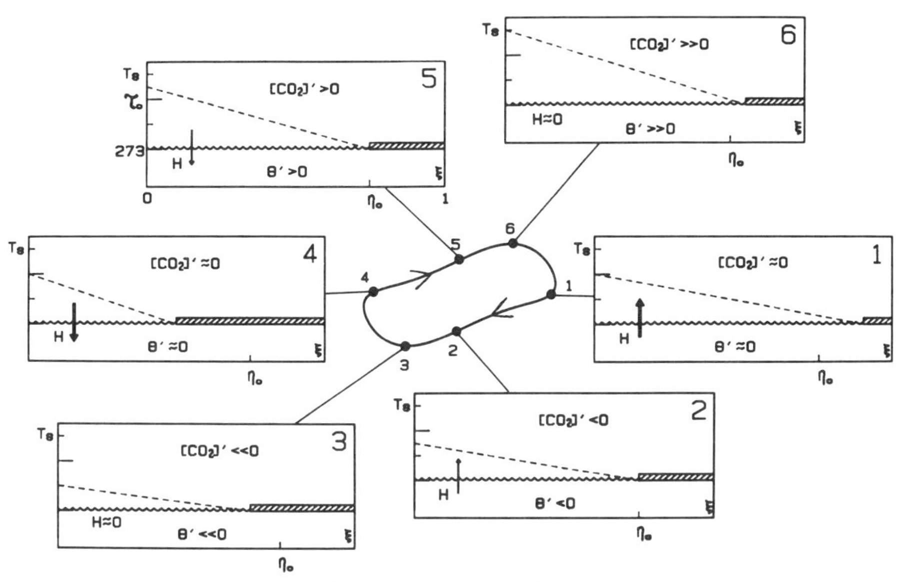

Figure 12. Schematic sequence of climatic states corresponding to the limit cycle solution obtained by Saltzman[112] and Saltzman et al.[129] τ_0 and η_0 are the equilibrium equatorial surface temperature, and sine of the ice-edge latitude, respectively; H denotes the total heat flux across the ocean surface, θ is the mean temperature of the whole ocean, and the stippled region denotes sea-ice coverage.

deep ocean temperature; this provides a basic source of *positive* feedback and instability. The possible relevance of such free variations of CO_2 is supported by recent chemical analyses of ice cores.[117,118]

With plausible estimates of the values of certain parameters in the model, one can deduce a variety of possible modes of behavior of the system. These include stable limit cycles, the periods of which could in principle range up to about 100,000 years for the thickest grounded forms of sea ice in an ocean whose thermal variations are at least partly controlled by semi-diffusive processes. In figure 12 we show the general pattern of evolution for such a deduced limit cycle.

Sea-ice formation and deep ocean thermal processes also play some role in the multi-variable system described by Sergin,[103] but because the assigned time constants for these domains are relatively small, they are not of dominant importance in achieving the greater than 10,000 year auto-oscillations reported.

Further comments on quasi-deductive models. It would appear that we are still a very long way from the kind of physical understanding necessary to adequately portray all the slow time-dependent processes taking place simultaneously in the climatic system—that is, the evolution of ice sheets, ice shelves, deep ocean, and trace-chemical concentrations. The above deterministic models involving these slow domains are but a few examples in which some quantitative efforts have been made. Many others are surely possible, such as a quantification of the role of salinity variations in the ocean.[119,120] A few of the areas in which further work is clearly needed are:

1. The determination and parameterization of the vertical fluxes of heat and matter at the surface and between the abysal and upper mixed layer of the ocean, including the salinity budget.
2. An understanding and dynamical representation of the *departures* from radiative equilibrium at the top of the atmosphere.
3. The inclusion of the release of the latent heat of fusion in models in which the dynamics of large ice masses is considered.[121]
4. The combining of the marine temperature/sea ice models of long-term climatic change with the continental ice sheet models (one suggestion in this regard has recently been formulated[2]).
5. The modeling of the *free,* non-anthropogenically forced evolution of atmospheric CO_2 in terms of the long-term prognostic variables. It remains to be determined whether such a relationship will be diagnostic or prognostic.

In the last analysis, all of the models that have been proposed or contemplated thus far will require some relatively arbitrary prescription of response time involving flux processes that are beyond our ability to observe or to parameterize from purely physical considerations. In this category are the representations now being used to parameterize snow accumulation/ablation and ice shelf edge growth in terms of the thermal state of the system. For example, the study by Oerlemans and Vernekar[122] gives ample testimony to the difficulty of establishing even the sign of the snowfall-temperature dependence.

More generally, it must be recognized that *it is unlikely that any amount of physical rigor regarding the faster behavior of the system will be able to generate with sufficient accuracy the extremely low net rates of flux of mass and energy involved in long-term climatic change (10^{-1} Wm^{-2}).* For this reason it will probably be a matter of necessity to adopt a more *inductive* line of attack (based on our gross understanding of the feedbacks that are likely to be involved) in which systems of equations are formulated that are capable of generating the known climatic variations without violating any *integral* conservation constraints. Many applications of this approach have already been made, most quite recently.

Inductive Deterministic Models

Numerous, but certainly not exhaustive, accounts have been given of the many possible feedbacks, both positive and negative, that are operative in the climatic system.[1,22,123,124] As we have indicated, one approach to paleoclimatic modeling (perhaps the only feasible one) is to assume some mathematical structure to represent these possibilities (based on incomplete observational estimates of the feedbacks, and on the suggestive results of deductive models of the type discussed in the previous section), with the aim of achieving solutions that are consistent with as many lines of observational evidence as possible.

In spite of its probable necessity, such an inductive approach is bound to be looked down upon as nothing more than "curve fitting"—a charge that is fundamentally difficult to refute. All we can probably say is that for an inductive model to be acceptable, we must demand that it continually satisfy all the constraints arising from (1) our present and growing knowledge of the physical processes and feedbacks at work in the climatic system (e.g., the response times implied by the model must be within plausible limits), and (2) our present and growing bank of data concerning both the variables specifically governed by the model and those that are not yet included explicitly. There is, indeed, some room for triumphant verification of such models by predicting what the paleoevolution of variables not yet adequately measured should be (e.g., the deep ocean or thermocline temperature, shelf ice mass, atmospheric CO_2 concentration); that is, while we may "fit" the output curve for one variable, this output may simultaneously entail variations in other, as yet unmeasured, variables which can serve as strong tests of the validity of the model. In this sense, an inductive model can still fulfill a major requirement of a "theory" of paleocclimate by predicting new phenomena and suggesting the search for new physically relevant processes and feedbacks. Beyond these considerations, as in all models we must continue to require a minimization of the "free parameters," and a "robustness" of the model with respect to small changes of these parameters, as well as to changes in forcing (both deterministic and stochastic) within the limits their amplitudes can be specified.

In any event, it would be natural to start such an inductive process with the simplest equations and gradually build up to greater complexity. In this hierarchy we could include such essentially diagnostic models as the offshoots of the Budyko–Sellers model postulated by Sutera[125] and MacAyeal,[126] for example, but we shall begin here with a model having only a single explicit *prognostic* variable and proceed to consider examples of time-dependent dynamical systems comprised of two or more prognostic variables.

A Model Involving One Explicitly Prognostic Variable. Since all physical systems must be dissipative to some extent, the simplest governing equation one could assume, say for the total ice mass M_i, is of the form

$$\frac{dM_i}{dt} = -K_i M_i + F(t) \tag{20}$$

Where K_i^{-1} is the "time constant" and $F(t)$ is the forcing of M_i, respectively.

This general form was adopted by Imbrie and Imbrie[7] in a first attempt to account for the Pleistocene ice mass record considered as a response to earth orbital forcing represented by[7,99]

$$\begin{aligned} F(t) &= F(e,\epsilon,\omega;t) \\ &\approx F_0(e_0,\epsilon_0,\omega_0) + \alpha \left\{ \frac{\epsilon'(t)}{\sqrt{\epsilon'^2}} + \beta \frac{e'(t)}{\sqrt{e'^2}} \sin\left[\omega(t) - \gamma\right] \right\} \end{aligned} \tag{21}$$

where $e_0(=0)$, ϵ_0, and ω_0 are standard values of the eccentricity, obliquity, and longitude of perihelion and the primes denote departures from these standard values. Moreover, to account for the "sawtooth" structure of the observed ice mass variation, showing relatively slow growth and more rapid decline of ice mass[127] (see figure 1) a non-linearity is introduced in the form of a bimodal value of K_i depending on whether dM_i/dt is positive or negative. (This can be viewed as the introduction of a second implicit prognostic variable.) These two values, K_{i+} and K_{i-}, plus the three additional constants in the representation of F (i.e., α, β, and γ) provide the degrees of freedom for tuning the output to the observed ice mass fluctuations. A "best fit" for the past 150,000 years is obtained for time constants $K_{i+}^{-1} = 42{,}000$ years and $K_{i-}^{-1} = 10{,}600$ years, a precessional phase lag γ corresponding to about 2000 years, and $\beta = -2$. The quantity α is an arbitrary positive number setting the scale of the ice mass variations.

Although the duplication of the observed record is fairly good over this 100,000 year period (see figure 13), the model shows only moderate to poor success in its ability to duplicate the entire observed ice variation as depicted in figure 1*A*. A need for adding further complexity to the dynamical system treated is indicated. One direction for adding this complexity is to increase the number of prognostic variables treated; if forcing alone cannot account for the record, then at least one other internal variable besides total ice mass must also be varying on the same time scale in order to account for times of growth versus times of decay for the same value of M_i. As we noted, this Imbrie-Imbrie model does implicitly contain such a second variable in the form of the specification of different values of K for growth and decay of M_i. However, there is no connection between the variation of K and any explicit second dependent climatic variable. Some likely candidates for such a second variable are (1) the *spatial distribution* of the continental ice mass (figure 11B), (2) the mean ocean temperature and its spatial distribution, (3) the CO_2 content of the atmosphere, (4) the salinity distribution in the ocean, and (5) the marine ice mass in the form of shelves, icebergs and sea-ice, all of which can in principle be different in periods of growth and decay of total ice mass.

Two-Prognostic Variable Models. One of the earliest general discussions of inductive climate models containing two prognostic variables was given by Eriksson.[128] Considering global mean surface temperature (T_s) and fraction of the earth covered by ice (x) as the two variables, he described many of the possible features of such systems including the possibility for internal self-sustained oscillations (i.e., limit-cycle or auto-oscillatory behavior). Further general discussions of this possibility were made in a series of papers by Sergin[103], followed by the previously mentioned "deductive" studies of Källén et al.[90] (governing continental ice and a mean temperature), and Saltzman[1,112] and Saltzman and Moritz[114] (governing marine ice and deep ocean temperature).

It was recognized that although much effort was expended in these latter deductive models by Saltzman et al. in trying to properly parameterize the relevant fluxes, there were still large uncertainties in these parameterizations and the assigned constants (see, e.g., figure 7 and table 2 of reference 114). It would therefore be no less accurate to form equivalent models that have the *qualitative* properties of these so-called "deductive" ones, in which the coefficients are assigned values that are within certain "plausibility" ranges suggested by observations and the more detailed deductive model. This essentially "inductive" attitude has been applied in reference 129, where the limit cycle behavior is modeled more simply by a coupled

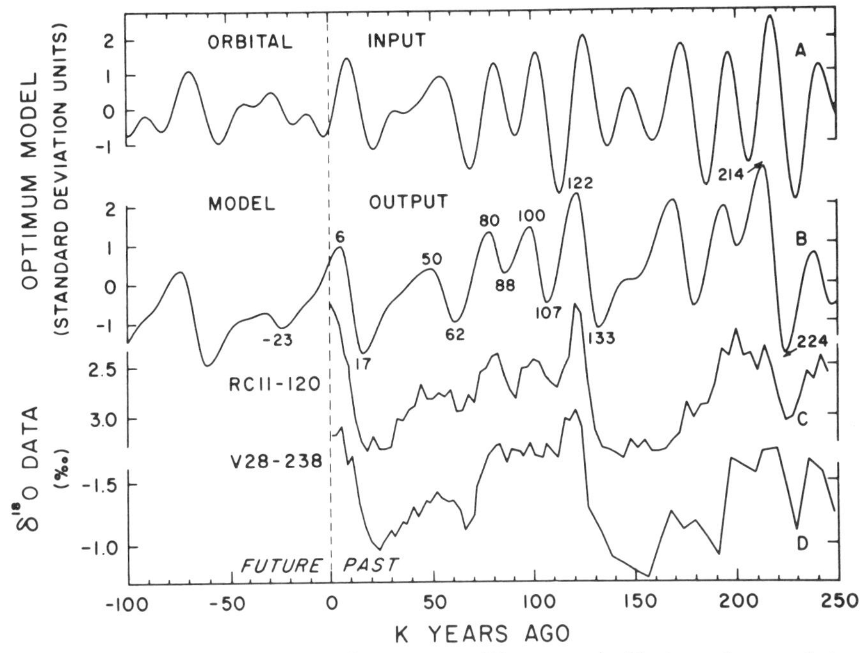

Figure 13. Orbital input (A), and ice volume response (B), compared with observed oxygen isotope derived ice-volume curves (C and D). From Imbrie and Imbrie,[7] reprinted by permission from *Science,* Vol. 207, Fig. 7, copyright 1980 by the American Association for the Advancement of Science.

system of equations governing a modified form of the van der Pol relaxation oscillation,

$$\frac{dx'}{dt} = -\phi_1,\theta' - \phi_2 x' \tag{22}$$

$$\frac{d\theta'}{dt} = \psi_1 x' + \psi_2 \theta' - \psi_3 x'^2 \theta' \tag{23}$$

where the primes denote a departure from an unstable equilibrium at (0,0), and ϕ_1, ϕ_2, ψ_1, ψ_2 and ψ_3 are positive constants. The spiral instability necessary for a limit cycle is generated by the $\psi_2\theta'$ term in eq. (23) which is physically identified with the positive $[CO_2] - \theta$ feedback previously postulated.[114] This system leads to a solution of the form depicted in figure 12 where $\eta = (1 - x)$. The possibility for auto-oscillatory behavior within the framework of this model is taken to the limit,[130] wherein values of the coefficients possibly appropriate for extremely thick marine ice shelf formation (cf., Denton and Hughes[94]) are assigned in order to generate a free 100,000 year limit cycle. This would represent a marine analogue of the Oerlemans[92] suggestion of a free 100,000 period involving *continental* ice sheets only. The influence of periodic earth-orbital forcing on relaxation oscillation climate models of the type just described have been considered.[109,151]

As we have noted, such two-dimensional limit cycle behavior demands the presence of an unstable sprial equilibrium, with strong stabilizing restoring mechanisms for large displacements from the equilibrium. Another possible scenario for climatic change that is admissable in a two-prognostic variable system involves the existence of a *stable* equilibrium toward which trajectories can spiral in a damped harmonic fashion.[1] The possibility exists that real climatic change is occurring near such a stable equilibrium, where the system would permenantly reside were it not for changes in external forcing and stochastic forcing that continually "kick" the system. This latter scenario, with earth-orbital forcing as the "kicker," would seem to play some role in the aforementioned study described by Oerlemans.[92] An assortment of other deterministic possibilities including the presence of multiple equilibria, have been discussed.[112]

Three or More Prognostic Variable Models. To represent the complete dynamical behavior of the climatic system will, in principle, require a multi-dimensional system of the form (6). In particular, a *three*-component system ($j = 1, 2, 3$) already contains so much dynamical richness beyond that representable by a two-component model, that this system should become the focus of a great deal of further paleoclimatological modeling. A major example of this richness is provided by Lorenz's[131] strange attractor system in which ergodic "almost intransitive" behavior is generated deterministically. Such "almost-intransivity" is not possible in a one- or two-variable dynamical system without the additional specification of stochastic forcing.

Examples of three-component models are provided in the work of Sergin,[103] Ghil and LeTreut,[89] Saltzman et al.[150] Saltzman and Sutera,[149] and Saltzman et al.[152] In these latter studies, it is shown that a nonlinear feedback system involving continental ice mass ζ, marine ice mass χ, and mean ocean temperature θ, can yield *free* and orbitally forced solutions having many features in common with the $^{18}0/^{16}0$ derived records of ζ (see figure 14). In particular, the solution shows a near 100,000 year period with rapid deglaciation, and predicts concomitant variability in χ and θ that can serve as a check on the model. The distributions of ζ, χ, and θ corresponding to six consecutive points of interest in the free solution cycle are shown in figure 15.

It is clear that to validate multi-variable models it will be necessary to obtain synchronous long-term records of at least one significant variable in addition to ice volume, the most important of which is probably the ocean temperature (see chapter 5). A highly promising possibility for estimating the evolution of θ is to examine deep-sea stratigraphic cores with high sedimentation rates for evidence of the O_2 content of the oceans, high (aerobic) values signifying generally cool conditions, and low (anoxic) values signifying warm conditions.[132] The sedimentary signatures of these states have been discussed.[133] For example, major anoxic periods can be clearly identified by black shale laminae.

As we noted, there have been some exciting reports recently concerning the possible variations of atmospheric CO_2 in conjunction with the ice ages, as determined by chemical analyses of glacial ice cores.[117,118] The results seem to indicate that substantially reduced amounts (of the order of 50 percent) of CO_2 were present

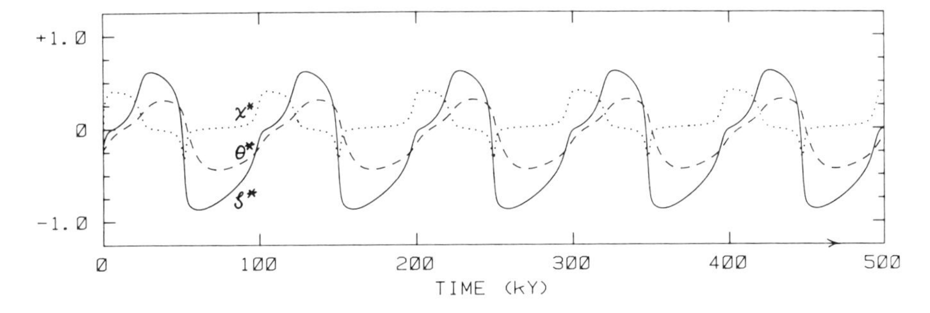

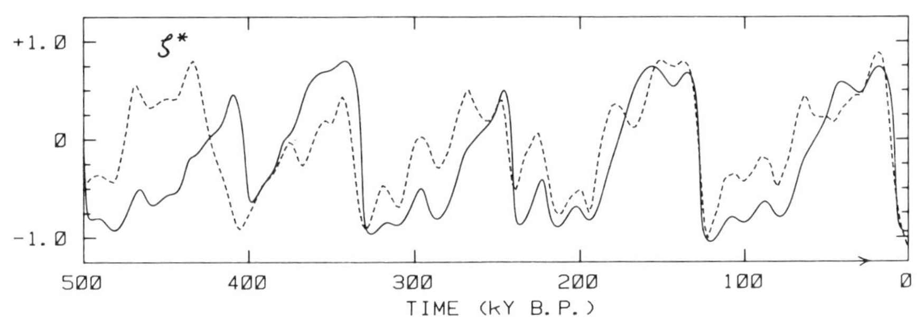

Figure 14. Top: Free solution of the (ζ, χ, θ) system, in nondimensional units (from ref. 149). Bottom: Orbitally-forced solution of the (ζ, χ, θ) system for the ice mass 3 (solid curve), compared with the SPECMAP $\delta^{18}O$ estimate of the true ζ-variations (dashed curve) (from ref. 152).

in the atmosphere at the time of large ice volume, findings that are not inconsistent with the dynamical climatic system discussed above.

Further comments on inductive models. In viewing these inductive models we give substance to our remarks in the Introduction to the effect that "completely new mathematical forms, bearing little resemblance to the continuum equations (1)–(5), may be required to express the conservation laws governing very long-term climatic change." It is also clear from the relatively primitive state of the models proposed thus far that this development is just beginning. The ultimate model of paleoclimatic variability will undoubtedly involve some combination of the deductive and inductive procedure[153] in which the forms of parameterization adopted will be designed to conform simultaneously (1) to our physical understanding of the flux process as determined by spatially and temporally "local" studies, and (2) to the known output desired to be achieved by means of the parameterization. The "tuning" of coefficients within "physically plausible" limits is an example of this latter process. In reference 153 an attempt is made to show how the deductive-equilibrium models of the fast response variables and the inductive-nonequilibrium models of the slow response variables can be harmonized into a coherent theory for the evolving 3-dimensional climatic pattern.

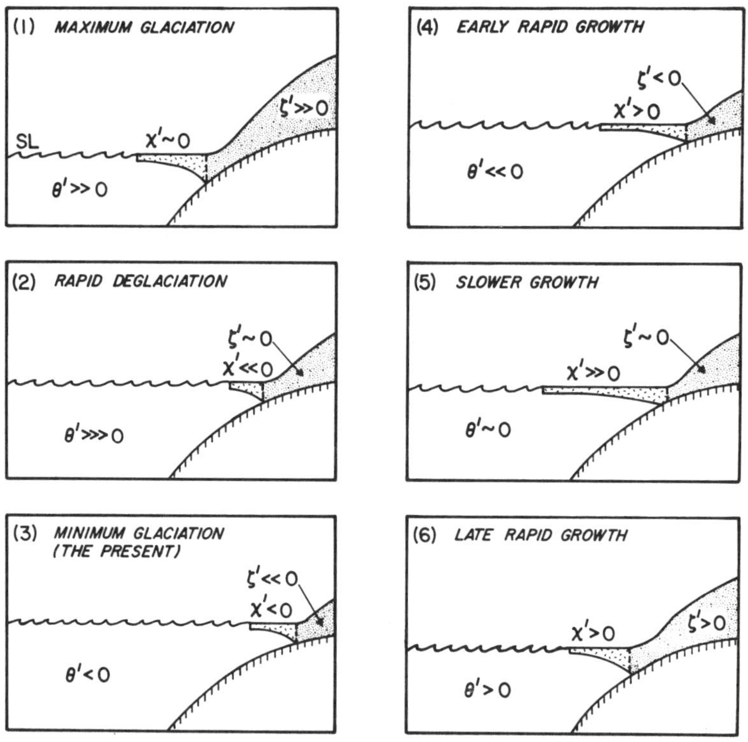

Figure 15. Schematic snapshot representations of the solution shown in figure 14 at selected consecutive points in time starting with maximum continental glaciation. The first three points (1, 2, 3) represent rapid deglaciation and the next three points (4, 5, 6) represent the glacial buildup (from ref. 149).

As a final point, we also note that within the domain of inductive methods resides the often used (and abused) statistical methodology of pure curve fitting or "modeling" by regression procedures. With this methodology no attempt is made to link the statistical model formulas to any dynamical-physical system from which they can be generated. Nonetheless, it is conceivable that such formulas account for a good deal of variance, have predictive capabilities, and play a useful role in suggesting the form of more dynamically-based models.

Stochastic Effects. The deterministic systems discussed in previous sections are, at best, highly speculative and simplified models of some of the feedbacks that are likely to be operating in the real climatic system. Clearly, there is much room for improvement of the content and fidelity of the models. However, it is unlikely that any amount of added rigor would enable us to represent the true behavior of the climatically averaged variables as a purely *deterministic* system.[112,134] The most we could hope for is that with such added rigor the residual effects owing to inadequate representation of phenomena of higher frequency than the climatic averaging period can be considered "random," that is representable by a white-noise process of suitable amplitude. This constitutes the *stochastic* component of any climatic dynamical system.

In essence this stochastic component represents the effects of determinism on scales that we cannot, or choose not to, consider, the sources of which are threefold:

1. The impossibility of fulfilling the so-called Reynolds' conditions for the climatic average because of aperiodic fluctuations on shorter time scales.
2. The difficulty of adequately parameterizing the nonlinear effects of the higher frequency phenomena (i.e., all parameterizations should have error terms added to them, at least a part of which is "random").
3. The impossibility of specifying initial conditions and boundary conditions (i.e., external forcing) without some level of uncertainty and error.

We must therefore conclude that *the variations of climate derived from any model must be imprecise to some extent and hence be describable only in some probabilistic form.* A schematic flowchart showing the components of a theoretical model of climatic variability, including the stochastic components, is given, in figure 4. An appropriate way to add to the completeness of the deterministic feedback systems discussed in the last two subsections is to include the effects of a white-noise stochastic forcing process by writing the equations as a "stochastic-dynamical" system in the so-called Ito form.[135]

$$\delta x_j = f_j\ (x_j,F_j)\delta t + \epsilon_j^{1/2}\ \delta W \tag{24}$$

where δx_j are increments of x_j over a finite time step δt, f_j are the deterministic functions for which solutions were obtained as in the previous subsections, $(\epsilon_j^{1/2}/\delta t)$ is the standard deviation (or, typical amplitude) of the random fluctuations, and δW is the incremental white-noise function that can be identified with the output of random number generator of variance equal to unity.

By solving the system (equation 24) numerically, subject at each time step to the random "kick" denoted by $\epsilon_j^{1/2}\delta W$, a realization of the stochastic-dynamic process can be obtained. By a sequence of such realizations one can determine the "residence density" of the solution in x_j-space, which can be considered to asymptotically approach the "probability density" for an infinite sequence. This numerical "Langevin-type" approach is essentially the equivalent of solving the Fokker-Planck equation explicitly governing the probability distribution,[134] a task that can be formidable for a complex determinism.

The first, and perhaps most critical requirement for an analysis of the stochastic properties of the model is to estimate a reasonable reference amplitude of the noise as measured by $\epsilon_j^{1/2}$. The simplest approximation is to assume that the past record of aperiodic variability about the running average climatic variations gives an acceptable measure of the combined effects. For example, for $\delta = 10$ y the interannual variability of the annual mean ice edge should give a good first approximation $\epsilon_\eta^{1/2}$ for pack sea ice.

A fundamental question is the degree to which the deterministic properties of a model can survive the presence of a reasonable level of stochastic noise, and to what extent the stochastic noise introduces new features when interacting with the deterministic solutions that would never appear in its absence (e.g., "tunneling" or quasi-periodic "exit" from one stable equilibrium to another). As a general rule, stochastic noise is a source of instability in a system tending to prevent it from ever settling down to a fixed stable point or limit cycle.

Over the last few years there have been an increasing number of studies of the effects of stochastic noise within the framework of paleoclimatic models. Some of the main results are the following:

1. In a pioneering study, Hasselman[134] pointed out that, because of the great separation in time scales between "weather" and "climate," one could, as the very simplest first approximation, assume the climatic mean state is representable as a linear, forced, dissipative system subject to "Brownian" perturbations due to the random component of weather. The departure of the climatic mean state, x', from the single, stable, equilibrium of such a system is governed by

$$\frac{dx'}{dt} = -Kx' + \epsilon_x^{1/2}\,\delta W,$$

the well-known solution for which represents a red noise spectrum similar to that observed for the climatic record. This idea was applied by Fraedrich[51] and Lemke[136] to zero and one-dimensional energy balance models of the Budyko type, respectively, showing how random fluctuations of climate can occur near a stable equilibrium at which the probability density is a maximum. A further application of this methodology to an ice sheet model with planetary wave feedback was made by Oerlemans.[38]

2. In addition to applying stochastic perturbations as a separate forcing term, it is equally appropriate to assume that the coefficients of a deterministic model are subject to stochastic variation. This "multiplicative coupling" of stochastic noise with the dependent variable was applied to a zero-dimensional Budyko-Sellers model by Nicolis and Nicolis,[137] showing the great sensitivity of the thermal equilibria of the model to such perturbations.

3. In an important extension of the above studies, Sutera[125] showed that when stochastic perturbations are applied to a Budyko type energy balance model containing *two* stable equilibria separated by an unstable equilibrium (i.e., a *double-well* potential system) a new time scale is introduced into the problem of climatic change, namely, the "exit time" for the climatic state to surmount the potential barrier and exhibit a new mode of behavior. With such a flipping between bimodal states, the system becomes "almost-intransitive" in the terminology of Lorenz.[138] Although the range of exit times can be quite broad, for reasonable values of the parameters their *mean* value can plausibly be near 100,000 years. In a significant further step, if one introduces some *forcing* on the scale of 100,000 years (e.g., the earth-orbital eccentricity variations), the potential barrier can fluctuate on this scale allowing for a "resonance" with this mean stochastic exit time such that a higher amplitude output of this periodicity can be achieved.[139,140] It is only by this sort of stochastic scenario, involving multiple equilibria, that a one-variable, diagnostic, model of the Budyko-Sellers type can be made to exhibit free internal climatic variability.

4. A great many new possibilities arise when stochastic perturbations are introduced in two and three-prognostic variable systems. Some of these have been explored in the context of climate modeling in a series of papers by Saltzman[2,112] and Saltzman et al.[129] wherein a mapping of the "residence density" of the solution in the phase plane is employed to reveal the probabilistic nature of climatic solutions. A few of the main results are the following:

a) As one would expect, if a two variable system has a single, spiral equilibrium, the addition of stochastic perturbations can lead to sustained quasi-periodic fluxtuations with a symmetric residence density portrait centered on the equilibrium point. However, if a second, unstable, equilibrium is also present, the stochastic trajectories will tend to be driven away from this unstable fixed point, leading to an asymmetric residence density portrait with respect to the stable fixed point. Thus, if such multiple equilibria are present under the influence of stochastic perturbations, a climatic solution not only will not permanently coincide with the stable equilibrium point but *may not even have a maximum probability of being at that equilibrium point.* (For a single-variable system containing a stable equilibrium and a nearby unstable equilibrium, as in the Budyko-Sellers model, the variable has a maximum probability of being at the stable equilibrium but the shape of the residence density curve may be skewed away from the unstable point.)

b) For a stable limit cycle under the influence of stochastic perturbations, the residence density portrait can be essentially similar to that for the stable spiral equilibrium from which it can evolve through a Hopf bifurcation. (An analogous result was obtained by Sutera[141] in considering the three-prognostic variable Lorenz attractor system). Thus, residence density signatures obtained from climatic records do not uniquely reveal the nature of the underlying deterministic solution.

c) By adding stochastic perturbations to a relaxation oscillation of the type studied in,[129] characterized by a bimodal trajectory speed, we obtain an excellent example of an almost intransitive climatic system that tends to reside for relatively long periods in the two regions of slowest trajectory flow around the limit cycle. As an illustrative example of this, we show in figure 16 the results of applying a reasonable level of white noise to the limit cycle portrayed in figure 12.

d) Stochastic perturbations acting strongly on one variable of the dynamical system can surpress the amplitude of the variations of the other variable to the point where its signature may be difficult to observe. For example, in the model treated the mean ocean temperature fluctuations can be severely damped by noisy fluctuations of the marine ice extent

e) Stochastic perturbations can alter the period of some cycles of a nonlinear oscillation.

All of these two-prognostic variable results with stochastic forcing are of potential relevance for modeling climatic change. In view of the general reluctance of climate dynamicists to include such stochastic effects in climate models, we reiterate our statement that these *stochastic processes are a necessary part of any climate model without which it is incomplete.* Our hope must be that the amplitude of stochastic noise will be small enough not to obscure the basic determinism of the system.

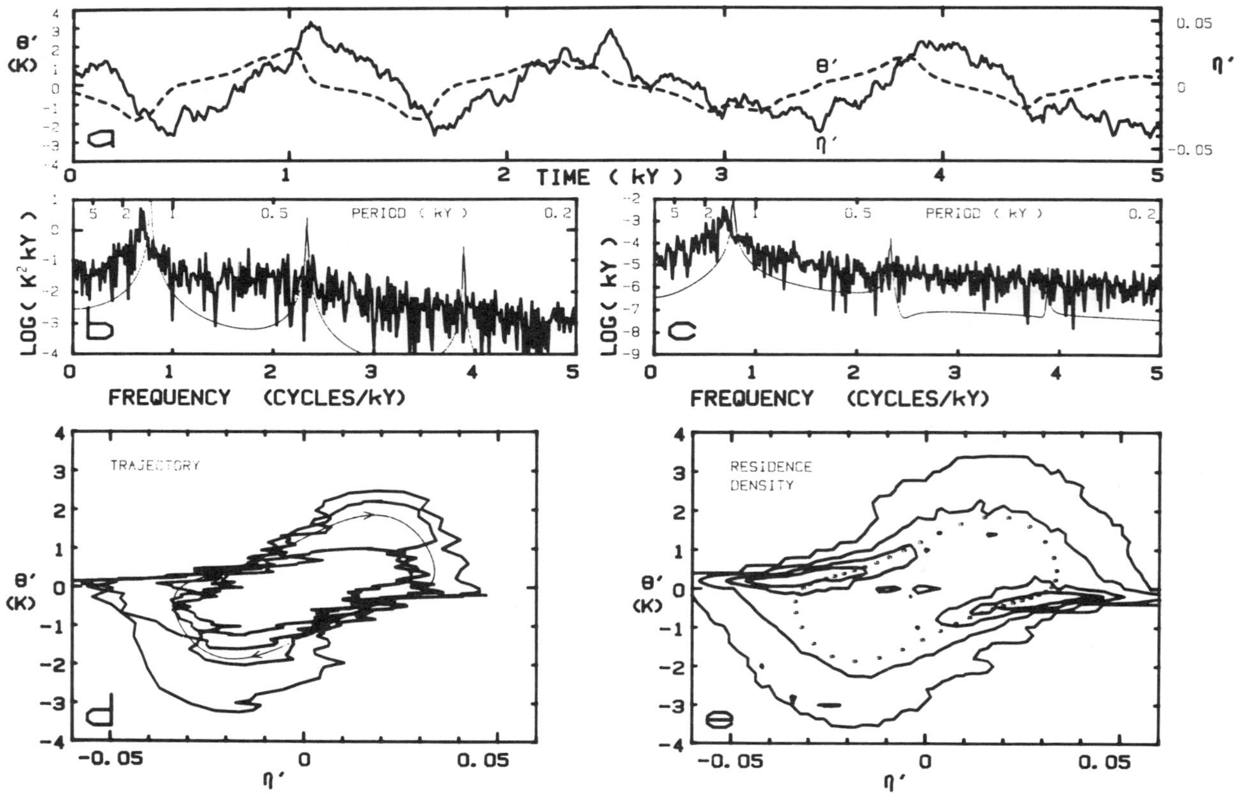

Figure 16. Sample stochastic solutions near the stable limit cycle portrayed in figure 12, obtained by perturbing η with random forcing of an amplitude corresponding to observed interannual variations. The sample evolution is shown in *a*, the variance spectra for η and θ are illustrated in *b* and *c*, the sample phase plane trajectory is depicted in *d*, and the "residence density" of the solution is given in *e*. Light curves in *B*, *C*, and *D*, and dots in *E* represent the deterministic solution. From Saltzman et al.[129]

EPILOGUE

An assessment of what we have summarized in this chapter can only leave one somewhat disappointed in the status and accomplishments of "paleoclimatic modeling." It is clear that we have barely begun the development of a theory of climatic variations, with even our philosophical and physical underpinnings yet to be secured. Bluntly stated, at this time we do not have an adequate model for climatic change on any time scale. An acceptable explanation for even so enormous a change in the climatic state as the last ice age still seems beyond our grasp.

It is logical that our first concern should be with the most dramatic changes in climate, such as the ice ages, which represent peaks in the climatic spectrum. Most of what we have discussed here has, in fact, dealt with these events which appear to have recurred in a near-100,000 year "cycle" over the past 700,000 years, with evidence of lower amplitude fluctuations at the precessional and obliquity periods of 20,000 and 40,000 years. For these fluctuations we have only some plausible scenerios involving earth-orbital forcing applied to a complex nonlinear system in which high inertia components such as continental ice and its underlying bedrock, thick marine ice shelves, and the thermal state of the deep ocean are likely to play a dynamical role. At the least, the evidence seems to point to the relevance of earth orbital variations in forcing fluctuations on this time scale of 10,000 to 100,000 years.

However, from what we know about paleoclimatic variations, the climatic state has been varying over an extremely broad and continuous band of time scales ranging from millions of years to hundreds of years. A good deal of this variability is in the nature of a red-noise "turbulence" that may not be capable of being modeled except by the crudest stochastic-dynamic system. The longest periods of this variability undoubtedly involve tectonic changes in the surface of the earth and chemical changes in the composition of the atomosphere, the former of which must be considered a major external forcing of the system. On the other end of the spectrum there is a rich variety of fluctuations of periods less than 10,000 years[142] that are almost within the historical experience of mankind (e.g., the "Little Ice Age"). Aside from the discussions of forcing due to anthropogenic CO_2, very little physical modeling of this scale has been attempted though it has been noted that free periods in the 1000 year range may be possible.

At this time the best opportunities for advancing our modeling capabilities from the physical viewpoint seems to lie in improving our knowledge of (1) the deep ocean and its interaction with the upper mixed layer including all thermo-haline processes, (2) the carboncycle in all its detail with the ultimate objective of developing a dynamical theory of atmospheric CO_2 as a *free* variable, and (3) the dynamical and thermal processes at the interface of marine ice and ocean water, and continental ice and atmosphere.

As we have noted, however, the uncertainties in all these and other physical processes may be so great as to render our parameterizations and representations of them relatively useless for paleoclimatic modeling. As we convince ourselves of this state of affairs, our effort must shift even more to the development of inductive dynamical systems (containing the richness of three or more prognostic variables) in the hope that we can implicitly capture all of the above physical effects in a "working" model from which an ever growing set of observational facts can be derived.

ACKNOWLEDGMENTS

I am grateful to Alfonso Sutera for many useful and interesting discussions on the questions treated in this chapter.

This material is based upon work supported by the National Science Foundation under Grants ATM-7925013 and ATM-8411195 (Climate Dynamics Program, Division of Atmospheric Sciences).

REFERENCES

1. B. Saltzman, A survey of statistical-dynamical models of the terrestrial climate, "*Adv. Geophys*, **20,** 183-304 (1978).
2. B. Saltzman, "Climatic systems analysis," *Adv. Geophys.*, **25,** 173-233 (1983).
3. A. S. Monin and A. M. Yaglom, *Statistical Fluid Mechanics*, M.I.T. Press, Cambridge, 1971.
4. G. K. Batchelor, *The Theory of Homogeneous Turbulence*, Cambridge Press, 1953.
5. J. W. Kim, J. T. Chang, N. L. Baker, and W. L. Gates, "The climate inversion problem: Determination of the relationship between local and large-scale climate." Climate Research Institute Report 22, Oregon State University, (1981).
6. M. I. Budyko, "The effect of solar radiation variations on the climate of the earth," *Tellus*, **21,** 611-619 (1969).
7. J. Imbrie and J. Z. Imbrie, "Modeling the climatic response to orbital variations," *Science*, **207,** 943-953 (1980).
8. W. D. Sellers, "A global climatic model based on the energy balance of the earth-atmosphere system," *J. Appl. Meteorol.*, **8,** 392-400 (1969).
9. M. J. Suarez, and I. M. Held, "Modeling climatic response to orbital parameter variations," *Nature*, **263,** 46-47 (1976).
10. J. Chang, ed., "General circulation models of the atmosphere, *Methods Comput. Phys.*, **17,** (1977).
11. W. L. Gates, "The climate system and its portrayal by climate models: A review of the basic principles. "*Climatic Variations and Variability: Facts and theories.*" In A. Berger, ed., D. Reidel, 1981, pp. 435-459.
12. J. Smagorinsky, "Global atmospheric modeling and the numerical simulation of climate." In W. N. Hess, *Weather and Climate Modification*, Wiley, New York, 1974, pp. 633-686.
13. F. N. Alyea, "Numerical simulation of an ice age paleoclimate," Atmos. Sci. Paper 193, Colorado State, 1972.
14. J. Williams, R. G. Barry, and W. M. Washington, "Simulation of the atmospheric circulation using the NCAR global circulation model with ice age boundary conditions," *J. Appl. Meteor.*, **13,** 305-317 (1974).
15. W. L. Gates, "Modeling the ice-age climate," *Science*, **191,** 1138-1144 (1976).
16. W. L. Gates, "The numerical simulation of ice-age climate with a global general circulation model," *J. Atmos. Sci.*, **33,** 1844-1873 (1976).
17. S. Manabe and D. G. Hahn, "Simulation of the tropical climate of an ice age," *J. Geophys. Res.*, **82,** 3889-3911 (1977).

18. J. Williams, "The use of numerical models in studying climatic change." In J. Gribbin, ed., *Climatic Change*, Cambridge Press, 1978, pp. 178-190.

19. J. Williams, "A brief comparison of model simulations of glacial period maximum atmospheric circulation," *Paleogeogr., Paleoclimatol., Paleoecol.*, **25**, 191-198 (1978).

20. J. E. Kutzbach, "Monsoon climate of the early holocene: Climate experiment with the earth's orbital parameters for 9000 years ago," *Science*, **214**, 59-61 (1981).

21. B. L. Otto-Bliesner, G. W. Branstator, and D. D. Houghton, "A global low-order general circulation model, Part I: Formulation and seasonal climatology," *J. Atmos. Sci.*, **39**, 929-948 (1982).

22. S. H. Schneider and R. E. Dickinson, "Climate modeling," *Rev. Geophys. Space Phys.*, **12**, 447-493 (1974).

23. G. K. Vallis, "A statistical-dynamical climate model with a simple hydrology cycle," *Tellus*, **34**, 211-227 (1982).

24. B. Saltzman and A. D. Vernekar, "An equilibrium solution for the axially-symmetric component of the earth's macroclimate," *J. Geophys. Res.*, **76**, 1498-1524 (1971).

25. B. Saltzman and A. D. Vernekar, "Global equilibrium solutions for the zonally-averaged macroclimate," *J. Geophys. Res.*, **77**, 3936-3942 (1972).

26. B. Saltzman and A. D. Vernekar, "A solution for the northern hemisphere climatic zonation during a glacial maximum, *Quat. Res.*, **5**, 307-320 (1975).

27. A. Gruber, "Satellite estimates of the earth-atmosphere radiation balance." In H. J. Bolle, ed., *Radiation in the Atmosphere*, Science Press, Princeton, 1977, pp. 477-480

28. W. L. Smith, J. Hickey, H. B. Howell, H. Jacobowitz, D. T. Hilleary, and A. J. Drummond, "Nimbus—6 earth radiation budget experiment, *Appl. Opt.*, **16**, 306-318 (1977).

29. R. E. Newell, "Changes in the poleward energy flux by the atmosphere and ocean as a possible cause for ice ages," *Quat. Res.*, **4**, 117-127 (1974).

30. CLIMAP project members, "The surface of the ice-age earth," *Science*, **191**, 1131-1137 (1976).

31. A. D. Vernekar, "A calculation of normal temperature at the earth's surface," *J. Atmos. Sci.*, **32**, 2067-2081 (1975).

32. R. E. Dickinson, "Planetary waves: Theory and observation," *GARP Publ. Ser. No. 23.* (Orographic Effects in Planetary Flows), 51-84. (1980).

33. S. Manabe and T. B. Terpstra, "The effects of mountains on the general circulation of the atmosphere as identified by numerical experiments, *J. Atmos. Sci.*, **31**, 3-42 (1974).

34. J. D. Opsteegh and A. D. Vernekar, "A simulation of the January standing wave pattern including the effects of the transient eddies," *J. Atmos. Sci.*, **39**, 734-744. (1982).

35. B. Saltzman, "Surface boundary effects on the general circulation and macroclimate: A review of the theory of the quasi-stationary perturbations in the atmosphere," *Meteorol. Monogr.*, **30**, 4-19 (1968).

36. S. Ashe, "A nonlinear model of the time-average axial asymmetric flow induced by topography and diabatic heating," *J. Atmos. Sci.*, **36**, 109-126 (1979).

37. A. D. Vernekar and H. D. Chang, "A statistical-dynamical model for stationary perturbations in the atmosphere," *J. Atmos. Sci.*, **35**, 433-444 (1978).

38. J. Oerlemans, "A model of a stochastically driven ice sheet with planetary wave feedback," *Tellus*, **31**, 469-477 (1979).

39. W. D. Sellers, "A two-dimensional global climatic model." *Mon. Weather Rev.*, **104**, 233-248 (1976).

40. P. J. Webster and K. M. W. Lau, "A simple ocean-atmosphere climate model: Basic model and a simple experiment," *J. Atmos. Sci.*, **34**, 1063-1084 (1977).

41. V. Ramanathan and J. A. Coakley, Jr., "Climate modeling through radiative-convective models," *Rev. Geophys. Space Phys.*, **16**, 465-489 (1978).

42. I. M. Held, "On the height of the tropopause and the static stability of the troposphere," *J. Atmos. Sci.*, **39**, 412-417 (1982).

43. J. Adem, "Experiments aiming at monthly and seasonal numerical weather predicition," *Mon. Weather Rev.*, **93,** 495-503 (1965).
44. I. Kubota, "Calculation of seasonal variation in the lower tropospheric temperature with heat budget equations," *J. Meteor. Soc. Jpn.*, **50,** 18-34 (1972).
45. B. Saltzman, "Steady state solutions for axially-symmetric climatic variables," *Pure Appl. Geophys.*, **69,** 237-259 (1968).
46. W. L. Donn and D. M. Shaw, "Model of climate evolution based on continental drift and polar wandering," *Geol. Soc. Am. Bull.* **88,** 390-396 (1977).
47. J. Adem, "Numerical experiments on ice age climates," *Climatic Change*, **3,** 155-171 (1981).
48. J. Adem, "Numerical simulation of the annual cycle of climate during the ice ages," *J. Geophys. Res.*, **86,** 12, 015—12, 034 (1981).
49. J. E. Kutzbach, "Estimates of past climate at paleolake Chad, North Africa, based on a hydrological and energy-balance model," *Quat. Res.*, **14,** 210-223 (1980).
50. E. J. Barron, S. L. Thompson, and S. H. Schneider, "An ice-free cretaceous? Results from climate model simulations," *Science,* **212,** 501-508 (1981).
51. K. Fraedrich, "Structural and stochastic analysis of a zero dimensional climate system," *Q. J. R. Meteorol. Soc.*, **104,** 461-474 (1978).
52. B. Saltzman and A. D. Vernekar, "Note on the effect of earth orbital radiation variations on climate," *J. Geophys. Res.*, **76,** 4195-4197 (1971).
53. D. M. Shaw and W. L. Donn, "Milankovich radiation variations: A quantitative evaluation," *Science*, **162,** 1270-1272 (1968).
54. M. Ewing and W. L. Donn, "A theory of ice ages," *Science*, **123,** 1061-1066 (1956).
55. M Ghil, "Energy-balance models: An introduction." *Climatic Variations and Variability: Facts and Theories, In A. Berger, ed.*, D. Reidel, Dordecht, 1981, pp. 461-480.
56. G. R. North, R. F. Cahalan, and J. A. Coakley, Jr., "Energy balance climate models." *Rev. Geophys. Space Phys.*, **19,** 91-121 (1981).
57. G. North and J. A. Coakley, "Differences between seasonal and mean annual energy balance model calculations of climate and climate sensitivity," *J. Atmos. Sci.*, **36,** 1189-1204 (1979).
58. E. N. Lorenz, "Climatic predictability." In The Physical Basis of Climate and Climate Modelling," *GARP Publ. No. 16*, ICSU/WMO, 132-136 (1975).
59. S. H. Schneider and C. Mass," Volcanic dust, sunspots, and temperature trends," *Science*, **190,** 741-746 (1975).
60. R. A. Bryson and G. J. Dittberner, "A non-equilibrium model of hemispheric mean surface temperature," *J. Atmos. Sci.*, **33,** 2094-2106 (1976).
61. S. F. Woronko, "Comments on A non-equilibrium model of hemispheric mean surface temperature." *J. Atmos. Sci.*, **34,** 1820-1821 (1977).
62. J. F. Adhémar, *Les Revolutions de la Mer Deluges Perodiques*, Paris, 1842.
63. J. Croll, *Climate and time in their geological relations: A theory of secular changes of the earth's climate.* Edw. Stanford, London, 1875.
64. M. Milankovitch, *Théorie mathématique des phénoménes thermiques produits par la radiation solaire*, Gauthier-Villars, Paris, 1920.
65. M. Milankovitch, "Mathematische klimalehre in köppen-geiger," *Handbuch der Klimatologie*, Gebrüder Bormträger, 1, A, Berlin, 1930.
66. M. Milankovitch, *Kanon der Erdbestrahlung und seine Anvending auf des Eiszeitproblem*, Royal Serbian Academy, Belgrad 1941. English transl., U.S. Commerce, Clearinghouse for Fed. Scientific and Technical Inf., Springfield, VA 1969.
67. A. D. Vernekar, "Long-period global variations of incoming solar radiation," *Meteor. Monogr.*, **12,** (34), (1972).
68. A. L. Berger, "Obliquity and precession of the last 5,000,000 years," *Astr. Astrophys.*, **51,** 127-135 (1976).

69. A. L. Berger, "Power and limitation of an energy-balance climate model as applied to the astronomical theory of paleoclimates," *Palaeogeogr., Paleoclimatol., Palaeolcol.*, **21,** 227-235 (1977).

70. A. L. Berger, "Long-term variation of the earth's orbital elements, *Celestial Mech.*, (1977).

71. A. L. Berger, "Support for the astronomical theory of climatic change, *Nature*, **269,** 44-45 (1977).

72. A. L. Berger, "Long-term variations of caloric insolation resulting from the earth's orbital elements, *Quat. Res.*, **9,** 139-167 (1978).

73. S. A. Bowling, Comments on "The effect of changes in the earth's obliquity on the distribution of mean annual sea level temperature," *J. Appl. Meteor.*, **9,** (1971).

74. G. J. Kukla, "Missing link between Milankovitch and climate," *Nature*, **253,** 600-603 (1975).

75. J. D. Hays, J. Imbrie, and J. J. Shackleton, "Variations in the earth's orbit: Pacemaker of the ice ages," *Science*, **194,** 1121-1132 (1976).

76. J. Weertman, "Milankovitch solar radiation variations and ice age sheet sizes, *Nature*, 17-20 (1976).

77. G. E. Birchfield, "A study of the stability of a model continental ice sheet subject to periodic variations in heat input, *J. Geophys. Res.*, **82,** 4909-4913 (1977).

78. E. J. Öpik, "Climatic change in cosmic perspective, *Icarus*, **4,** 289-307 (1965).

79. R. J. Talbot, D. M. Butler, and M. J. Newman, "Climatic effects during passage of the solar system through interstellar clouds, *Nature*, **262,** 561-563 (1976).

80. G. J. F. MacDonald, "Tidal friction," *Rev. Geophys.*, **2,** 467-541 (1964).

81. J. W. Wells, "Coral growth and geochronometry," *Nature*, **167,** 948-950 (1963).

82. B. G. Hunt, "The effects of past variations of the earth's rotation rate on climate," *Nature*, **281,** 188-191 (1979).

83. G. S. Golitsyn, "A similarity approach to the general circulation of planetary atmospheres," *Icarus*, **13,** 1-24 (1970).

84. J. R. Bray, "Volcanic triggering of glaciation," *Nature*, **260,** 414-415 (1976).

85. A. T. Wilson, "Origin of ice ages: An ice shelf theory for Pleistocene glaciation," *Nature*, **201,** 147-149 (1964).

86. J. Weertman, "Rate of growth or shrinkage of nonequilibrium ice sheets," *J. Glaciol.*, **6,** 145-158 (1964).

87. G. E. Birchfield and J. Weertman, "A note on the spectral response of a model continental ice sheet," *J. Geophys. Res.*, **83,** 4123-4125 (1978).

88. G. E. Birchfield, J. Weertman, and A. T. Lunde, "A paleoclimate model of northern hemisphere ice sheets," *Quat. Res.*, **15,** 126-142 (1981).

89. M. Ghil, and H. LeTreut, "A climate model with cryodynamics and geodynamics," *J. Geophys. Res.*, **86,** 5262-5270 (1981).

90. E. Källén, C. Crafoord, and M. Ghil, "Free oscillations in a climate model with ice-sheet dynamics," *J. Atmos. Sci.*, **36,** 2292-2303 (1979).

91. J. Oerlemans, "Continental ice sheets and the planetary radiation budget," *Quat. Res.*, **14,** 349-359 (1980a).

92. J. Oerlemans, "Model experiments on the 100,000-yr. glacial cycle," *Nature*, **287,** 430-432 (1980b).

93. D. Pollard, A. P. Ingersoll, and J. G. Lockwood, "Response of a zonal climate ice sheet model to the orbital perturbations during the Quaternary ice ages," *Tellus*, **32,** 301-319 (1980).

94. G. H. Denton and T. J. Hughes, "The Arctic ice sheet: An outrageous hypothesis." In G. H. Denton and T. J. Hughes, eds., *The Last Great Ice Sheets*, Wiley, New York, 1981, pp. 437-467.

95. T. J. Hughes, G. H. Denton, and M. G. Grosswald, "Was there a late Würm Arctic Ice Sheet?" *Nature*, **266,** 596-602 (1977).

96. P. Fong, "Latent heat of melting and its importance for glaciation cycles," *Climatic Change*, **4,** 199–206 (1982).
97. R. F. Flint, *Glacial and Quaternary Geology*, Wiley, New York, 1971.
98. H. Pollack, "The heat flow from the earth: A review. In P. A. Davies and S. K. Runcorn, eds., Academic Press, New York, 1980, pp. 183–192.
99. I. M. Held, "Climate models and the astronomical theory of the ice ages," *Icarus*, **50,** 408–422 (1983).
100. M. A. W. Mahaffy, "A numerical three-dimensional ice flow model," *J. Geophys. Res.*, **81,** 1059–1066 (1976).
101. W. F. Budd and I. N. Smith, "The growth and retreat of ice sheets in response to orbital radiation changes," In Sea Level, Ice, and Climatic Change," *IAHS Publ. No. 131*, 369–409. (1981).
102. G. E. Birchfield, J. Weertman, and A. T. Lunde, A model study of the role of high-latitude topography in the climatic response to orbital insolation anomalies," *J. Atmos. Sci.*, **39,** 71–87 (1982).
103. V. Ya Sergin, "Numerical modeling of the glaciers—ocean—atmosphere global system, *J. Geophys. Res.*, **84,** 3191–3204 (1979).
104. D. Pollard, "Simple ice sheet model yields realistic 100 kyr glacial cycles," *Nature*, **296,** 334–338 (1982).
105. D. Pollard, "An investigation of the astronomical theory of the ice ages using a simple climate—icesheet model, *Nature*, **272,** 233–235. (1978).
106. J. Oerlemans, "Modelling of Pleistocene European ice-sheets: Some experiments with simple mass-balance parameterizations," *Quat. Res.*, **15,** 77–85 (1981).
107. J. Oerlemans, "Some basic experiments with a vertically-integrated ice sheet model, *Tellus*, **33,** 1–11 (1981).
108. J. Oerlemans, "A model of the Antarctic ice sheet," *Nature*, **297,** 550–553 (1982).
109. H. Le Treut and M. Ghil, "Orbital forcing, climatic interactions, and glaciation cycles," *J. Geophys. Res.*, **88,** 5167–5190 (1983).
110. T. M. L. Wigley, "Spectral analysis and the astronomical theory of climatic change, *Nature*, **264,** 629–631 (1976).
111. W. R. Peltier, "Dynamics of the ice age earth," *Adv. Geophys.*, **24,** (1982).
112. B. Saltzman, "Stochastically-driven climatic fluctuations in the sea-ice, ocean temperature, CO_2 feedback system," *Tellus*, **34,** 97–112 (1982).
113. R. E. Moritz, "Nonlinear analysis of a simple sea-ice ocean temperature oscillator model," *J. Geophys. Res.*, **84,** 4916–4920
114. B. Saltzman, and R. E. Moritz, "A time-dependent climatic feed-back system involving sea-ice extent, ocean temperature, and CO_2," *Tellus*, **32,** 93–118 (1980).
115. W. D. Hibler, "Sea ice growth, drift, and decay." In S. C. Colbeck, ed., *Dynamics of Snow and Ice Masses*, Academic Press, New York, 1980.
116. W. S. B. Paterson, "Ice sheets and ice shelves." In S. C. Colbeck, ed., *Dynamics of Snow and Ice Masses.* Academic Press, New York 1980.
117. W. Berner, H. Oeschger and B. Stauffer, "Information on the CO_2 cycle from ice core studies," *Radiocarbon*, **22,** 227–236 (1980).
118. R. J. Delmas, J. M. Ascencio, and M. Legrand, "Polar ice evidence that atmospheric CO_2 20,000 yr. B.P. was 50% of present," *Nature*, **284,** 115–157 (1980).
119. G. W. Brass, J. R. Southam, and W. H. Peterson, "Warm saline bottom water in the ancient ocean," *Nature*, **296,** 620–623 (1982).
120. P. K. Weyl, "The role of the oceans in climatic change: A theory of the ice ages," *Meteorol. Monogr.*, **30,** 37–62 (1968).
121. B. Saltzman, "Global mass and energy requirements for glacial oscillations and their implications

for mean ocean temperature oscillations," *Tellus*, **29**, 205–212 (1977). (Reply to J. G. Lockwood, ibid., **30**, 190–191 (1978)).

122. J. Oerlemans and A. D. Vernekar, "A model study of the relation between northern hemisphere glaciation and precipitation, *Contrib. to Atmos. Phys.*, **54**, 352–361 (1981).
123. W. W. Kellogg, "Climatic feedback mechanisms involving the polar regions." In G. Weller and S. A. Bowling, eds., *Climate of the Arctic*, Geophysics Institute, University of Alaska, 1975, pp. 111–116.
124. W. F. Ruddiman and A. McIntyre, "The mode and mechanism of the last deglaciation: Oceanic evidence," *Quat. Res.*, **16**, 125–134 (1981).
125. A. Sutera, "On stochastic perturbation and long-term climate behavior," *Q. J. R. Meteorol. Soc.*, **107**, 137–153 (1981).
126. D. R. MacAyeal, "A catastrophe model of the paleoclimate," *J. Glaciol.*, **24**, 245–257 (1979).
127. W. S. Broecker and J. van Donk, "Insolation changes, ice volumes, and the O^{18} record in deep-sea cores," *Rev. Geophys. Space Phys.*, **8**, 169–198 (1970).
128. E. Eriksson, "Air–ocean-icecap interactions in relation to climatic fluctuations and glaciation cycles," *Meteor Monogr.*, **8**, (30), 68–92 (1968).
129. B. Saltzman, A. Sutera, and A. Evenson, "Structural stochastic stability of a simple auto-oscillatory climatic feedback system," *J. Atmos. Sci.*, **38**, 494–503 (1981).
130. B. Saltzman, A. Sutera, and A. R. Hansen, "A possible marine mechanism for an internally generated 100,000 year climate cycle, *J. Atmos. Sci.*, **39**, 2634–2637 (1982).
131. E. N. Lorenz, "Deterministic non-periodic flow," *J. Atmos. Sci.*, **20**, 130–141 (1963).
132. A. G. Fischer and M. A. Arthur, "Secular variations in the pelagic realm," In H. E. Cook and P. Enos, eds., Deep-water Carbonate Environments, Spec. Publ. Soc. Econ. Paleontol. Miner., 19–50 (1977).
133. D. C. Rhoads, and J. W. Morse, "Evolutionary and ecologic significance of oxygen-deficient marine basins," *Lethaia*, **4**, 413–428 (1971).
134. K. Hasselmann, "Stochastic climate models, Part 1: Theory," *Tellus*, **28**, 473–485 (1976).
135. Z. Schuss, "Singular perturbation methods in stochastic differential equations of mathematical physics," *SIAM Rev.*, **22**, 119–155 (1980).
136. P. Lemke, "Stochastic climate models, Part 3: Application to zonally averaged energy models, *Tellus*, **29**, 385–392 (1977).
137. C. Nicolis and G. Nicolis, "Environmental fluctuation effects on the global energy balance, *Nature*, **281**, 132–134 (1979).
138. E. N. Lorenz, "Climatic determinism," *Meteorol. Monogr.*, **5**, 1–3 (1968).
139. R. Benzi, G. Parisi, A. Sutera, and A. Vulpiani, "Stochastic resonance in climatic change," *Tellus*, **34**, 10–16 (1982).
140. C. Nicolis, "Stochastic aspects of climatic transitions—response to a periodic forcing," *Tellus*, **34**, 1–9 (1982).
141. A. Sutera, "Stochastic perturbation of a pure convective motion," *J. Atmos. Sci.*, **37**, 245–249 (1980).
142. J. E. Kutzbach and R. A. Bryson, "Variance spectrum of Holocene climatic fluctuations in the North Atlantic sector, *J. Atmos. Sci.*, **31**, 1958–1963 (1974).
143. W. Dansgaard, S. J. Johnson, H. B. Clausen, and C. C. Langway, Jr., "Climatic record revealed by the Camp Century ice core." K. K. Turekian, ed., *The Late Cenozoic Glacial Ages*, Yale University Press, New Haven, 1971, pp. 37–56.
144. B. J. Mason, "Towards the understanding and prediction of climatic variations, *Q. J. Meteorol. Soc.*, **102**, 473–498 (1976).
145. J. M. Mitchell, "An overview of climatic variability and its causal mechanisms," *Quat. Res.*, **6**, 481–493. (1976).

146. M. I. Hoffert, A. J. Callegari, and C. T. Hsieh, "The role of deep sea heat storage in the secular response to climatic forcing," *J. Geophys. Res.*, **85,** 6667–6679 (1980).

147. R. L. Nace, "World water inventory and control." In R. J. Chorley, ed., *Water, Earth and Man.* Methuen London, 1969 pp. 31–42.

148. C. H. B. Priestley, *Turbulent Transfer in the Lower Atmosphere*, University of Chicago Press, 1959.

149. B. Saltzman and A. Sutera, "A model of the internal feedback system involved in the late Quaternary climatic variations, *J. Atmos. Sci.*, **41,** 736–745 (1984).

150. B. Saltzman, A. Sutera, and A. R. Hansen, "Long period free oscillations in a three–component climate model." In A. Berger, ed., *New Perspectives in Climate Modelling*, Elsevier, Amsterdam, 1984, pp. 289–298.

151. B. Saltzman, A. Sutera, and A. R. Hansen, "Earth–orbital eccentricity variations and climatic change" In A. Berger, J. Imbrie, J. Hays, G. Kukla, and B. Saltzman, eds., *Milankovitch and Climate*, D. Reidel, Dordrecht, 1984, 615–636.

152. B. Saltzman, A. R. Hansen, and K. Maasch, "The late Quaternary glaciations as a response of a three component feedback system to earth-orbital forcing," *J. Atmos. Sci.* in press 1984.

153. B. Saltzman, "On the role of equilibrium atmospheric climate models in the theory of long period glacial variations," *J. Atmos. Sci., 41,* 2263–2266 (1984).

9

CLIMATE MODELS: APPLICATIONS FOR THE PRE-PLEISTOCENE

Eric J. Barron

National Center for Atmospheric Research, Boulder, Colorado

The simulation of paleoclimates requires: (1) specification of the external climatic forcing factors which influenced the climate, (2) a mathematical model of the climate system which incorporates all the processes and feedback mechanisms necessary to predict the basic features of climatic change, and (3) the results must be verifiable through comparison with the paleoclimatic record. Particularly for the pre-Pleistocene, each of these three aspects is characterized by a number of problems. Because of these problems a comprehensive "simulation" of a pre-Pleistocene climate is, as yet, unfeasible. However, there is tremendous potential in a quantitative approach to paleoclimatology based on a set of physical laws which can be applied to the study of hypothetical climate systems. Initially, a quantitative approach to pre-Pleistocene climate modeling must be based on the method of sensitivity analysis.

In a sensitivity study, the model is perturbed by making a change in a boundary condition or internal process and the results are compared with a "control" or unperturbed simulation. By modifying one model characteristic at a time, cause and effect can be isolated. In this manner, one hopes to gain insight into the model and to identify which mechanisms may have the most significance in the real climate system. At the very least this research approach can identify those areas in need of further investigation.

The study of geography as a potential climatic forcing factor is one of the few existing examples of the application of models and sensitivity analysis to the study of pre-Pleistocene, Phanerozoic climates. The changes in the surface of the earth on 10^7 year time scales are relatively large and therefore have potential as an important external climatic influence. For the most part, the surface of the earth can also be well specified as a boundary condition in climate models. There are a large number of sensitivity experiments, using a diverse group of climate models, which have attempted to examine the climatic importance of paleogeography. Combined with a brief description of other sensitivity studies which have application to pre-Pleistocene climates, these examples provide a "source book" of experiments, model limitations, and results for the pre-Pleistocene.

PRE-PLEISTOCENE CLIMATES: THE NATURE OF THE PROBLEM

The pre-Pleistocene climate problem consists of three components: (1) determination of the climatic state over any specified interval of geologic time, (2) determination of the controlling factors which influenced the climate, and (3) determination of the climatic response to any change in these controlling factors.

The first component consists of three problems: reconstruction of the surface of the earth, the interpretation of paleoclimatic indicators, and the ability to stratigraphically subdivide the geologic record. Paleoclimatic indicators must be correctly reconstructed with respect to the spin axis of the earth (latitude), geographic position (coastal, continental interior, etc.), and elevation. The climatic state must be determined from a variety of evidence. In some cases this evidence may not have a clear quantitative relationship to climate, and the relationship often becomes increasingly obscure the older the record. Many time stratigraphic units represent intervals of a few million years. Within these units areal synchroneity cannot be demonstrated. The climatic record may consist of both instantaneous events, such as the record of a single subtropical cyclone, and long-term (million year) averages. Each of these factors result in sufficient ambiguity that Phanerozoic climates can be reconstructed only within broad limits. Yet, within these limits the geologic record poses a number of challenging climatic problems, and in many cases the data are sufficient to develop and test hypotheses.

The number of external climatic forcing factors which may have been important during the pre-Pleistocene is the second major component of the problem. Orography, sea level, continental positions, CO_2, volcanism, orbital elements, solar luminosity, and galactic clouds have all been proposed as important climatic forcing factors. These factors may have jointly or independently influenced paleoclimates. For instance, as yet it is unclear whether a single cause (e.g., high latitude land area) may serve as an explanation of Phanerozoic glaciations. Unfortunately, for most of geologic time the climatic forcing factors are only poorly known. Only the surface of the earth can be accurately reconstructed, but this becomes increasingly uncertain prior to 100 million years ago. Although it is not yet feasible to specify all the external climatic forcing factors which may have been important during a specific interval of geologic time, the significance of a number of specific mechanisms can be investigated.

The third major component is to demonstrate that climate is sensitive to the change in a specific forcing factor and to determine the climatic response to any variations in that factor. To some extent the climate sensitivity may be demonstrated by correlation of "causes" and "effects" as reconstructed from the geologic record.

However, a qualitative understanding of the climate system is insufficient to solve most paleoclimatic problems. Yet it is infeasible to incorporate every detail of the atmosphere, hydrosphere, and cryosphere system in a model and the natural system must be simplified. Hopefully the simplified models will include all the important processes necessary to predict climatic change. Primarily, most models can be verified only against observations of the present-day climate. For these reasons, our confidence in the ability of a model to simulate a climate very different from the present-day must be tempered by knowledge of the limitations of the various physical representations of the climate system.

There is considerable potential in studying various mechanisms quantitatively by performing simulations with climate models despite these limitations. However, the research approach must be based on sensitivity analyses rather than attempts to produce a verified comprehensive simulation of a pre-Pleistocene climate.

PERSPECTIVE ON PRE-PLEISTOCENE CLIMATE MODELING

In pre-Pleistocene research, the term "model" often refers to a qualitative description of a process or processes which serves to explain some aspect of the geologic record. Following the basic geologic tenet of uniformitarianism, models are commonly based on qualitative analogy with modern processes and the recent geologic record. A major goal in paleoclimatology should be to replace the qualitative "modern analogy" with a set of physical laws (e.g., the equation of motion, the first law of thermodynamics, etc.) as discussed in the previous chapter. These physical laws can then be assumed to govern all processes within the climatic system. The laws must, in practice, be simplified or approximated and then applied to a hypothetical climate system. Regardless of the level of complexity, this model contains a mathematical representation of the physical processes thought to be important in determining climate. In this case, the contribution to paleoclimatology will be quantitative, reproducible, and improved upon more easily.

The Primary Task of Climate Modeling

The primary task of climate modeling is to replace the complex natural system by a hierarchy of simplified ones which can be used as quantitative tools for insight. There are a number of ways in which this can be accomplished, although pre-Pleistocene climate modeling presents many specific problems.

Rather than incorporate processes which would explicitly take into account all the possible time scales over which climatic change might take place, a climate model is divided into an "internal" system which is described by variables and an "external" system which is described by specified quantities. This is relatively straightforward for present-day simulations. For example, for a present-day January simulation, the ocean is part of the "external" system and sea surface temperatures can be specified as observed values.

For the pre-Pleistocene, some temporal averaging can also reduce the complexity of the climate system. The problem is that the geologic data may be insufficient to specify many of the "external" components which are typically specified in present-day simulations. The stratigraphic resolution is often limited to a few million years, where even the lithosphere is part of the "internal" climate system. Initially, it would appear that solutions to pre-Pleistocene climate problems would require comprehensive models (e.g. coupled ocean-atmosphere-lithosphere three-dimensional models) and large computer resources.[1]

There are a number of other ways in which the natural climate system may be simplified. There is a clear trade-off between spatial resolution and computational time. Typically, the computational time varies as a function of the cube of the horizontal spatial resolution.[1] A number of models take advantage of spatial averaging. For instance, models designed to investigate the change in the latitudinal position of the "snow line" for a change in solar input are typically one-dimen sional.[2] These Energy Balance Models (EBMs) are zonally averaged, vertically integrated, and typically the meridional resolution is restricted to 10° latitude zones.

Many processes which, in theory, can be computed from general physical laws can often be parameterized (i.e., approximated). For instance, a low resolution model must either ignore or parameterize factors which operate on a scale less than the model resolution, such as approximating eddies or molecular processes by a simple diffusion law applied to zonal averaged variables. Observational data may also be used to derive an empirical relationship which can be substituted for a more complex calculation. For instance, as a result of a series of aircraft flights, Kung et al.[3] determined the approximate latitudinal distribution of winter and summer surface albedo for North America. Using these data, it is possible to relate the seasonal change at each latitude to a change in zonally averaged surface temperature, if the albedo variations are corrected for other factors (e.g., solar zenith angle). In this manner, the complex natural system may also be simplified by restricting the number of variables. A primary limitation of parameterizations is that it may restrict the model's ability to predict a large excursion from the present-day climate. This may be of critical importance in investigations of paleoclimates.

These simplifications introduce a number of uncertainties in climate modeling, particularly for the pre-Pleistocene. Parameterizations or specified conditions may artificially constrain the climate system. The nature of the boundary conditions (e.g., solar luminosity, land-sea distribution, etc.) are also subject to error. Consequently, the extent to which any climate model is capable of accurately simulating climatic change is uncertain.

Application of Climate Models to Hypothetical Systems

Given the uncertainties in applying climate models to the pre-Pleistocene, a comprehensive simulation of a particular geologic period, like that described in chapter 5 for maximum Wisconsin at 18,000 years B.P., is not yet a reasonable goal. Consider, for example, the scientific goal of understanding the atmospheric circulation for a specific Paleozoic Period. The oceans are practically unknown and even oceanic boundary conditions, such as bathymetry for an ocean model, cannot be well specified. In addition to some ocean specification, an atmospheric simulation would require paleocontinental positions, elevations, sea level, and surface vegetation characteristics even if orbital elements and solar luminosity were assumed to be similar to that of the present day. The specification of each of these characteristics is associated with substantial error. In addition the simulation would be difficult to verify. Even if all these characteristics could be accurately specified, by including so many variables, cause and effect would be difficult to interpret. The result would be a climatic snapshot, of uncertain accuracy, which would give little or no insight into the climate of any other time period.

Considerably more insight can be gained by performing sensitivity experiments for each important geologic variable. The most information will be gained, if initially the sensitivity experiments are based on relatively large perturbations. For instance, if the scientific goal is to investigate the climatic importance of large-scale eustatic sea level variations, considerably more insight will be gained by examining the largest, well-defined sea level variation. First, a large perturbation will give the clearest model response. Second, it will clarify whether the suite of smaller sea level variations are likely to be significant. Third, a clear model response will also facilitate any comparisons with the geologic record. In contrast to an attempt at a comprehensive Paleozoic atmospheric simulation, the sensitivity experiment may have considerable significance for the interpretation of pre-Pleistocene climates in general.

In addition, the use of a hierarchy of models, from simple to complex, is an effective research approach because each model is characterized by different assumptions and limitations. Simple models are usually highly parameterized but have the advantage of being computationally efficient and relatively easy to interpret. More complex models are characterized by higher vertical and horizontal resolution and include physical processes more explicitly. Extensive sensitivity experiments can usually be completed with relatively simple models with comparable resources to that required for a single experiment using a complex model. Sensitivity experiments with simple models may indicate those processes which might deserve further refined investigations. For instance, a coupled ocean-atmosphere model experiment may not be justifiable if there is large uncertainty as to whether the changes in boundary conditions are important. In almost every case, the most effective research approach is to start simply and then to proceed by utilizing a hierarchy of models, from simple to complex.

Finally, the application of models to hypothetical climate systems requires observational checks. As closely as possible, these checks should be consistent with the spatial and temporal resolution of the model. Each of the limitations associated with the determination of the climatic state over a specified interval of time enters into our ability to interpret model experiments.

Summary

The primary task in producing a mathematical model is to replace the complex natural system by a simplified one which can be used as a quantitative tool. Because of the requirement to simplify the natural system, it is unlikely that any model will include all the processes and feedback mechanisms which may be required to predict climatic change. In particular for the pre-Pleistocene, inadequacies of models and inaccuracies in specified conditions may artificially constrain the climate system. Therefore the key to model interpretation and insight into the climate system is the completion of sensitivity experiments, modifying one model characteristic (boundary condition or parameterization) at a time. This approach will be the most effective if the sensitivity experiments utilize the largest well-defined perturbations which are likely to have occurred, and if relatively simple models are utilized before proceeding up the model hierarchy.

APPLICATION OF CLIMATE MODELING TO THE PRE-PLEISTOCENE

There is considerable evidence for substantial climatic variation throughout the Phanerozoic. This evidence can be used to construct a schematic temperature history for the earth (figure 1). One major problem which can be addressed with climate models is the causes of these climatic variations. A number of explanations have been proposed for the variations between glacial and warm equable climatic extremes.[4,5] In a sensitivity study approach these factors must be investigated one at a time.

Paleogeography is a logical starting point. This mechanism has been frequently cited as an explanation of Phanerozoic climatic change.[6–11] The changes in the surface of the earth are demonstrably large and can be well specified over the last 100 million years. The last 100 million years was characterized by a number of

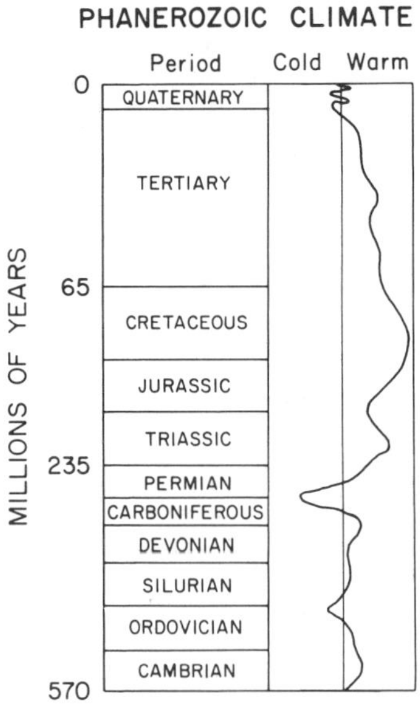

Figure 1. A schematic global temperature history of the earth for the Phanerozoic.[36]

distinct paleogeographic trends[12] and the two end points, Pleistocene glaciations and the warm, equable Cretaceous, were considerable climatic contrasts. Consequently, the majority of pre-Pleistocene climate sensitivity studies have attempted to examine the importance of geography as a climatic forcing factor. In particular a diverse group of climate models have been utilized to examine the importance of geography as an explanation of the warm, equable Cretaceous.

In addition to examining the causes of Phanerozoic climatic change, a number of other important scientific questions can be examined through the application of climate models. For example, one major problem is how the change in a particular boundary condition may modify the nature of the circulation of the atmosphere and the oceans. Both changes in geography and the nature of the planetary warming may alter the characteristics of the circulation. These problems, as well as many others, can be addressed in sensitivity studies using the more comprehensive climate models.

The applications of climate modeling to pre-Pleistocene climate problems will be examined by comparing experiments using a hierarchy of models, from simple to complex, that examine the scientific problems mentioned above. These applications will be described by outlining previous concepts, the goals of the model experiments, the model characteristics, results of experiments, verification of the results, and the limits and problems of the models and their application. This will provide a source book of pre-Pleistocene, Phanerozoic climate model applications.

Surface Albedo Models

One of the most frequently cited causes of Phanerozoic glaciation is the occurrence of land at high latitudes available for accumulation of high albedo snow.[8,11] The differences between the albedo of land, ocean, snow and ice, and the functional relationship between latitude and incoming solar energy suggest that changes in the distribution of continents and land area may play a role in climate evolution. A simple surface albedo calculation may be used to examine this qualitative hypothesis. One of the simplest types of calculations can be based on the area of land in each latitude zone through time as measured from continental reconstructions. Reasonable surface albedos, α, for land and ocean can be assigned to each latitude zone. By assuming present-day mean annual solar radiation, Q, the surface energy budget term, $Q(1 - \alpha)$ can be calculated for a series of time intervals. Without additional assumptions or calculations, this surface energy budget component cannot be related to temperature variations. The goal is to determine if the land-sea differences result in a climatic forcing which is in the correct sense to explain the paleoclimatic record.

Cogley[13] considered the albedo contrast due to continental paleopositions as measured from the reconstructions of Smith and Briden.[14] The Smith and Briden maps do not include changes in total land area due to sea level variations. In a calculation which assumed no snow cover, Cogley[13] determined that low latitude changes in continental distribution contributed more to the radiation balance than the increase in land area at high latitudes throughout the Cenozoic. Based on the sensitivity of other climate models, Cogley estimated that continental positions would have resulted in only a 0.3°K difference in surface temperature over the last 240 million years. He speculated that calculations with snow might be considerably more sensitive.

Barron et al.[15] planimetered the area of land and sea in 10° latitude zones from a series of paleogeographic maps (taking into account sea level variations) at 20 million year increments from 180 million years ago to the present. Land was assigned an albedo of 0.15, and ocean between 0–30° latitude an albedo of 0.06 increasing to 0.13 at 60° latitude. Continental snow was assumed poleward of 60° latitude and was assigned an albedo of 0.65. Sea ice was assigned an albedo of 0.35. This calculation resulted in several conclusions. Both the changes in land at high latitudes and at low latitudes were suggestive of a global cooling from 100 million years ago to the present. However, the relative decrease caused by land-sea changes at low latitudes exceeded that at high latitudes. The importance of land-sea changes at low latitudes was accentuated if desert albedos (0.35) were assigned for land area between 10 and 30° in latitude. There was a strong inverse correlation between total land area and $Q(1 - \alpha)$, indicating that sea level variations may have been a more important climatic forcing factor than continental paleopositions over the last 180 million years.

The results of the surface albedo calculation are probably only indicative of the long-term paleoclimatic trends (all other factors being the same). As such, the results may be compared with a number of long-term records of specific paleocli matic indicators. The isotopic paleotemperature data from late Mesozoic and Cenozoic benthic foraminifera are a useful record of the temperature history. As shown in figure 2, the data of Savin[16] compare favorably with the trends in $Q(1 - \alpha)$.

The above calculations should not be over interpreted. Not a single climatic feedback mechanism was taken into consideration. One calculation assumed no snow cover and the other snow cover on all land above a specific latitude. Only the surface albedo was taken into account and heat transport across latitudes and infrared radiation loss to space were not included. The conclusions are still of interest. Paleogeography appears to operate in the correct sense to explain the paleoclimatic data. These simple calculations have indicated that more detailed model studies of the role of paleogeography as a climatic forcing factor are warranted.

Planetary Albedo Models

The next step in the model hierarchy is a more rigorous calculation of the planetary albedo as a function of changes in the surface of the earth. This calculation will provide an estimate of the magnitude of the difference in total absorbed solar energy between the mid-Cretaceous and the present day. In this manner, it can be determined whether the increase in absorbed solar energy during the Cretaceous is of the right order to help maintain the warm global climate.

Thompson and Barron[17] developed a zonally averaged planetary albedo model specifically to examine the influence of a number of geologic variables (land-sea

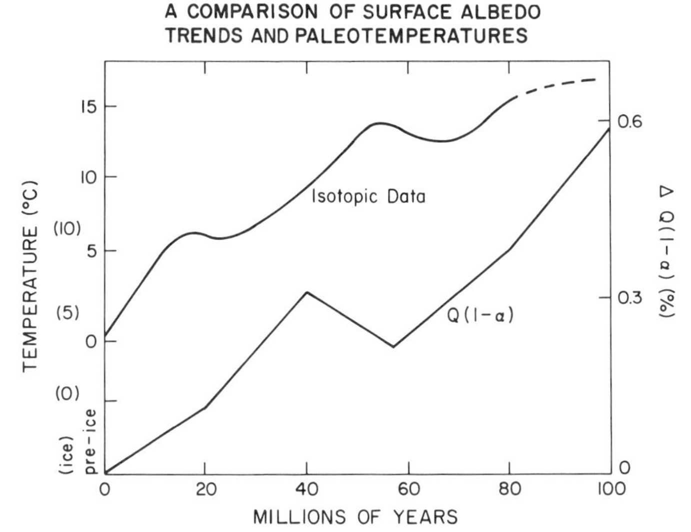

Figure 2. A comparison of a simple surface albedo calculation as a function of geography[15] with a bottom water temperature history based on isotopic temperatures determined from benthic foraminifera.[16] Note the temperature scale is given for both present-day and ice-free conditions.

distribution, temperature, sea ice, vegetation, and cloud cover) on the planetary albedo. The planetary albedo model takes into account the duration and intensity of the insolation, the solar zenith angle, and the mechanisms of modifying the incident solar beam (absorption, scattering, and reflection). The surface albedo is a function of surface type, solar zenith angle and the nature of the radiation (direct or diffuse). Snow cover albedo effects are included as a function of surface temperature. This model includes both cloud albedo as a linear function of solar zenith angle and the effects of multiple cloud-surface reflections. For the present day the model calculations compare favorably with the mean annual and seasonal variation of planetary albedo at each latitude from satellite observations.

For the Cretaceous calculation a number of boundary conditions must be specified, including cloud fraction, land-sea distribution, surface temperatures, surface vegetation characteristics and sea ice extent. Given the limitations of the geologic record, it is clear that model results which are dependent on these assumptions may be suspect. Therefore, Thompson and Barron[17] performed a series of sensitivity experiments to determine the importance of each assumption.

The following assumptions were made for the Cretaceous case. Equatorial temperatures were very similar to the present day and polar temperatures were near or above freezing. The initial land albedos were assumed to range from 0.12 in a wide tropical belt to 0.16 at higher latitudes. The sea ice fraction was assumed to be zero. Cloud fraction was specified as present-day zonal cloud cover. The present-day global mean cloudiness is approximately 0.55. Additional cloud cover sensitivity studies were completed for uniform 0.5 cloud fraction. Such a decrease in cloud fraction may be a plausible assumption based on dynamical model sensitivity studies with increased surface temperatures.[18,19] By changing variables sequentially, Thompson and Barron[17] determined the contribution of each variable to the difference between present day and Cretaceous planetary albedos.

The use of Cretaceous geography and surface albedo in the model increased the absorbed solar energy approximately 1 percent compared to the control. The temperature and sea ice assumptions resulted in a 1.3 percent increase in absorbed solar energy. The decrease in global mean cloudiness of 0.05 increased the absorbed solar energy by 1.6 percent. These results have a number of interesting implications. The global change in absorbed solar energy is large. A plausible value for the change in global mean surface temperature for a 1 percent change in absorbed solar energy is 1.5°C.[20] Given the uncertainties in determining the actual climate sensitivity, a 2–4 percent increase in solar energy could give a 2–8°C increase in global mean surface temperature. For comparison, the Cretaceous temperature assumption, of present-day equatorial temperatures and polar temperatures of 0°C in the mean, is a 6°C increase in global mean surface temperature for the Cretaceous compared to the present day.

This result implies that paleogeography, through the radiation budget alone, has been an important climatic forcing factor. It appears that the increase in total absorbed solar energy is of the correct order of magnitude to maintain a warm Cretaceous climate. However, a substantial amount of the increase in absorbed solar energy resulted from the surface temperature, sea ice, and cloud cover assumptions. These factors were specified as plausible values rather than calculated as true climatic feedbacks. The paleogeographic mechanism must be tested by using land-sea distribution as a specified variable in a more complete climate model, which is not dependent on assumptions of surface temperature. This requires a model which includes heat transport across latitudes, infrared radiation loss to space, and the planetary albedo.

Energy Balance Climate Models

The basis of energy balance climate models is that over long time scales (short relative to the geologic record) incoming solar energy and outgoing infrared radiation should balance. A hierarchy of energy balance models exists, from globally averaged equilibrium climate models to time-dependent zonally averaged energy balance models, such as developed by Thompson and Schneider,[21] which include seasonal heat storage. The energy balance equation of Thompson and Schneider is

$$\frac{\partial}{\partial t}[R(\phi)T_s(\phi,t)] = Q(\phi,t)[1 - \alpha(\phi,t)] - F_{\mathrm{IR}}^{\uparrow}(\phi,t) - \mathrm{div}F(\phi,t) \qquad (1)$$

On the right-hand side are the energy balance components as a function of latitude, ϕ, and time, t. Q is the incoming solar energy, α is the albedo, $F_{\mathrm{IR}}\uparrow$ the outgoing infrared radiation and div F is the divergence of the heat flux. On the left-hand side, R is the thermal inertia (vertically integrated mean heat capacity per unit area) and T_s is the surface temperature. This equation will be a good approximation of the climate system if each of the terms can be approximated adequately. With each term in the equation either known (e.g., Q) or a specified function of T_s (such as F_{IR}), the equation can be solved for surface temperature. A series of sensitivity experiments can then be performed by changing an external or internal condition to investigate the model response to specific forcing factors or assumed physical processes.

Sellers and Meadows[22] used an energy balance-type model to estimate the average surface temperature of the earth for two extreme cases: (1) where the continents were grouped at the pole and (2) where the continents were grouped at the equator. The difference in albedo between the two cases resulted in a 12°C change in surface temperature. At least in this limit, land at high latitudes resulted in the coldest surface temperatures. Although the details of the parameterizations were not given, the results appear to contribute to the hypothesis that continental drift is capable of producing significant climatic changes.

Donn and Shaw[23,24] applied an energy balance climate model to examine the importance of Northern Hemisphere changes in continental positions from the Triassic to the present day. The model calculates temperature at 512 grid points for the Northern Hemisphere. Thus the resolution of the model is much greater than in the previously described studies.

However, there are a number of other limitations. First, the surface albedo was prescribed based on present-day characteristics. The high albedo due to snow was therefore a function of land fraction at high latitudes and not the temperature. Second, the horizontal eddy diffusion of heat was prescribed as a time-dependent term, whereby the coefficient was arbitrarily increased from the Triassic to the present day. Donn and Shaw also utilized an early reconstruction of continental paleopositions with fixed continental area.

From this model, Donn and Shaw determined that the area of land at high latitudes was the major cause of global cooling since the Mesozoic. It is interesting to note that the model estimates less than a 2°C increase in Northern Hemispheric mean temperature. Noting that the temperatures are far too cool to explain the paleoclimatic data, Donn[25] suggests that the enigma of high-latitude paleoclimates may result from a long-term bias between the rotational and dipole (paleomagnetic pole) locations. The limitations of the model may make such strong conclusions suspect. None of the assumptions in the model are necessarily unreasonable, but if one factor at a time (e.g., geography, eddy coefficient) had been modified, in a true sensitivity experiment approach, these conclusions would be easier to interpret. The surface albedo specification and prescribed heat transport coefficient may have limited the model's ability to predict climatic change.

Barron et al.[26] used the Thompson-Schneider[21] version of a zonally averaged energy balance model with the identical planetary albedo model as described by Thompson and Barron.[17] The outgoing infrared radiation was assumed to be a linear function of T_s, with a correction for cloud cover.[27,28] The energy transported poleward by the atmosphere was approximated as a diffusion process. Sensible plus potential energy transport was proportional to the meridional gradient of T_s. Latent heat transport was proportional to the meridional gradient of atmospheric water vapor concentration (proportional to saturation vapor pressure of T_s). Outside the tropics the diffusion coefficients were assumed to be proportional to the absolute value of the temperature gradient, thus the diffusion was nonlinear.

Barron et al.[26] calculated the change in surface temperatures from present-day surface conditions and cloud cover to a case with prescribed Cretaceous geography and present-day cloud cover. For this sensitivity experiment the change in geography resulted in a 1.6°C increase in globally averaged surface temperature. The 1.2 percent increase in total absorbed solar energy was only slightly greater than the calculation from the planetary albedo model.[17]

A 6°C increase in globally averaged surface temperature is a conservative estimate of the warming needed to satisfy the Cretaceous record of paleoclimatic indicators.[12] This is the case with equatorial temperatures similar to the present-day and polar temperatures of 0°C in the mean. Figure 3 clearly shows that the simulations do not reproduce even this conservative estimate of Cretaceous temperatures. Cretaceous globally averaged surface temperatures may have been as much as 14°C higher than at present. Barron[12] bases this limit on estimates of polar mean annual temperatures as high as 10°C.

The discrepancy between the model simulations and the climatic record may be explained by any of three factors: (1) the model may be inadequate, (2) the paleoclimatic record may be misinterpreted, or (3) other forcing factors may be required, in addition to paleogeography. This is precisely the nature of the problem of pre-

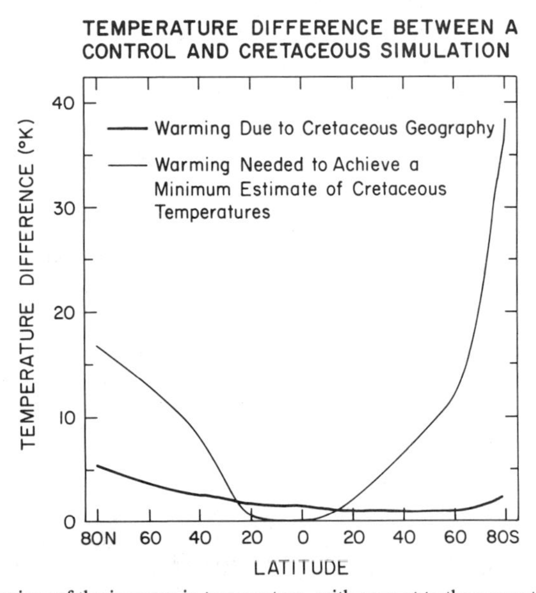

Figure 3. A comparison of the increase in temperature, with respect to the present day, at each latitude required to achieve a minimal estimate of Cretaceous temperatures with the increase generated in an Energy Balance Climate Model[26] as a result of specifying Cretaceous geography as a surface boundary condition.

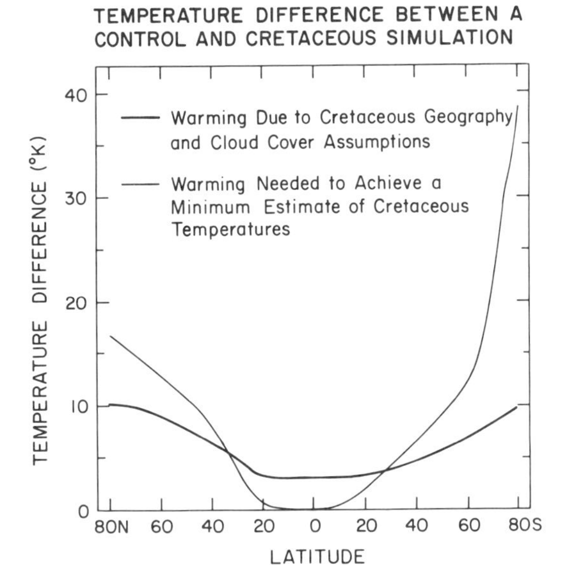

Figure 4. A comparison of the increase in temperature, with respect to the present day, at each latitude required to achieve a minimal estimate of Cretaceous temperatures with the increase generated in an Energy Balance Climate Model[26] as a result of specifying Cretaceous geography and cloud-climate feedbacks.

Pleistocene climates as previously discussed. At this point it is necessary to modify the model approach to some degree. Rather than arbitrarily including other forcing factors or rejecting the paleoclimatic interpretation, the model may be used to determine whether the conservative estimate of Cretaceous temperatures may be achieved by plausible changes in internal model conditions. Barron et al.[26] determined that plausible cloud-climate feedbacks due to the increased surface temperature (similar to those mentioned in the planetary albedo studies), combined with Cretaceous geography, could result in a 5°C increase in globally averaged surface temperature compared to the present day. However, the tropics were too warm and the poles remained too cold (figure 4).

Although factors such as geography and cloud have been modified nonuniformly with respect to latitude in the Cretaceous experiments, the diffusive heat transport formulation tends to smooth out temperature differences between latitudes. Consequently, the distribution of temperatures estimated from the paleoclimatic data may be difficult to explain using the diffusive heat transport formulation. Stated simply, as the equator-to-pole surface temperature gradient decreases, poleward heat transport decreases and thus the poles remain too cold in the model. Barron et al.[26] calculated the heat transport required to achieve the Cretaceous distribution of temperatures given Cretaceous geography and cloud-climate feedbacks (i.e. a global increase in absorbed solar energy sufficient to account for the globally averaged increase in hypothesized Cretaceous surface temperature). Of particular interest, the heat transport convergence required to maintain the Cretaceous distribution of temperatures was almost identical to the present-day heat transport convergence. The model results suggested that if present-day heat transport can be maintained despite the large decrease in equator-to-pole surface temperature gradient, the conservative estimate of Cretaceous temperatures can be achieved with a change in geography and plausible cloud-climate feedbacks. However, if Cretaceous temperatures were actually warmer, then other external climatic forcing factors (like a CO_2 increase) or major departures in internal factors may be required to explain the Cretaceous paleoclimatic record.

A number of the problems apparent as a result of the energy balance climate model studies may be examined using more comprehensive climate models. In particular, the problem requires better models in terms of atmospheric and oceanic heat transport and cloud parameterizations. High resolution models may be used to predict the geographic distribution of temperature, thereby facilitating a direct comparison of the model results with the geologic record. A more direct comparison of model temperatures with the geologic record may reduce the uncertainty in whether the model has achieved a sufficient warming to explain the Cretaceous climate.

General Circulation Models

General Circulation Models (GCMs) are three-dimensional time dependent climate models. These models describe the evolution of the dynamic and thermodynamic state of the atmosphere or ocean. The resolution of GCMs is typically on the order of 5° in latitude and longitude and 2 to 24 levels in the vertical.

Sensitivity experiments with GCMs can be used to examine a number of pre-Pleistocene paleoclimatic problems, in addition to estimating the change in the distribution of temperature with a change in geography. In particular, paleoclimatologists have hypothesized that the equator-to-pole surface temperature gradient is a major control on the nature of the atmospheric circulation.[29,30] The hypothesis suggests that during periods of reduced equator-to-pole surface temperature gradient there was a poleward displacement of large-scale circulation features and a "sluggish" atmospheric and oceanic circulation in the sense of weaker winds and wind-driven ocean surface currents. These concepts have been applied to a wide range of geologic studies.[31] Sensitivity experiments with GCMs can address both the problem of Cretaceous warming as well as how geography and the nature of the surface warming affect the characteristics of the circulation.

Barron and Washington[32] performed a number of sensitivity experiments with a version of the National Center for Atmospheric Research (NCAR) General Circulation Model of the atmosphere.[33] The model consists of an atmospheric GCM coupled with a simple mixed layer ocean model with uniform thickness. The vertical extent of the model is 24 km with eight layers of equal thickness and the horizontal resolution was 5° in latitude and longitude. The model includes an explicit hydrologic cycle with predictive equations for soil moisture and snow cover. Cloudiness is diagnostically determined from the relative humidity. The ocean portion of the model is a simple mixed layer with uniform temperature at each ocean grid point calculated as a function of the surface heat balance. If the ocean temperature drops to −2°C, sea ice can form and grow according to the thermodynamic sea ice model of Semtner.[34] Barron and Washington[32] performed simulations with specified, time invariant insolation (March, January, July) until the model approached equilibrium.

The GCM experiments with a coupled mixed layer ocean by Barron and Washington[31,32] were designed to examine quantitatively the relationships between paleogeography, surface temperature gradient, and the nature of the atmospheric circulation by performing model sensitivity experiments. Three simulations were completed with March (Equinox) present-day insolation: (1) "control" simulation with present-day geography, (2) a Cretaceous experimental simulation (experiment 1) with specified mid-Cretaceous geography, and (3) the Cretaceous experiment repeated but with the restriction that the minimum sea surface temperature was constrained to be 10°C (experiment 2). By comparing a present-day control simulation with two Cretaceous cases with identical geography and insolation but different surface temperature characteristics, possible relationships between geography, surface temperature gradients, and the nature of the atmospheric circulation can be examined.

Neither the poleward displacement of circulation features nor the reduced intensity of the atmospheric winds hypothesized by paleoclimatologists for periods of reduced equator-to-pole surface temperature gradients occurred in these Cretaceous experiments. For instance, the position of the subtropical high shifted equatorward in the Cretaceous experiments to a position associated with the zonal Tethys Ocean. In both Cretaceous cases the poles were characterized by high surface pressure and easterly winds, the opposite of the previously held hypothesis for periods of reduced surface temperature gradients. In fact, the general pattern of equator-to-pole surface pressure characteristics was very similar in the experimental cases although the surface temperature gradient was very different (figure 5). The differences between the Cretaceous and present-day control experiments implies that paleogeography exerts a major influence on the nature of the circulation. The major difference between the two experimental cases involves the zonality of the pressure characteristics. Cretaceous experimental case 2, with stronger land-sea thermal contrasts, has a less zonal surface pressure pattern.

Despite the decrease in equator-to-pole surface temperature gradient, the intensity of the atmospheric circulation did not decrease in the Cretaceous cases. Barron and Washington[31] determined that although the meridional surface temperature gradient decreased, the vertically integrated meridional temperature gradient was maintained. While polar regions warmed substantially, there was also a small increase in tropical sea surface temperatures (temperature of the mixed layer) in the Cretaceous experiments. Because the saturation vapor pressure increases as a nonlinear function of increasing temperature, even a small tropical surface warming (or cooling) may be significant. The heat of condensation of water increases the potential temperature of the air, and the change in sea surface temperature is amplified in the upper atmosphere. Because of this effect, the vertically integrated meridional temperature gradient may be maintained despite the decrease in surface temperature gradient. Kraus[35] discussed the importance of these effects for the atmospheric circulation during glacial and interglacial time periods.

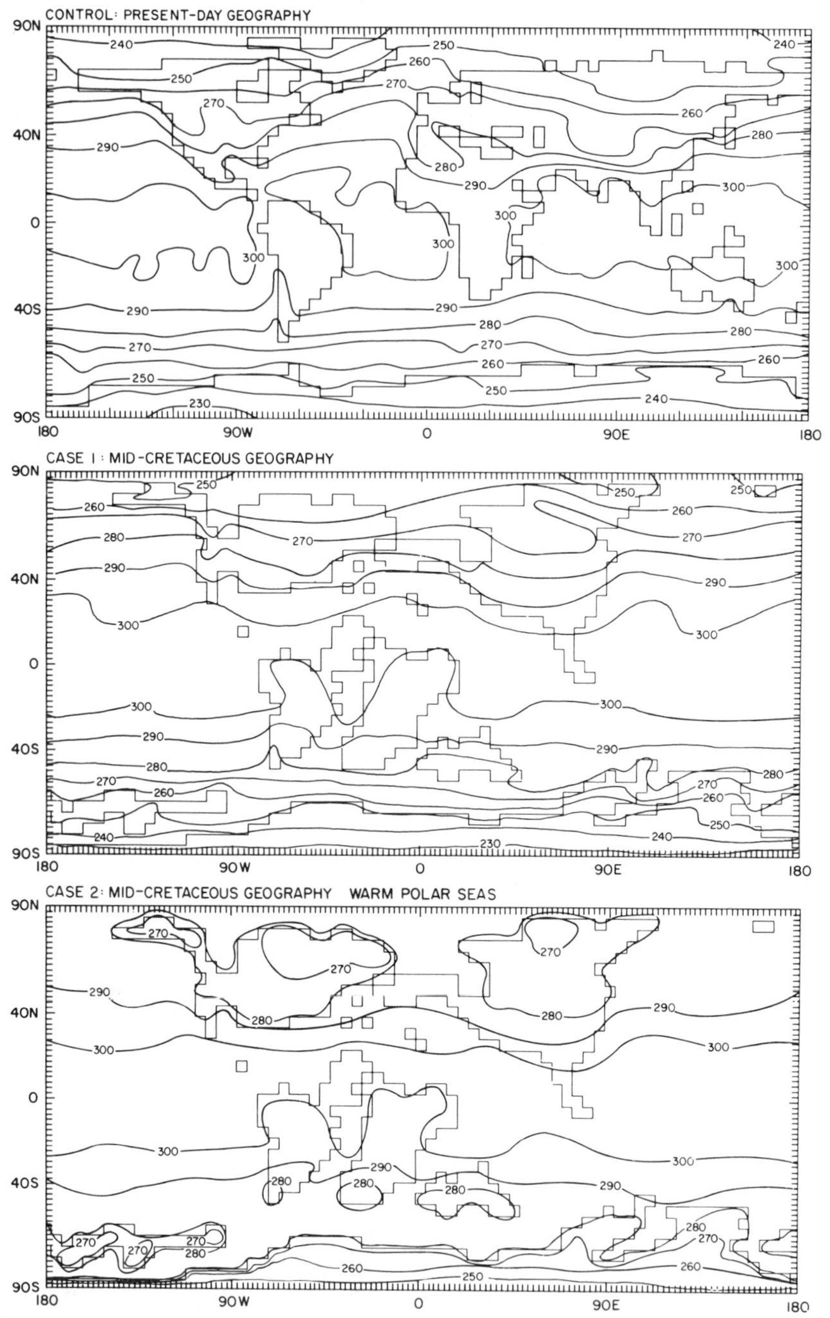

Figure 5. A comparison of time mean (30-day average) temperature at the ground (mixed layer temperatures at ocean grid points) and sea level pressure for three General Circulation Model experiments with March insolation: (1) present-day control, (2) Cretaceous case 1 and (3) Cretaceous case 2. Cretaceous case 1 has specified mid-Cretaceous paleogeography and Cretaceous case 2 has specified mid-Cretaceous paleogeography and sea surface temperatures are constrained such that the minimum value is 10°C.

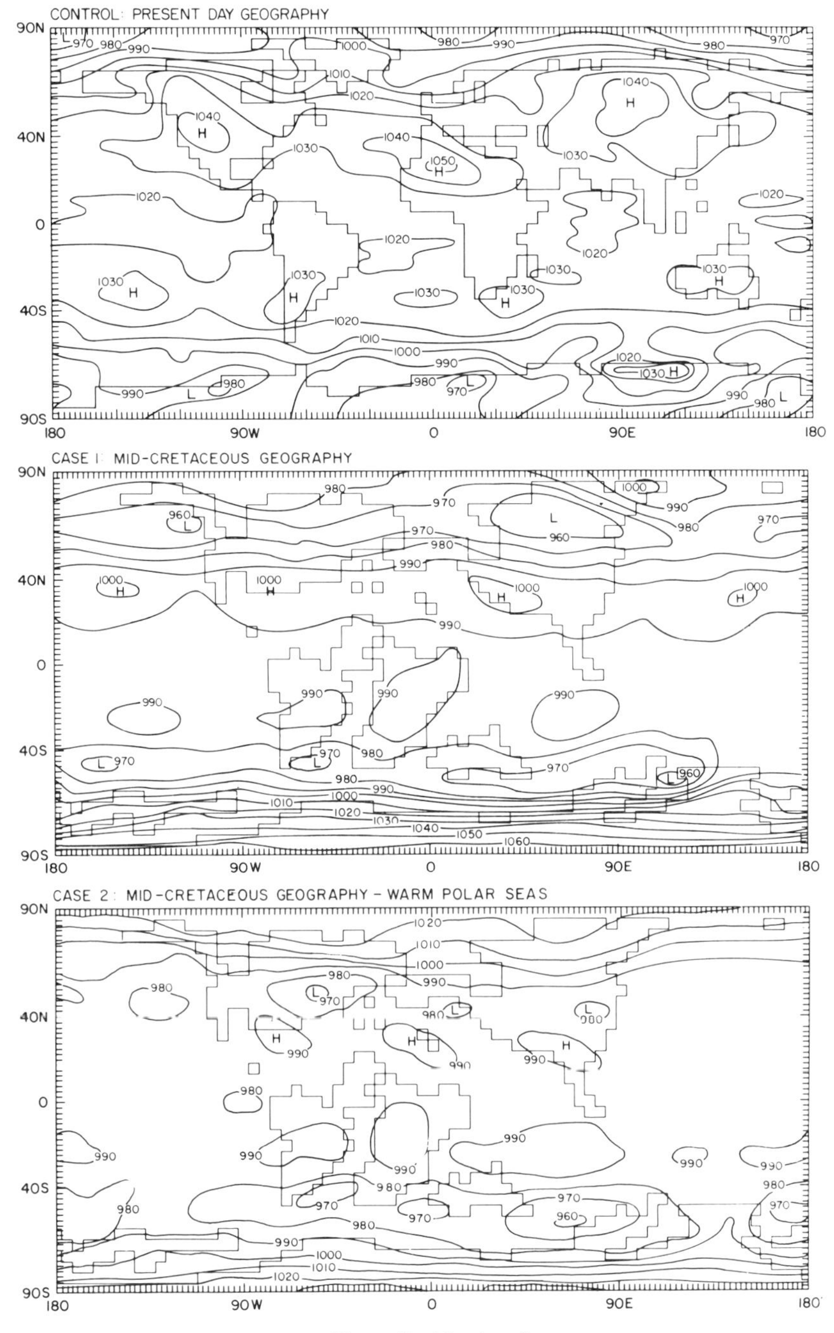

Figure 5. *(Continued)*

The sensitivity experiments with the GCM with a coupled mixed layer are also useful to compare the nature of the planetary warming with Cretaceous paleoclimatic indicators. For example, the interpretation of paleofloral data from Asia distinguish between tropical-subtropical floras, and floras which have experienced subfreezing winter conditions. These data can be compared directly with a January insolation mixed layer experiment (using the identical model) with the minimum sea surface temperatures constrained to be 10°C. This case, with very warm polar ocean temperatures in winter, is much closer to the "warmest" estimates of Cretaceous polar temperatures. This is qualitatively the same as prescribing a large oceanic heat transport. A greatly increased oceanic heat transport for the Cretaceous, compred to the present day, has been hypothesized by numerous authors[36] as an important element in explaining the warmth of Cretaceous polar regions.

Even with warm oceans surrounding the high latitude continents, the continental interiors remain below freezing in winter. Apparently, the model atmosphere's ability to advect warm air from surrounding oceanic regions into the continental interiors is limited. This result implies that unless other factors are operating, large continents at high latitudes are likely to be characterized by cold interior winter conditions, even if polar ocean temperatures are warm. A comparison with the paleofloral data of Vakhrameev[37] from Asia, for the mid-Cretaceous, suggests that much of the data can be explained by a warm adjacent sea. However one point in Mongolia is an exception. The model calculated mean January surface temperature of approximately −10°C is associated with a diverse flora of subtropical character (figure 6). If these data are correct, if the biota cannot withstand seasonally subfreezing conditions, and if the paleogeography and orography has been adequately specified, then neither geography nor increased oceanic heat transport are sufficient to explain the Cretaceous data. Unfortunately, such continental interior localities which can be used for verification of model results are rare. Barron and Washington[32] make a number of comparisons between the model simulations and the geologic record, in addition to the point described above.

In a second set of experiments, Barron and Washington[38] utilized a spectral GCM with nine levels which employs a rhomboidal truncation at wave number 15. The model includes interactive clouds and radiative processes as described by Ramanathan et al.[39] Pitcher et al.[40] describe the results of January and July simulations with fixed sea surface temperatures and sea ice distribution. Barron and Washington[38] calculate the ocean surface temperatures based on the surface energy balance for a layer with no heat capacity ("Swamp" model). Sea-ice forms when the ocean surface temperature falls below −2°C. With this model the sensitivity experiments with Cretaceous geography employed mean annual insolation.

The sensitivity experiments with the spectral GCM and mean annual insolation offer an additional estimate of the planetary warming due to Cretaceous geography. Barron and Washington[38] performed a series of sensitivity experiments specifically to examine the importance of each of the Cretaceous geographic variables (topography, continental area or sea level, and continental positions). Other than a change in a single geographic variable, both the present control and the sensitivity experiments started with identical initial surface boundary conditions of a uniform sea surface temperature, uniform land surface temperatures, and uniform surface al-

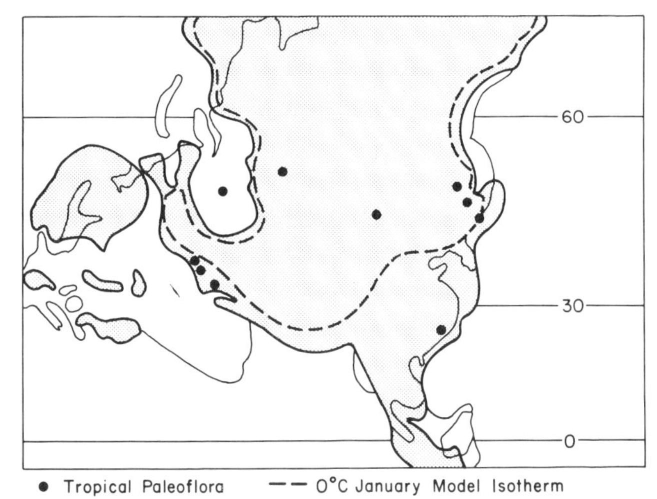

Figure 6. A comparison of the model predicted 273°K surface isotherm from a GCM experiment[32] with ocean temperatures constrained to be greater than 10°C for present-day January insolation with the locations of tropical-subtropical paleofloras on a mid-Cretaceous paleogeographic map.[32]

bedo and soil moisture. The difference between the final "realistic" Cretaceous experiment and the present-day control is a 4.8°C increase in globally averaged surface temperature with a 15–30°C surface warming at each pole. For the case of mean annual insolation the regions of subfreezing conditions are restricted geographically. Continental position was the most important variable in the Northern Hemisphere warming while Antarctic topography and ice (both ice related) were more important in the Southern Hemisphere warming.

A 4.8°C increase in globally averaged surface temperature for the Cretaceous case compared to the present day may be a reasonable estimate of the planetary warming due to geography alone, including the feedbacks incorporated in the spectral GCM. This series of geographic sensitivity experiments designed to investigate the importance of continental positions, sea level, and elevation serves as a foundation for full seasonal simulations, simulations which examine the importance of other forcing factors (e.g., CO_2), and ocean heat transport experiments. A seasonal simulation would allow a much better comparison with paleotemperature data to determine if the warming is sufficient to explain the Cretaceous climate.

It is important to note that the above results may be model dependent. In particular, neither of the GCM experiments adequately treat the surface or the deep ocean. However, the results are of interest for a number of reasons. All the simulations suggest that paleogeography is an important mechanism of climatic change in terms of planetary temperature modifications. The GCM results also suggest that paleogeography influences the nature of the circulation. The model results bring into question the previously held qualitative hypotheses concerning the importance of the equator-to-pole surface temperature gradient as a major control of the nature of the circulation. The results may also guide areas of future paleoclimatic research. They raise the question of whether even the combined effects of geography and increased oceanic heat transport are sufficient to explain the paleoclimatic record. Research into the nature of the paleoclimatic record of continental interiors may prove crucial in answering the question of whether other climatic forcing factors may be necessary to explain Cretaceous climates.

An Overview of Pre-Pleistocene Applications

A hierarchy of climate models, characterized by different limitations, resolution, and representations of physical processes, has been applied to the problem of explaining the contrast between the warm, equable Cretaceous and the present glacial climate. The estimates of the planetary surface temperature warming due to Cretaceous geography, as compared to present-day geography, range from $\approx 1.5°C$ to 8°C. Barron[12] estimates that the Cretaceous globally averaged surface temperature was 6 to 14°C higher than at present. A number of Cretacous simulations resulted in a warming similar to the conservative estimate of Cretaceous globally averaged surface temperature. In every case, the models quantitatively demonstrated that paleogeography was an important climatic forcing factor.

A second problem is whether the warming due to a change in geography is sufficient to explain the paleoclimatic data. At least one simulation by Barron and Washington[32] questions whether the change in geography plus increased oceanic heat transport was sufficient to explain warm winters in the continental interior of Asia. Although this result has not yet been confirmed by other model simulations, it is important for a number of reasons. It focuses our attention back on the nature of the pre-Pleistocene climate problem discussed earlier. Is the discrepancy between the record and the model due to inadequacies of physical models? Are other climatic forcing factors in addition to geography required to explain the paleoclimatic record? Has the geologic record been misinterpreted?

The model studies have also served to reexamine the early qualitative arguments which have been applied to time periods greatly different from the present day. In one case a qualitative hypothesis, that geography was an important climatic forcing factor, has been demonstrated quantitatively. In another case, the applicability of a qualitative hypothesis, that the equator-to-pole surface temperature gradient was a major influence on the nature of the circulation, has been questioned.

A number of additional climatic sensitivity experiments have been performed based on forcing factors which may be important in explaining the geologic record. Hunt[41,42] used a general circulation model to examine the effects of past variations of the earth's rotation rate on climate. A number of models both simple[43] and complex[44] have examined the importance of changing the solar constant. Pollard et al.,[45] Birchfield,[46] and Schneider and Thompson,[47] in addition to several other authors have modeled the effects of orbital element variations. The importance of CO_2 variations in climate has been examined with a whole hierarchy of models.[48–55] The consequences of volcanic eruptions has also been addressed through climate models by Pollack et. al.[56] and Hunt.[57] The above list is by no means exhaustive.

These simulations may give some insight into the importance of other geologic variables as climatic forcing factors in addition to paleogeography. For instance, a doubling of CO_2 above today's values, all other factors being the same, results in a 2 to 3°C increase in globally averaged surface temperature in most model studies. However, these factors are only beginning to be considered as forcing factors in pre-Pleistocene Phanerozoic climate experiments.

As yet, it is not possible to simulate accurately the climate of any past geologic period, because of the limitations in models, the geologic record and in knowledge of the important climatic forcing factors. The application of climate models must be based on sensitivity model studies. The sensitivity experiments also provide the foundation for the interpretation of future simulations.

FUTURE RESEARCH

The pre-Pleistocene climate model sensitivity experiments are of extremely limited scope. There are two directions for future research. First, additional sensitivity experiments are required to understand the Cretaceous climate, and second, there are a whole suite of Phanerozoic problems in addition to the problem of a warm, equable Cretaceous.

There remains considerable additional sensitivity experiments required to examine the Cretaceous climate. Many of the geography sensitivity studies included more than a change in a single geographic variable. Continental positions, sea level variations, and orography must be considered in detail to understand fully cause and effect. Highly idealized geographic sensitivity experiments might contribute even more effectively to the understanding of the role of geography in climatic change. To date the treatment of the oceans has been wholly inadequate. Lacking the ability to specify sea surface temperatures, coupled ocean-atmosphere experiments, or simple ocean heat transport parameterizations for mixed layer atmospheric GCMs should provide considerable insight. Ocean circulation model sensitivity studies for different specified atmospheric boundary conditions are a prerequisite for coupled ocean-atmosphere experiments. As in paleoclimatology there are a large number of qualitative hypotheses in paleoceanography which may be addressed by the application of quantitative models. Through intensive research on the climatic extremes of the Pleistocene and the Cretaceous we will gain confidence in our ability to understand the entire spectrum of Phanerozoic climatic change.

However, there is no guarantee that the planetary warming which resulted from Cretaceous geography may be used to interpret any other warm geologic period illustrated in figure 1. At least one major avenue of research should be to investigate the entire range of Phanerozoic geographies starting at the simple end of the model hierarchy. There are three possible results of this type of investigation, which are illustrated by means of hypothetical model results in figure 7. First, there may exist a clear correlation between geography and warm and cool geologic periods (figure 7*a*). If previous models are indicative, the models may not produce variations in surface temperature which are of the "correct" magnitude to explain the geologic record. This case will illustrate that geography is an important climatic forcing

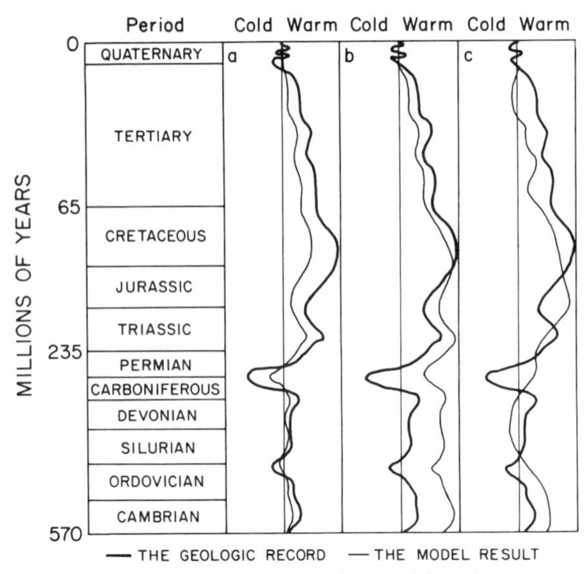

Figure 7. Schematic Phanerozoic climate model scenarios. (*a*) A hypothetical case where models with specified Phanerozoic geography result in warming or cooling in the correct sense but not of the correct magnitude. (*b*) A hypothetical case where models with specified Phanerozoic geography result in warming or cooling in the correct sense with a superimposed longer term trend. (*c*) A hypothetical case where models with specified Phanerozoic geography result in a Cretaceous warming, but other periods have little correlation.

factor, and that additional research must concentrate on feedbacks and qualifying the paleoclimatic record. A second possibility is that there will exist some correlation between the model results and the paleoclimatic record, but that there will also exist systematic shifts between the two (figure 7*b*). This case would suggest that other long-term climatic forcing factors are operating (e.g., variations in solar luminosity) in addition to geography. A third possibility (figure 7*c*) is that there will be no correlation between geography and the paleoclimatic record. Although neither the geography nor the paleoclimatic interpretation were described in detail, Burrett[58] suggests the latter case may be correct. If this is true, then each geologic period may have been influenced by different combinations or by unique climatic forcing factors. Or conversely, we may have a great deal to learn about continental reconstructions, data interpretation, and climate modeling.

The advances in climate modeling have been achieved by a decade of examining model performance in comparison with observations, the inter-comparison of a diverse group of models each with different limitations, and through sensitivity studies to understand the relationship between climate and changes in external and internal factors. Although the Phanerozoic is characterized by a number of special problems and limitations, the same research approach holds considerable promise for paleoclimatology and paleoceanography.

REFERENCES

1. W. L. Gates, *Paleoclimatic Modeling-A Review of Problems and Prospects for the Pre-Pleistocene, Climate in Earth History*, National Academy of Science, Washington, D.C., 1982.
2. S. H. Schneider and T. Gal-Chen, "Numerical experiments in climate stability," *J. Geophys. Res.*, **78,** 6182-6194 (1973).
3. E. C. Kung, R. A. Bryson, and D. H. Lenschow, "Study of a continental surface albedo on the basis of flight measurements and structure of the earth's surface over North America," *Mon. Weather Rev.*, **92,** 543-564 (1964).
4. D. H. Tarling, "The Geological-Geophysical Framework of Ice Ages." In J. Gribben, ed., *Climatic Change*, Cambridge University Press, Cambridge, 1978.
5. T. M. L. Wigley, "Climate and paleoclimate: What can we learn about solar luminosity variations?," *Sol. Phys.*, **74,** 435-471 (1981).
6. P. Damon, "The relationship between terrestrial factors and climate," *Meteorol. Monogr.*, **8,** 106-111 (1968).
7. E. Irving and W. Robertson, "The Distribution of Continental Crust and Its Relation to Ice Ages." In R. Phinney, ed., *The History of the Earth's Crust*, Princeton University Press, Princeton, 1968.
8. J. Crowell and L. Frakes, "Phanerozoic glaciation and the causes of the ice ages," *Am. J. Sci.*, **268,** 193-224 (1970).
9. R. Fairbridge, "Glaciation and Plate Migration." In D. Tarling and S. Runcorn, eds., *Implications of Continental Drift to the Earth Sciences*, Academic Press, New York, 1973.
10. L. Frakes and E. Kemp, "Influence of continental positions on early Tertiary climates," *Nature*, **240,** 97-100 (1972).
11. C. Beaty, "The causes of glaciation," *Am. Sci.*, **66,** 452-459 (1978).
12. E. J. Barron, "A warm, equable Cretaceous: The nature of the problem," *Earth-Sci. Rev.*, **19,** 305-338 (1983).
13. J. Cogley, Albedo contrast of glaciation due to continental drift," *Nature*, **279,** 712-713 (1979).
14. A. Smith and J. Briden, *Cenozoic Paleocontinental Maps*, Cambridge University Press, Cambridge, 1977.
15. E. J. Barron, J. L. Sloan, and C. G. A. Harrison, "Potential significance of land-sea distribution and surface albedo variations as a climatic forcing factor; 180 M.Y. to the present," *Palaeogeogr. Palaeoclimatol., Palaeoecol.*, **30,** 17-40 (1980).
16. S. M. Savin, "The history of the earth's surface temperature during the past 100 million years," *Ann. Rev. Earth Planet. Sci.*, **5,** 319-355 (1977).
17. S. L. Thompson and E. J. Barron, "Comparison of Cretaceous and present earth albedos: Implications for the causes of paleoclimates," *J. Geol.*, **89,** 143-167 (1981).
18. J. Roads, "Numerical experiments on the climatic sensitivity of an atmospheric hydrologic cycle, *J. Atmos. Sci.*, **35,** 753-773 (1978).
19. S. H. Schneider, W. M. Washington, and R. M. Chervin, "Cloudiness as a climatic feedback mechanism: Effects on cloud amounts of prescribed global and regional surface changes in the NCAR GCM," *J. Atmos. Sci.*, **35,** 2207-2221 (1978).
20. S. H. Schneider and S. L. Thompson, "Cosmic conclusions from climate models: Can they be justified?," *Icarus*, **41,** 456-469 (1980).
21. S. L. Thompson and S. H. Schneider, "A seasonal zonal energy balance model with an interactive lower layer," *J. Geophys. Res.*, **84,** 2401-2414 (1979).
22. A. Sellers and A. Meadows, "Long-term variations in the albedo and surface temperature of the Earth," *Nature*, **254,** (1975).
23. W. Donn and D. Shaw, "Model of climate evolution based on continental drift and polar wandering, *Bull. Geol. Soc. Am.*, **88,** 390-396 (1977).

24. W. Donn and D. Shaw, "Continental Drift and Arctic Climate." In M. Dunbar, ed., *Polar Oceans*, Arctic Institute of North America, 1977.
25. W. Donn, "The enigma of high-latitude paleoclimate," *Palaeogeogr. Palaeoclimatol. Paleoecol.*, **40,** 199–212 (1982).
26. E. J. Barron, S. L. Thompson, and S. H. Schneider, "An ice-free Cretaceous? Results from climate model simulations," *Science*, **212,** 501–508 (1981).
27. S. G. Warren and S. H. Schneider, "Seasonal simulation as a test for uncertainties in the parameterizations of a Budyko–Sellers zonal climate model," *J. Atmos. Sci.*, **36,** 1377–1391 (1979).
28. V. Ramanathan, "Interactions between ice–albedo, lapse rate, and cloud-top feedbacks: An analysis of the nonlinear response of a GCM climate model," *J. Atmos. Sci.*, **34,** 1885–1897 (1977).
29. C. E. P. Brooks, *Climate Through the Ages*, Yale University Press, New Haven, Connecticut, 1928.
30. H. Flohn, "Grundfragen du paläoklimatologie in lichte einer theoretischem klimatologie," *Geol. Rundsch.*, **54,** 504–515 (1964).
31. E. J. Barron and W. M. Washington, "Atmospheric circulation during warm geologic periods: Is the equator-to-pole surface temperature gradient the controlling factor?," *Geology*, **10,** 633–636 (1982).
32. E. J. Barron and W. M. Washington, "Cretaceous climate: A comparison of atmospheric simulations with the geologic record," *Palaeogeogr. Palaeoclimatol. Palaeoecol.*, **40,** 103–134 (1982).
33. W. M. Washington and D. L. Williamson, "A description of the NCAR global circulation models." In J. Chang, ed., *Methods in Computational Physics*, Vol. 17, Academic Press, New York, 1977.
34. A. J. Semtner, "A model for the thermodynamic growth of sea ice in numerical investigations of climate," *J. Phys. Oceanogr.*, **6,** 379–389 (1976).
35. E. B. Kraus, "Comparison between ice age and present general circulations," *Nature*, **245,** 129–133 (1973).
36. L. A. Frakes, *Climates Throughout Geologic Time*, Elsevier, Amsterdam, 1979.
37. V. A. Vakhrameev, "Main features of phytogeography of the globe in Jurassic and Early Cretaceous time," *Paleontol. J.*, **2,** 123–133 (1975).
38. E. J. Barron and W. M. Washington, "The role of geographic variables in explaining paleoclimates: Results from Cretaceous Climate Model Sensitivity Studies," *J. Geophys. Res.*, **89,** 1267–1279 (1984).
39. V. Ramanathan, E. J. Pitcher, R. C. Malone, and M. L. Blackmon," The response of a spectral general circulation model to refinements in radiative processes," *J. Atmos. Sci.*, **40,** 605–630 (1983).
40. E. J. Pitcher, R. C. Malone, V. Ramanathan, M. L. Blackmon, K. Puri, and W. Bourke, "January and July simulations with a spectral general circulation model," J. Atmos. Sci., **40,** 580–604 (1983).
41. B. G. Hunt, "The effects of past variations of the earth's rotation rate on climate," *Nature*, **281,** 188–191 (1979).
42. B. G. Hunt, "The influence of the earth's rotation rate on the general circulation of the atmosphere," *J. Atmos. Sci.*, **36,** 1392–1408 (1979).
43. T. Gal-Chen and S. H. Schneider, "Energy balance climate modeling: Comparison of radiative and dynamic feedback mechanisms," *Tellus*, **28,** 108–121 (1976).
44. R. Wetherald and S. Manabe, "The effects of changing the solar constant on the climate of a general circulation model," *J. Atmos. Sci.*, **32,** 2044–2059 (1975).
45. D. Pollard, A. P. Ingersol, and J. G. Lockwood, "Response of a zonal climate-ice sheet model to the orbital perturbations during the Quaternary ice ages," *Tellus*, **32,** 301–319 (1980).
46. G. E. Birchfield, J. Weertman, and A. T. Lunde, "A model study of the role of high-latitude

topography in the climatic response to orbital insolation anomalies," *J. Atmos. Sci.*, **39,** 71–87 (1982).

47. S. H. Schneider and S. L. Thompson, "Ice ages and orbital variations: Some simple theory and modeling," *Quat. Res.*, **12,** 188–203 (1979).
48. S. H. Schneider, "On the carbon-dioxide-climate confusion." *J. Atmos. Sci.*, **32,** 2060–2066 (1975).
49. S. Manabe and R. Wetherald, "Thermal equilibrium of the atmosphere with a given distribution of relative humidity," *J. Atmos. Sci.*, **24,** 241–259 (1967).
50. S. Manabe and R. Wetherald, "The effects of doubling the CO_2 concentration on the climate of a general circulation model," *J. Atmos. Sci.*, **32,** 3–15 (1975).
51. S. Manabe and R. Wetherald, "On the distribution of climate change resulting from an increase in CO_2 content of the atmosphere," *J. Atmos. Sci.*, **37,** 99–118 (1980).
52. S. Manabe and R. J. Stouffer, "Sensitivity of a global climate model to an increase of CO_2 concentration in the atmosphere," *J. Geophys. Res.*, **85,** 5529–5554 (1980).
53. V. Ramanathan, M. Lian, and R. Cess, "Increased atmospheric CO_2: Zonal and seasonal estimates of the effect on the radiation energy balance and surface temperature," *J. Geophys. Res.*, **84,** 4949–4958 (1979).
54. T. Owen, R. D. Cess, and V. Ramanathan, "Enhanced CO_2 greenhouse to compensate for reduced solar luminosity on early earth," *Nature*, **277,** 640–642 (1979).
55. S. H. Schneider and S. L. Thompson, "Atmospheric CO_2 and climate: Importance of the transient response," *J. Geophys. Res.*, **86,** 3135–3147 (1981).
56. J. B. Pollack, O. B. Toon, C. Sagan, A. Summers, B. Baldwin, and W. VanCamp, "Volcanic explosions and climatic change: A theoretical assessment," *J. Geophys. Res.*, **81,** 1071–1083 (1976).
57. B. G. Hunt, "A simulation of the possible consequences of a volcanic eruption on the general circulation of the atmosphere," *Mon. Weather Rev.*, **105,** 247–260 (1977).
58. C. F. Burrett, "Phanerozoic land–sea and albedo variations as climate controls," *Nature*, **296,** 54–56 (1982).

10

THE FUTURE OF PALEOCLIMATOLOGY

John Imbrie

Department of Geological Sciences, Brown University, Providence, Rhode Island

As summarized in chapter 1, the past 20 years have witnessed remarkable progress in our ability to reconstruct and understand past climates. In this chapter I attempt to look forward and imagine what the future will bring—not over the next 20 years (for such forecasts have little more than curiosity value for future historians), but over the next few years, when we may have some confidence that strategies now in place will still be influential in guiding research.

The Narrative of Past Climate

Paleoclimatic data. Like all sciences, paleoclimatology is partly empirical and partly theoretical. As a way of organizing ideas about the empirical side of this discipline we will find it useful to consider its basic observations as a multivariate function of time and space

$$X_i = \phi\ (T,S)$$

where X_i is a set of n different kinds of observations ($i = 1, 2, \ldots, n$) which reflect in n different ways the state of the climate system at time T and place S. If the sampling has been carried out at t stratigraphic horizons and s places, the observations form a three-dimensional array of size n by t by s. Consider, for example, the $n = 60$ observations that might be made on a sample of deep-sea sediment deposited $T = 120{,}000$ years ago at some site S in the South Atlantic. This set of observations forms a one-dimensional array $X_1, X_2 \ldots X_{60}$, and might include measurements of $\delta^{18}O$ and $\delta^{13}C$ on a planktonic foram species (X_1, X_2); the same measurements, plus measurements of Cd/Ca, on a benthic species ($X_3 - X_5$);

measurements of percent Ca CO_3, percent quartz on two different grain sizes, and percent opal ($X_6 - X_9$); and species census data on coccoliths ($X_{10} - X_{19}$), radiolaria ($X_{20} - X_{39}$), planktonic forams ($X_{40} - X_{55}$), and benthic forams ($X_{56} - X_{60}$).

Paleoclimatic information Having assembled these basic observations X_i, our next step will be to transform these *climatic data* into *climatic information* about the state of the climatic system. Generally, these transformations will take the form of *point reconstructions*

$$\hat{C}_j = \phi\ (T,S)$$

where $i = 1, 2, \ldots, m$ ($m \langle\langle n$), and $\hat{C}_j$ are estimates of *m indices of local climate* C_j which describe, in compact form, the actual state of the climate system at time T and place S. In the illustration from marine paleoclimatology given above, the point reconstruction $\hat{C}_j$ might, for example, include estimates of $m = 8$ indices of ocean climate: summer sea-surface temperature, winter sea-surface temperature, mean surface salinity, mean productivity, mean bottom water temperature, calcium carbonate compensation depth, distance from the site of the Subtropical Convergence, and the flux of North Atlantic Deep Water. Real examples of transformations of this kind are found in end-of-chapter references.[1,2,3,] In the field of terrestrial paleoclimatology, point reconstruction would typically be based on observations made on samples of soils, lake sediments, or ice cores, and might include estimates of $m = 8$ indices of atmospheric climate such as mean summer air temperature, mean winter air temperature, extreme winter air temperature, mean rainfall, mean evaporation-precipitation ratio, snowfall, wind speed and direction, and the mean frequency of given airmasses over the site.[4,5,6]

It is perhaps worth emphasizing that the procedure of transforming (and condensing) a large set of observations into estimates of small sets of climatic indices is not a peculiarity of the field of paleoclimatology alone, but a basic part of modern meteorology, oceanography, and climatology as well. In these fields, as in paleoclimatology, the definition of a useful set of climatic indices, as well as the development of procedures for estimating them, often represents an important scientific advance. As an illustration of this point, consider the recent progress made in understanding the origin of climatic anomalies known as El Niño and Southern Oscillation (ENSO) events. These anomalies, which have a recurrence time on the order of several years, involve changes in the tropical atmosphere and ocean over a very large sector of the globe. The changes are so complex, and involve such large fields of atmospheric and oceanic variables, that the mere description of what happens during an ENSO event is itself a formidable task—not to mention the even more challenging task of understanding why the phenomenon occurs. As recent reviews by Rasmusson and Carpenter[7] and Philander[8] make clear, the progress made in understanding ENSO oscillations can be attributed in part to the identification of a small number of climatic indices which provide, in effect, empirical handles by which the investigator can get a grip on a large and slippery problem. One of these indices is the well known Southern Oscillation Index, which is the difference in atmospheric pressure between locations on the opposite sides of the Indonesian-

South Pacific seesaw (say between Tahiti and Darwin). Other indices include some measure of Indonesian rainfall, the sea-surface temperature off the coast of Peru, and the position of the Intertropical and South Pacific Convergence Zones.

Recent progress in the field of paleoclimatology has also been marked by efforts to transform a large set of observations about the climate at a particular place and time into a small number of informative indices. In some cases, these transformations take the form of quantitative algorithms (sometimes called transfer functions or calibration functions) which have been calibrated on Recent data.[9] Examples include estimates of sea-surface temperature,[1] sea-surface salinity,[10] air temperature,[4] and air-mass frequency.[11] In other cases, semi-quantitative procedures have been employed to estimate the position of zones of convergence in the surface ocean. Examples include the position of the Polar Front in the North Atlantic,[12] the Antarctic Convergence,[13] the Subtropical Convergence in the Indian Ocean,[14] and the position of the Oyashio-Kuroshio convergence in the North Pacific.[15]

These tranformations of raw data into useful information illustrate the process of making point reconstructions of past climate, that is, reconstructions at a particular site or in a small, well-sampled area. In addition, methods are now available for transformating observations made at a particular site into climatic conditions integrated over large areas of the globe. Such estimates can be referred to as *areal reconstructions*. Paramount among paleoclimatic indices of this kind is the oxygen isotope ratio, $\delta^{18}O$, in marine foraminifera. As shown by Shackleton and Opdyke,[16] variations in this ratio are highly correlated with changes in the global volume of glacial ice. Subsequent work has shown that this correlation is particularly strong if measurements are made on planktonic foraminifera at open-ocean sites in low- and mid-latitudes, provided that accumulation rates exceed 2 cm/1000 years.

Several other climatic indices in addition to $\delta^{18}O$ may provide climatic information integrated over large areas of the globe. Among these are (1) the δ^{18} record in polar ice cores; (2) measures of the dust content of polar ice cores; (3) biotic measures of the position of the Antarctic Convergence; and (4) chemical, isotopic, or other measures of the flux of North Atlantic Deep Water.

The Science of Paleoclimatology

Our discussion so far has centered on the empirical side of paleoclimatology. As defined here, this effort includes gathering raw data and then transforming it into reconstructions of past climate, either at some particular place and time (point reconstructions) or integrated over some segment of the globe (areal reconstructions). If investigation stops at this point, the result will be a narrative description of past climate. But such a narrative, fascinating and useful as it may be, does not constitute a science, for it does not (by itself) address questions of cause. Why was the reconstructed climate different from the climate today? How did the climate system balance the radiation budget at the time of interest?

To approach these questions, two additional procedures are normally followed. First, the climatic information is organized into a useful *monitoring array*: either a *time series* of information at some particular site, a *transect* of time series along a linear array of sites, or a *map* of information at some particular time. Second, the climatic information displayed in the array is compared with a *theoretical model* having a comparable time-space structure. Normally, the initial comparison will involve a physically based *conceptual model*. If the results of this comparison are promising, the investigator will proceed as rapidly as possible to compare the climatic information with the output of a *numerical model*.

The work of Ruddiman and McIntyre[17] illustrates the use of a conceptual model (water-mass migration driven by Milankovitch forcing) to explain a transect of information in the Pleistocene North Atlantic. Studies by Manabe and Hahn,[18] Gates,[19] and Kutzbach and Otto Bliesner[20] illustrate the use of numerical models of the general circulation to explain maps of past climate. Attempts to model climatic time series are illustrated (in the frequency domain) by Hays et al.[21] and Kominz and Pisias,[22] and in the time domain by Ruddiman and McIntyre[23] and Imbrie and Imbrie.[24]

Trends in Paleoclimatic Research

Several recent developments in paleoclimatology seem likely to influence future research. These include (1) the development of improved, multivariate procedures for reconstructing past climate; (2) the use of high-resolution stratigraphic frameworks to take advantage of more accurate geologic time scales; and (3) the conduct of numerical experiments, that is, the organization of climatic information into time-space arrays that are suitable for comparison with the output of numerical models.

Reconstruction Procedures As discussed above, the empirical side of paleoclimatology depends fundamentally on making observations X_i which reflect the state of the climate system at some particular time in the past; and then transforming these observations into estimates of indices $\hat{C}_j$ which describe what that state was. In recent years notable progress has been made both in finding useful new properties to measure, and in developing more accurate transformation procedures. Many examples of progress in this area might be cited, but I will limit myself to two which illustrate the growing importance of chemical and physical techiques: use of Cd/Ca ratios in deep sea foraminifera to obtain information about the phosphate content of oceanic deep water,[3] and measurement of the CO_2 content in polar ice cores to obtain information on atmospheric CO_2 levels.[6] Progress on the transformation problem has been marked by an increased use of *multivariate* procedures. By this I mean the simultaneous use of a number of different observations to estimate a climatic parameter of interest. Although the procedures used may or may not involve formal multivariate regression, the important point is not the particular statistical methodology, but the use of multiple observations to partial a particular effect out of a complex web of influences. A notable example of this trend is the simultaneous use of carbon and oxygen isotope measurements on multiple species of planktonic forams—in addition to species census data on the foram community—to obtain information on the structure of the water column.[25] Another example is the simultaneous use of species census data on three different groups of marine plankton (foraminifera, radiolaria, and coccolithophorids) to obtain information about ice-age sea-surface temperatures.[2] It seems likely that future reconstructions of past climate will, increasingly, be based on multivariate sets of data in which isotopic, chemical, and paleontologic observations are combined.

Stratigraphy and Chronology Over the past 20 years our ability to correlate and date sedimentary sequences has improved significantly. Nowhere is this improvement more evident than in Cenozoic marine sediments, where the various magnetic reversals can be distinguished from each other with the aid of planktonic biostratigraphic zones and then dated (indirectly) by correlation with the terrestrial succession of K/A-dated reversals. The result is a magneto-biostratigraphic framework that provides a *first order* stratigraphic zonation scheme having a correlation and dating accuracy on the order of the duration of a typical zone, that is, one or two million years.[26] For the Quaternary sequence, the benchmarks of this first order succession provide the starting point for *second order* stratigraphic subdivisions based on oxygen-isotope zones,[16] and a dating scheme for these subdivisions based on the Milankovitch theory.[21] For the late Quaternary, the result is a magneto-bio-isotopic framework with a correlation and dating accuracy of about 5,000 years.[27]

These improvements in the geologic time scale—and in the ability to tie many sedimentary sequences into that time scale stratigraphically—have already had a major impact on the field of paleoclimatology. In the future, as the second-order scheme is extended to older sequences, and as geologic observations are made at an appropriately fine spacing, this impact will certainly be even greater. This impact of a better stratigraphy and chronology are already seen in two developments that are of fundamental importance to the science of paleoclimatology. First, the refined stratigraphic schemes make it possible to reconstruct the spatial pattern of climate over a large portion of the globe and during a short interval of time. This, in turn, makes it possible to attempt to explain the *geographic pattern* of a particular past climate as an *equilibrium response* of the climate system to boundary conditions imposed during (and prior to) that particular interval.[1] Second, the absolute time scale can be used to estimate the *rate of climatic change*. This, in turn, makes it possible to attempt to explain the change in climate as a *dynamic response* of the climate system. Part of this response will represent a reaction to changing external boundary conditions (deterministic forcing);[24] and part will reflect the interaction of processes within the climate system (stochastic forcing).[22,28,29]

Modeling Experiments As reviewed above, research conducted over the past 20 years has given us a remarkable ability to monitor temporal and geographic changes in climate. As a result, important opportunities now exist to further our understanding of the causes of climatic change. These opportunities are best exploited by conducting numerical modeling experiments, that is, by comparing the output of a physically based model of the climate system with an appropriate array of climatic observations. To conduct such experiments, it is necessary that the geologist design and acquire an observational array that will match the output of a particular climate model. In general, three kinds of experiment are possible: those based on arrays in the *space domain*, *time domain*, and *frequency domain*.

Experiments in the Space Domain For the most general models, climatic arrays in the space domain are specified in three dimensions (u, v, z); but experiments so far conducted have been limited to a comparison of mapped climatic estimates ($\hat{C}_{uv}$) with model output (M_{uv}) for a given vertical level z and time t. Such experiments take the form ($M_{uv} - \hat{C}_{uv}$). In the future, we can expect experiments with vertical arrays of the form $\hat{C}_{uz}$ or $\hat{C}_{vz}$ to assume greater importance, particularly as models of ocean circulation in three dimensions are improved, and as geological information about deep water becomes more reliable. So far, numerical experiments in paleoclimatology have been conducted by using models in which the response of only one part of the climate system (either the atmosphere or the ocean) is calculated, while the state of the other parts (including the cryosphere) is fixed. Hopefully, future experimental designs will be able to make use of coupled models, in which the atmosphere and ocean (and eventually the atmosphere, ocean, *and* cryosphere) will interact.

Experiments in the Time Domain Considering that climatic estimates in the time domain $\hat{C}_t$ are relatively easy to acquire at particular sites, while the construction of maps is a far more difficult and time-consuming procedure, it might seem surprising that, so far, relatively few modeling experiments in the time domain ($M_t - \hat{C}_t$) domain have been conducted. The explanation of this paradox is that the development of physically based theories of *climate change* (particularly change on geologic time scales) has lagged far behind the development of theories of *equilibrium climate*. One opportunity for work of this kind has been provided by the Milankovitch theory; but even here, only very simplified time domain models have been used so far.[24,30,31] More sophisticated experiments in this field can be anticipated.

Another opportunity for time domain modeling of Quaternary climate is provided by the oxygen-isotope record. Taking that record as an indirect measure of fluctuations in global sea level, Broecker[32] proposed that changes in sea level should be accompanied by variations in the phosphorous content of the ocean, and that these variations in turn would cause changes in atmospheric CO_2 concentration. Keir and Berger[33] have developed a time-dependent model for exploring this hypothesis. Because changes in CO_2 have such important consequences for atmospheric climate and for ocean chemistry, it is likely that the matter will be actively pursued over the next several years.

Experiments in the Frequency Domain Some of the difficulties involved in predicting how the climate system should respond in the time domain can be circumvented by transforming the estimates $\hat{C}_t$ into the frequency domain $\hat{C}_f$. More precisely, a time series of climatic information, $\hat{C} = \phi\ (t)$, where the data is specified as a function of time, is transformed into a spectrum, $\sigma^2/f = \phi\ (f)$, where the variance density of the data is specified as a function of frequency. Or, if two time series $\hat{C}_1$ and $\hat{C}_2$ are studied simultaneously, the average (linear) correlation and average phase relationships between the two signals can be analyzed into contributions from different frequency bands.[34] The result of this cross-sprectral analysis is a coherency spectrum, $k = \phi\ (f)$, and a phase spectrum, $\Theta = \phi(f)$.

At first glance, the transformation of a "perfectly good time series" into its rather abstract, frequency domain counterpart may strike many investigators as a muddying of waters that are already sufficiently opaque. But there may be good reasons for doing this. In the first place, the (Fourier) transformation may actually simplify the modeling effort by allowing the investigator to focus on one particular frequency band at a time, where the effects of one particular physical process may be modeled and studied in isolation from the effects of other processes operating at other frequencies. In effect, the procedures of formal time series analysis allow the investigator to break up a large and difficult theoretical problem into smaller, tractable units. Underlying this strategy is the widely recognized principle that every physical process operates at a characteristic rate.[28] Since changes in climate result from many physical processes, each operating at its own rate, the virtues of focusing on one frequency band at a time are apparent.

Indeed, recent improvements in geologic chronology now make it possible to use cross-spectral methods to estimate the *rate* at which important climatic processes actually operate. To illustrate this point, consider the set of processes by which various parts of the climate system respond to changes in the intensity of incoming solar radiation (Milankovitch forcing). Given a time series of some particular climatic response $\hat{C}(t)$, say of oxygen-isotope ratios, and a time series of insolation intensity $Q^*(t)$, one procedure for estimating the response rate begins by making some very generalized assumptions about the system under investigation. Normally, the assumption is made that climate behaves as a linear, single-exponential system. In such a system, any given value of Q^* will, after a suitably long interval of time, give rise to an equilibrium response C that is linearly dependent on Q^*. For convenience, therefore, the forcing function Q^* may be transformed to yield a function $Q(t)$ having the same scale as $C(t)$. Given a change in Q, the system will move towards an equilibrium value at a rate dC/dt that is proportional to the departure from equilibrium at any time t. This model[34] is expressed in the equation

$$dC/dt = 1/K\,[Q(t) - C(t)] \tag{1}$$

where the rate constant $1/K$ is expressed in terms of a *time constant* K. If K is large, the system responds slowly; if $K = 0$, the response is instantaneous.

Clearly, it would be desirable to obtain a general solution to equation (1) so that the *output* of the system $C(t)$ could be specified given any input $Q(t)$. Although a general analytical solution does not exist, solutions for important special cases are available. One such case involves an instantaneous, step-function change in Q from one long-term value to another. If the amount of the change in Q is unity at time $t = 0$, the response is

$$C(t) = 1 - e^{-t/K} \tag{2}$$

at any time t. After an elapsed time $t = K$, for example, $C = 1 - 1/e = 0.63$, and C has moved 63 percent of the way towards equilibrium. The time constant K is therefore often referred to as the "$1/e$ time" or "e-folding time" of the system. However, many geologists prefer to think in terms of half response times, $K_{1/2} =$

$(\log_e 2)\ K$. An analytical solution to (1) is also available for the case of a sinusoidal input having a given amplitude, a given phase, and a given frequency f. The corresponding output is a sinusoid having the same frequency as the input, but exhibiting a reduced amplitude and a certain phase lag Θ given by

$$\Theta = -\arctan(2\pi f K) \tag{3}$$

where f is the frequency of the forcing, K the system time constant, and the minus sign indicates that the response lags the input by the specified phase angle. It is this relationship that makes it possible to estimate K from observations on the output of a system driven at a known frequency. For example, observations on the phase difference between $\delta^{18}O$ and orbital forcing at the frequencies of variation in precession (1/21,000 year) and obliquity (1/41,000 year) yield an estimate of the time constant of the radiation-ice-sheet system as 17,000 years $\pm$ 3,000 years.[24]

Using equation (3), this value of K can be used to predict the phase of the system's response at *any* frequency. Moreover, these predictions should hold for a *family* of climatic, biologic, and geologic responses that are directly related to the volume of glacial ice. For example, one member of this family is the sea surface temperature in the North Atlantic Subpolar Gyre.[23] Climatic responses having distinctly different phase lags must represent fundamentally different physical processes having different time constants. For example, changes in the surface temperature in high southern latitudes[21] have a considerably shorter lag than $\delta^{18}O$ and high latitude North Atlantic temperatures; and changes in the intensity of the Indian Monsoon[35] and in the Atlantic Subtropical Gyre[23] exhibit no lag behind precessional forcing. Although serious research on phase spectra is only beginning, this approach offers much promise for sorting out and identifying families of climatic responses that are controlled by different time constants. The power of this approach offers one justification for transforming observations into the frequency domain.

Another reason for displaying climatic information in the frequency domain is that it is often easy to predict the *frequencies* over which a particular cause will operate, whereas it is often difficult to predict what the precise *form* (i.e., the *amplitude* and *phase*) of the response to that cause will be. A familiar example of this principle is the seasonal cycle. Here it is easy to predict from first principles that the climatic response in many areas will be dominated by oscillations having a 1 year period; but it is no easy matter to predict what the amplitude and phase of the seasonal cycle will be at a particular site. Another example may be cited from the field of paleoclimatology, where the Milankovitch theory makes it easy to predict that climatic oscillations having a 41,000 year period will be caused by variations in orbital obliquity.[21] It is, however, no trivial matter to predict from first principles what the amplitude and phase of the climatic response to obliquity forcing will be at a particular site.

Thus the first step in unravelling the chain of cause and effect in climatology is often the identification of a fundamental cause by finding a predicted peak in the climate spectrum. And the second step—identification of the mechanism by which the climate system responds to the fundamental cause—is often aided by observing what the amplitude and phase of the response (in the relevant frequency band) is at different sites.[7,8,21,23,32,33]

REFERENCES

1. CLIMAP, "Seasonal reconstructions of the Earth's surface at the last glacial maximum," *Geol. Soc. Am. Map Chart Ser. MC*, **36,** (1981).
2. B. Molfino, N. G. Kipp, and J. J. Morley. "Comparison of foraminiferal, coccolithophorid, and radiolarian paleotemperature equations: Assemblage coherency and estimate concordancy," *Quat. Res.*, **17,** 279-313 (1982).
3. E. A. Boyle. "Deep circulation of the ocean during the past 150,000 years: Geochemical evidence," *Trans. Am. Geophys. Union*, **63,** 71–72 (1982).
4. T. Webb III. "Reconstructing climatic sequences from botanical data," *J. Interdisc. His.*, **10,** 749–772 (1980).
5. T. Webb III and D. R. Clark. "Calibrating micropaleontological data in climatic terms: A critical review," *Ann. N. Y. Acad. Sci.*, **288,** 93–118 (1977).
6. G. de Q. Robin. *The Climatic Record in Polar Ice Sheets*, Cambridge University Press, Cambridge, 1983.
7. E. M. Rasmusson and T. H. Carpenter, "Variations in tropical sea surface temperature and surface wind fields associated with the Southern Oscillation/El Nino," *Mon. Weather Rev.*, **110,** 354–384 (1982).
8. S. G. H. Philander, "El Nino Southern Oscillation phenomena," *Nature*, **302,** 295–301 (1983).
9. J. Imbrie and T. Webb III, "Transfer Functions: Calibrating Micropaleontological Data in Climatic Terms." In A. Berger ed., D. Reidel, Dordecht, *Climatic Variations and Variability: Facts and Theories*, 1981, pp. 527-538.
10. J. L. Cullen, "Microfossil evidence for changing salinity patterns in the Bay of Bengal over the last 20,000 years," *Palaeogeogr., Palaeoclimatol.*, and *Palaeoecol.*, **35,** 315–356 (1981).
11. T. Webb III and R. A. Bryson, "Late- and postglacial climatic change in the northern midwest, U.S.A.: Quantitative estimates derived from fossil pollen spectra by multivariate statistical analysis," *Quat. Res.*, **2,** 70–115 (1972).
12. W. F. Ruddiman and A. McIntyre, "Time–transgressive deglacial retreat of polar waters from the North Atlantic," *Quat. Res.*, **3,** 117–130 (1973).
13. J. D. Hays, J. A. Lozano, N. J. Shackleton, and G. Irving, "Reconstruction of the Atlantic and western Indian Ocean sectors of the 18,000 B.P. Antarctic Ocean." In R. M. Cline and J. D. Hays eds., Investigation of late Quaternary paleoceanography and paleoclimatology, Geological Society American Memoir 145, 337–372 (1976).
14. W. L. Prell, W. H. Hutson, and D. F. Williams, "The subtropical convergence and late Quaternary circulation in the southern Indian Ocean," *Mar. Micro.*, **4,** 225–234 (1979).
15. T. C. Moore, W. H. Hutson, N. G. Kipp, J. D. Hays, W. L. Prell, P. Thompson, and G. Boden, "The biological record of the ice-age ocean," *Palaeogeogr, Palaeoclimatol., Palaeoecol.*, **35,** 357-370 (1981).
16. N. J. Shackleton and N. D. Opdyke, "Oxygen isotope and paleomagnetic stratigraphy of equatorial Pacific core V28-238: Oxygen isotope temperatures and ice volumes on a 10^5 and 10^6 year scale," *Quat. Res.*, **3,** 39–55 (1973).
17. W. F. Ruddiman and A. McIntyre. "Northeast Atlantic paleoclimatic changes over the past 600,000 years." In R. M. Cline and J. D. Hays eds., *Quaternary paleoceanography and paleoclimatology*, Geology Society American Memoir 145, 111–146 (1976).
18. S. Manabe and D. C. Hahn, "Simulation of tropical climate of an ice age." *J. Geophys. Res.*, **82,** 3889-3912 (1977).

19. W. L. Gates, "The numerical simulation of ice age climate with a general circulation model," *J. Atmos. Sci.*, **33,** 1844–1873 (1976).

20. J. E. Kutzbach and B. L. Otto–Bliesner, "The sensitivity of the African-Asian monsoonal climate to orbital parameter changes for 9000 years B.P. in a low-resolution general circulation model," *J. Atmos. Sci.*, **39,** 1177–1188 (1982).

21. J. D. Hays, J. Imbrie, and N. J. Shackleton, "Variations in the earth's orbit: Pacemaker of the ice ages," *Science*, **194,** 1121–1132 (1976).

22. M. A. Kominz and N. G. Pisias, "Pleistocene climate: Deterministic or stochastic?" *Science*, **204,** 171–173 (1979).

23. W. F. Ruddiman and A. McIntyre, "Oceanic mechanisms for amplification of the 23,000-year ice volume cycle," *Science*, **212,** 617–627 (1976).

24. J. Imbrie and J. Z. Imbrie, "Modeling the climatic response to orbital variations," *Science*, **207,** 943–953 (1980).

25. R. G. Fairbanks, M. Sverdlove, R. Free, P.H. Wiebe, and A. W. Be, "Vertical distribution and isotopic fractionation of living planktonic foraminifera from the Panama Basin," *Nature*, **298,** 841–844 (1982).

26. W. A. Berggren. "Tertiary boundaries." In B. F. Funnel and W. R. Riedel, eds. *Micropalenontology of the Oceans*, Cambridge University Press, 1971, pp. 693–809.

27. J. Imbrie, J. D. Hays, D. G. Martinson, A. McIntyre, A. C. Mix, J. J. Morley, N. G. Pisias, W. L. Prell, and N. J. Shackleton, "The orbital theory of Pleistocene climate: Support from a revised chronology of the marine $\delta^{18}O$ record." In A. Berger et al., eds., *Milankovitch and Climate*, D. Reidel, Dordrecht, 1984, pp. 269–305.

28. J. M. Mitchell, Jr., "An overview of climatic variability and its causal mechanisms," *Quat. Res.*, **6,** 481–493 (1976).

29. K. Hasselmann, "Stochastic climate models, Part 1: Theory," *Tellus*, **28,** 473–484 (1976).

30. D. A. Pollard, A. P. Ingersol, and J. G. Lockwood, "Response of a zonal climate–ice sheet model to the orbital perturbations during the Quaternary ice ages," *Tellus*, **32,** 301–319 (1980).

31. G. E. Birchfield, J. Weertman, and A. T. Lunde, "A paleoclimate model of northern hemisphere ice sheets," *Quat. Res.*, **15,** 126–142 (1981).

32. W. S. Broecker, "Glacial to interglacial changes in ocean chemistry," *Progr. Oceanogr.*, **11,** 151–197 (1982).

33. R. S. Keir and W. J. Berger, "Atmospheric CO_2 content in the last 120,000 years: The phosphate-extraction model," *J. Geophys. Res.*, **88,** 6027–6038 (1983).

34. G. M. Jenkins and D. G. Watts, *Spectral Analysis and Its Applications*, Holden-Day, San Francisco, 1968.

35. W. L. Prell, "Monsoonal climate of the Arabian Sea during the Late Quaternary: A response to changing solar radiation." In A. Berger, et al. eds., *Milankovitch and Climate*, D. Reidel, Dordrecht, 1984, pp. 349–366.

INDEX

INDUSTRIAL LOCATION AND AIR QUALITY CONTROL: A Planning Approach
Jean-Michel Guldmann and Daniel Shefer

PLANT DISEASE CONTROL: Resistance and Susceptibility
Richard C. Staples and Gary H. Toenniessen, Editors

AQUATIC POLLUTION
Edward A. Laws

MODELING WASTEWATER RENOVATION: Land Treatment
I. K. Iskandar, Editor

AIR AND WATER POLLUTION CONTROL: A Benefit Cost Assessment
A. Myrick Freeman, III

SYSTEMS ECOLOGY: An Introduction
Howard T. Odum

INDOOR AIR POLLUTION: Characterization, Prediction, and Control
Richard A. Wadden and Peter A. Scheff

INTRODUCTION TO INSECT PEST MANAGEMENT, Second Edition
Robert L. Metcalf and William H. Luckman, Editors

WASTES IN THE OCEAN—Volume 1: Industrial and Sewage Wastes in the Ocean
Iver W. Duedall, Bostwick H. Ketchum, P. Kilho Park, and Dana R. Kester, Editors

WASTES IN THE OCEAN—Volume 2: Dredged Material Disposal In the Ocean
Dana R. Kester, Bostwick H. Ketchum, Iver W. Duedall and P. Kilho Park, Editors

WASTES IN THE OCEAN—Volume 3: Radioactive Wastes and the Ocean
P. Kilho Park, Dana R. Kester, Iver W. Duedall, and Bostwick H. Ketchum, Editors

LEAD AND LEAD POISONING IN ANTIQUITY
Jerome O. Nriagu

INTEGRATED MANAGEMENT OF INSECT PESTS OF POME AND STONE FRUITS
B. A. Croft and S. C. Hoyt

PRINCIPLES OF ANIMAL EXTRAPOLATION
Edward J. Calabrese

NONCONVENTIONAL ENERGY RESOURCES
Philip R. Pryde

VIBRIOS IN THE ENVIRONMENT
Rita R. Colwell, Editor

WATER RESOURCES: Distribution, Use and Management
John R. Mather